Finite Element Analysis with Error Estimators

An Introduction to the FEM and Adaptive Error Analysis Engineering Students

Finite Element Analysis with Error Estimators

An Introduction to the FEM and Adaptive Error Analysis for Engineering Students

J. E. Akin

ELSEVIER
BUTTERWORTH HEINEMANN

AMSTERDAM • BOSTON • HEIDELBERG • LONDON • NEW YORK • OXFORD
PARIS • SAN DIEGO • SAN FRANCISCO • SINGAPORE • SYDNEY • TOKYO

Elsevier Butterworth-Heinemann
Linacre House, Jordan Hill, Oxford OX2 8DP
30 Corporate Drive, Burlington, MA 01803

First published 2005

British Library Cataloguing in Publication Data
A catalogue record for this book is available from the British Library

Library of Congress Cataloguing in Publication Data
A catalogue record for this book is available from the Library of Congress

ISBN 0 7506 6722 2

For information on all Elsevier Butterworth-Heinemann publications visit our website at http://books.elsevier.com

Printed and bound in Great Britain

Contents

* Denotes sections or chapters that can be omitted for a first reading or shorter course.

Preface

There are many good texts on the application of finite element analysis techniques. Most do not address the concept and implementation of error estimation. Now that computers are so powerful there is no reason not to carry out a re-analysis until the error levels reach the point that the user is satisfied. Having an error estimation is critical to automating the adaptation of the finite element analysis process. Today several commercial programs include automatic adaptation, based on an error analysis. The user of such programs should have a clear concept of the theory and limitations of such tools. Thus, this text includes the basic finite element theory and its mathematical foundations, the error estimation processes, and the associated computational procedures, as well as several example applications.

This book is primarily intended for advanced undergraduate engineering students and beginning graduate students. The text contains more material than could be covered in a single quarter or semester course. Therefore, a number of chapters or sections that could be omitted in a first course have been marked with an asterisk ($*$). Most of the subject matter deals with linear heat transfer and elementary stress analysis.

The future of finite element analysis will probably heavily involve adaptive analysis methods. One should have a course in Functional Analysis to best understand those techniques. Most undergraduate curriculums do not contain such courses. Therefore, a chapter on mathematical preliminaries is included.

All the Fortran 95 source programs for the general finite element library (called MODEL), and the corresponding application and supporting data file can be downloaded from the World Wide Web (for non-commercial use only). They can be found at the Elsevier site http://www.books.elsevier.com/companions/. The same is true of a large library of small Matlab plotting scripts that display the input and output results shown in the text.

I would like to thank many current and former students at Rice University for their constructive criticisms and comments during the evolution of this book. Special thanks go to Prof. R. L. Taylor, of the University of California at Berkeley for his many detailed and constructive suggestions. Mr. Don Schroder helped with the preparation of a large part of the manuscript. Finally, this book would not have been completed without the support and patience of my wife Kimberly.

Ed Akin
Houston, Texas
2005

Features of the text and accompanying resources

End of chapter exercises
Each chapter ends with a range of exercises that are suitable for homework and assignment work, as well as for private study.

Worked solutions to the exercises are freely available to teachers who adopt or recommend the text to their students. For details on accessing this material please visit http://books.elsevier.com/manuals and follow the registration instructions on screen.

Fortran 95 source programs
Source programs for the general finite element library, and the corresponding application and supporting data file can be freely downloaded from the accompanying website. Go to http://books.elsevier.com/companions and follow the instructions on screen. This material is presented for non-commercial use only.

Matlab plotting scripts library
A library of Matlab plotting scripts that display the input and output results shown in the text are also available for free download from the accompanying website. Go to http://books.elsevier.com/companions and follow the instructions on screen. This material is presented for non-commercial use only.

Notation

The symbols most commonly used throughout the book are defined below. When appearing in the text matrices, tensors, and vectors are identified by boldface type.

Mathematical symbols

$(\hat{.})$	Based on element gradient
$(.^*)$	Based on nodally continuous gradient
$\{.\}$	Column vector, n by 1
$\lvert . \rvert$	Determinant of a matrix
$\mathbf{\Delta}^T$	Divergence operator
$\varnothing$	Empty set
$\mathbf{\nabla}$	Gradient operator
$\in$	In
$\bigcap$	Intersection
$[.]^{-1}$	Inverse of a square matrix
$\square$	Non-dimensional parametric space
$\lVert . \rVert$	Norm of a matrix or vector
$].,.[$	Open one-dimensional domain
$\lfloor\ \rfloor^T \lfloor\ \rfloor$	Outer product square matrix, m by m
$,_{(.)}$	Partial differentiation with respect to (.)
$\boldsymbol{\partial}_G, \boldsymbol{\partial}_\Omega$	Partial derivatives in global Cartesian space
$\boldsymbol{\partial}_L, \boldsymbol{\partial}_\square$	Partial derivatives in local parametric space
$\propto$	Proportional to
$[.]$	Rectangular, m by n, or square matrix
$\lfloor . \rfloor$	Row vector, 1 by m
$\subset$	Subset
$[.]^T$	Transpose of a matrix
$\bigcup$	Union

Latin Symbols

A	Area
$\mathbf{a}$	Acceleration vector
a, b, c	Natural coordinates on −1 to +1
$(.)^b$	Relating to a boundary domain
$\mathbf{B}$	Differential operator acting on interpolation matrix **H** or **N**
$\mathbf{b}$	Differential operator acting on global interpolation matrix **h**
C^n	Field continuity of degree n
$\mathbf{C}$	System source vector
$\mathbf{C}^b$	Source vector from a boundary segment
$\mathbf{C}^e$	Source vector from an element
$\mathbf{D}$	System degrees of freedom vector
D	Differential operator.
$\mathbf{D}^b$	Boundary segment degrees of freedom vector
$\mathbf{D}^e$	Element degrees of freedom vector
$\mathbf{d}$	Cartesian gradient of **H**
$\mathbf{d}_x$	First row of **d**, etc. for y, z
dof	Degree(s) of freedom
E	Modulus of elasticity of a material
$\mathbf{E}$	Constitutive law (stress-strain) matrix
e	Error
$(.)^e$	Relating to an element domain
$\mathbf{F}$	Resultant force vector
G	Shear modulus of a material
$\mathbf{G}$	Geometry interpolation row matrix (usually $\mathbf{G} = \mathbf{H}$)
$\mathbf{H}^b$	Boundary interpolation row matrix for a scalar
$\mathbf{H}^e$	Element interpolation row matrix for a scalar
h	Characteristic length. Convection coefficient
$\mathbf{h}$	Global interpolation matrix
$I^e, \mathbf{I}^e$	Integral of a scalar or matrix, respectively, on an element
$\mathbf{I}$	Identity matrix
$\mathbf{J}$	Jacobian matrix of a geometric transformation
$\mathbf{K}$	Stiffness matrix
k	Thermal conductivity of a material, or spring stiffness
$\mathbf{L}$	Differential operator
L	Length
L_k	Barycentric coordinates, $\sum L_k = 1$
$\mathbf{M}$	Mass matrix of the system
$\mathbf{m}^e$	Mass matrix, or thermal capacity matrix of an element
m	Mass
$\mathbf{N}$	Interpolation matrix for generalized degrees of freedom (often $\mathbf{N} = \mathbf{H}$)
$\mathbf{n}$	Unit normal vector
n_a	Number of adjacent elements, *NEIGH_L*
n_b	Number of boundary segments, *N_MIXED* + *N_SEG*
n_c	Number of constraint equations, *N_CEQ*

n_d	Number of system degrees of freedom ($n_m \times n_g$), *N_D_FRE*
n_e	Number of elements in the system, *N_ELEMS*
n_f	Maximum number of flux components, *N_G_FLUX*
n_g	Number of generalized dof per node, *N_G_DOF*
n_h	Number of scalar interpolations in **H**, *LT_FREE*
n_i	Number of element equation index terms ($n_n \times n_g$), *LT_FREE*
n_l	Number of elements in a patch, *L_IN_PATCH*
n_m	Maximum node number in the system, *MAX_NP*
n_n	Maximum number of nodes per element, *NOD_PER_EL*
n_o	Number of mixed or Robin BC segments, *N_MIXED*
n_p	Dimension of the parametric space, *N_PARM*
n_q	Number of quadrature points, *N_QP*
n_r	Number of rows in the **B** matrix, *N_R_B*
n_s	Dimension of the physical space, *N_SPACE*
n_t	Number of different element types, *N_L_TYPE*
n_v	Number of vector interpolations in **V**, *LT_FREE*
n_x	Number of element geometry definition nodes, *N_GEOM*
$\mathbf{P}$	Polynomial row matrix. Reaction vector
p	Pressure
$\mathbf{Q}$	Source per unit volume
$\mathbf{Q}^e$	Source per unit volume at element node points
q	Source per unit length
q_n	Heat flux normal to boundary ($\mathbf{q}_n = q_n \mathbf{n}$)
$\mathbf{q}$	Heat flux vector at a point
$\mathbf{R}$	Matrix of position vectors, $\mathbf{R} = [\mathbf{x}\,\mathbf{y}\,\mathbf{z}]$
R	Residual error in Ω^e
r, s, t	Unit coordinates on 0 to 1
$\mathbf{S}$	Square matrix of the system
$\mathbf{S}^b$	Square matrix from a boundary segment
$\mathbf{S}^e$	Square matrix from an element
t	Thickness, time
$\mathbf{T}$	Transformation matrix, or boundary traction matrix
U	Strain energy
$\mathbf{u}$	Displacement vector. Velocity vector
u, v, w	Components of displacement vector
V	Volume
$\mathbf{v}$	Velocity vector
W	Mechanical work
x, y, z	Cartesian coordinates
$\mathbf{X}$	Body force vector
$\mathbf{x}$	Vector of x-coordinates
$\mathbf{x}^e$	Vector of x-coordinates of the element nodes
$\mathbf{y}$	Vector of y-coordinates
$\mathbf{z}$	Vector of z-coordinates

Greek symbols

α	Coefficient of thermal expansion
$\boldsymbol{\beta}$	Boolean gather matrix
$\boldsymbol{\beta}^T$	Boolean scatter matrix
$\sum_e \boldsymbol{\beta}^{eT} \mathbf{C}^e$	Column vector element assembly process
$\sum_e \boldsymbol{\beta}^{eT} \mathbf{S}^e \boldsymbol{\beta}^e$	Square matrix element assembly process
$\boldsymbol{\Gamma}$	Boundary of a domain, Ω
$\boldsymbol{\Gamma}^b$	Segment of the boundary $\boldsymbol{\Gamma}$
$\boldsymbol{\Gamma}^e$	Boundary of an element domain, Ω^e
γ	Weight per unit volume
$\boldsymbol{\Delta}$	Local derivatives of the interpolation matrix **H** or **N**
$\boldsymbol{\delta}$	Element or boundary segment dof.
$\boldsymbol{\varepsilon}$	Strain or gradient
ζ	Refinement parameter
η	Allowed percentage error
θ	Temperature, or angle
Θ	Effectivity index
λ	Direction cosine wrt x. Lame' constant.
μ	Direction cosine wrt y. Lame' constant.
ν	Poisson's ratio of a material. Direction cosine wrt z.
$\boldsymbol{\Pi}$	Total potential energy, $\boldsymbol{\Pi} = U - W$
π	Mathematical constant 3.14159...
ρ	Mass density of a material
$\boldsymbol{\rho}$	Position vector to a point, $\boldsymbol{\rho} = [x, y, z]$
$\boldsymbol{\sigma}$	Flux or stress
$\boldsymbol{\sigma}^*$	Smoothed flux or stress approximation
$\hat{\boldsymbol{\sigma}}$	Finite element flux or stress approximation
τ	Stabilization parameter
$\boldsymbol{\tau}$	Shear stress
$\boldsymbol{\Phi}$	System degrees of freedom vector
$\boldsymbol{\Phi}_k$	k-th unknown
ϕ	Scalar unknown. Velocity potential
ψ	Stream function
ω	Angular velocity
$\boldsymbol{\Omega}$	Domain
$\boldsymbol{\Omega}^e$	Element domain

Selected program notation (Array sizes follow in parentheses.)

AJ	Jacobian matrix: (N_SPACE, N_SPACE)
AVE	Average quantities at a system node: (N_R_B + 2, MAX_NP)
B	Gradient versus dof matrix: (N_R_B, LT_FREE)
C	Element column matrix: (LT_FREE)
CC	Column matrix of system equations: (N_D_FRE)
COORD	Coordinates of all nodes on an element: (LT_N, N_SPACE)
C_B	Boundary segment column matrix: (LT_FREE)
D	Nodal parameters associated with an element: (LT_FREE)
DD	System list of nodal parameters: (N_D_FRE)
DGH	Global derivatives of scalar functions **H** : (N_SPACE, LT_N)
DGV	Global derivatives of vector functions **V** : (N_SPACE, LT_FREE)
DLG	Local derivatives of geometry functions **G** : (LT_PARM, LT_GEOM)
DLH	Local derivatives of scalar functions **H** : (LT_PARM, LT_N)
E	Constitutive matrix: (N_R_B, N_R_B)
EL_M	Element mass matrix: (LT_FREE, LT_FREE)
FLUX_LT	Flux at element nodes from a SCP: (SCP_FIT, LT_N
G	Interpolation functions for geometry: (LT_GEOM)
GLOBAL	Global derivatives of scalar interpolation functions **H**
H	Interpolation functions for an element scalar: (LT_N)
H_INTG	Integral of scalar interpolation functions **H** : (LT_N)
H_QP	Interpolation for **H** at quadrature point: (LT_N, LT_QP)
INDEX	System degree of freedom numbers array: (LT_FREE)
L_B_N	Maximum number of nodes on an element boundary segment
LT	Element type number
LT_FREE	Number of degrees of freedom per element
LT_GEOM	Number of geometric nodes per element
LT_N	Maximum number of nodes for element type
LT_PARM	Dimension of parametric space for element type
LT_QP	Number of quadrature points for element type
LT_SHAP	Current element type shape flag number
L_B_N	Number of nodes on an element boundary segment
L_SHAPE	Shape: 0=Point 1=Line 2=Triangle 3=Quadrilateral 4=Hexahedron 5=Tetrahedron etc.
L_TYPE	Type number array of all elements: (L_S_TOT)
MAT_FLO	Number of real material properties
MAX_NP	Number of system nodes
MISC_FL	Number of miscellaneous floating point (real) system properties
MISC_FX	Number of miscellaneous fixed point (integer) system properties
M_B_N	Number of nodes on a mixed boundagy condition segment
NODES	Node incidences of all elements: (L_S_TOT, NOD_PER_EL)
NOD_PER_EL	Maximum number of nodes per element
N_BS_FIX	Number of boundary segment integer properties
N_BS_FLO	Number of boundary segment real properties
N_CEQ	Number of system constraint equations
N_D_FLUX	Maximum number of flux segment dof = L_B_N * N_G_DOF

N_D_FRE	Total number of system degrees of freedom
N_ELEMS	Number of elements in the system
N_EL_FRE	Maximum number of degrees of freedom per element
N_GEOM	Maximum number of element geometry nodes
N_G_DOF	Number of generalized parameters (dof) per node
N_G_FLUX	Number of flux components per segment node
N_LP_FIX	Number of integer element properties
N_LP_FLO	Number of floating point (real) element properties
N_MAT	Number of materail types
N_MX_FIX	Number of fixed point (integer) mixed segment properties
N_MX_FLO	Number of floating point (real) mixed segment properties
N_NP_FIX	Number of fixed point (integer) nodal properties
N_NP_FLO	Number of floating point (real) nodal properties
N_PARM	Dimension of parametric space
N_PATCH	Number of SCP patches = MAX_NP or N_ELEMS
N_QP	Maximum number of element quadrature points
N_R_B	Number of rows in $\mathbf{B}$ and $\mathbf{E}$ matrices
N_SEG	Number of element boundary segments with given flux
N_SPACE	Dimension of space
PATCH_FIT	Local patch flux values at its nodes: (SCP_N, SCP_FIT)
PT	Quadrature coordinates: (LT_PARM, LT_QP)
S	Element square matrix: (LT_FREE, LT_FREE)
SCP_COUNTS	Number of patches used for each nodal averages: (MAX_NP)
SCP_FIT	Number of terms being fit in a patch, N_R_B usually
SCP_GEOM	Number of patch geometry nodes
SCP_H	Interpolation functions for patch, usually is $\mathbf{H}$ (SCP_N)
SCP_LT	Patch type number
SCP_N	Number of nodes per patch
SCP_PARM	Number of parametric spaces for patch
SCP_QP	Number of quadrature points needed in a SCP patch
SIGMA_HAT	Flux components at a point in original element: (SCP_FIT)
SIGMA_SCP	Flux components at a point in smoothed SCP: (SCP_FIT)
SS	Square matrix of system equations: (N_D_FREE, N_D_FREE)
STRAIN	Strain or gradient vector: (N_R_B + 2)
STRAIN_0	Initial strain or gradient vector, if any: (N_R_B)
STRESS	Stress vector at a point: (N_R_B + 2)
S_B	Boundary segment square matrix, if any: (LT_FREE, LT_FREE)
THIS_EL	Current element number
THIS_LT	Current element type number
THIS_STEP	Current time step number
TIME	Current time in dynamic or transient solution
V	Interpolation functions for vectors: (LT_FREE)
WT	Quadrature weights: (LT_QP)
X	Coordinates of all system nodes: (MAX_NP, N_SPACE)
XYZ	Spatial coordinates at a point: (N_SPACE)

Chapter 1

Introduction

1.1 Finite element methods

The goal of this text is to introduce finite element methods from a rather broad perspective. We will consider the basic theory of finite element methods as utilized as an engineering tool. Likewise, example engineering applications will be presented to illustrate practical concepts of heat transfer, stress analysis, and other fields. Today the subject of error analysis for adaptivity of finite element methods has reached the point that it is both economical and reliable and should be considered in an engineering analysis. Finally, we will consider in some detail the typical computational procedures required to apply modern finite element analysis, and the associated error analysis. In this chapter we will begin with an overview of the finite element method. We close it with consideration of modern programming approaches and a discussion of how the software provided differs from the author's previous implementations of finite element computational procedures.

In modern engineering analysis it is rare to find a project that does not require some type of finite element analysis (FEA). The practical advantages of FEA in stress analysis and structural dynamics have made it the accepted tool for the last two decades. It is also heavily employed in thermal analysis, especially for thermal stress analysis.

Clearly, the greatest advantage of FEA is its ability to handle truly arbitrary geometry. Probably its next most important features are the ability to deal with general boundary conditions and to include nonhomogeneous and anisotropic materials. These features alone mean that we can treat systems of arbitrary shape that are made up of numerous different material regions. Each material could have constant properties or the properties could vary with spatial location. To these very desirable features we can add a large amount of freedom in prescribing the loading conditions and in the post-processing of items such as the stresses and strains. For elliptical boundary value problems the FEA procedures offer significant computational and storage efficiencies that further enhance its use. That class of problems include stress analysis, heat conduction, electrical fields, magnetic fields, ideal fluid flow, etc. FEA also gives us an important solution technique for other problem classes such as the nonlinear Navier-Stokes equations for fluid dynamics, and for plasticity in nonlinear solids.

Here we will show what FEA has to offer and illustrate some of its theoretical formulations and practical applications. A design engineer should study finite element methods in more detail than we can consider here. It is still an active area of research. The current trends are toward the use of error estimators and automatic adaptive FEA procedures that give the maximum accuracy for the minimum computational cost. This is also closely tied to shape modification and optimization procedures.

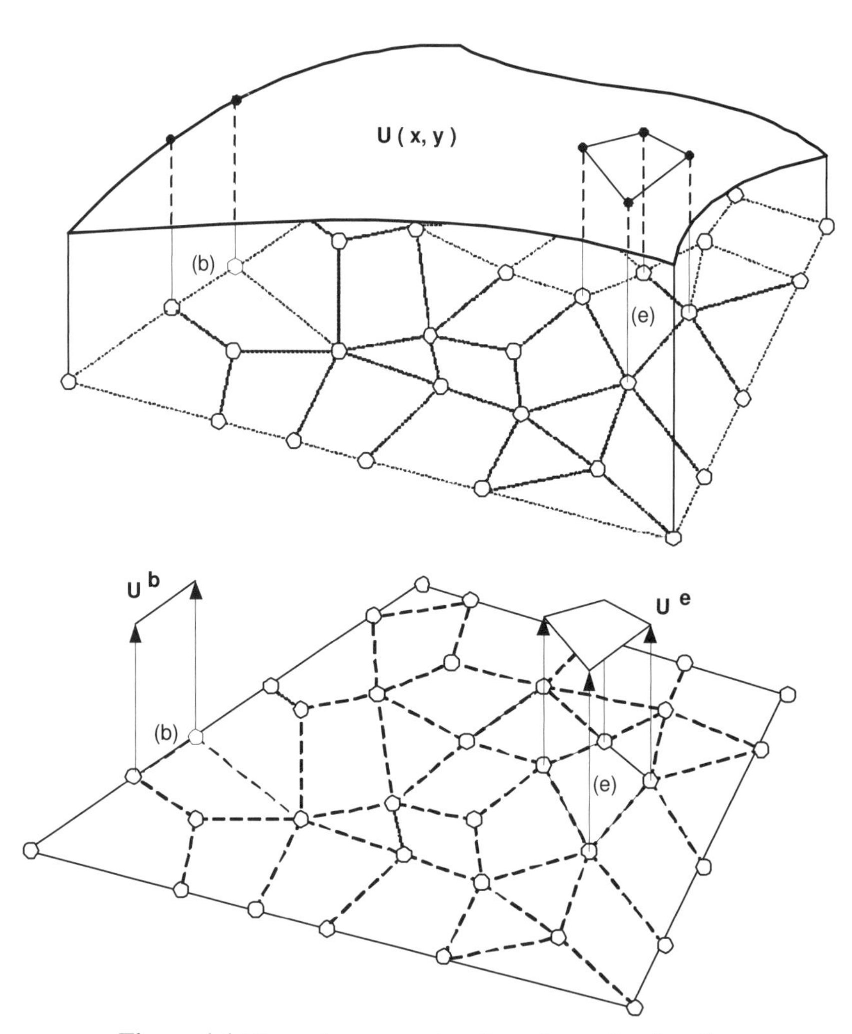

Figure 1.1 *Piecewise approximation of a scalar function*

1.2 Capabilities of FEA

There are many commercial and public-domain finite element systems that are available today. To summarize the typical capabilities, several of the most widely used software systems have been compared to identify what they have in common. Often we find about 90 percent of the options are available in all the systems. Some offer very specialized capabilities such as aeroelastic flutter or hydroelastic lubrication. The mainstream capabilities to be listed here are found to be included in the majority of the commercial systems. The newer adaptive systems may have fewer options installed but they are rapidly adding features common to those given above. Most of these systems are available on engineering workstations and personal computers as well as mainframes and supercomputers. The extent of the usefulness of an FEA system is directly related to the extent of its element library. The typical elements found within a single system usually include membrane, solid, and axisymmetric elements that offer linear, quadratic, and cubic approximations with a fixed number of unknowns per node. The new hierarchical elements have relatively few basic shapes but they do offer a potentially large number of unknowns per node (more than 80). Thus, the actual effective element library size is extremely large.

In the finite element method, the boundary and interior of the region are subdivided by lines (or surfaces) into a finite number of discrete sized subregions or finite elements. A number of nodal points are established with the mesh. The size of an element is usually associated with a reference length denoted by h. It, for example, may be the diameter of the smallest sphere that can enclose the element. These nodal points can lie anywhere along, or inside, the subdividing mesh, but they are usually located at intersecting mesh lines (or surfaces). The elements may have straight boundaries and thus, some geometric approximations will be introduced in the geometric idealization if the actual region of interest has curvilinear boundaries. These concepts are graphically represented in Fig. 1.1.

The nodal points and elements are assigned identifying integer numbers beginning with unity and ranging to some maximum value. The assignment of the nodal numbers and element numbers will have a significant effect on the solution time and storage requirements. The analyst assigns a number of generalized degrees of freedom to each and every node. These are the unknown nodal parameters that have been chosen by the analyst to govern the formulation of the problem of interest. Common nodal parameters are displacement components, temperatures, and velocity components. The nodal parameters do not have to have a physical meaning, although they usually do. For example, the hierarchical elements typically use the derivatives up to order six as the midside nodal parameters. This idealization procedure defines the total number of degrees of freedom associated with a typical node, a typical element, and the total system. Data must be supplied to define the spatial coordinates of each nodal point. It is common to associate some code to each node to indicate which, if any, of the parameters at the node have boundary constraints specified. In the new adaptive systems the number of nodes, elements, and parameters per node usually all change with each new iteration.

Another important concept is that of *element connectivity*, (or topology) i.e., the list of global node numbers that are attached to an element. The element connectivity data defines the topology of the (initial) mesh, which is used, in turn, to assemble the system

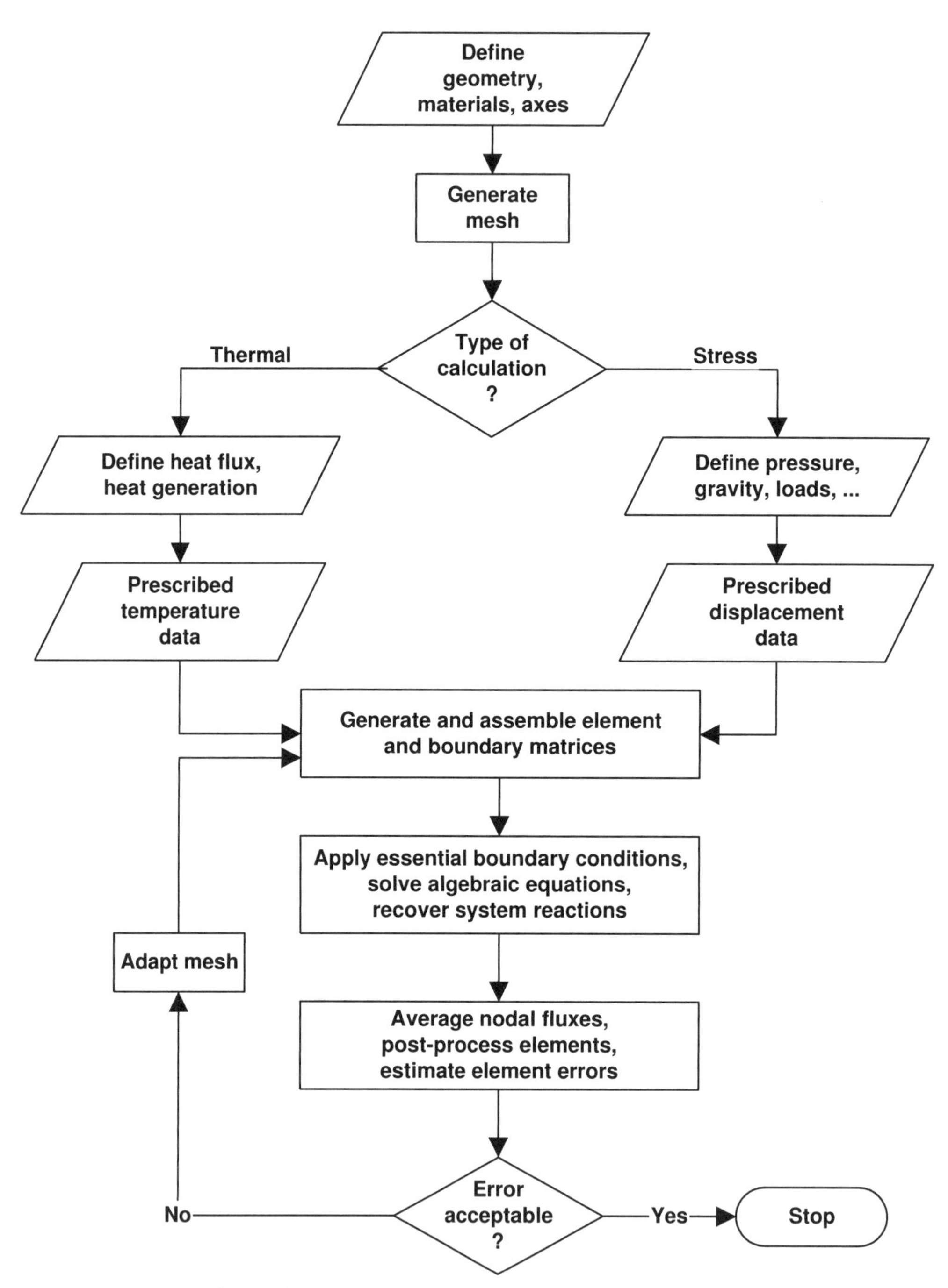

Figure 1.2 *Typical stages in a finite element analysis*

algebraic equations. Thus, for each element it is necessary to input, in some consistent order, the node numbers that are associated with that particular element. The list of node numbers connected to a particular element is usually referred to as the element incident list for that element. We usually assign a material code, or properties, to each element.

Finite element analysis can require very large amounts of input data. Thus, most FEA systems offer the user significant data generation or supplemental capabilities. The common data generation and validation options include the generation and/or replication of coordinate systems, node locations, element connectivity, loading sets, restraint conditions, etc. The verification of such extensive amounts of input and generated data is greatly enhanced by the use of computer graphics.

In the adaptive methods we must also compute the error indicators, error estimators, and various energy norms. All these quantities are usually output at 1 to 27 points in each of thousands of elements. The most commonly needed information in an engineering analysis is the state of temperatures, or displacements and stresses. Thus, almost every system offers linear static stress analysis capabilities, and linear thermal analysis capabilities for conduction and convection that are often needed to provide temperature distributions for thermal stress analysis. Usually the same mesh geometry is used for the temperature analysis and the thermal stress analysis. Of course, some studies require information on the natural frequencies of vibration or the response to dynamic forces or the effect of frequency driven excitations. Thus, dynamic analysis options are usually available. The efficient utilization of materials often requires us to employ nonlinear material properties and/or nonlinear equations. Such resources require a more experienced and sophisticated user. The usual nonlinear stress analysis features in large commercial FEA systems include buckling, creep, large deflections, and plasticity. Those advanced methods will not be considered here.

There are certain features of finite element systems which are so important from a practical point of view that, essentially, we cannot get along without them. Basically we

Table 1.1 *Typical unknown variables in finite element analysis*

Application	Primary	Associated	Secondary
Stress analysis	Displacement, Rotation	Force, Moment	Stress, Failure criterion Error estimates
Heat transfer	Temperature	Flux	Interior flux Error estimates
Potential flow	Potential function	Normal velocity	Interior velocity Error estimates
Navier-Stokes	Velocity	Pressure	Error estimates
Eigen-problem	Eigenvalues	Eigenvectors	Error estimates

Table 1.2 *Typical given variables and corresponding reactions*

Application	Given	Reaction
Stress analysis	Displacement	Force
	Rotation	Moment
	Force	Displacement
	Couple	Rotation
Heat transfer	Temperature	Heat flux
	Heat flux	Temperature
Potential flow	Potential	Normal velocity
	Normal velocity	Potential
Navier-Stokes	Velocity	Force

have the ability to handle completely arbitrary geometries, which is essential to practical engineering analysis. Almost all the structural analysis, whether static, dynamic, linear or nonlinear, is done by finite element techniques on large problems. The other abilities provide a lot of flexibility in specifying loading and restraints (support capabilities). Typically, we will have several different materials at different arbitrary locations within an object and we automatically have the capability to handle these nonhomogeneous materials. Just as importantly, the boundary conditions that attach one material to another are usually automatic, and we don't have to do anything to describe them unless it is possible for gaps to open between materials. Most important, or practical, engineering components are made up of more than one material, and we need an easy way to handle that. What takes place less often is the fact that we have *anisotropic materials* (one whose properties vary with direction, instead of being the same in all directions). There is a great wealth of materials that have this behavior, although at the undergraduate level, anisotropic materials are rarely mentioned. Many materials, such as reinforced concrete, plywood, any filament-wound material, and composite materials, are essentially anisotropic. Likewise, for heat-transfer problems, we will have thermal conductivities that are directionally dependent and, therefore, we would have to enter two or three thermal conductivities that indicate how this material is directionally dependent. These advantages mean that for practical use finite element analysis is very important to us. The biggest disadvantage of the finite element method is that it has so much power that large amounts of data and computation will be required.

All real objects are three-dimensional but several common special cases have been defined that allow two-dimensional studies to provide useful insight. The most common examples in solid mechanics are the states of *plane stress* (covered in undergraduate mechanics of materials) and *plane strain*, the *axisymmetric solid* model, the *thin-plate* model, and the *thin-shell* model. The latter is defined in terms of two parametric surface coordinates even though the shell exists in three dimensions. The *thin beam* can be thought of as a degenerate case of the thin-plate model. Even though today's solid modelers can generate three-dimensional meshes relatively easily one should learn to approach such problems carefully. A well planned series of two-dimensional approximations can provide important insight into planning a good three-dimensional model. They also provide good 'ball-park' checks on the three-dimensional answers. Of course, the use of basic handbook calculations in estimating the answer before approaching an FEA system is also highly recommended.

The typical unknown variables in a finite element analysis are listed in Table 1.1 and a list of related action-reaction variables are cited in Table 1.2. Figure 1.2 outlines as a flow chart the major steps needed for either a thermal analysis or stress analysis. Note that these segments are very similar. One of the benefits of developing a finite element approach is that most of the changes related to a new field of application occur at the element level and usually represent less than 5 percent of the total coding.

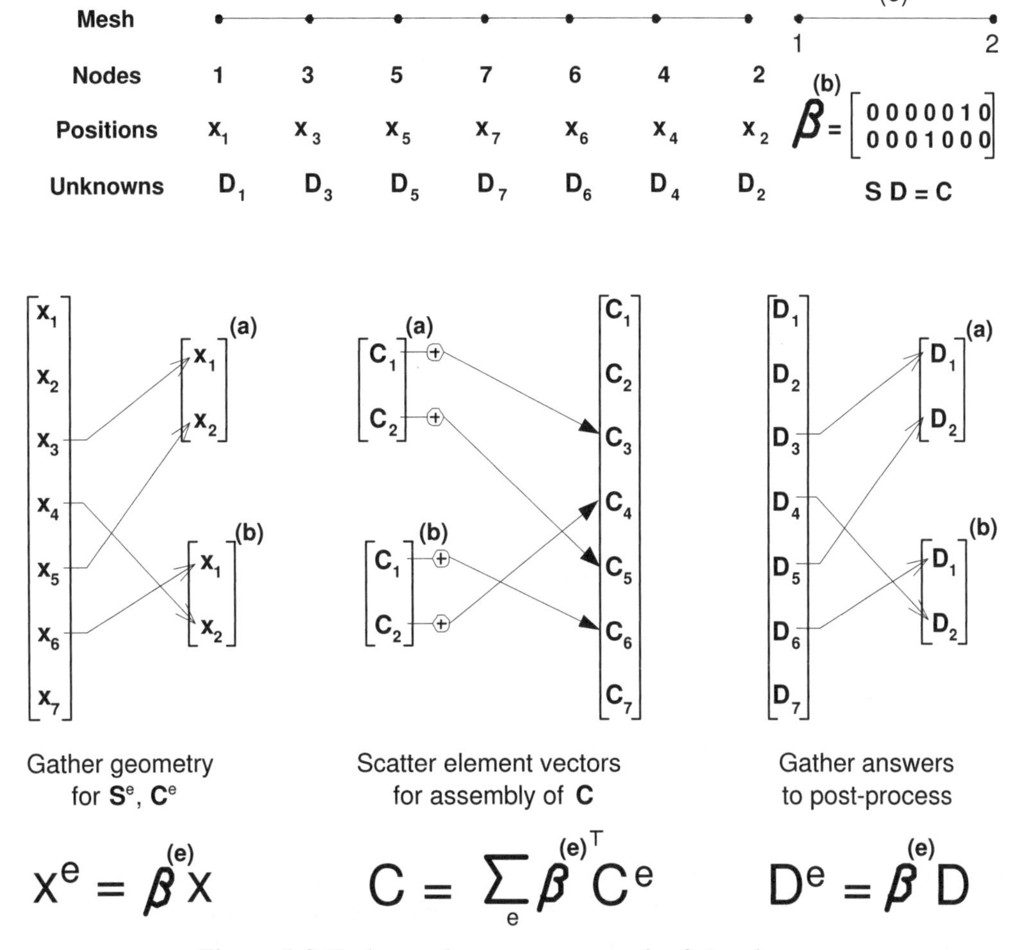

Figure 1.3 *Gather and scatter concepts for finite elements*

1.3 Outline of finite element procedures

From the mathematical point of view the finite element method is an integral formulation. Modern finite element integral formulations are usually obtained by either of two different procedures: *weighted residual* or *variational formulations.* The following sections briefly outline the common procedures for establishing finite element models. It is fortunate that all these techniques use the same bookkeeping operations to generate the final assembly of algebraic equations that must be solved for the unknowns.

The generation of finite element models by the utilization of weighted residual techniques is increasingly important in the solution of differential equations for non-structural applications. The weighted residual method starts with the governing differential equation to be satisfied in a domain Ω:

$$L(\phi) = Q, \tag{1.1}$$

where L denotes a differential operator acting on the primary unknown, ϕ, and Q is a source term. Generally we assume an approximate solution, say ϕ^*, for the spatial distribution of the unknown, say

$$\phi(x) \approx \phi^* = \sum_i^n h_i(x)\, \Phi_i^*, \tag{1.2}$$

where the $h_i(x)$ are spatial distributions associated with the coefficient Φ_i^*. That assumption leads to a corresponding assumption for the spatial gradient of the assumed behavior. Next we substitute these assumptions on spatial distributions into the differential equation. Since the assumption is approximate, this operation defines a residual error term, R, in the differential equation

$$L(\phi^*) - Q = R \neq 0. \tag{1.3}$$

Although we cannot force the residual term to vanish, it is possible to force a weighted integral, over the solution domain, of the residual to vanish. That is, the integral of the product of the residual term and some weighting function is set equal to zero, so that

$$I_i = \int_\Omega R\, w_i\, d\Omega = 0 \tag{1.4}$$

leads to the same number of equations as there are unknown Φ_i^* values. Most of the time we will find it very useful to employ integration by parts on this governing integral.

Substituting an assumed spatial behavior for the approximate solution, ϕ^*, and the weighting function, w, results in a set of algebraic equations that can be solved for the unknown nodal coefficients in the approximate solution. This is because the unknown coefficients can be pulled out of the spatial integrals involved in the assembly process.

The choice of weighting function defines the type of weighted residual technique being utilized. The Galerkin criterion selects

$$w_i = h_i(x), \tag{1.5}$$

to make the residual error orthogonal to the approximate solution. Use of integration by parts with the Galerkin procedure (i.e., the Divergence Theorem) reduces the continuity requirements of the approximating functions. If a Euler variational procedure exists, the Galerkin criterion will lead to the same element matrices.

A spatial interpolation, or blending function is assumed for the purpose of relating the quantity of interest within the element in terms of the values of the nodal parameters at the nodes connected to that particular element. For both weighted residual and variational formulations, the following restrictions are accepted for establishing convergence of the finite element model as the mesh refinement increases:

1. The element interpolation functions must be capable of modeling any constant values of the dependent variable or its derivatives, to the order present in the defining integral statement, in the limit as the element size decreases.
2. The element interpolation functions should be chosen so that at element interfaces the dependent variable and its derivatives, of one order less than those occurring in the defining integral statement, are continuous.

Through the assumption of the spatial interpolations, the variables of interest and their derivatives are uniquely specified throughout the solution domain by the nodal parameters associated with the nodal points of the system. The parameters at a particular node directly influence only the elements connected to that particular node. The domain will be split into a mesh. That will require that we establish some bookkeeping processes to keep up with data going to, or coming from a node or element. Those processes are commonly called gather and scatter, respectively. Figure 1.3 shows some of these processes for a simple mesh with one generalized scalar unknown per node, $n_g = 1$, in a one-dimensional physical space. There the system node numbers are shown numbered in an arbitrary fashion. To establish the local element space domain we must usually gather the coordinates of each of its nodes. For example, for element $b(= 5)$ it gathers the data for system node numbers 6 and 4, respectively, so that the element length, $L^{(5)} = x_6 - x_4$, can be computed. Usually we also have to gather some data on the coefficients in the differential equation (material properties). If the coefficients vary over space they may be supplied as data at the nodes that must also be gathered to form the element matrices.

After the element behavior has been described by spatial assumptions, then the derivatives of the space functions are used to approximate the spatial derivatives required in the integral form. The remaining fundamental problem is to establish the element matrices, $\mathbf{S}^e$ and $\mathbf{C}^e$. This involves substituting the approximation space functions and their derivatives into the governing integral form and moving the unknown coefficients, $\mathbf{D}^e$, outside the integrals. Historically, the resulting matrices have been called the element stiffness matrix and load vector, respectively.

Once the element equations have been established the contribution of each element is added, using its topology (or connectivity), to form the system equations. The system of algebraic equations resulting from FEA (of a linear system) will be of the form $\mathbf{S}\,\mathbf{D} = \mathbf{C}$. The vector $\mathbf{D}$ contains the unknown nodal parameters, and the matrices $\mathbf{S}$ and $\mathbf{C}$ are obtained by assembling the known element matrices, $\mathbf{S}^e$ and $\mathbf{C}^e$, respectively. Figure 1.3 shows how the local coefficients of the element source vector, $\mathbf{C}^e$, are scattered and added into the resultant system source, $\mathbf{C}$. That illustration shows a conversion of local row numbers to the corresponding system row numbers (by using the element connectivity data). An identical conversion is used to convert the local and system column numbers needed in assembling each $\mathbf{S}^e$ into $\mathbf{S}$. In the majority of problems $\mathbf{S}^e$, and thus, $\mathbf{S}$, will be symmetric. Also, the system square matrix, $\mathbf{S}$, is usually banded about the diagonal or at least *sparse*. If **S** is unsymmetric its upper and lower triangles have the same sparsity.

After the system equations have been assembled, it is necessary to apply the *essential boundary constraints* before solving for the unknown nodal parameters. The most common types of essential boundary conditions (EBC) are (1) defining explicit values of the parameter at a node and (2) defining constraint equations that are linear combinations of the unknown nodal quantities. The latter constraints are often referred to in the literature as *multi-point constraints* (MPC). An essential boundary condition should not be confused with a forcing condition of the type that involves a flux or traction on the boundary of one or more elements. These element boundary source, or forcing, terms contribute additional terms to the governing integral form and thus to the element square and/or column matrices for the elements on which the sources were applied. Thus, although these (*Neumann-type*, and *Robin* or *mixed-type*) conditions do enter into the system equations, their presence may not be obvious at the system level. Wherever essential boundary conditions do not act on part of the boundary, then at such locations, source terms from a lower order differential equation automatically apply. If one does not supply data for the source terms, then they default to zero. Such portions of the boundary are said to be subject to natural boundary conditions (NBC). The natural boundary condition varies with the integral form, and typical examples will appear later.

The initial sparseness (the relative percentage of zero entries) of the square matrix, **S**, is an important consideration since we only want to store non-zero terms. If we employ a direct solver then many initially zero terms will become non-zero during the solution process and the assigned storage must allow for that. The 'fill-in' depends on the numbering of the nodes. If the FEA system being employed does not have an automatic renumbering system to increase sparseness, then the user must learn how to number nodes (or elements) efficiently. After the system algebraic equations have been solved for the unknown nodal parameters, it is usually necessary to output the parameters, **D**. For every essential boundary condition on **D**, there is a corresponding unknown *reaction* term in **C** that can be computed after **D** is known. These usually have physical meanings and should be output to help check the results.

In rare cases the problem would be considered completed at this point, but in most cases it is necessary to use the calculated values of the nodal parameters to calculate other quantities of interest. For example, in stress analysis we use the calculated nodal displacements to solve for the strains and stresses. All adaptive programs must do a very large amount of post-processing to be sure that the solution, **D**, has been obtained to the level of accuracy specified by the analyst. Figure 1.3 also shows that the gather operation is needed again for extracting the local results, $\mathbf{D}^e$, from the total results, **D**, so they can be employed in special element post-processing and/or error estimates.

Usually the post-processing calculations involve determining the spatial derivatives of the solution throughout the mesh. Those gradients are continuous within each element domain, but are discontinuous across the inter-element boundaries. The true solution usually has continuous derivatives so it is desirable to somehow average the individual element gradient estimates to create continuous gradient estimate values that can be reported at the nodes. Fortunately, this addition gradient averaging process also provides new information that allows the estimate of the problem error norm to be calculated. That gradient averaging process will be presented in Chapters 2 and 6.

In the next chapter we will review the historical approach of the method of weighted residuals and its extension to finite element analysis. The earliest formulations for finite element models were based on variational techniques. This is especially true in the areas of structural mechanics and stress analysis. Modern analysis in these areas has come to rely on FEA almost exclusively. Variational models find the nodal parameters that yield a minimum (or stationary) value of an integral known as a functional. In most cases it is possible to assign a physical meaning to the integral. For example, in solid mechanics the integral represents the *total potential energy*, whereas in a fluid mechanics problem it may correspond to the rate of entropy production. Most physical problems with variational formulations result in quadratic forms that yield algebraic equations for the system which are symmetric and positive definite. The solution that yields a minimum value of the integral functional and satisfies the essential boundary conditions is equivalent to the solution of an associated differential equation. This is known as the Euler theorem.

Compared to the method of weighted residuals, where we start with the differential equation, it may seem strange to start a variational formulation with an integral form and then check to see if it corresponds to the differential equation we want. However, from Euler's work more than two centuries ago we know the variational forms of most even order differential equations that appear in science, engineering, and applied mathematics. This is especially true for elliptical equations. Euler's Theorem of Variational Calculus states that the solution, u, that satisfies the essential boundary conditions and renders stationary the functional

$$I = \int_{\Omega} f\left(x, y, z, \phi, \frac{\partial \phi}{\partial x}, \frac{\partial \phi}{\partial y}, \frac{\partial \phi}{\partial z} \right) d\Omega + \int_{\Gamma} \left(q\phi + a\phi^2/2 \right) d\Gamma \tag{1.6}$$

also satisfies the partial differential equation

$$\frac{\partial f}{\partial \phi} - \frac{\partial}{\partial x} \frac{\partial f}{\partial(\partial \phi/\partial x)} - \frac{\partial}{\partial y} \frac{\partial f}{\partial(\partial \phi/\partial y)} - \frac{\partial}{\partial z} \frac{\partial f}{\partial(\partial \phi/\partial z)} = 0 \tag{1.7}$$

in Ω, and satisfies the natural boundary condition that

$$n_x \frac{\partial f}{\partial(\partial \phi/\partial x)} + n_y \frac{\partial f}{\partial(\partial \phi/\partial y)} + n_z \frac{\partial f}{\partial(\partial \phi/\partial z)} + q + a\phi = 0 \tag{1.8}$$

on Γ that is not subject to an essential boundary. Here n_x, n_y, n_z are the components of the normal vector on the boundary, Γ. Note that this theorem also defines the natural boundary condition, as well as the corresponding differential equation. In Chapter 7 we will examine some common Euler variational forms for finite element analysis.

1.4 Assembly into the system equations

1.4.1 Introduction

An important but often misunderstood topic is the procedure for *assembling* the system equations from the element equations and any boundary contributions. Here assemblying is defined as the operation of adding the coefficients of the element equations into the proper locations in the system equations. There are various methods for accomplishing this but most are numerically inefficient. The numerically efficient *direct assembly* technique will be described here in some detail. We begin by reviewing the simple but important relationship between a set of local (nodal point, or element) degree of freedom numbers and the corresponding system degree of freedom numbers.

The assembly process, introduced in part in Fig. 1.3, is graphically illustrated in Fig. 1.4 for a mesh consisting of six nodes ($n_m = 6$), three elements ($n_e = 3$). It has a four-node quadrilateral and two three-node triangles, with one generalized parameter per node ($n_g = 1$). The top of the figure shows the nodal connectivity of the three elements and a cross-hatching to define the source of the various coefficients that are occurring in the matrices assembled in the lower part of the figure. The assembly of the system **S** and **C** matrices is graphically coded to denote the sources of the contributing terms but not their values. A hatched area indicates a term that was added in from an element that has the same hash code. For example, the load vector term C(6), coming from the only parameter at node 6, is seen to be the sum of contributions from elements 1 and 2, which are hatched with horizontal (-) and vertical (|) lines, respectively. The connectivity table implies the same thing since node 6 is only connected to those two elements. By way of comparison, the term C(1) has a contribution only from element 2. The connectivity table shows only that element is connected to that corner node.

Note that we have to set **S** = **0** to begin the summation. Referring to Fig. 1.4 we see that 10 of the coefficients in **S** remain initially zero. So that example is initially about 27 percent sparse. (This will changed if a direct solution process is used.) In practical problems the assembled matrix may initially be 90 percent sparse, or more. Special equation solving techniques take advantage of this feature to save on memory and operation counts.

1.4.2 Computing the equation index

There are a number of ways to assign the equation numbers of the algebraic system that results from a finite element analysis procedure. Here we will select one that has a simple equation that is valid for most applications. Consider a typical nodal point in the system and assume that there are n_g parameters associated with each node. Thus, at a typical node there will be n_g local degree of freedom numbers ($1 \le J \le n_g$) and a corresponding set of system degree of freedom numbers. If I denotes the system node number of the point, then the n_g corresponding system degrees of freedom, Φ_k have their equation number, k assigned as

$$k(I, J) = n_g * (I - 1) + J \qquad 1 \le I \le n_m, \qquad 1 \le J \le n_g, \tag{1.9}$$

where n_m is the maximum node number in the system. That is, they start at 1 and range to n_g at the first system node then at the second node they range from $(n_g + 1)$ to $(2\, n_g)$

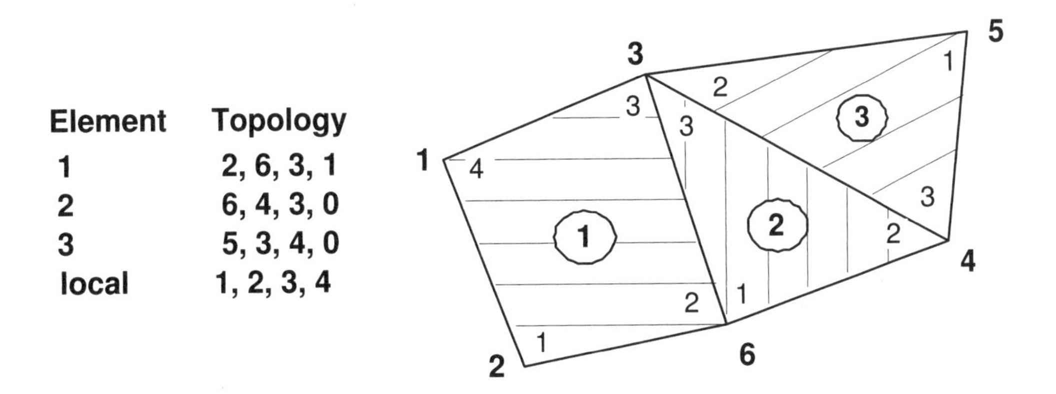

Global assembly: S * D = C

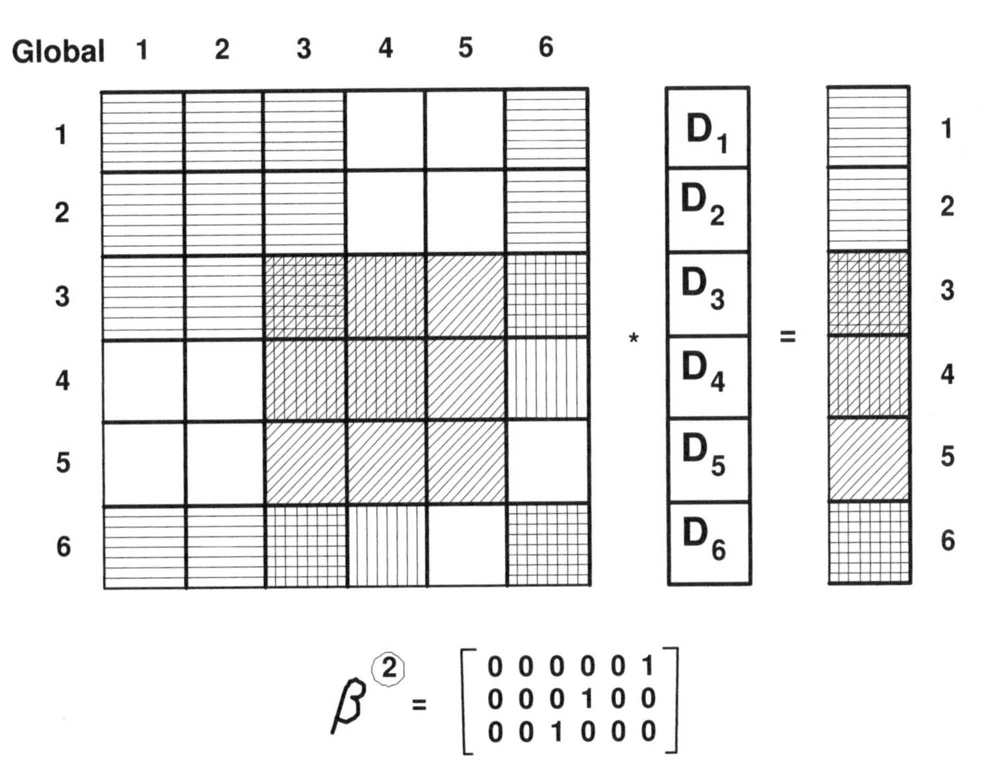

$$\beta^{(2)} = \begin{bmatrix} 0 & 0 & 0 & 0 & 0 & 1 \\ 0 & 0 & 0 & 1 & 0 & 0 \\ 0 & 0 & 1 & 0 & 0 & 0 \end{bmatrix}$$

Figure 1.4 *Graphical illustration of matrix assembly*

```
FUNCTION GET_INDEX_AT_PT (I_PT) RESULT (INDEX)                      ! 1
! * * * * * * * * * * * * * * * * * * * * * * * * * * *             ! 2
!  DETERMINE DEGREES OF FREEDOM NUMBERS AT A NODE                   ! 3
! * * * * * * * * * * * * * * * * * * * * * * * * * * *             ! 4
Use System_Constants ! for N_G_DOF                                  ! 5
 IMPLICIT NONE                                                      ! 6
 INTEGER, INTENT(IN)  :: I_PT                                       ! 7
 INTEGER              :: INDEX (N_G_DOF)                            ! 8
 INTEGER :: J ! implied loop                                        ! 9
                                                                    !10
! N_G_DOF = NUMBER OF PARAMETERS (DOF) PER NODE                     !11
! I_PT    = SYSTEM NODE NUMBER                                      !12
! INDEX   = SYSTEM DOF NOS OF NODAL DOF                             !13
!           INDEX (J) = N_G_DOF*(I_PT - 1) + J                      !14
                                                                    !15
  INDEX = (/ (N_G_DOF*(I_PT - 1) + J, J = 1, N_G_DOF) /)            !16
END FUNCTION GET_INDEX_AT_PT                                        !17

FUNCTION GET_ELEM_INDEX (LT_N, ELEM_NODES) RESULT(INDEX)            ! 1
! * * * * * * * * * * * * * * * * * * * * * * * * * * * *           ! 2
!   DETERMINE DEGREES OF FREEDOM NUMBERS OF ELEMENT                 ! 3
! * * * * * * * * * * * * * * * * * * * * * * * * * * * *           ! 4
Use System_Constants    ! for N_G_DOF                               ! 5
 IMPLICIT NONE                                                      ! 6
 INTEGER, INTENT(IN) :: LT_N, ELEM_NODES (LT_N)                     ! 7
 INTEGER             :: INDEX (LT_N * N_G_DOF)    ! OUT             ! 8
 INTEGER :: EQ_ELEM, EQ_SYS, IG, K, SYS_K         ! LOOPS           ! 9
                                                                    !10
! ELEMNODES = NODAL INCIDENCES OF THE ELEMENT                       !11
! EQ_ELEM   = LOCAL  EQUATION NUMBER                                !12
! EQ_SYS    = SYSTEM EQUATION NUMBER                                !13
! INDEX     = SYSTEM DOF NUMBERS OF ELEMENT DOF NUMBERS             !14
! INDEX (N_G_DOF*(K-1)+IG) = N_G_DOF*(ELEM_NODES(K)-1) + IG         !15
! LT_N      = NUMBER OF NODES PER ELEMENT                           !16
! N_G_DOF   = NUMBER OF GENERAL PARAMETERS (DOF) PER NODE           !17
                                                                    !18
  DO K = 1, LT_N                  ! LOOP OVER NODES OF ELEMENT      !19
    SYS_K = ELEM_NODES (K)        ! SYSTEM NODE NUMBER              !20
    DO IG = 1, N_G_DOF            ! LOOP OVER GENERALIZED DOF       !21
      EQ_ELEM = IG + N_G_DOF * (K     - 1)      ! LOCAL  EQ         !22
      EQ_SYS  = IG + N_G_DOF * (SYS_K - 1)      ! SYSTEM EQ         !23
      IF ( SYS_K > 0 ) THEN              ! VALID NODE               !24
        INDEX (EQ_ELEM) = EQ_SYS                                    !25
      ELSE                               ! ALLOW MISSING NODE       !26
        INDEX (EQ_ELEM) = 0                                         !27
      END IF ! MISSING NODE                                         !28
    END DO ! OVER DOF                                               !29
  END DO ! OVER LOCAL NODES                                         !30
END FUNCTION GET_ELEM_INDEX                                         !31
```

Figure 1.5 *Computing equation numbers for homogeneous nodal dof*

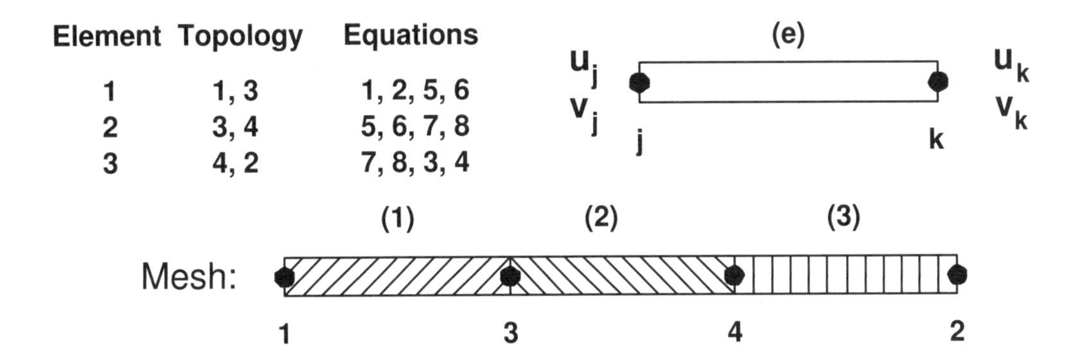

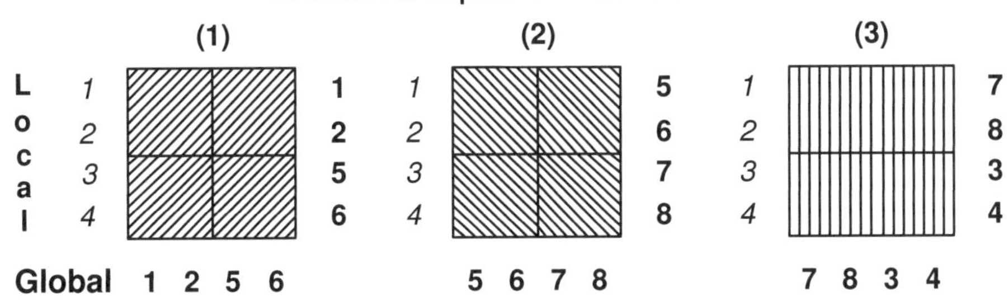

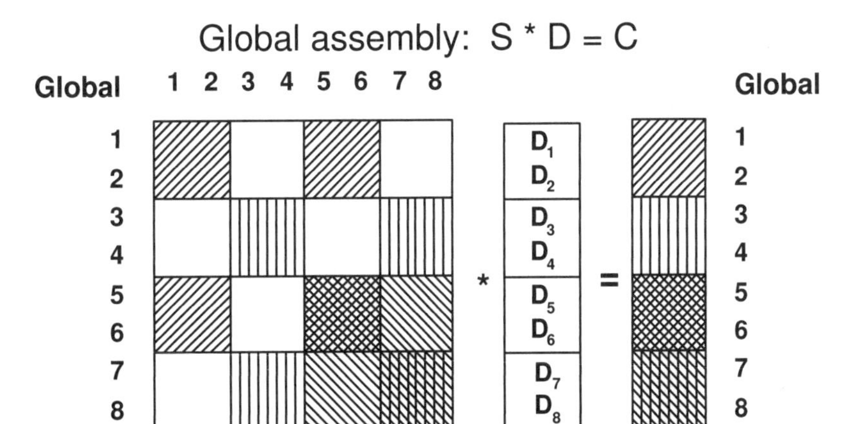

Figure 1.6 *Assembling two unknowns per node*

```
SUBROUTINE STORE_COLUMN (N_D_FRE, N_EL_FRE, INDEX, C, CC)        ! 1
! * * * * * * * * * * * * * * * * * * * * * * * * * * * *        ! 2
! STORE ELEMENT COLUMN MATRIX IN SYSTEM COLUMN MATRIX             ! 3
! * * * * * * * * * * * * * * * * * * * * * * * * * * * *        ! 4
Use Precision_Module ! Defines DP for double precision           ! 5
 IMPLICIT NONE                                                   ! 6
 INTEGER,  INTENT(IN)    :: N_D_FRE, N_EL_FRE                    ! 7
 INTEGER,  INTENT(IN)    :: INDEX (N_EL_FRE)                     ! 8
 REAL(DP), INTENT(IN)    :: C      (N_EL_FRE)                    ! 9
 REAL(DP), INTENT(INOUT) :: CC     (N_D_FRE)                     !10
 INTEGER :: I, J                                                 !11
                                                                 !12
! N_D_FRE  = NO DEGREES OF FREEDOM IN THE SYSTEM                 !13
! N_EL_FRE = NUMBER OF DEGREES OF FREEDOM PER ELEMENT            !14
! INDEX    = SYSTEM DOF NOS OF THE ELEMENT DOF                   !15
! C        = ELEMENT COLUMN MATRIX                               !16
! CC       = SYSTEM COLUMN MATRIX                                !17
                                                                 !18
  DO I = 1, N_EL_FRE                       ! ELEMENT ROW         !19
    J = INDEX (I)                          ! SYSTEM ROW NUMBER   !20
    IF ( J > 0 ) CC (J) = CC (J) + C (I)   ! SKIP INACTIVE ROW   !21
  END DO ! OVER ROWS                                             !22
END SUBROUTINE STORE_COLUMN                                      !23

SUBROUTINE STORE_FULL_SQUARE (N_D_FRE, N_EL_FRE, S, SS, INDEX)   ! 1
! * * * * * * * * * * * * * * * * * * * * * * * * * * * * * * *  ! 2
!     STORE ELEMENT SQ MATRIX IN FULL SYSTEM SQ MATRIX           ! 3
! * * * * * * * * * * * * * * * * * * * * * * * * * * * * * * *  ! 4
Use Precision_Module ! Defines DP for double precision           ! 5
 IMPLICIT NONE                                                   ! 6
 INTEGER,  INTENT(IN)    :: N_D_FRE, N_EL_FRE                    ! 7
 INTEGER,  INTENT(IN)    :: INDEX (N_EL_FRE)                     ! 8
 REAL(DP), INTENT(IN)    :: S      (N_EL_FRE, N_EL_FRE)          ! 9
 REAL(DP), INTENT(INOUT) :: SS     (N_D_FRE,  N_D_FRE)           !10
 INTEGER   :: I, II, J, JJ                                       !11
                                                                 !12
! N_D_FRE  = TOTAL NO OF SYSTEM DEGREES OF FREEDOM               !13
! N_EL_FRE = NO DEGREES OF FREEDOM PER ELEMENT                   !14
! INDEX    = SYSTEM DOF NOS OF ELEMENT PARAMETERS                !15
! S        = FULL ELEMENT SQUARE MATRIX                          !16
! SS       = FULL SYSTEM SQUARE MATRIX                           !17
                                                                 !18
  DO I = 1, N_EL_FRE                       ! ELEMENT ROW         !19
    II = INDEX (I)                         ! SYSTEM ROW NUMBER   !20
    IF ( II > 0 ) THEN                     ! SKIP INACTIVE ROW   !21
      DO J = 1, N_EL_FRE                   ! ELEMENT COLUMN      !22
        JJ = INDEX (J)                     ! SYSTEM COLUMN       !23
        IF ( JJ > 0 ) SS (II, JJ) = SS (II, JJ) + S (I, J)       !24
      END DO ! OVER COLUMNS                                      !25
    END IF                                                       !26
  END DO ! OVER ROWS                                             !27
END SUBROUTINE STORE_FULL_SQUARE                                 !28
```

Figure 1.7 *Assembly of element arrays into system arrays*

and so on through the mesh. These elementary calculations are carried out by subroutine *GET_INDEX_AT_PT*. The program assigns n_g storage locations for the vector, say *INDEX*, containing the system degree of freedom numbers associated with the specified nodal point — see Tables 1.3 and 1.4 for the related details.

A similar expression defines the local equation numbers in an element or on a boundary segment. The difference is that then I corresponds to a local node number and has an upper limit of n_n or n_b, respectively. In the latter two cases the local equation number is the subscript for the *INDEX* array and the corresponding system equation number is the integer value stored in *INDEX* at that position. In other words, Eq. 1.9 is used to find local or system equation numbers and J always refers to the specific dof of interest and I corresponds to the type of node number (I_S for a system node, I_E for a local element node, or I_B for a local boundary segment node). For a typical element type subroutine GET_ELEM_INDEX, Fig. 1.5, fills the above element *INDEX* vector for any standard or boundary element. In that subroutine storage locations are likewise established for the n_n element incidences (extracted by subroutine GET_ELEM_NODES) and the corresponding $n_i = n_n \times n_g$ system degree of freedom numbers associated with the element, in vector *INDEX*.

Figure 1.6 illustrates the use of Eq. 1.9 for calculating the system equation numbers for $n_g = 2$ and $n_m = 4$. The **D** vector in the bottom portion shows that at each node we count its dof before moving to the next node. In the middle section the cross-hatched element matrices show the 4 local equation numbers to the left of the square matrix, and the corresponding system equation numbers are shown to the right of the square matrix, in bold font. Noting that there are $n_g = 2$ dof per node explains why the top left topology list (element connectivity with $n_n = 2$) is expanded to the system equation number list with 4 columns.

Once the system degree of freedom numbers for the element have been stored in a vector, say *INDEX*, then the subscripts of a coefficient in the element equation can be directly converted to the subscripts of the corresponding system coefficient to which it is to be added. This correspondence between local and system subscripts is illustrated in Fig. 1.6. The expressions for these assembly, or scatter, operations are generally of the form

$$C_I = C_I + C_i^e, \qquad S_{I,J} = S_{I,J} + S_{i,j}^e \tag{1.10}$$

where i and j are the local subscripts of a coefficient in the element square matrix, $\mathbf{S}^e$, and I, J are the corresponding subscripts of the system equation coefficient, in **S**, to which the element contribution is to be added. The direct conversions are given by $I = INDEX(i)$, $J = INDEX(j)$, where the *INDEX* array for element, e, is generated from Eq. 1.1 by subroutine GET_ELEM_INDEX.

Figure 1.5 shows how that index could be computed for a node or element for the common case where the number of generalized degrees of freedom per node is everywhere constant. For a single unknown per node ($n_g = 1$), as shown in Fig. 1.4, then the nodal degree of freedom loop (at lines 16 and 21 in Fig. 1.5) simply equates the equation number to the global node number. An example where there are two unknowns per node is illustrated in Fig. 1.6. That figure shows a line element mesh with two nodes per element and two dof per node (such as a standard beam element). In that case it is similar to the assembly of Fig. 1.4, but instead of a single coefficient we are adding a set of smaller square sub-matrices into **S**. Figure 1.7 shows how the assembly can be implemented for column matrices (subroutine STORE_COLUMN) and full (non-sparse)

Table 1.3 *Degree of freedom numbers at system node* $\boldsymbol{I_s}$

Local	System[†]
1	$INDEX(1)$
2	$INDEX(2)$
•	•
•	•
•	•
J	$INDEX(J)$
•	•
•	•
•	•
n_g	$INDEX(n_g)$

[†] $INDEX(J) = n_g * (I_s - 1) + J$

Table 1.4 *Relating local and system equation numbers*

Local node	Parameter number	System node	Element degree of freedom numbers	
			Local	System
I_L	J	$I_S = node(I_L)$	$n_g * (I_L - 1) + J$	$n_g * (I_S - 1) + J$
1	1	$node(1)$	1	$n_g * [node(1)-1] + 1$
1	2	$node(1)$	2	
•	•	•		
•	•	•	•	•
1	n_g	$node(1)$	•	•
2	1	$node(2)$	•	•
•	•	•		
•	•	•		
•	•	•		
K	J_g	$node(K)$	$n_g * (K-1) + J_g$	$n_g * [node(K)-1] + J_g$
•	•	•		
•	•	•	•	•
n_n	1	$node(n_n)$	•	•
•	•	•	•	•
•	•	•		
n_n	n_g	$node(n_n)$	$n_n * n_g$	$n_g * [node(n_n)-1] + n_g$

square matrices (STORE_FULL_SQUARE) if one has an integer index that relates the local element degrees of freedom to the system dof.

1.4.3 Example equation numbers

Consider a two-dimensional problem ($n_s = 2$) involving 400 nodal points ($n_m = 400$) and 35 elements ($n_e = 35$). Assume two parameters per node ($n_g = 2$) and let these parameters represent the horizontal and vertical components of some vector. In a stress analysis problem, the vector could represent the displacement vector of the node, whereas in a fluid flow problem it could represent the velocity vector at the nodal point. Assume the elements to be triangular with three corner nodes ($n_n = 3$). The local numbers of these nodes will be defined in some consistent manner, e.g., by numbering counter-clockwise from one corner. This mesh is illustrated in Fig. 1.8.

By utilizing the above control parameters, it is easy to determine the total number of degrees of freedom in the system, n_d, and associated with a typical element, n_i are: $n_d = n_m * n_g = 400 * 2 = 800$, and $n_i = n_n * n_g = 3 * 2 = 6$, respectively. In addition to the total number of degrees of freedom in the system, it is important to be able to identify the system degree of freedom number that is associated with any parameter in the system. Table 1.4, or subroutine GET_DOF_INDEX, provides this information. This relation has many practical uses. For example, when one specifies that the first parameter ($J = 1$) at system node 50 ($I_S = 50$) has some given value what one is indirectly saying is that system degree of freedom number $DOF = 2 * (50 - 1) + 1 = 99$ has a given value. In a

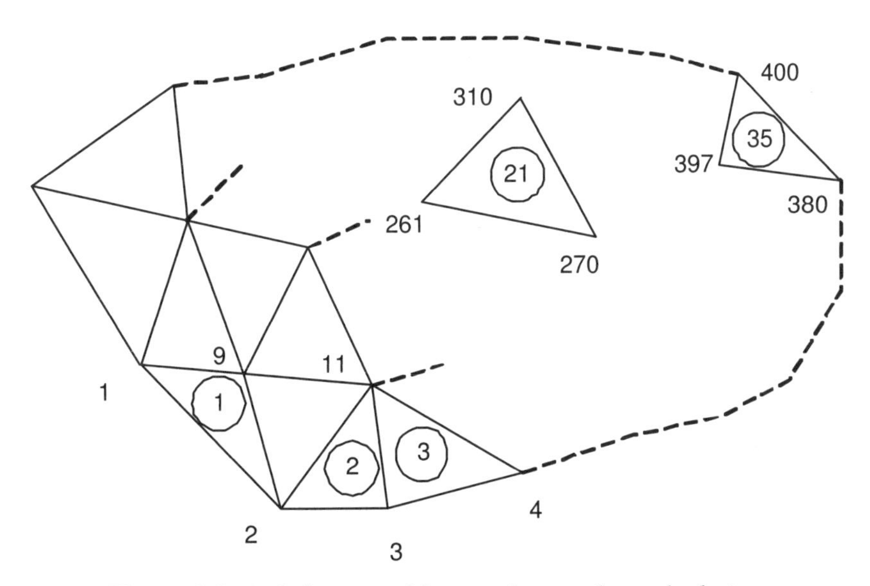

Figure 1.8 *Mesh for assembly equation number calculations*

Table 1.5 *Example mesh connectivity*

Element	Element indices		
1	1	2	9
2	2	3	11
3	3	4	11
⋮	⋮	⋮	⋮
21	261	270	310
⋮	⋮	⋮	⋮
35	397	398	400
	Array NODES		

Table 1.6 *Example element and system equation numbers*

Node local i_L	Number system I_S	Parameter number J	Degree of Freedom System DOF_S	Degree of Freedom Local DOF_L
1	261	1	521	1
1	261	2	522	2
2	270	1	539	3
2	270	2	540	4
3	310	1	619	5
3	310	2	620	6
	Array *ELEM_NODES*		Array *INDEX*	

similar manner, we often need to identify the system degree of freedom numbers that correspond to the n_i local degrees of freedom of the element. In order to utilize Eq. 1.9 to do this, one must be able to identify the n_n node numbers associated with the element of interest. This is relatively easy to accomplish since those data are part of the input data (element incidences). For example from Table 1.5, for element number 21 we find the three element incidences (row 21 of the system connectivity array) to be

System	261	270	310	← Array ELEM_NODES
Local	1	2	3	$(1 \times n_n)$

Therefore, by applying Eq. 1.9, we find the degree of freedom numbers in Table 1.6. The element array *INDEX* has many programming uses. Its most important application is to aid in the assembly (scatter) of the element equations to form the governing system equations. We see from Fig. 1.5 that the element equations are expressed in terms of local degree of freedom numbers. In order to add these element coefficients into the system equations one must identify the relation between the local degree of freedom numbers and the corresponding system degree of freedom numbers. Array *INDEX* provides this information for a specific element. In practice, the assembly procedure is as follows. First the system matrices **S** and **C** are set equal to zero. Then a loop over all the elements if performed. For each element, the element matrices are generated in terms of the local degrees of freedom. The coefficients of the element matrices are added to the corresponding coefficients in the system matrices. Before the addition is carried out, the element array *INDEX* is used to convert the local subscripts of the coefficient to the system subscripts of the term in the system equations to which the coefficient is to be added. That is, we scatter

$$S^e_{i,j} \overset{+}{\rightarrow} S_{I,J}, \qquad C^e_i \overset{+}{\rightarrow} C_I \tag{1.11}$$

where $I_S = INDEX(i_L)$ and $J_S = INDEX(j_L)$ are the corresponding row and column numbers in the system equations, i_L, j_L are the subscripts of the coefficients in terms of the local degrees of freedom, and the symbol $\overset{+}{\rightarrow}$ reads as 'is added to'. Considering all of the terms in the element matrices for element 21 in the previous example, one finds six typical scatters from the $\mathbf{S}^e$ and $\mathbf{C}^e$ arrays are

$$\begin{array}{ll}
S^e_{1,1} \overset{+}{\rightarrow} S_{521,521} & C^e_1 \overset{+}{\rightarrow} C_{521} \\
S^e_{2,3} \overset{+}{\rightarrow} S_{522,539} & C^e_2 \overset{+}{\rightarrow} C_{522} \\
S^e_{3,4} \overset{+}{\rightarrow} S_{539,540} & C^e_3 \overset{+}{\rightarrow} C_{539} \\
S^e_{4,5} \overset{+}{\rightarrow} S_{540,620} & C^e_4 \overset{+}{\rightarrow} C_{540} \\
S^e_{5,6} \overset{+}{\rightarrow} S_{619,620} & C^e_5 \overset{+}{\rightarrow} C_{619} \\
S^e_{1,6} \overset{+}{\rightarrow} S_{521,620} & C^e_6 \overset{+}{\rightarrow} C_{620}\,.
\end{array}$$

1.5 Error concepts

Part of the emphasis of this book will be on error analysis in finite element studies. Thus this will be a good point to mention some of the items that will be of interest to us later. We will always employ integral forms. Denote the highest derivative occurring in the integral by the integer m. Assume all elements have the same shape and use the same interpolation polynomial. Let the characteristic element length size be the real value h, and assume that we are using a complete polynomial of integer degree p. Later we will

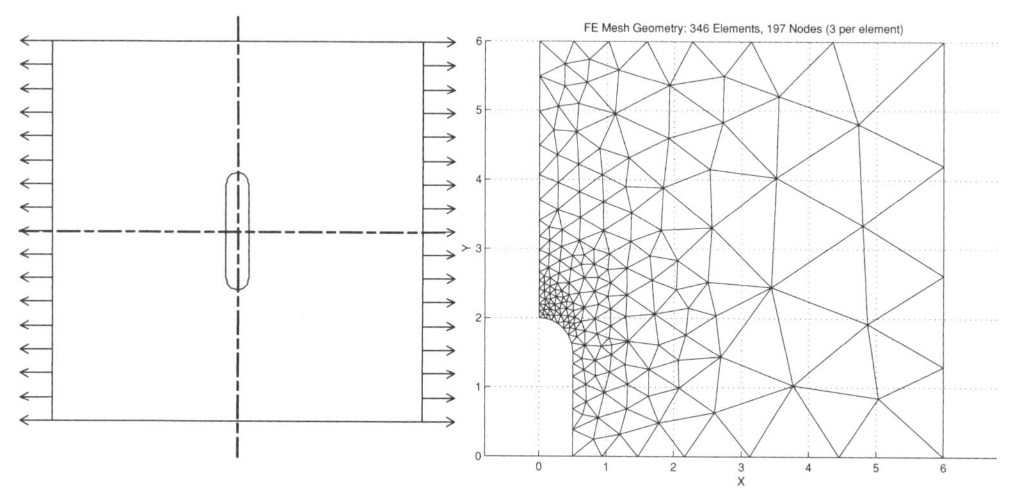

Figure 1.9 *Relating element size to expected gradients*

be interested in the asymptotic convergence rate, in some norm, as the element size approaches zero, $h \rightarrow 0$. Here we will just mention the point wise error that provides the insight into creating a good manual mesh or a reasonable starting point for an adaptive mesh. For a problem with a smooth solution the local finite element error is proportional to the product of the $m-th$ derivative at the point and the element size, h, raised to the $p-th$ power. That is,

$$e(x) \propto h^p \, \partial^m u(\mathbf{x}) / \partial \mathbf{x}^m \, . \qquad (1.12)$$

This provides some engineering judgement in establishing an initial mesh. Where you expect the gradients (the $m-th$ derivative) to be high then make the elements very small. Conversely, where the gradients will be zero or small we can have large elements to reduce the cost. These concepts are illustrated in Fig. 1.9 where the stresses around a hole in a large flat plate are shown. There we see linear three noded triangles (so $p = 1$) in a quarter symmetry mesh. Later we will show that the integral form contains the first derivatives (gradient, so $m = 1$). Undergraduate studies refer to this as a stress concentration problem and show that the gradients rapidly increase by a factor of about 3 over a very small region near the top of the hole. Thus we need small elements there. At the far boundaries the tractions are constant so the gradient of the displacements are nearly zero there and the elements can be big. Later we will automate estimating local error calculations and the associated element size changes needed for an accurate and cost effective solution.

1.6 Exercises

1. Assume (unrealistically) that all the entries in an element square matrix and column vector are equal to the element number. Carry out the assembly of the system in

Fig. 1.4 to obtain the final numerical values for each coefficient in the **S** and **C** matrices. Hint: manually loop over each element and carry out the line by line steps given in *GET_ELEM_INDEX*, and then those in *STORE_COLUMN*, and finally those in *STORE_FULL_SQUARE* before going to the next element.

2. Assume (unrealistically) that all the entries in an element square matrix and column vector are equal to the element number. Carry out the assembly of the system in Fig. 1.6 to obtain the final numerical values for each coefficient in the **S** and **C** matrices.

3. In Fig. 1.6 assume that the global nodes are numbered consecutively from 1 to 4 (from left to right). Write the element index vector for each of the three elements.

4. List the topology (connectivity data) for the six elements in Fig. 1.3.

5. Why does the β Boolean array in Fig. 1.3 have two rows and seven columns?

6. What is the Boolean array, β, for element $a(= 2)$ in Fig. 1.3?

7. What is the percent of sparsity of the **S** matrix in Fig. 1.6?

8. What is the Boolean array, β, for element 3 in Fig. 1.4?

9. What is the size of the Boolean array, β, for any element in Fig. 1.6? Explain why.

10. What is the Boolean array, β, for element 1 in Fig. 1.6?

11. Referring to Fig. 1.4, multiply the given 6×1 **D** array by the 3×6 Boolean array, β, for the second element to gather its corresponding $\mathbf{D}^e$ local dof.

12. In an FEA stress analysis where a translational displacement is prescribed the reaction necessary for equilibrium is _____: a) heat flux, b) force vector, c) pressure, d) temperature, e) moment (couple) vector.

13. In an FEA stress analysis where a rotational displacement is prescribed the reaction necessary for equilibrium is ____: a) heat flux, b) force vector, c) pressure, d) temperature, e) moment (couple) vector.

14. In an FEA thermal analysis where a temperature is prescribed the reaction necessary for equilibrium is _____: a) heat flux, b) force vector, c) pressure, d) temperature, e) moment (couple) vector.

15. A material that is the same at all points is ___: a) homogeneous, b) non-homogeneous, c) isotropic, d) anisotropic, e) orthotropic.

16. A material that is the same in all directions at a point is ___: a) homogeneous, b) non-homogeneous, c) isotropic, d) anisotropic, e) orthotropic.

17. A material that has at least 2 preferred directions is ___: a) homogeneous, b) non-homogeneous, c) isotropic, d) anisotropic, e) orthotropic.

18. A material with the most general directional dependence is ___: a) homogeneous, b) non-homogeneous, c) isotropic d) anisotropic, e) orthotropic.

19. Define a scalar, vector, and tensor quantity to have zero, one, and two subscripts, respectively. Identify which of the above describe the following items: _____ mass, _____ time, _____ position, _____ centroid, _____ volume, _____ surface area, _____ displacement, _____ temperature, _____ heat flux, _____ heat source, _____ stress, _____ moment of inertia, _____ force, _____ moment, _____ velocity.
20. In a finite element solution at a node, the _____: a) primary variable is most accurate, b) primary variable is least accurate, c) secondary variable is most accurate, d) secondary variable is least accurate.
21. Interior to an element in an FEA solution, the _____: a) primary variable is most accurate, b) primary variable is least accurate, c) secondary variable is most accurate, d) secondary variable is least accurate.
22. An eigen-problem will define two independent square matrices. Can you still use Eqs. 1.9-10 to assemble each of them? If so why?
23. A transient problem will define two independent square matrices. Can you still use Eqs. 1.9-10 to assemble each of them? If so why?
24. A dynamics problem defines two or three independent square matrices. Can you still use Eqs. 1.9-10 to assemble each of them? If so why?
25. If we allow the number of generalized unknowns, n_g, to vary at each node rather than being a global constant then Eq. 1.9 is invalid. Assume that the number of unknowns at each node are available as an input array, say $DOF\_PT(1:n_m)$. Create a new array, $DOF\_PT\_SUM(0:n_m)$, and use it to develop an alternate to Eq. 1.9 to find the equation number for the $j-th$ unknown at the $i-th$ node.

1.7 Bibliography

[1] Adams, V. and Askenazi, A., *Building Better Products with Finite Element Analysis*, Santa Fe: Onword Press (1999).

[2] Akin, J.E., *Finite Elements for Analysis and Design*, London: Academic Press (1994).

[3] Akin, J.E. and Singh, M., "Object-Oriented Fortran 90 P-Adaptive Finite Element Method," pp. 141–149 in *Developments in Engineering Computational Technology*, ed. B.H.V. Topping, Edinburgh: Civil_Comp Press (2000).

[4] Axelsson, O. and Baker, V.A., *Finite Element Solution of Boundary Value Problems*, Philadelphia, PA: SIAM (2001).

[5] Bathe, K.J., *Finite Element Procedures*, Englewood Cliffs: Prentice Hall (1996).

[6] Becker, E.B., Carey, G.F., and Oden, J.T., *Finite Elements – An Introduction*, Englewood Cliffs: Prentice Hall (1981).

[7] Cook, R.D., Malkus, D.S., and Plesha, N.E., *Concepts and Applications of Finite Element Analysis*, New York: John Wiley (1989).

[8] Cook, R.D., Malkus, D.S., Plesha, N.E., and Witt, R.J., *Concepts and Applications of Finite Element Analysis*, New York: John Wiley (2002).

[9] Desai, C.S. and Kundu, T., *Introduction to the Finite Element Method*, Boca Raton: CRC Press (2001).

[10] Gupta, K.K. and Meek, J.L., *Finite Element Multidisciplinary Analysis*, Reston: AIAA (2000).

[11] Huebner, K.H., Thornton, E.A., and Byrom, T.G., *Finite Element Method for Engineers*, New York: John Wiley (1994).

[12] Hughes, T.J.R., *The Finite Element Method*, Mineola: Dover Publications (2003).

[13] Kwon, Y.W. and Bang, H., *The Finite Element Method using Matlab*, Boca Raton: CRC Press (1997).

[14] Logan, D.L., *A First Course in the Finite Element Method,* 3rd Edition, Pacific Grove: Brooks Cole (2002).

[15] Nassehi, V., *Practical Aspects of Finite Element Modeling of Polymer Processing*, New York: John Wiley (2002).

[16] Norrie, D.H. and DeVries, G., *Finite Element Bibliography*, New York: Plenum Press (1976).

[17] Pironneau, O., *Finite Element Methods for Fluids*, New York: John Wiley (1991).

[18] Portela, A. and Charafi, A., *Finite Elements Using Maple*, Berlin: Springer (2002).

[19] Rao, S.S., *The Finite Element Method in Engineering*, Boston: Butterworth Heinemann (1999).

[20] Reddy, J.N., *An Introduction to the Finite Element Method,* 2nd Edition, McGraw-Hill (1993).

[21] Segerlind, L.J., *Applied Finite Element Analysis*, New York: John Wiley (1987).

[22] Silvester, P.P. and Ferrari, R.L., *Finite Elements for Electrical Engineers*, Cambridge: Cambridge University Press (1996).

[23] Smith, I.M. and Griffiths, D.V., *Programming the Finite Element Method,* 3rd Edition, Chichester: John Wiley (1998).

[24] Szabo, B. and Babuska, I., *Finite Element Analysis*, New York: John Wiley (1991).

[25] Zienkiewicz, O.C. and Taylor, R.L., *The Finite Element Method,* 4th Edition, New York: McGraw-Hill (1991).

[26] Zienkiewicz, O.C. and Taylor, R.L., *The Finite Element Method,* 5th Edition, London: Butterworth-Heinemann (2000).

Chapter 2

Mathematical preliminaries

2.1 Introduction

The earliest forms of finite element analysis were based on physical intuition with little recourse to higher mathematics. As the range of applications expanded, for example to the theory of plates and shells, some physical approaches failed and some succeeded. The use of higher mathematics such as variational calculus explained why the successful methods worked. At the same time the mathematicians were attracted by this new field of study. In the last few years the mathematical theory of finite element analysis has grown quite large. Since the state of the art now depends heavily on error estimators and error indicators it is necessary for an engineer to be aware of some basic mathematical topics of finite element analysis. We will consider load vectors and solution vectors, and residuals of various weak forms. All of these require us to define some method to 'measure' these entities. For the above linear vectors with discrete coefficients, $\mathbf{V}^T = [\, V_1 \; V_2 \cdots V_n \,]$, we might want to use a measure like the root mean square, *RMS*:

$$RMS^2 = \frac{1}{n} \sum_{i=1}^{n} V_i^2 = \frac{1}{n} \mathbf{V}^T \mathbf{V}$$

which we will come to call a norm of the linear vector space. Other quantities vary with spatial position and appear in integrals over the solution domain and/or its boundaries. We will introduce various other norms to measure these integral quantities.

The finite element method always involves integrals so it is useful to review some integral identities such as Gauss' Theorem (Divergence Theorem):

$$\int_\Omega \nabla \cdot \mathbf{u} \, d\Omega = \int_\Gamma \mathbf{u} \cdot \mathbf{n} \, d\Gamma = \int_\Gamma \frac{\partial \mathrm{u}}{\partial \mathrm{n}} \, d\Gamma$$

which is expressed in Cartesian tensor form as

$$\int_\Omega u_{i,i} \, d\Omega = \int_\Gamma u_i \, n_i \, d\Gamma$$

where there is an implied summation over subscripts that occurs an even number of times and a comma denotes partial differentiation with respect to the directions that follow it. That is, $(\,)_{,i} = \partial(\,)/\partial x_i$. The above theorem can be generalized to a tensor with any number of subscripts:

$$\int_{\Omega} A_{ijk...q,r}\, d\Omega = \int_{\Gamma} A_{ijk...q}\, n_r\, d\Gamma.$$

We will often have need for one of the Green's Theorems:

$$\int_{\Omega} (\nabla A \cdot \nabla B + A\nabla^2 B)\, d\Omega = \int_{\Gamma} A \frac{\partial B}{\partial n}\, d\Gamma$$

and

$$\int_{\Gamma} (A\nabla^2 B - B\nabla^2 A)\, d\Omega = \int_{\Gamma} (A\nabla B - B\nabla A) \cdot \mathbf{n}\, d\Gamma$$

which in Cartesian tensor form are

$$\int_{\Omega} (A_{,i} B_{,i} + AB_{,ii})\, d\Omega = \int_{\Gamma} AB_{,i}\, n_i\, d\Gamma$$

and

$$\int_{\Omega} (AB_{,ii} - BA_{,ii})\, d\Omega = \int_{\Gamma} (AB_{,i} - BA_{,i})\, n_i\, d\Gamma .$$

We need these relations to derive the Galerkin weak form statements and to manipulate the associated error estimators. Usually, we are interested in removing the highest derivative term in an integral and use the second from last equation in the form

$$\int_{\Omega} AB_{,ii}\, d\Omega = \int_{\Gamma} AB_{,i}\, n_i\, d\Gamma - \int_{\Omega} A_{,i} B_{,i}\, d\Omega . \tag{2.1}$$

In one-dimensional applications this process is called integration by parts:

$$\int_a^b p\, dq = pq \Big|_a^b - \int_a^b q\, dp.$$

Error estimator proofs utilize inequalities like the Schwarz inequality

$$|\mathbf{a} \cdot \mathbf{b}| \leq |\mathbf{a}|\, |\mathbf{b}| \tag{2.2}$$

and the triangle inequality

$$|\mathbf{a} + \mathbf{b}| \leq |\mathbf{a}| + |\mathbf{b}|. \tag{2.3}$$

Finite element error estimates often use the Minkowski inequality

$$\left[\sum_{i=1}^{n} |x_i \pm y_i|^p \right]^{1/p} \leq \left[\sum_{i=1}^{n} |x_i|^p \right]^{1/p} + \left[\sum_{i=1}^{n} |y_i|^p \right]^{1/p}, \quad 1 < p < \infty, \tag{2.4}$$

and the corresponding integral inequality

$$\left[\int_{\Omega} |x \pm y|^p\, d\Omega \right]^{1/p} \leq \left[\int_{\Omega} |x|^p\, d\Omega \right]^{1/p} + \left[\int_{\Omega} |y|^p\, d\Omega \right]^{1/p}, \quad 1 < p < \infty. \tag{2.5}$$

We begin the preliminary concepts by introducing linear spaces. These are a collection of objects for which the operations of addition and scalar multiplication are defined in a simple and logical fashion.

2.2 Linear spaces and norms

The increased practical importance of error estimates and adaptive methods makes the use of *functional analysis* a necessary tool in finite element analysis. Today's student should consider taking a course in functional analysis, or studying texts such as those of Liusternik [10], Nowinski [12], or Oden [14]. This chapter will only cover certain basic topics. Other related advanced works, such as that of Hughes [9], should also be consulted. We are usually seeking to approximate a more complicated solution by a finite element solution. To develop a feel for the 'closeness' or 'distance between' these solutions, we need to have some basic mathematical tools. Since the approximation and the true solution vary throughout the spatial domain of interest, we are not interested in examining their difference at every point. The error at specific points is important and methods for estimating such an error are given by Ainsworth and Oden [1] but will not be considered here. Instead, we will want to examine integrals of the solutions, or integrals of differences between the solutions. This leads us naturally into the concepts of linear spaces and norms. We will also be interested in integrals of the derivatives of the solution. That will lead us to the Sobolev norm which includes both the function and its derivatives. Consider a set of functions $\phi_1(x), \ \phi_2(x), \ \cdots \ \phi_n(x)$. If the functions can be linearly combined they are called elements of a *linear space*. The following properties hold for the space of real numbers, R:

$$\begin{aligned} &\alpha, \ \beta \in R \\ &\phi_1 + \phi_2 = \phi_2 + \phi_1 \\ &(\alpha + \beta)\,\phi = \alpha\,\phi + \beta\,\phi \\ &\alpha\,(\phi_1 + \phi_2) = \alpha\,\phi_1 + \alpha\,\phi_2\,. \end{aligned} \tag{2.6}$$

An *inner product,* $< \bullet\,, \bullet >$, on a real linear space A is a map that assigns to an ordered pair $x, y \in A$ a real number R denoted by $< x, y >$. This process is often represented by the symbolic notation: $< \bullet, \ \bullet >: A \times A \rightarrow R$. It has the following properties

i. $< x, y > \; = \; < y, x >$ symmetry

ii. $< \alpha\, x, y > \; = \alpha < y, x >$
iii. $< (x + y), z > \; = \; < x, z > + < y, z >$ } linearity

iv. $< x, x > \; \geq 0$ and
$< x, x > \; = 0$ iff $x = 0$ } positive-definiteness.

The pair $x, \ y \in A$ are said to be orthogonal if $< x, y > \; = 0$. Another useful property is the Schwarz inequality: $< x, y >^2 \; \leq \; < x, x >< y, y >$. An inner product also represents an operation such as

$$< u, v > \; = \int_{x_1}^{x_2} u(x)\, v(x)\, dx. \tag{2.7}$$

Note that when the inner product operations is an integration the symbol $< u, v >$ is often replaced by the symbol (u, v) and may be called the *bilinear form*. A *norm*, $\| \bullet \|$, on a linear space A is a map of the function to a real number, $\| \bullet \|$: $A \rightarrow R$, with the properties (for $x, y \in A$ and $\alpha \in R$)

$$\left.\begin{array}{lll} \text{i.} & \| x \| \geq 0 \text{ and} \\ & \| x \| = 0 \text{ iff } x = 0 \end{array}\right\} \text{positive-definiteness}$$

$$\begin{array}{ll} \text{ii.} & \| \alpha x \| = | \alpha | \, \| x \| \\ \text{iii.} & \| x + y \| \leq \| x \| + \| y \|, \quad \text{triangle inequality.} \end{array} \tag{2.8}$$

A *semi-norm*, $|x|$, is defined in a similar manner except that it is positive semi-definite. That is, condition i is weakened so we can have $|x| = 0$ for x not zero. A *measure* or *natural norm* of a function x can be taken as the square root of the inner product with itself. This is denoted as

$$\| x \| = < x, x >^{\frac{1}{2}} \tag{2.9}$$

2.3 Sobolev norms *

The $L_2(\Omega)$ inner product norm involves only the inner product of the functions, and no derivatives: $(u, v) = \int_\Omega uv \, d\Omega$ where $\Omega \subset R^n$, $n \geq 1$. Then the norm is

$$\| u \|_{L_2} = \| u \|_0 = (u, u)^{\frac{1}{2}} = [\int_\Omega u^2 \, d\Omega]^{\frac{1}{2}}. \tag{2.10}$$

The $H^1(\Omega)$ inner product and norm includes both the functions and their first derivatives

$$(u, v)_1 = \int_\Omega [uv + \sum_{k=1}^{n} u_{,k} \, v_{,k}] \, d\Omega$$

where $(\ )_{,k} = \partial(\ )/\partial x_k$, and

$$\| u \|_H^1 = \| u \|_1 = (u, u)_1^{\frac{1}{2}} = \left[\int_\Omega \left(u^2 + \sum_{k=1}^{n} u_{,k}^2 \right) d\Omega \right]^{\frac{1}{2}}. \tag{2.11}$$

Note $H^0 = L_2$. Likewise, we can extend $H^s(\Omega)$ to include the S-*th* order derivatives.

2.4 Dual problem, self-adjointness

One often hears references to a boundary condition as either being an essential or a natural condition. Usually an essential boundary condition simply specifies a value of the primary unknown at a point. However, there is an established mathematical definition of these terms. Consider a homogeneous differential operator represented as

$$L(u) = 0 \in \Omega. \tag{2.12}$$

We form the inner product of $L(u)$ with another function, say v, to get

$$< L(u), v > = \int_0^1 u \frac{d^2 v}{dx^2} \, dx. \tag{2.13}$$

If we integrate by parts (sometimes repeatedly) we obtain the alternate form

$$< L(u), v > = < u, L^*(v) > + \int_\Omega [F(v) \, G(u) - F(u) \, G^*(v)] \, d\Omega, \tag{2.14}$$

where F and G are differential operators whose forms follow naturally from integration by parts. The operator L^* is the *adjoint* of L. If $L^* = L$ then L is *self-adjoint* and $G^* = G$, also. The $F(u)$ are called the *essential boundary conditions* and $G(u)$ are the *natural boundary conditions*. When $L^* = L$, then $F(u)$ is prescribed on Γ_1, and $G(u)$ is prescribed on Γ_2 where $\Gamma = \Gamma_1 \cup \Gamma_2$, $\Gamma_1 \cap \Gamma_2 = \varnothing$. We say that $< L(u), u > \; > 0$ is positive definite *iff* $L^* = L$, and $u \neq 0$. A self-adjoint problem will lead to a set of symmetric bilinear forms and a corresponding set of symmetric algebraic equations for the unknown coefficients in the problem. The weak form given by Eq. 2.14 is also referred to as the *dual problem*. If both the original weak form and the dual problem are solved it is possible to compute both an upper bound and a lower bound of the error in the approximation. Having both bounds is not always worth the extra computational cost.

To illustrate how to classify the boundary conditions, or to establish a dual problem, consider the model differential equation

$$L(u) = \frac{d^2u}{dx^2} \qquad x \in]0, 1[$$

has the inner product

$$< v, L(u) > = \int_0^1 v\, L(u)\, dx = \int_0^1 v\, \frac{d^2 u}{dx^2}\, dx.$$

Using integration by parts: $\int_a^b p dq = pq \Big|_a^b - \int_a^b q\, dp$. Here we let $p = v$ so that its derivative is $dp = (dv/dx)\, dx$, and $dq = (d^2u/dx^2)\, dx$, so $q = du/dx$, such that

$$< v, L(u) > = v \frac{du}{dx} \Big|_0^1 - \int_0^1 \frac{du}{dx} \frac{dv}{dx}\, dx.$$

Integrate by parts again

$$< v, L(u) > = v \frac{du}{dx} \Big|_0^1 - [\, u \frac{dv}{dx} \Big|_0^1 - \int_0^1 u \frac{d^2 v}{dx^2}\, dx\,]$$

$$= \; < L^*(v),\, u > + [\, v \frac{du}{dx} - u \frac{dv}{dx}\,] \Big|_0^1 .$$

Comparing this result to the definitions in Eq. 2.14 we see that the adjoint operator is $L^* = L = d^2()/dx^2$, the essential boundary condition involves $F(v) = 1 * v$ so it applies to the primary variable. The natural boundary condition assigns $G() = G^*() = d()/dx$, which is the gradient or slope of the primary variable. The original ordinary differential equation requires two boundary conditions. Our usual options are: a) give u at $x = 0$ and $x = 1$ and recover du/dx at $x = 0$ and $x = 1$ from the solution, b) give u at $x = 0$ and du/dx at $x = 1$ (or vice versa). We compute u for all x and recover du/dx at $x = 0$, c) give du/dx at $x = 0$ and $x = 1$. This determines u to within an arbitrary constant.

There are some other general observations about the types of boundary conditions and solution continuity that are associated with even order differential equations. Let the highest order derivative be $2m$. Then the essential boundary conditions involve derivatives of order zero (i.e., the solution itself) through $(m - 1)$. The non-essential boundary conditions involve the remaining derivatives of order m through $(2m - 1)$. The approximation must maintain continuity of the zero-th through $(m - 1)$ derivatives.

2.5 Weighted residuals

Here we will introduce the concept of approximating the solution to a differential equation by the *method of weighted residuals* (MWR) as it was originally used: on a global basis. That approach requires that we guess the solution over the entire domain and that our guess exactly satisfy the boundary conditions. Then we will introduce the simple but important change that the finite element approach adds to the MWR process. Guessing a solution that satisfies the boundary conditions is very difficult in two- and three-dimensional space, but it is relatively easy in one-dimension. To illustrate a global (or single element solution) consider the following model equation:

$$L(u) = \frac{d^2 u}{dx^2} + u + Q(x) = 0, \quad x \in]0, 1[\tag{2.15}$$

with a spatially varying source term $Q(x) = x$, essential boundary conditions of $u = 0$ at $x = 0$ and $u = 0$ at $x = 1$ so that the exact solution to this problem is $u = \text{Sin } x / \text{Sin } 1 - x$. We want to find a global approximate solution involving constants Φ_i, $1 \le i \le n$ that will lead to a set of n simultaneous equations. For homogeneous essential boundary conditions we usually pick a global product approximation of the form

$$u^* = g(x) \, f(x, \Phi_i) \tag{2.16}$$

where $g(x) \equiv 0$ on Γ. Here the boundary is $x = 0$ and $x - 1 = 0$ so we select a form such as $g_1(x) = x(1 - x)$, or $g_2(x) = x - \text{Sin } x / \text{Sin } 1$. We could pick $f(x, \Phi_i)$ as a polynomial $f(x) = \Phi_1 + \Phi_2 x + \cdots \Phi_n x^{(n-1)}$. For simplicity, select $n = 2$ and use $g_1(x)$ so the approximate solution is

$$u^*(x) = x(1 - x)(\Phi_1 + \Phi_2 x) = \mathbf{h}(x) \mathbf{\Phi}. \tag{2.17}$$

Expanding, this gives:

$$u^*(x) = (x - x^2) \Phi_1 + (x^2 - x^3) \Phi_2 = h_1(x) \Phi_1 + h_2(x) \Phi_2.$$

Here we will employ the MWR to find the $\mathbf{\Phi}$'s. From them we will know the value of $u^*(x)$ at all points and compute the error in the solution, $e = u(x) - u^*(x)$, and its norm, $\|u\|$. Here, however, we will focus on the residual error in the governing differential equation. From Eqs. 2.15 and 17 we see that the residual error in the differential equation at any point is $R(x) = u^{*\prime\prime} + u^* + Q(x)$, or in expanded form:

$$R(x) = Q(x) + [\frac{d^2}{dx^2} + 1] \mathbf{h}(x) \mathbf{\Phi}$$

$$R(x) = Q(x) + [\mathbf{h}'' + \mathbf{h}] \mathbf{\Phi} = Q(x) + \mathbf{b}(x) \mathbf{\Phi}$$

$$R(x) = Q(x) + (-2 + x - x^2) \Phi_1 + (2 - 6x + x^2 - x^3) \Phi_2 \neq 0. \tag{2.18}$$

where $b_1 = h_1'' + h_1(x) = (0 - 2) + (x - x^2)$. For an approximate solution with n constants we can split the residual R into parts including and independent of the Φ_j, say

$$R(x) = R(x)_0 + \sum_{j=1}^{n} b_j(x) \Phi_j = R_0 + \mathbf{b}(x) \mathbf{\Phi} \tag{2.19}$$

where **b** is a row matrix and **Φ** is a column vector. Usually R_0 is associated with the source term in the differential equation. Note for future reference that the partial derivatives of the residual with respect to the unknown degrees of freedom are:

$$\partial R / \partial \Phi_1 = (-2 + x - x^2), \quad \partial R / \partial \Phi_2 = (2 - 6x + x^2 - x^3),$$

or in general $\partial R / \partial \Phi_j = b_j(x)$. The residual error will vanish everywhere only if we guess the exact solution. Since that is usually not possible the method of weighted residuals requires that a weighted integral of the residual vanish instead;

$$\int_0^1 R(x)\, w(x)\, dx \equiv 0 \tag{2.20}$$

where $w(x)$ is a weighting function. We use n weights to get the necessary system of algebraic equations to find the unknown Φ_j. Substituting Eq. 2.19 gives

$$\int_\Omega R\, w_k\, d\Omega = \int_\Omega \left(R_0 + \sum_{j=1}^{n} b_j(x)\, \Phi_j \right) w_k\, d\Omega = 0_k, \quad 1 \le k \le n$$

or

$$\sum_{j=1}^{n} \int_\Omega b_j(x)\, w_k(x)\, \Phi_j\, d\Omega = -\int_\Omega R_0(x)\, w_k(x)\, d\Omega, \quad 1 \le k \le n. \tag{2.21}$$

In matrix form this system of equations is written as:

$$\underset{n \times n}{[\mathbf{S}]} \quad \underset{n \times 1}{\{\mathbf{D}\}} = \underset{n \times 1.}{\{\mathbf{C}\}} \tag{2.22}$$

Usually we call **S** and **C** the stiffness matrix and source vector, respectively. Clearly, there are many ways to pick the weighting functions, w_k. Mathematical analysis and engineering experience have lead to the following five most common choices of the weights used in various weighted residual methods:

A) Collocation Method: For this method we force the residual error to vanish at n arbitrarily selected points. Thus, we select

$$w_k(x) = \delta(x - x_k), \quad 1 \le k \le n \tag{2.23}$$

where the Dirac Delta distribution $\delta(x - x_k)$ which has the properties

$$\delta(x - x_k) = \begin{cases} 0 & x \ne x_k \\ \infty & x = x_k \end{cases}$$

$$\int_{-\infty}^{\infty} \delta(x - x_k)\, dx = \int_{x_k - a}^{x_k + a} \delta(x - x_k)\, dx = 1$$

and for any function $f(x)$ continuous at x_k

$$\int_{-\infty}^{\infty} \delta(x - x_k)\, f(x)\, dx = \int_{x_k - a}^{x_k + a} \delta(x - x_k)\, f(x)\, dx = f(x_k). \tag{2.24}$$

By inspection this reduces Eq. 2.21 to simply

$$\sum_{j=1}^{n} b_j(x_k)\,\Phi_j = -R_0(x_k), \qquad 1 \le k \le n.$$

Our problem is that we have an infinite number of choices for the collocation points, x_k. For $n = 2$, we could pick two points where R is large, or the third point, or the Gaussian quadrature points that are used in numerical integration, etc. Pick the two collocation points as $x_1 = 1/4$ and $x_2 = 1/2$; then

$$\begin{bmatrix} \frac{29}{16} & -\frac{35}{64} \\ \frac{7}{4} & \frac{7}{8} \end{bmatrix} \begin{Bmatrix} \Phi_1 \\ \Phi_2 \end{Bmatrix} = \begin{Bmatrix} \frac{1}{4} \\ \frac{1}{2} \end{Bmatrix}$$

is our unsymmetric algebraic system. Since the essential boundary conditions have already been satisfied by the assumed solution we can solve these equations without additional modifications. Here we obtain $\Phi_1 = 6/31$ and $\Phi_2 = 40/217$ so that our first approximate solution is given by $u^* = x(1-x)\,(42 + 40\,x)/217$. Selected interior results compared to the exact solution are:

x	u	u^*
1/4	0.044	0.045
1/2	0.070	0.071
3/4	0.060	0.062

Note that $u(x_k) - u^*(x_k) \neq 0$ even though $R(x_k) = 0$. That is, the error in the differential equation is zero at these collocation points, but the error in the solution is not zero. This can be viewed as similar to a finite difference solution.

B) Least Squares Method: For the n equations pick

$$\int_0^1 R(x)\, w_i(x)\, dx = 0, \qquad 1 \le i \le n$$

with the weights defined as

$$w_i(x) = \frac{\partial R(x)}{\partial \Phi_i} = b_i(x), \tag{2.25}$$

from Eq. 2.19. This choice is equivalent to solving the minimization problem:

$$\frac{1}{2}\int_0^1 R^2(x)\, dx \;\rightarrow\; \text{stationary (minimum)}. \tag{2.26}$$

Equation 2.26 means in this case Eq. 2.21 becomes

$$\sum_{j=1}^{n} \int_{\Omega} b_j(x)\, b_i(x)\, \Phi_j\, d\Omega = -\int_{\Omega} R_0(x)\, b_i(x)\, d\Omega, \qquad 1 \le i \le n.$$

For this example

$$\int_0^1 R(x)\frac{\partial R}{\partial \Phi_1}\, dx = 0, \quad \int_0^1 R(x)\frac{\partial R}{\partial \Phi_2}\, dx = 0$$

and substitutions from Eq. 2.18 gives

$$\frac{202}{60}\,\Phi_1 + \frac{101}{60}\,\Phi_2 = \frac{55}{60}$$

$$\frac{101}{60}\,\Phi_1 + \frac{393}{105}\,\Phi_2 = \frac{57}{60}.$$

It should be noted from Eqs. 2.19, 21, 25 that this procedure yields a square matrix which is always symmetric. Solving gives $\Phi_1 = 0.188$, $\Phi_2 = 0.170$ and selected results at the three interior points of: 0.043, 0.068, and 0.059, respectively.

C) Galerkin Method: The concept here is to make the residual error orthogonal to the functions associated with the spatial influence of the constants. That is, let

$$u^*(x) = g(x)\, f(x, \Phi_i) = \sum_{i=1}^{n} h_i(x)\,\Phi_i.$$

Here the h_i term defines how we have assumed the contribution from Φ_i will vary over space. Here for $n = 2$ and $h_1 = (x - x^2)$ and $h_2 = (x^2 - x^3)$, we set

$$w_i(x) \equiv h_i(x) \tag{2.27}$$

so Eq. 2.21 simplifies to

$$\sum_{j=1}^{n} \int_{\Omega} b_j(x)\, h_i(x)\,\Phi_j\, d\Omega = -\int_{\Omega} R_0(x)\, h_i(x)\, d\Omega, \quad 1 \le i \le n. \tag{2.28}$$

and for this specific example we require

$$\int_0^1 R(x)\, h_1(x)\, dx = 0, \quad \int_0^1 R(x)\, h_2(x)\, dx = 0$$

and Eq. 2.18 yields

$$\frac{3}{10}\,\Phi_1 + \frac{3}{20}\,\Phi_2 = \frac{1}{12}$$

$$\frac{3}{20}\,\Phi_1 + \frac{13}{105}\,\Phi_2 = \frac{1}{20}$$

which is again symmetric (for the self-adjoint equation). Solving gives degree of freedom values of $\Phi_1 = 71/369$, $\Phi_2 = 7/41$ and selected results at the three interior points of: 0.044, 0.070, and 0.060, respectively.

D) Method of Moments: Pick a spatial coordinate lever arm as a weight:

$$w_i(x) \equiv x^{(i-1)} \tag{2.29}$$

so that in the current one-dimensional example

$$\int_0^1 R(x)\, x^0\, dx = 0, \quad \int_0^1 R(x)\, x^1\, dx = 0 \tag{2.30}$$

gives the algebraic system

$$\frac{11}{6}\,\Phi_1 + \frac{11}{12}\,\Phi_2 = \frac{1}{2}$$

$$\frac{11}{12}\,\Phi_1 + \frac{19}{20}\,\Phi_2 = \frac{1}{3}$$

with the solution $\Phi_1 = 122/649$, $\Phi_2 = 110/649$ and selected results at the three interior points of: 0.043, 0.068, and 0.059, respectively. This method usually yields an unsymmetrical system. It is popular in certain physics applications.

E) Subdomain Method: For this final method we split the solution domain, Ω, into n arbitrary non-overlapping subdomains, Ω_k, that completely fill the space such that

$$\Omega = \bigcup_{k=1}^{n} \Omega_k \tag{2.31}$$

Then we define

$$w_k(x) \equiv 1 \qquad \text{for} \quad x \in \Omega_k \tag{2.32}$$

and it is zero elsewhere. This makes the residual error vanish on each of n different regions. Here $n = 2$, so we arbitrarily pick $\Omega_1 =]0, \frac{1}{2}[$ and $\Omega_2 =]\frac{1}{2}, 1[$. Then

$$\int_{\Omega_1} R(x)\, dx = 0, \quad \int_{\Omega_2} R(x)\, dx = 0 \tag{2.33}$$

yields the unsymmetric algebraic system

$$\begin{bmatrix} \frac{11}{12} & \frac{-53}{192} \\ \frac{11}{12} & \frac{229}{192} \end{bmatrix} \begin{Bmatrix} \Phi_1 \\ \Phi_2 \end{Bmatrix} = \begin{Bmatrix} \frac{1}{8} \\ \frac{3}{8} \end{Bmatrix}.$$

This results in $\Phi^T = [388 \;\; 352] / 2068$ and selected results at the three interior points of: 0.043, 0.068, and 0.059, respectively.

These examples show how analytical approximations can be obtained for differential equations. These approximate methods offer some practical advantages. Instead of solving a differential equation we are now presented with the easier problem of solving an algebraic problem, resulting from an integral relation, for a set of coefficients that define the approximation. The weighted residual procedure is valid of any number of spatial dimensions. The procedure is valid for any shaped domain Ω. It allows non-homogeneous coefficients. That is, the coefficient multiplying the derivatives in the differential operator L can vary with location. Note that so far we have not yet made any references to finite element methods. Later, you may look back on these examples as special cases of a single element solution. These simple examples could have been solved with matrix inversion routines. In practice, inversions are much too computationally expressive, and one must solve the equations by iterative methods or by a factorization process such as the process outlined in Fig. 2.1. By starting with a triangular matrix the substitution processes have only one unknown per row. The factored triangular arrays are stored in the locations of the original square matrix. Practical implementations of direct solvers must account for sparse array storage options and the fact that the factorization operations increases the 'fill-in' and thus the total storage requirement.

2.6 Boundary condition terms

If a boundary condition involves a non-zero value then we must extend the assumed approximate solution to include additional constants to be used to satisfy the essential boundary conditions. Usually these conditions are invoked prior to or during the solution

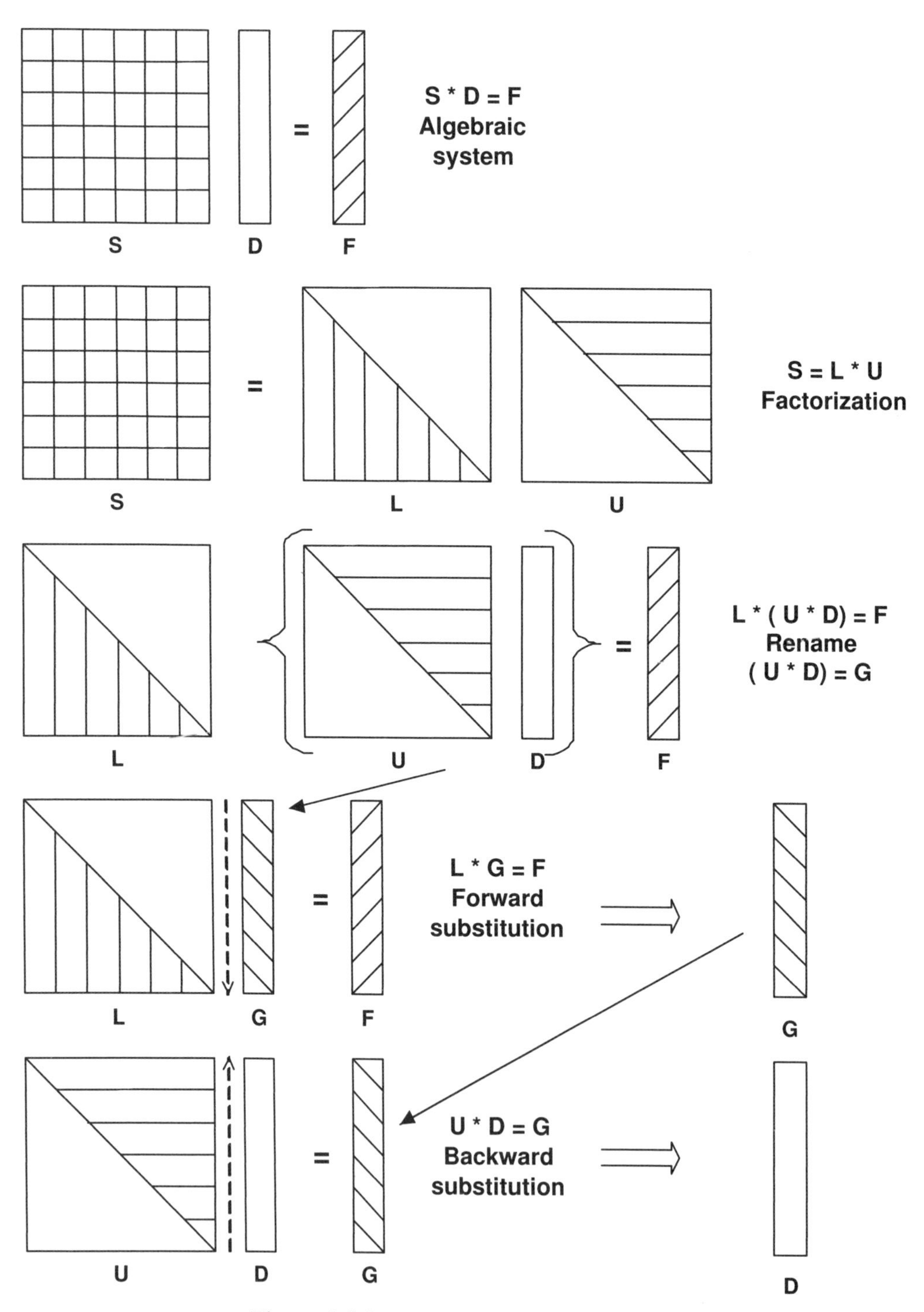

Figure 2.1 *Steps in the factorization process*

of the corresponding algebraic equations. Of course, since we are going to satisfy both the differential equation and the boundary conditions th total number of algebraic equations developed must be equal to the number of unknown parameters. To illustrate the algebraic procedure that is usually used we will solve the ODE in Eq. 2.15 with an approximation that allows any value to be assigned as the boundary condition at $x = 0$, say $u(0) = \Phi_1$. To apply the boundary condition after we have selected an approximate solution we will use Eq. 2.16 and pick $g(x) = (1 - x)$ so that only the boundary condition at $x = 1$ is satisfied in advance. We can add another constant to $f(x)$ to allow any boundary condition at $x = 0$: $f(x) = \Phi_1 + \Phi_2 x + \Phi_3 x^2$. Then the residual error is

$$R = x + (1 - x)\,\Phi_1 + (x + 2 - x^2)\,\Phi_2 + (x^2 + 2 - 6x - x^3)\,\Phi_3.$$

Since we now have three unknown degrees of freedom, Φ, we must have three weighted residual equations. For simplicity we will choose the collocation method and pick three equally spaced collocation points. Evaluating the residual at the quarter points and multiplying by the common denominator gives the three equations

$$\begin{bmatrix} -48 & 116 & -35 \\ -32 & 112 & 56 \\ -16 & 116 & 151 \end{bmatrix} \begin{Bmatrix} \Phi_1 \\ \Phi_2 \\ \Phi_3 \end{Bmatrix} = \begin{Bmatrix} 16 \\ 32 \\ 48 \end{Bmatrix}.$$

Note that since we know u at $x = 0$ these unknowns are not independent. Substituting $x = 0$ into our approximate solution and equating it to the assign boundary value there gives $u(0) = \Phi_1$. We call this an *essential boundary condition* on Φ_1. There are an infinite number of possible boundary conditions and we gain flexibility by allowing extra constants to satisfy them. Since Φ_1 will be a known number only the last two rows are independent for determining the remaining terms in Φ. Note that the first column of numbers, in the last two rows, is now multiplied by a known value and thus they can be carried to the right hand side to give the reduced algebraic system for the independent Φ:

$$\begin{bmatrix} 112 & 56 \\ 116 & 151 \end{bmatrix} \begin{Bmatrix} \Phi_2 \\ \Phi_3 \end{Bmatrix} = \begin{Bmatrix} 32 \\ 48 \end{Bmatrix} + \begin{Bmatrix} 32 \\ 16 \end{Bmatrix} \Phi_1.$$

An equivalent matrix modification routine in the *MODEL* code deletes the redundant coefficients, but keeps the matrix the same size to avoid re-ordering all the coefficients as done above. For the common original boundary condition of $u(0) = 0$, we have $\Phi_1 = 0$ and the changes to the right hand side (RHS) are not necessary. But the above form also allows us the option of specifying any non-zero boundary condition we need. Using the zero value gives a solution of $\Phi_2 = 0.2058$ and $\Phi_3 = 0.1598$. The resulting values at the interior quarter points are 0.046, 0.071, and 0.061, respectively. These compare well with the previous results.

If the second boundary condition had been applied other than at $x = 0$ then we would have a more complicated relation between the Φ. For example, assume we move the boundary condition to $x = 0.5$. Then evaluating the approximate solution there yields

$$u(0.5) = 0.5\,\Phi_1 + 0.25\,\Phi_2 + 0.125\,\Phi_3$$

which is called a *linear constraint equation* on Φ, or a *multipoint constraint* (MPC). In

Application dependent software

Term / Process	Required (or use keyword example)	Optional
Generate matrices		
Differential operator, Save for flux averages $\mathbf{S}_K^E$, $\mathbf{M}^E$	ELEM_SQ_MATRIX my_el_sq_inc	APPLICATION_B_MATRIX my_b_matrix_inc APPLICATION_E_MATRIX my_e_matrix_inc
Volumetric source $\mathbf{C}_Q^E$		ELEM_COL_MATRIX my_el_col_inc or ELEM_SQ_MATRIX my_el_sq_inc
Mixed or Robin BC $\mathbf{S}_h^B$, $\mathbf{C}_h^B$	MIXED_SQ_MATRIX my_mixed_sq_inc	
Boundary flux $\mathbf{C}_F^B$	SEG_COL_MATRIX my_seg_col_inc	EXACT_NORMAL_FLUX my_exact_normal_flux_inc
Save for post-processing		ELEM_SQ_MATRIX my_el_sq_inc or ELEM_POST_DATA my_el_post_inc
Post-process element		POST_PROCESS_ELEM my_post_el_inc
Energy norm error estimate	APPLICATION_B_MATRIX APPLICATION_E_MATRIX	my_b_matrix_inc my_e_matrix_inc
Use an exact solution		
Exact essential BC		EXACT_SOLUTION my_exact_inc
List exact solution		EXACT_SOLUTION my_exact_inc
List exact fluxes		EXACT_SOLUTION_FLUX my_exact_flux_inc
Use exact source		EXACT_SOURCE my_exact_source_inc

Figure 2.2 *User software interfaces in MODEL*

other words we would have to solve the weighted residual algebraic system subject to a linear constraint. This is a fairly common situation in practical design problems and adaptive analysis procedures. The computational details for enforcing the above essential boundary conditions are discussed in detail later.

2.7 Adding more unknowns

Since the exact solution of the model problem is not a polynomial our global polynomial approach can never yield an exact solution. However, we can significantly improve the accuracy by adding more unknown coefficients to the expansion in Eq. 2.17. In matrix notation the original global Galerkin matrices become

$$\mathbf{S}^e = \int_L \mathbf{h}^T \, \mathbf{b} \, dx, \qquad \mathbf{C}^e = -\int_L \mathbf{h}^T \, Q(x) \, dx$$

where $Q(x) = x$ is the source term, $\mathbf{b} = \mathbf{h}'' + \mathbf{h}$ comes from the differential operator acting on u, and where a prime denotes a derivative. Likewise, using Least Squares:

$$\mathbf{S}^e = \int_L \mathbf{b}^T \, \mathbf{b} \, dx, \qquad \mathbf{C}^e = \int_L \mathbf{b}^T \, Q(x) \, dx$$

Here we see that the Least Squares square matrix will always be symmetric, but the Galerkin form may not be. As we add more unknown coefficients we just increase the size of the functions in **h**, and thus in **b**, and increase the number of integration points to account for the higher degree polynomials occurring in the matrices. A disadvantage of adding more unknowns to a global solution is that the unknown parameters are fully coupled to each other. That means the algebraic equations to be solved are fully populated, and thus very expensive to solve. The finite element method will lead to very sparse equations that are efficient to solve.

2.8 Numerical integration

Since numerical integration simply replaces an integral with a special summation this approach has the potential for automating all the above integrals required by the MWR. Then we can include thousands of unknown coefficients, Φ_i, in our test solution. Here we are dealing with polynomials. It is well known that in one-dimension Gaussian quadrature with n_q terms will exactly integrate a polynomial of order $(2\,n_q - 1)$. Gauss proved that this is the minimum number of points that can be used in a summation to yield the exact results. Therefore, it is the most efficient method available for integrating polynomials. Thus, we could replace the above integrals with a two-point Gauss rule. (This will be considered in full detail later in Sec. 4.4, and Table 4.2.) For example, the Galerkin source term is

$$C_1^e = \int_0^1 x \, h_1(x) \, dx = \sum_{j=1}^{n_q} x_j \, h_1(x_j) \, w_j \tag{2.34}$$

where the x_j and w_j are tabulated data. For $n_q = 2$ on the domain $\Omega =]0, 1[$ we have $w_1 = w_2 = 1/2$ and $x_j = (1 \pm 1/\sqrt{3})/2$, or $x_1 = 0.2113325$ and $x_2 = 0.788675$. So

$$\int_0^1 x(x - x^2)\, dx = \sum_{j=1}^{n_q} (x_j^2 - x_j^3)\, w_j = \sum_{j=1}^{2} x_j^2 (1 - x_j)\, w_j$$

$$= [\,(0.2113248)^2 (0.7886751)\, 1/2 + (0.7886751)^2 (0.2113248)\, 1/2\,]$$

$$= (0.16666667)\, 1/2 = 0.083333\,.$$

If we had an infinite word length machine this process would yield the exact value of 1/12 which was previously found in Eq. 2.28.

```
                 Interface from MODEL to ELEM_SQ_MATRIX, 1
Note: MODEL requires strong typing (implicit none).  Any item not
defined here (and later) in the interface must have its variable
type and size defined by the user.

Type    Status Name    Remarks

INTEGER (IN) DP        Double precision kind for this hardware
INTEGER (IN) LT_GEOM Number of element type geometric nodes
INTEGER (IN) LT_FREE Number of element type unknowns
INTEGER (IN) LT_N      Number of element type solution nodes
INTEGER (IN) LT_PARM Parametric dimension of element type
INTEGER (IN) LT_QP     Number of element type quadrature points
INTEGER (IN) N_SPACE Physical space dimension of problem (space)

REAL(DP) (IN) COORD (LT_N,N_SPACE)    Element type coordinates
REAL(DP) (IN) PT    (LT_PARM,LT_QP)   Quadrature parametric points
REAL(DP) (IN) WT    (LT_QP)           Quadrature parametric weights
REAL(DP) (IN) X     (MAX_NP,N_SPACE)  All nodal coordinates

REAL(DP) (OUT) C    (LT_FREE)          Element column matrix
REAL(DP) (OUT) DGH  (N_SPACE,LT_N)     Global derivatives of H
REAL(DP) (OUT) DLH  (LT_PARM,LT_N)     Local  derivatives of H
REAL(DP) (OUT) G    (LT_GEOM)          Geometry interpolation array
REAL(DP) (OUT) H    (LT_N)             Solution interpolation array
REAL(DP) (OUT) S    (LT_FREE,LT_FREE)  Element square matrix
REAL(DP) (OUT) EL_M (LT_FREE,LT_FREE)  Element square matrix

GET_G_AT_QP     Form G array at quadrature point
GET_H_AT_QP     Form H array at quadrature point
GET_DLH_AT_QP   Form DLH array at quadrature point
```

Figure 2.3 *User interface to ELEM_SQ_MATRIX (part 1)*

A typical partial implementation of these global Galerkin and Least Squares procedures will be illustrated with the *MODEL* program. Only a very small part of it changes for each application. Every application requires that we formulate a square matrix. That is done in subroutine *ELEM_SQ_MATRIX*, which also allows the optional calculation of an element column matrix. A number of prior applications are supplied in a library form and will be discussed later. The coding for a totally new application is usually supplied by an 'include file' that the compiler inserts into the necessary subprogram. Figure 2.2 shows all of the user subroutines and *INCLUDE* files that we will use in this book. Note that for educational purposes it includes access to selected exact solutions so they can be compared to the finite element model solution and the error estimator to be consider later. By using the keyword controls in a data file *MODEL* allocates space for the most commonly needed items in a finite element analysis. As we find need for such items we will declare how they interface to subroutine

ELEM_SQ_MATRIX, and others. Figure 2.3 lists the portion of the interface arguments that will be used here. The coding of the above problems, by numerical integration, is shown in Figs. 2.4 and 2.6, respectively. The results agree well, as do all of our weighted residual solutions. The global Galerkin and Least Squares results are listed in Fig. 2.5. Plotting the resulting solutions shows very similar curves from all five approaches to the methods of weighted residuals.

2.9 Integration by parts

The use of integration by parts will be very important in most finite element Galerkin methods. From the matrix definitions in the previous sections we see that the Least Square process involves the same order derivative in both terms in the matrix product in the square matrix and thus can not benefit from integration by parts. However, in the Galerkin square matrix since $\mathbf{b} = \mathbf{h}'' + \mathbf{h}$, the first product involves $\mathbf{h}$ and $\mathbf{h}''$ so integration by parts can be applied to that one matrix product. Returning to the original scalar form causing that term we see:

$$\int_L w\, u''\, dx = w\, u' \Big|_0^L - \int_L w' u'\, dx. \tag{2.35}$$

Here the assumed solution is zero at the two ends so the appearance of the boundary terms is not clearly important in this global analysis. But in finite element analysis, where we will have extra unknown coefficients at the end points, they will be very important and yield physically significant reaction recovery data. When we utilize the above integration by parts, and change all the signs, the previous coding in Fig. 2.4 changes to that in Fig. 2.7. Note that the square matrix is now clearly symmetric (see lines 52 & 56 of Fig. 2.7), and we no longer need the storage array for the second derivative of h (see line 49 of Figs 2.4 & 5). The numerical results are identical to the original ones given in Fig. 2.6. The full cubic approximation is seen in Fig. 2.8. If we had only one degree of freedom ($\Phi_2 = 0$) this would reduce to a quadratic approximation with much higher error as seen in Fig. 2.8. Increasing the number of degrees of freedom quickly decreases the error to the point that it can not be seen, but can be computed by an error estimator.

2.10 Finite element model problem

In order to extend the previous introductory concepts on the global MWR to the more powerful finite element method consider the same one-dimensional model problem as our first example. The differential equation of interest, Eq. 2.15, is

$$L(u) = \frac{d^2 u}{dx^2} + u + Q(x) = 0, \qquad x \in]0, L[$$

on the closed domain, $x \in]0, L[$, and is subjected to two boundary conditions to yield a unique solution. Here $Q(x) = x$ denotes a source term per unit length, as before. The corresponding governing integral statement to be used for the finite element model is obtained from the *Galerkin weighted residual method*, followed by integration by parts which introduces the term $du/dx = -q$, which we will define as the flux. In higher

```
! ...  Partial Global Access Arrays                                          ! 1
 REAL(DP) :: C (LT_FREE), S (LT_FREE, LT_FREE) ! Results                     ! 2
 REAL(DP) :: PT (LT_PARM, LT_QP), WT (LT_QP)   ! Quadratures                 ! 3
 REAL(DP) :: H (LT_N), DGH (N_SPACE, LT_N)     ! Solution                    ! 4
 REAL(DP) :: G (LT_GEOM)                       ! Geometry                    ! 5
 REAL(DP) :: COORD (LT_N, N_SPACE)             ! Coordinates                 ! 6
                                                                             ! 7
! ...  Partial Notations List                                                ! 8
! COORD     = SPATIAL COORDINATES OF ELEMENT'S NODES                         ! 9
! DGH       = GLOBAL DERIVATIVES OF INTERPOLATION FUNCTIONS                  !10
! G         = GEOMETRIC INTERPOLATION FUNCTIONS                              !11
! H         = SCALAR INTERPOLATION FUNCTIONS                                 !12
! LT_FREE   = NUMBER OF DEGREES OF FREEDOM                                   !13
! LT_GEOM   = NUMBER OF GEOMETRY NODES                                       !14
! LT_PARM   = DIMENSION OF PARAMETRIC SPACE                                  !15
! LT_QP     = NUMBER OF QUADRATURE POINTS                                    !16
! LT_N      = NUMBER OF NODES PER ELEMENT                                    !17
! N_SPACE   = DIMENSION OF PHYSICAL SPACE                                    !18
! PT        = QUADRATURE COORDINATES                                         !19
! WT        = QUADRATURE WEIGHTS                                             !20
! ...         see full notation file                                         !21
                                                                             !22
! .........................................................                  !23
! **  ELEM_SQ_MATRIX PROBLEM DEPENDENT STATEMENTS FOLLOW **                  !24
! .........................................................                  !25
! Define any new local array or variable types                               !26
!                                                                            !27
!      GLOBAL (SINGLE ELEMENT) Galerkin MWR FOR ODE                          !28
!   U,xx + U + X = 0, U(0)=0=U(1), U = sin(x)/sin(1) - x                     !29
!             Without integration by parts                                   !30
                                                                             !31
 REAL(DP) :: D2GH (1, 1:2)     ! Second global derivative                    !32
 REAL(DP) :: DL, DX_DN, X_Q    ! Length, Jacobian, Position                  !33
 INTEGER  :: IQ                ! Loops                                       !34
                                                                             !35
 DL    = COORD (LT_N, 1) - COORD (1, 1)  ! LENGTH                            !36
 DX_DN = DL / 2.                         ! CONSTANT JACOBIAN                 !37
 S = 0.d0; C = 0.d0                      ! ZERO SUMS                         !38
                                                                             !39
 DO IQ = 1, LT_QP           ! LOOP OVER QUADRATURES                          !40
                                                                             !41
!        GET GEOMETRIC INTERPOLATION FUNCTIONS, AND X-COORD                  !42
  G   = GET_G_AT_QP (IQ)                          ! parametric               !43
  X_Q = DOT_PRODUCT (G, COORD (1:LT_GEOM, 1)) ! x at point                   !44
                                                                             !45
!    GLOBAL INTERPOLATION, 1st & 2nd GLOBAL DERIVATIVES                      !46
  H     (:)   = (/ (X_Q - X_Q**2), (X_Q**2 - X_Q**3)  /)                     !47
  DGH   (1,:) = (/ (1 - 2*X_Q),    (2*X_Q - 3*X_Q**2) /)                     !48
  D2GH  (1,:) = (/ (-2),           (2 - 3*X_Q)        /)                     !49
                                                                             !50
  C = C - H * X_Q * WT (IQ) * DX_DN  ! SOURCE, from Q(x)                     !51
                                                                             !52
!           SQUARE MATRIX ( ? SYMMETRIC ? )                                  !53
  S = S + ( MATMUL (TRANSPOSE(D2GH), H)             & ! from u"              !54
          + OUTER_PRODUCT (H, H)) * WT (IQ) * DX_DN ! from u                 !55
 END DO  ! QUADRATURE                                                        !56
!  Outer product C_sub_jk = A_sub_j * B_sub_k                                !57
!  End of application dependent code                                         !58
```

Figure 2.4 *A global Galerkin implementation*

```
! ...  Partial Global Access Arrays                              ! 1
 REAL(DP) :: C (LT_FREE), S (LT_FREE, LT_FREE) ! Results         ! 2
 REAL(DP) :: PT (LT_PARM, LT_QP), WT (LT_QP)   ! Quadratures     ! 3
 REAL(DP) :: H (LT_N), DGH (N_SPACE, LT_N)     ! Solution        ! 4
 REAL(DP) :: G (LT_GEOM)                       ! Geometry        ! 5
 REAL(DP) :: COORD (LT_N, N_SPACE)             ! Coordinates     ! 6
                                                                 ! 7
! ...  Partial Notations List                                    ! 8
! COORD   = SPATIAL COORDINATES OF ELEMENT'S NODES               ! 9
! DGH     = GLOBAL DERIVATIVES OF INTERPOLATION FUNCTIONS        !10
! G       = GEOMETRIC INTERPOLATION FUNCTIONS                    !11
! H       = SCALAR INTERPOLATION FUNCTIONS                       !12
! LT_FREE = NUMBER OF DEGREES OF FREEDOM                         !13
! LT_GEOM = NUMBER OF GEOMETRY NODES                             !14
! LT_PARM = DIMENSION OF PARAMETRIC SPACE                        !15
! LT_QP   = NUMBER OF QUADRATURE POINTS                          !16
! LT_N    = NUMBER OF NODES PER ELEMENT                          !17
! N_SPACE = DIMENSION OF PHYSICAL SPACE                          !18
! PT      = QUADRATURE COORDINATES                               !19
! WT      = QUADRATURE WEIGHTS                                   !20
! ...       see full notation file                               !21
                                                                 !22
! ..............................................................  !23
! **  ELEM_SQ_MATRIX PROBLEM DEPENDENT STATEMENTS FOLLOW **      !24
! ..............................................................  !25
! Define new local array or variable types, then statements      !26
!                                                                !27
!    GLOBAL (SINGLE ELEMENT) LEAST SQUARE METHOD FOR ODE         !28
!    U,xx + U + X = 0, U(0)=0=U(1), U = sin(x)/sin(1) - x        !29
                                                                 !30
 REAL(DP) :: DL, DX_DN, X_IQ ! Length, Jacobian, Position        !31
 REAL(DP) :: D2GH (1, 2)     ! Second derivative                 !32
 REAL(DP) :: F (2)           ! H'' + H, Work space               !33
 INTEGER  :: IQ              ! Loops                             !34
                                                                 !35
 DL    = COORD (LT_N, 1) - COORD (1, 1) ! LENGTH                 !36
 DX_DN = DL / 2.                        ! CONSTANT JACOBIAN      !37
 S = 0.d0; C = 0.d0                     ! ZERO SUMS              !38
                                                                 !39
 DO IQ = 1, LT_QP        ! LOOP OVER QUADRATURES                 !40
                                                                 !41
!       GET GEOMETRIC INTERPOLATION FUNCTIONS, AND X-COORD       !42
   G   = GET_G_AT_QP (IQ)                      ! parametric      !43
   X_Q = DOT_PRODUCT (G, COORD (1:LT_GEOM, 1)) ! x at point      !44
                                                                 !45
!       GLOBAL INTERPOLATION, 1st & 2nd GLOBAL DERIVATIVES       !46
   H    (:)   = (/ (X_Q - X_Q**2), (X_Q**2 - X_Q**3)  /)         !47
   DGH  (1,:) = (/ (1 - 2*X_Q),    (2*X_Q - 3*X_Q**2) /)         !48
   D2GH (1,:) = (/ (-2),           (2 - 6*X_Q)        /)         !49
   F    (:)   = D2GH (1,:) + H (:)                               !50
                                                                 !51
   C = C - F * X_Q * WT (IQ) * DX_DN ! SOURCE, from Q(x)         !52
                                                                 !53
!        SQUARE MATRIX, from U" and U, SYMMETRIC                 !54
   S = S + OUTER_PRODUCT (F, F) * WT (IQ) * DX_DN                !55
 END DO  ! QUADRATURE                                            !56
! Outer product C_sub_jk = A_sub_j * B_sub_k                     !57
! End of application dependent code                              !58
```

Figure 2.5 *A global least squares implementation*

```
U,xx + U + X = 0, U(0)=0=U(1), Exact U = sin(x)/sin(1) - x           ! 1
                                                                      ! 2
Single cubic element (Global) Galerkin Solution:                      ! 3
------------------------------------------------                      ! 4
                                                                      ! 5
***  OUTPUT OF RESULTS IN NODAL ORDER  ***                            ! 6
  NODE, 1 COORDINATES,  1 PARAMETERS.                                 ! 7
     1  0.00000E+00     1.92412E-01                                   ! 8
     2  1.00000E+00     1.70732E-01                                   ! 9
                                                                      !10
** ELEMENT GAUSS POINT  RESULTS **                                    !11
ELEM    X       EXACT          FEA       GRADIENT    FE_GRADIENT      !12
1    0.000  0.0000E+00  0.0000E+00   1.88395E-01   1.92412E-01        !13
1    0.069  1.3014E-02  1.3198E-02   1.85532E-01   1.86932E-01        !14
1    0.330  5.5092E-02  5.5001E-02   1.24268E-01   1.22321E-01        !15
1    0.670  6.7974E-02  6.7835E-02  -6.85032E-02  -6.65570E-02        !16
1    0.931  2.2476E-02  2.2697E-02  -2.90078E-01  -2.91477E-01        !17
1    1.000  0.0000E+00  0.0000E+00  -3.57907E-01  -3.63144E-01        !18
                                                                      !19
Single cubic element (Global) Least square Solution:                  !20
----------------------------------------------------                  !21
                                                                      !22
***  OUTPUT OF RESULTS IN NODAL ORDER  ***                            !23
  NODE, 1 COORDINATES,  1 PARAMETERS.                                 !24
     1  0.00000E+00     1.87542E-01                                   !25
     2  1.00000E+00     1.69471E-01                                   !26
                                                                      !27
** ELEMENT GAUSS POINT  RESULTS **                                    !28
ELEM    X       EXACT          FEA       GRADIENT    FE_GRADIENT      !29
1    0.000  0.0000E+00  0.0000E+00   1.88395E-01   1.87542E-01        !30
1    0.034  6.3536E-03  6.3053E-03   1.87718E-01   1.85742E-01        !31
1    0.169  3.0952E-02  3.0426E-02   1.71385E-01   1.66831E-01        !32
1    0.381  6.0872E-02  5.9426E-02   1.03316E-01   1.00101E-01        !33
1    0.619  7.0522E-02  6.8960E-02  -3.23143E-02  -2.98400E-02        !34
1    0.831  4.6834E-02  4.6193E-02  -1.98511E-01  -1.93234E-01        !35
1    0.966  1.1519E-02  1.1461E-02  -3.24515E-01  -3.22039E-01        !36
1    1.000  0.0000E+00  0.0000E+00  -3.57907E-01  -3.57013E-01        !37
                                                                      !38
! Notes:                                                              !39
!   The "nodal parameters" above do not actually occur                !40
!   at the nodes for a global solution as they will later             !41
!   for all later finite element solutions.                           !42
!                                                                     !43
!   There must be as many "nodes" as global degrees of                !44
!   freedom to trick the MODEL code into doing a global               !45
!   solution.  Likewise, there needs to be one fake "element"         !46
!   connected to all the nodes.                                       !47
                                                                      !48
```

Figure 2.6 *Global MWR solutions and gradients*

```
! ...  Partial Global Access Arrays                                  ! 1
 REAL(DP) :: C (LT_FREE), S (LT_FREE, LT_FREE) ! Results             ! 2
 REAL(DP) :: PT (LT_PARM, LT_QP), WT (LT_QP)   ! Quadratures         ! 3
 REAL(DP) :: H (LT_N), DGH (N_SPACE, LT_N)     ! Solution            ! 4
 REAL(DP) :: G (LT_GEOM)                       ! Geometry            ! 5
 REAL(DP) :: COORD (LT_N, N_SPACE)             ! Coordinates         ! 6
                                                                     ! 7
! ...  Partial Notations List                                        ! 8
! COORD   = SPATIAL COORDINATES OF ELEMENT'S NODES                   ! 9
! DGH     = GLOBAL DERIVATIVES OF INTERPOLATION FUNCTIONS            !10
! G       = GEOMETRIC INTERPOLATION FUNCTIONS                        !11
! H       = SCALAR INTERPOLATION FUNCTIONS                           !12
! LT_FREE = NUMBER OF DEGREES OF FREEDOM                             !13
! LT_GEOM = NUMBER OF GEOMETRY NODES                                 !14
! LT_PARM = DIMENSION OF PARAMETRIC SPACE                            !15
! LT_QP   = NUMBER OF QUADRATURE POINTS                              !16
! LT_N    = NUMBER OF NODES PER ELEMENT                              !17
! N_SPACE = DIMENSION OF PHYSICAL SPACE                              !18
! PT      = QUADRATURE COORDINATES                                   !19
! WT      = QUADRATURE WEIGHTS                                       !20
! ...                                                                !21
                                                                     !22
! .......................................................            !23
! ** ELEM_SQ_MATRIX PROBLEM DEPENDENT STATEMENTS FOLLOW **           !24
! .......................................................            !25
! Define new local array or variable types, then statements          !26
!                                                                    !27
!      GLOBAL (SINGLE ELEMENT) Galerkin MWR FOR ODE                   !28
!   U,xx + U + X = 0, U(0)=0=U(1), U = sin(x)/sin(1) - x             !29
!               With integration by parts                            !30
                                                                     !31
 REAL(DP) :: DL, DX_DN, X_Q  ! Length, Jacobian, Position            !32
 INTEGER  :: IQ              ! Loops                                 !33
                                                                     !34
 DL    = COORD (LT_N, 1) - COORD (1, 1) ! LENGTH                     !35
 DX_DN = DL / 2.                        ! CONSTANT JACOBIAN          !36
 S = 0.d0; C = 0.d0                     ! ZERO SUMS                  !37
                                                                     !38
 DO IQ = 1, LT_QP          ! LOOP OVER QUADRATURES                   !39
                                                                     !40
!      GET GEOMETRIC INTERPOLATION FUNCTIONS, AND X-COORD            !41
   G   = GET_G_AT_QP (IQ)                       ! parametric         !42
   X_Q = DOT_PRODUCT (G, COORD (1:LT_GEOM, 1)) ! x at point          !43
                                                                     !44
!      GLOBAL INTERPOLATION AND GLOBAL DERIVATIVES (ONLY)            !45
   H   (:)   = (/ (X_Q - X_Q**2), (X_Q**2 - X_Q**3)  /)              !46
   DGH (1,:) = (/ (1 - 2*X_Q),    (2*X_Q - 3*X_Q**2) /)              !47
                                                                     !48
   C = C + H * X_Q * WT (IQ) * DX_DN ! SOURCE, from Q(x)             !49
                                                                     !50
!          SQUARE MATRIX ( SYMMETRIC )                               !51
   S = S + ( MATMUL (TRANSPOSE(DGH), DGH)          & ! from u"       !52
           - OUTER_PRODUCT (H, H)) * WT (IQ) * DX_DN ! from u        !53
                                                                     !54
 END DO  ! QUADRATURE                                                !55
!  Outer product C_sub_jk = A_sub_j * B_sub_k                        !56
!  End of application dependent code                                 !57
```

Figure 2.7 *Global Galerkin with integration by parts*

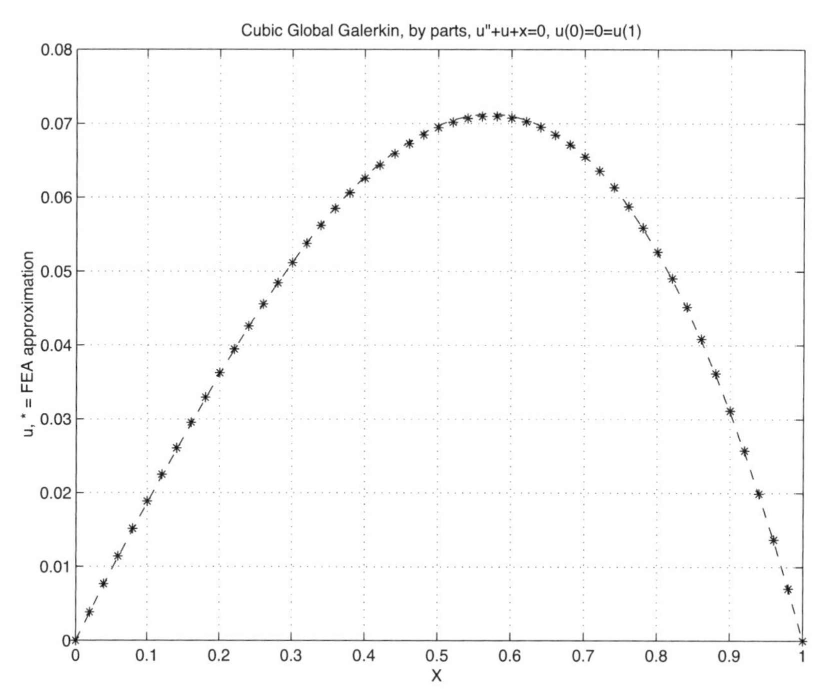

Figure 2.8 *Exact (-) and cubic global Galerkin (*) solutions*

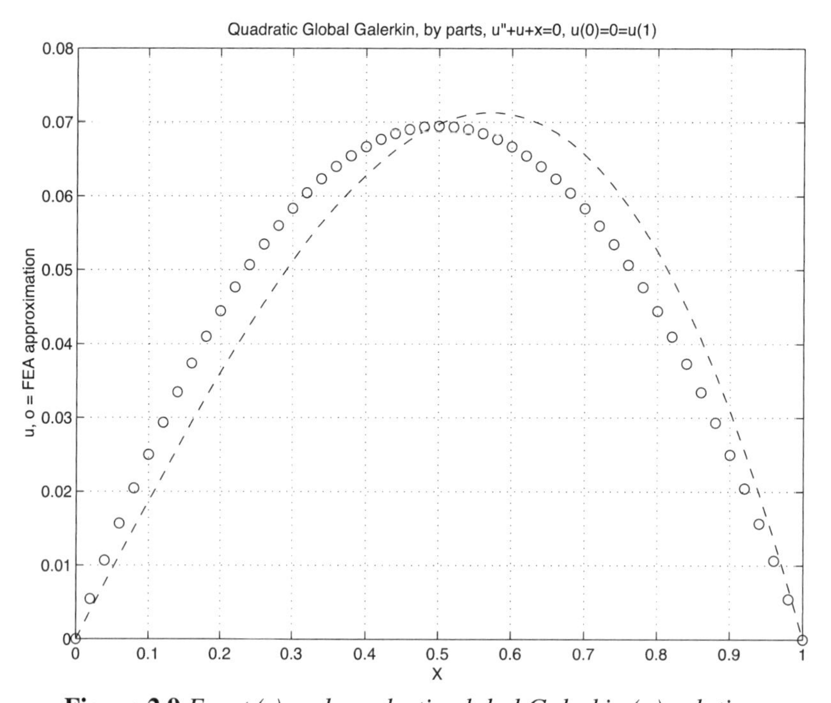

Figure 2.9 *Exact (-) and quadratic global Galerkin (o) solutions*

dimension problems it is the flux vector, $\mathbf{q}$. On a boundary we may be interested in a related scalar term, $q_n = \mathbf{q} \cdot \mathbf{n}$, which is the flux normal to the boundary defined by the unit normal vector $\mathbf{n}$. For the special case of the one-dimensional form being considered here we need to note that at the left limit of the domain $\mathbf{n} = -1\,\mathbf{i}$ while at the right limit it is $\mathbf{n} = +1\,\mathbf{i}$. In this case, the Galerkin method states that the function, w, that satisfies the boundary conditions and the integral form:

$$I = \int_0^L \left[\frac{dw}{dx}\frac{du}{dx} - w\,u - w\,Q\right] dx + u'(0)\,w(0) - u'(L)\,w(L) = 0, \quad (2.36)$$

also satisfies Eq. 2.15. For a finite element model we must generate a mesh that subdivides the domain and (usually) its boundary. The unknown coefficients in the finite element model, $\mathbf{D}$, will be assigned to the node points of the mesh. Within each element the solution will be approximated by an assumed local spatial behavior. That in turn defines the assumptions for spatial derivatives in an element domain. To illustrate this in one-dimension consider Fig. 2.10 which compares an exact solution (dashed) and a piecewise linear finite element model. The domains of influence of a typical element and a typical node are sketched there. In a finite element model, I is assumed to be the sum of the n_e element and n_b boundary segment contributions so that

$$I = \sum_{e=1}^{n_e} I^e + \sum_{b=1}^{n_b} I^b, \quad (2.37)$$

where here $n_b = 2$ and consists of the last two terms given in Eq. 2.36. A typical element term is

$$I^e = \int_{L^e} \left[(du^e / dx)^2 - (u^e)^2 - Q^e\, u^e \right] dx,$$

where L^e is the length of the element. To evaluate such a typical element contribution, it is necessary to introduce a set of interpolation functions, $\mathbf{H}$, so $u^e(x) = \mathbf{H}^e(x)\,\mathbf{D}^e$, and

$$du^e / dx = d\mathbf{H}^e / dx\,\mathbf{D}^e = \mathbf{D}^{e^T}\, d\mathbf{H}^{e^T} / dx, \quad (2.38)$$

where $\mathbf{D}^e$ denotes the nodal values of u for element e. One of the few standard notations in finite element analysis is to denote the result of the differential operator acting on the interpolation functions, $\mathbf{H}$, by the symbol $\mathbf{B}$. That is, $\mathbf{B}^e \equiv d\,\mathbf{H}^e / dx$. Thus, a typical element contribution is

$$I^e = \mathbf{D}^{e^T}\, \mathbf{S}^e\, \mathbf{D}^e - \mathbf{D}^{e^T}\, \mathbf{C}^e, \quad (2.39)$$

with $\mathbf{S}^e = (\mathbf{S}^e{}_1 - \mathbf{S}^e{}_2)$ and where the first contribution to the square matrix is

$$\mathbf{S}^e{}_1 \equiv \int_{L^e} \frac{d\mathbf{H}^{e^T}}{dx} \frac{d\mathbf{H}^e}{dx}\, dx = \int_{L^e} \mathbf{B}^{e^T}\, \mathbf{B}^e\, dx,$$

which, for this linear element, has a constant integrand and can be integrated by inspection. The second square matrix contribution and the resultant source vector are:

$$\mathbf{S}^e{}_2 \equiv \int_{L^e} \mathbf{H}^{e^T}\, \mathbf{H}^e\, dx, \quad \mathbf{C}^e \equiv \int_{L^e} Q^e\, \mathbf{H}^{e^T}\, dx.$$

Clearly, both the element degrees of freedom, $\mathbf{D}^e$, and the boundary degrees of freedom, $\mathbf{D}^b$, are subsets of the total vector of unknown parameters, $\mathbf{D}$. That is, $\mathbf{D}^e \subseteq \mathbf{D}$ and $\mathbf{D}^b \subset \mathbf{D}$. Of course, the $\mathbf{D}^b$ are usually a subset of the $\mathbf{D}^e$ (i.e., $\mathbf{D}^b \subset \mathbf{D}^e$ and in higher

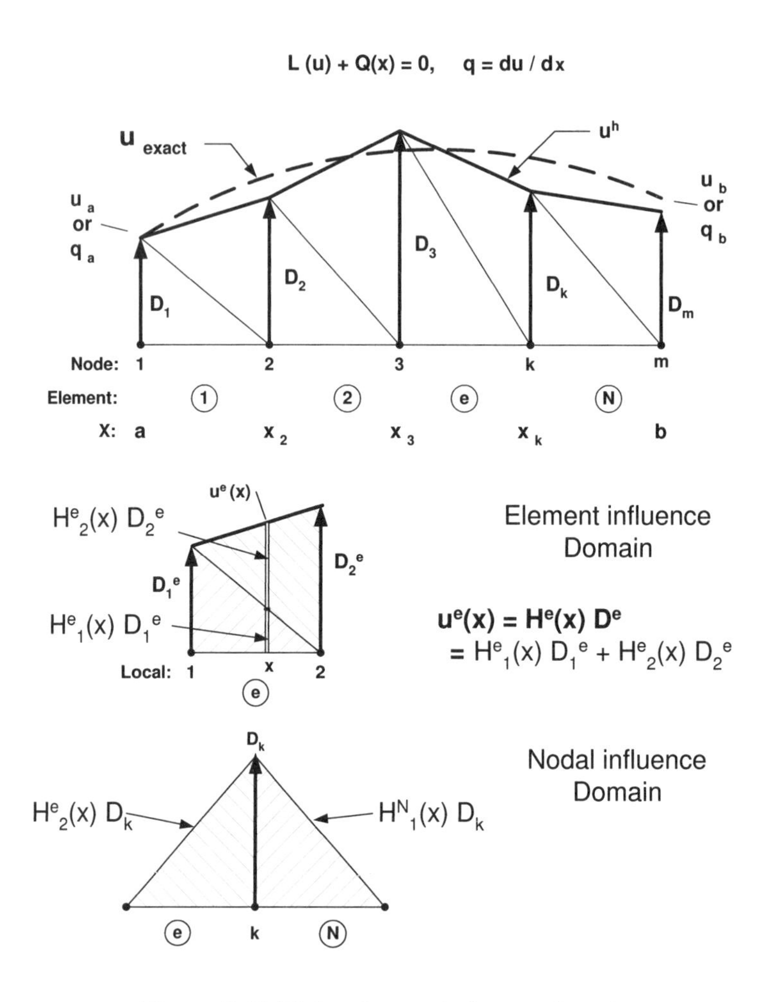

Figure 2.10 *Finite element influence domains*

dimensional problems $\mathbf{H}^b \subset \mathbf{H}^e$). The main point here is that $I = I(\mathbf{D})$, and that fact must be considered in the summation. The consideration of the subset relations is merely a bookkeeping problem. This allows Eq. 2.39 to be written as

$$I = \mathbf{D}^T \mathbf{S}\mathbf{D} - \mathbf{D}^T \mathbf{C} = \mathbf{D}^T (\mathbf{S}\mathbf{D} - \mathbf{C}) = 0$$

where

$$\mathbf{S} = \sum_{e=1}^{n_e} \boldsymbol{\beta}^{e^T} \mathbf{S}^e \boldsymbol{\beta}^e, \qquad \mathbf{C} = \sum_{e=1}^{n_e} \boldsymbol{\beta}^{e^T} \mathbf{C}^e + \sum_{b=1}^{n_b} \boldsymbol{\beta}^{b^T} \mathbf{C}^b, \tag{2.40}$$

and where $\boldsymbol{\beta}$ denotes a set of symbolic bookkeeping operations. The combination of the summations and bookkeeping is commonly referred to as the *assembly process*. Note that one set of operations acts on the rows of the column vector and the square matrix while a second set acts on the columns of the square matrix. This is often called 'scattering' the element contributions into the corresponding system coefficients.

It is easily shown that for a non-trivial solution, $\mathbf{D} \neq \mathbf{O}$, we must have $\mathbf{O} = \mathbf{S}\,\mathbf{D} - \mathbf{C}$, as the governing algebraic equations to be solved for the unknown nodal parameters, $\mathbf{D}$. To be specific, consider a linear interpolation element with two nodes per element, $(n_n = 2)$. If the element length is $L^e = (x_2 - x_1)^e$, then the element interpolation, written in physical coordinates, is

$$\mathbf{H}^e(x) = \left\lfloor \frac{(x_2^e - x)}{L^e} \quad \frac{(x - x_1^e)}{L^e} \right\rfloor,$$

so that

$$\mathbf{B}^e = \frac{d\mathbf{H}^e}{dx} = \left\lfloor -\frac{1}{L^e} \quad \frac{1}{L^e} \right\rfloor.$$

Therefore, the two parts (diffusion and convection) to the element square matrix are

$$\mathbf{S}^e{}_1 = \frac{1}{L^e}\begin{bmatrix} 1 & -1 \\ -1 & 1 \end{bmatrix}, \quad \mathbf{S}^e{}_2 = \frac{L^e}{6}\begin{bmatrix} 2 & 1 \\ 1 & 2 \end{bmatrix}, \tag{2.41}$$

while the element column (source) matrix is

$$\mathbf{C}^e = \int_{L^e} \mathbf{H}^{e^T} Q^e dx = \int_{L^e} \frac{Q^e}{L^e} \begin{Bmatrix} (x_2^e - x) \\ (x - x_1^e) \end{Bmatrix} dx.$$

If we were to assume that $Q = Q_0$, a constant, the constant source would simplify to $\mathbf{C}^{e^T} = \lfloor 1 \;\; 1 \rfloor Q_0 L^e/2$. That is, the finite element model would replace the constant source per unit length by lumping half its resultant, $Q_0\, L^e$, at each of the two nodes of the element. In the given case of $Q(x) = x$, the source vector reduces to

$$\mathbf{C}^e = \frac{L^e}{6}\begin{bmatrix} 2 & 1 \\ 1 & 2 \end{bmatrix}\begin{Bmatrix} Q_1 \\ Q_2 \end{Bmatrix}, \tag{2.42}$$

where $Q_1 = x_1$ and $Q_2 = x_2$ are the nodal values of the source. The latter form is what results if we make the usual assumption that spatially varying data are to be input at the system nodes and interpolated inside the element. In other words, it is common to interpolate from gathered nodal data to define $Q(x) = \mathbf{H}^e(x)\mathbf{Q}^e$ where $\mathbf{Q}^e$ are the local nodal values of the source. Then the resulting integral is the same as in $\mathbf{S}^e{}_2$, so $\mathbf{C}^e = \mathbf{S}^e{}_2\mathbf{Q}^e$ as given above. If we set $Q_1 = Q_2 = Q_0$ this agrees with the constant source resultant, as noted above.

These are all the arrays needed to carry out an analysis if no post-processing information is needed. Thus it is relatively easy to hard-code the source for this model problem. Figure 2.11 gives such an implementation as well as including comment statements that look ahead to saving data typically needed for post-processing operations to be introduced later. The element square matrices are defined at lines 18-20 and the matrix multiplication to form $\mathbf{C}^e$ is carried out at line 23, using the nodal values of x which happens in this example to be the nodal values of $Q(x)$. Two optional lines appear as comments at lines 15 and 28. They can be used to save data that can be used later in an error estimate or post-processing. Alternatively, those data could simply be re-computed in a later phase of the program.

```
! ..........................................................    ! 1
!  ***  ELEM_SQ_MATRIX PROBLEM DEPENDENT STATEMENTS FOLLOW ***   ! 2
! ..........................................................    ! 3
!   Hard Coded Galerkin MWR for the ODE (MODEL example 124)      ! 4
!  E * U,xx + U + x = 0, with E = 1 & boundary conditions like   ! 5
!  U(0)=0=U(1),      so U = sin(x)/sin(1) - x        or          ! 6
!  U(0)=0, U'(1)=0, so U = sin(x)/cos(1) - x        etc          ! 7
                                                                 ! 8
  REAL(DP) :: DL                              ! Length           ! 9
  REAL(DP) :: S_1 (2, 2), S_2 (2, 2)   ! stiffness, mass         !10
                                                                 !11
  DL = COORD (LT_N, 1) - COORD (1, 1) ! Length                   !12
                                                                 !13
! E  = 1.d0 ; LT_QP = 1        ! constitutive array, quadrature  !14
! CALL STORE_FLUX_POINT_COUNT ! Save LT_QP, for post-processing  !15
                                                                 !16
!           SQUARE MATRIX, CONDUCTION & CONVECTION               !17
  S_1 = RESHAPE ((/ 1, -1, -1, 1 /), (/2,2/)) / DL   ! stiffness !18
  S_2 = DL * RESHAPE ((/ 2,  1,  1, 2 /), (/2,2/)) / 6.d0 ! mass !19
  S   = S_1 - S_2                                     ! net      !20
                                                                 !21
!           INTERNAL SOURCE (EXACT INTEGRATION)                  !22
  C = MATMUL (S_2, COORD (1:2, 1))  ! linear source term         !23
                                                                 !24
!       SAVE FOR FLUX AVERAGING OR POST PROCESSING               !25
! B (1, :) = (/ -1, 1 /) / DL                    ! dH / dx       !26
! XYZ (1)   = (COORD (LT_N, 1) + COORD (1, 1))/2 ! center point  !27
! CALL STORE_FLUX_POINT_DATA (XYZ, E, DGH)      ! to postprocess !28
                                                                 !29
! End of application dependent code                              !30
```

Figure 2.11 *Exact integration linear element model for U" + U + X = 0*

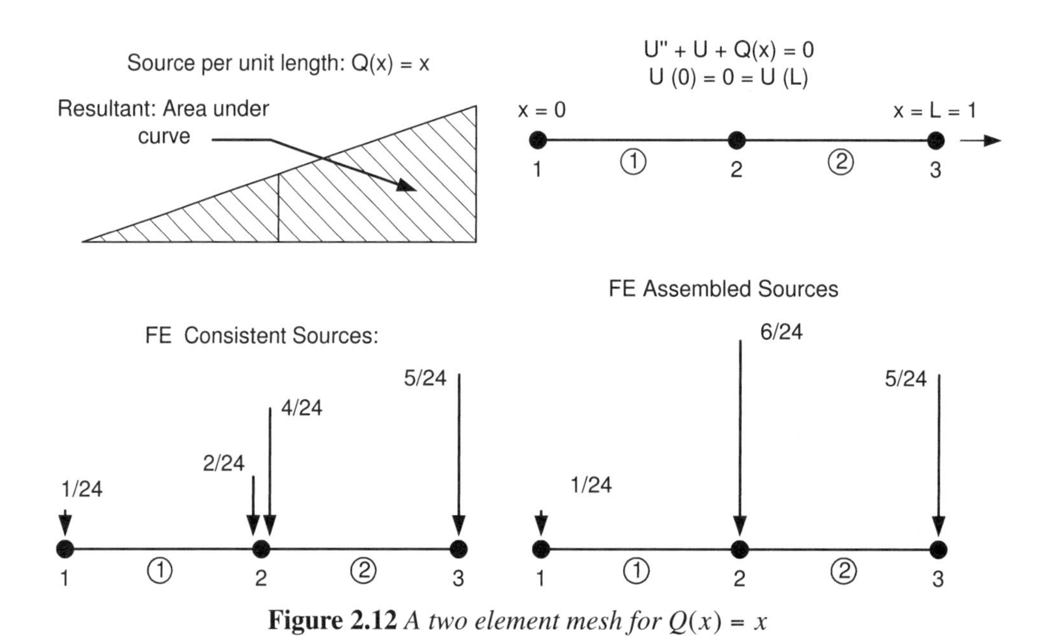

Figure 2.12 *A two element mesh for $Q(x) = x$*

To compare a finite element spatial approximation with the exact one, and the previous global MWR, select a two element model, as illustrated in Fig. 2.12 (which is stored as MODEL example 124). Let the elements be of equal length, $L^e = L/2$. Here we set $L = 1$ as in the previous global MWR. Then the element square matrices are the same for both elements and forming a common denominator gives the values:

$$\mathbf{S}^e \frac{1}{1/2}\begin{bmatrix} 1 & -1 \\ -1 & 1 \end{bmatrix} + \frac{1/2}{6}\begin{bmatrix} 2 & 1 \\ 1 & 2 \end{bmatrix} = \frac{1}{24}\begin{bmatrix} 44 & -50 \\ -50 & 44 \end{bmatrix}.$$

The two column matrices will differ because they occupy different regions of space, and thus different sources $Q(x)$. The reader should verify that their numerical values are:

$$\mathbf{C}_{(e=1)}^T = \lfloor 1 \quad 2 \rfloor / 24, \quad \mathbf{C}_{(e=2)}^T = \lfloor 4 \quad 5 \rfloor / 24.$$

Note from Fig. 2.12 that these two resultant element vectors account for the total applied source, $Q(x)$, because the sum of the coefficients of the above two vectors is $1/2$ which is the value of the integral of $Q(x)$ over the entire domain.

For the last two terms in Eq. 2.36 note that $u(0) = \phi_1$ and $u(L) = \phi_3$ so those terms become $\phi_1 u'(0) - \phi_3 u'(L)$. It is now clear that we can write the last two terms as the system level scalar (dot) product $\mathbf{D}^T \mathbf{C}_q$, where the only two non-zero entries in $\mathbf{C}_q$ are $u'(0)$ and $-u'(L)$, and they are carried to the RHS. The assembly process applied to the element matrices and the boundary matrices yields, $\mathbf{S}\,\mathbf{D} = \mathbf{C}$, as

$$\frac{1}{24}\begin{bmatrix} 44 & -50 & 0 \\ -50 & (44+44) & -50 \\ 0 & -50 & 44 \end{bmatrix}\begin{Bmatrix} \phi_1 \\ \phi_2 \\ \phi_3 \end{Bmatrix} = \frac{1}{24}\begin{Bmatrix} 1 \\ (2+4) \\ 5 \end{Bmatrix} + \begin{Bmatrix} -u'(0) \\ 0 \\ u'(L) \end{Bmatrix}. \qquad (2.43)$$

However, these equations do not yet satisfy the two essential boundary conditions of $u(0) = \phi_1 = u_0 = 0$, and $u(L) = \phi_3 = u_L = 0$. That is, the above system does not have a unique solution because it is a system of three equations for five unknowns $(\phi_1, \phi_2, \phi_3, u'(0), u'(L))$. Note that the essential boundary conditions have assigned values to the two end nodal values (ϕ_1, ϕ_3), so we move their columns (1 and 3) from $\mathbf{S}$ to the right hand side. After applying these conditions and simplifying:

$$\begin{bmatrix} 0 & -50 & 0 \\ 0 & 88 & 0 \\ 0 & -50 & 0 \end{bmatrix}\begin{Bmatrix} \phi_1 \\ \phi_2 \\ \phi_3 \end{Bmatrix} = \begin{Bmatrix} 1 \\ 6 \\ 5 \end{Bmatrix} + \begin{Bmatrix} -24\,u'(0) \\ 0 \\ 24\,u'(L) \end{Bmatrix} - \phi_1\begin{Bmatrix} 44 \\ -50 \\ 0 \end{Bmatrix} - \phi_3\begin{Bmatrix} 0 \\ -50 \\ 44 \end{Bmatrix}.$$

Now there are three unknowns $(\phi_2, u'(0), u'(L))$ and the system is non-singular. Retaining only the second row, which is the only independent sub-set of equations for a nodal value, and substituting the zero values for ϕ_1, ϕ_3 gives: $88\phi_2 = 6 + 0 + 0$, or $\phi_2 = 0.06818$ versus an exact value at that node of $u = 0.06975$.

Now it is possible to return to the remaining unused rows (1 and 3) in the algebraic system to recover the flux 'reactions' that are necessary to enforce the two essential boundary conditions. From the first row

```
title "Two L2 solution of U,xx + U + X = 0" ! begin keywords  ! 1
example    124 ! Application source code library number       ! 2
nodes        3 ! Number of nodes in the mesh                  ! 3
elems        2 ! Number of elements in the system             ! 4
dof          1 ! Number of unknowns per node                  ! 5
el_nodes     2 ! Maximum number of nodes per element          ! 6
bar_chart      ! Include bar chart printing in output         ! 7
exact_case   9 ! Analytic solution for list_exact, etc        ! 8
list_exact     ! List given exact answers at nodes, etc       ! 9
remarks      3 ! Number of user remarks                       !10
quit ! keyword input, remarks follow                          !11
1 U,xx + U + X = 0, U(0)=0=U(1), U = sin(x)/sin(1) - x        !12
2 Here we use two linear (L2) line elements.                  !13
3 Defaults to 1-D space, and line element                     !14
 1 1 0.             ! node, bc_flag, x                        !15
 2 0 0.5            ! node, bc_flag, x                        !16
 3 1 1.00           ! node, bc_flag, x                        !17
  1    1    2     ! elem, two nodes                           !18
  2    2    3     ! elem, two nodes                           !19
 1    1 0.        ! node, dof, essential BC value             !20
 3    1 0.        ! end of data                               !21
```

Figure 2.13 *Data for a two L2 element Galerkin model*

```
TITLE: "Two L2 solution of U,xx + U + x = 0"                         ! 1
                                                                     ! 2
*** INPUT SOURCE RESULTANTS ***                                      ! 3
ITEM        SUM            POSITIVE        NEGATIVE                  ! 4
    1    5.0000E-01      5.0000E-01      0.0000E+00                  ! 5
                                                                     ! 6
*** REACTION RECOVERY ***                                            ! 7
   NODE, PARAMETER,        REACTION, EQUATION                        ! 8
      1,  DOF_1,         1.8371E-01       1                          ! 9
      3,  DOF_1,        -3.5038E-01       3                          !10
                                                                     !11
***   RESULTS AND EXACT VALUES IN NODAL ORDER    ***                 !12
    NODE,  X-Coord,      DOF_1,        EXACT1,                       !13
       1  0.0000E+00  0.0000E+00  0.0000E+00                         !14
       2  5.0000E-01  6.8182E-02  6.9747E-02                         !15
       3  1.0000E+00  0.0000E+00  0.0000E+00                         !16
```

Figure 2.14 *Selected two L2 simple Galerkin model results*

$$0 - 50\,\phi_2 + 0 \;=\; 1 - 24u'(0) - 44\phi_1 - 0$$

or $-4.4091 = -24u'(0)$, so $u'(0) = 0.1837$ which compares to the exact flux (slope) of $u'(0) = 0.1884$ at $x = 0$. Likewise, the third row of the system yields the reaction $0 - 50\,\phi_2 + 0 \;=\; 5 + 24u'(L) + 0 - 44\phi_3$ so $u'(L) = -0.3504$ versus the exact $u'(L) = -0.3579$. Note the reduced equations would allow any values to be assigned to ϕ_1 and ϕ_3 and the required reaction flux values would change in proportion. Several finite element codes compute the boundary fluxes by computing the gradients in those elements that are adjacent to the boundary where the essential boundary conditions are applied. Getting those fluxes from the integral form, as done above, is usually much more accurate. This will be demonstrated in the typical post-processing steps where we recover the gradients in all the elements in the mesh.

We usually want to recover the element gradients at selected points inside the element. Here we have selected a linear interpolation, so the gradient is constant throughout each element. In such a case we usually report the gradient value at the centroid (center) of the element. Later we will show that location is the most accurate location for the gradient. Gathering each element's nodal values back from the solution, $\mathbf{D}^T = \lfloor 0. \quad 0.06818 \quad 0. \rfloor$, we compute the fluxes, say $\varepsilon = du/dx$, in the elements from Eq. 2.38 as

$$\varepsilon^e = \frac{1}{L^e} \lfloor -1 \quad 1 \rfloor \begin{Bmatrix} \phi_1^e \\ \phi_2^e \end{Bmatrix} \tag{2.44}$$

$$\varepsilon^{(1)} = \frac{1}{0.5} [-1 \quad 1] \begin{Bmatrix} 0.0 \\ 0.06818 \end{Bmatrix} = 0.1364$$

$$\varepsilon^{(2)} = \frac{1}{0.5} [-1 \quad 1] \begin{Bmatrix} 0.06818 \\ 0.0 \end{Bmatrix} = -0.1364$$

which are the two equal and opposite slopes of this crude approximate solution. Trying to extrapolate these fluxes from elements at the boundary to the point on the boundary would give much less accurate boundary fluxes (slopes) than those obtained above from the governing integral form.

The data for the two element model is shown in Fig. 2.13. They begin with a group of problem control words and are followed by the numerical data for the nodes, the elements, and the essential boundary conditions. The results from this crude approximation are listed in Fig. 2.14 and shown in Fig. 2.15 along with the exact solution (as a dashed line). While not exact, the function values are noticed to be most accurate at the nodes. Conversely, the approximate gradients are least accurate (and discontinuous) at the nodes. The poor function accuracy compared to the previous global MWR solution using three constants is due in part to the fact that two of the three constants have been used to satisfy the boundary conditions and that the prior solution was cubic while the local finite element solution is currently piecewise linear. If we simply increase the number of elements the quality of the approximation will increase as shown in Fig. 2.16 where six elements were employed.

The previous discussion of the model differential equation showed that to implement a numerical solution we must, as a minimum, code the calculation of an element square matrix, and often also need a column matrix due to a source term. The first six lines of Fig. 2.17 hint that there must be some sort of software interface to a routine that governs such calculations, and that interface provides the storage of the arrays that are generally required for interpolation, integration, position evaluation, etc., and access to any user supplied data. In the *MODEL* code the routine that is always required is called *ELEM_SQ_MATRIX*. Figure 2.2 summarized other optional and required routines contained within the software library. In order to carry out the above gradient recoveries we either have to recompute the **B** matrix or store it at each quadrature point. That is the

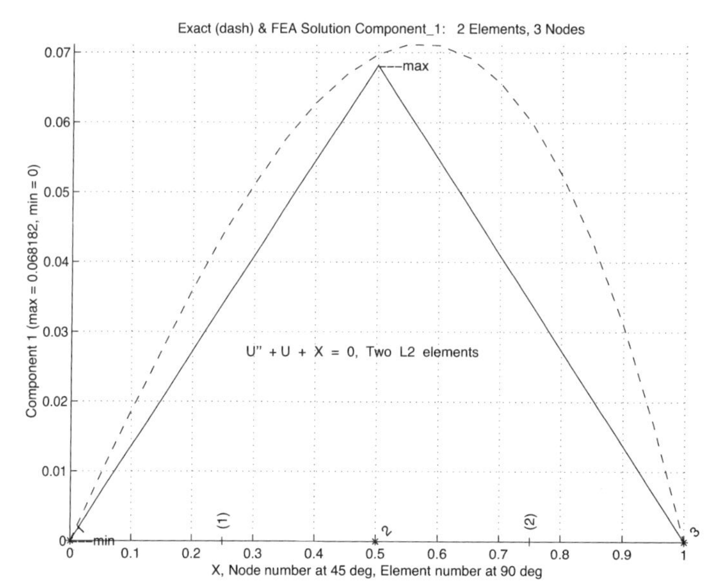

Figure 2.15 *Results from exact (-) and two linear elements (solid)*

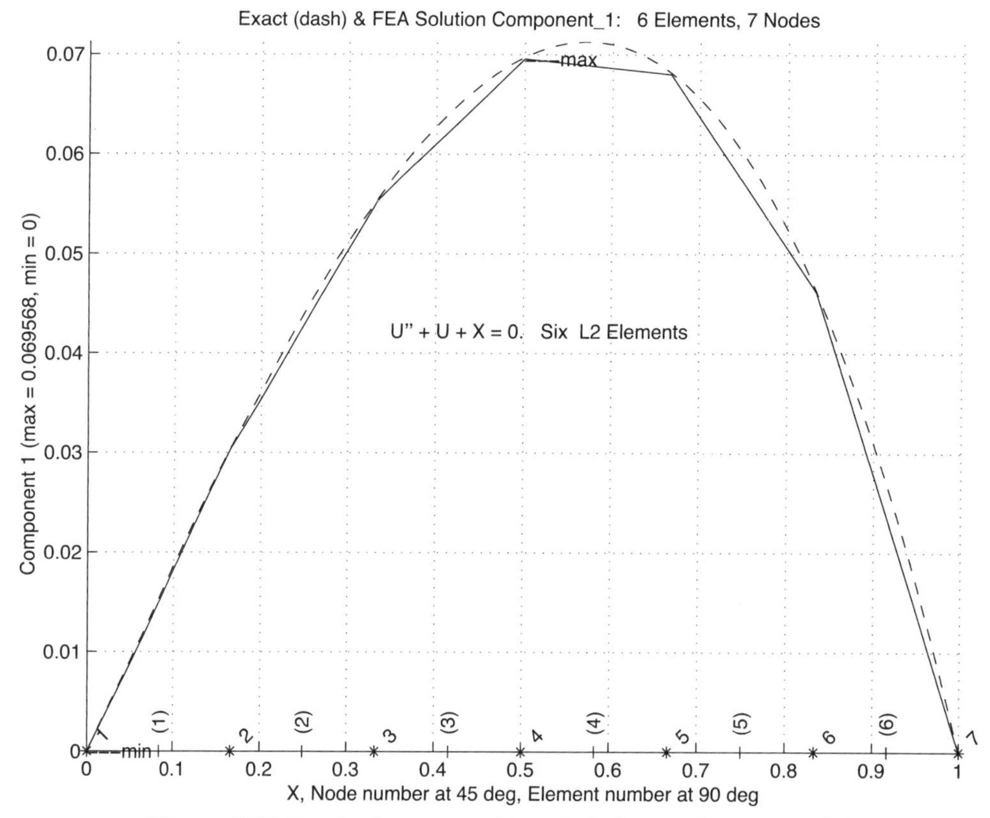

Figure 2.16 *Results from exact (-) and six linear elements (solid)*

```
! ...  Partial Global Access Arrays                              ! 1
  REAL(DP) :: C (LT_FREE), S (LT_FREE, LT_FREE) ! Results        ! 2
  REAL(DP) :: PT (LT_PARM, LT_QP), WT (LT_QP)   ! Quadratures    ! 3
  REAL(DP) :: H (LT_N), DGH (N_SPACE, LT_N)     ! Solution & deriv ! 4
  REAL(DP) :: COORD (LT_N, N_SPACE)             ! Elem coordinates ! 5
  REAL(DP) :: XYZ   (N_SPACE)                   ! Pt coordinates ! 6
                                                                 ! 7
! ...  Partial Notations List                                    ! 8
! COORD     = SPATIAL COORDINATES OF ELEMENT'S NODES             ! 9
! DGH       = GLOBAL DERIVATIVES SCALAR INTERPOLATION FUNCTIONS  !10
! H         = SCALAR INTERPOLATION FUNCTIONS                     !11
! LT_FREE   = NUMBER OF DEGREES OF FREEDOM                       !12
! LT_PARM   = DIMENSION OF PARAMETRIC SPACE                      !13
! LT_QP     = NUMBER OF QUADRATURE POINTS                        !14
! LT_N      = NUMBER OF NODES PER ELEMENT                        !15
! N_SPACE   = DIMENSION OF PHYSICAL SPACE                        !16
! PT        = QUADRATURE COORDINATES                             !17
! WT        = QUADRATURE WEIGHTS                                 !18
! XYZ       = PHYSICAL POINT                                     !19
                                                                 !20
! ..........................................................     !21
!  **  ELEM_SQ_MATRIX PROBLEM DEPENDENT STATEMENTS FOLLOW **     !22
! ..........................................................     !23
! (Stored as application example source file number 104.)       !24
!                                                                !25
! APPLICATION DEPENDENT Galerkin MWR via Gauss quadratures       !26
!   U,xx + U + X = 0, with boundary conditions like              !27
!   U(0)=0=U(1),     so U = sin(x)/sin(1) - x        or          !28
!   U(0)=0,U'(1)=0, so U = sin(x)/cos(1) - x        etc          !29
                                                                 !30
 REAL(DP) :: DL, DX_DN                     ! Length, Jacobian    !31
 INTEGER  :: IQ                            ! Loops               !32
                                                                 !33
 DL    = COORD (LT_N, 1) - COORD (1, 1)    ! LENGTH              !34
 DX_DN = DL / 2.                           ! CONSTANT JACOBIAN   !35
 E     = 1.d0                              ! CONSTANT E          !36
                                                                 !37
 CALL STORE_FLUX_POINT_COUNT ! Save LT_QP for post-process       !38
                                                                 !39
 DO IQ = 1, LT_QP          ! LOOP OVER QUADRATURES               !40
                                                                 !41
!       GET INTERPOLATION FUNCTIONS, AND X-COORD                 !42
   H     = GET_H_AT_QP (IQ)                                      !43
   XYZ   = MATMUL (H, COORD)   ! ISOPARAMETRIC                   !44
                                                                 !45
!       LOCAL AND GLOBAL DERIVATIVES, B = DGH                    !46
   DLH    = GET_DLH_AT_QP (IQ) ; DGH    = DLH / DX_DN            !47
                                                                 !48
!          SOURCE VECTOR WITH Q(X) = X = XYZ (1)                 !49
   C = C + H * XYZ (1) * WT (IQ) * DX_DN                         !50
                                                                 !51
!          SQUARE MATRIX                                         !52
   S = S + ( MATMUL (TRANSPOSE(DGH), DGH)                  &     !53
           - OUTER_PRODUCT (H, H)        ) * WT (IQ) * DX_DN     !54
                                                                 !55
!    SAVE FOR FLUX AVERAGING OR POST PROCESSING, B == DGH        !56
   CALL STORE_FLUX_POINT_DATA (XYZ, E, DGH) ! for post-proc      !57
 END DO  ! QUADRATURE                                            !58
!  End of application dependent code                             !59
```

Figure 2.17 *Element quadrature implementation for* $U'' + U + X = 0$

purpose of lines 38 and 57 in Fig. 2.11. The former declares how many quadrature points are being used by this element type, and the later line saves the **B** matrix and the spatial coordinate(s) of the point. (In this example, the 'constitutive data' are simply unity and are not really needed.) Thus, the post-processing loop has a similar pair of operations to gather those data and carry out the matrix products used above.

A typical subroutine segment for implementing any one-dimensional finite element by numerical integration is shown in Fig. 2.17. The coding is valid for any line element in the library of interpolation functions (currently linear through cubic) and the selection of element type is set in the control data, as noted later. For a linear element a one point quadrature rule would exactly integrate the first square matrix contribution, but a two point quadrature rule would be needed for the second square matrix contribution and for the column matrix. Clearly higher degree elements require a corresponding increase in the quadrature rule employed. The first 20 lines of that figure relate to an interface that has not yet been described. Only lines 26 and on change with each new application class. Lines 38 and 57 are optional for later post-processing uses. Line 36 accounts for the unit coefficient multiplying the d^2u/dx^2 term in the differential equation. Usually it has some other assigned user input value.

2.11 Continuous nodal flux recovery

Zienkiewicz and Zhu [20, 23, 24] developed the concept of utilizing a local patch of elements, sampled at their super-convergent points, to yield a smooth set of least square fit nodal gradients or fluxes. As noted earlier, the super-convergent points of an element are the special interior locations where the gradients of the element are most accurate. That is, those gradient locations match those of polynomials of one or more degrees higher. Numerous minor improvements to their original process have shown the SCP recovery process to be a practical way to get continuous nodal fluxes, $\boldsymbol{\sigma}^*$. They have demonstrated numerically that one can generate super-convergence estimates for $\boldsymbol{\sigma}^*$ at a node by employing patches of elements surrounding the node. These concepts are illustrated in Fig. 2.18.

A local least squares fit is generated over the patch of elements in the following way. Assume a polynomial approximation of the form

$$\boldsymbol{\sigma}^* = \mathbf{P}(\phi, \eta)\,\mathbf{a} \tag{2.45}$$

where **P** denotes a polynomial (in a local parametric coordinate system selected for each patch) that is of the same degree and completeness that was used to approximate the original solution, $\mathbf{u}_h$. That is, **P** is similar or identical to the solution interpolation array **H**. Here **a** represents a rectangular array that contains the nodal values of the flux. Since $\hat{\boldsymbol{\sigma}}$ was computed using the physical derivatives of **H**, $\hat{\boldsymbol{\sigma}} = \mathbf{E}^e\mathbf{B}^e\boldsymbol{\phi}^e$. To compute the estimate for $\boldsymbol{\sigma}^*$ at the nodes inside the patch, we minimize the function

$$F(\mathbf{a}) = \sum_{j=1}^{n} (\boldsymbol{\sigma}_j^* - \hat{\boldsymbol{\sigma}}_j)^2 \quad \rightarrow \quad \min$$

where n is the total number of integration points (or super-convergent points) used in the elements that define the patch and $\boldsymbol{\sigma}_j$ is the flux evaluated at point $\mathbf{x}_j$. Substituting the two different interpolation functions gives

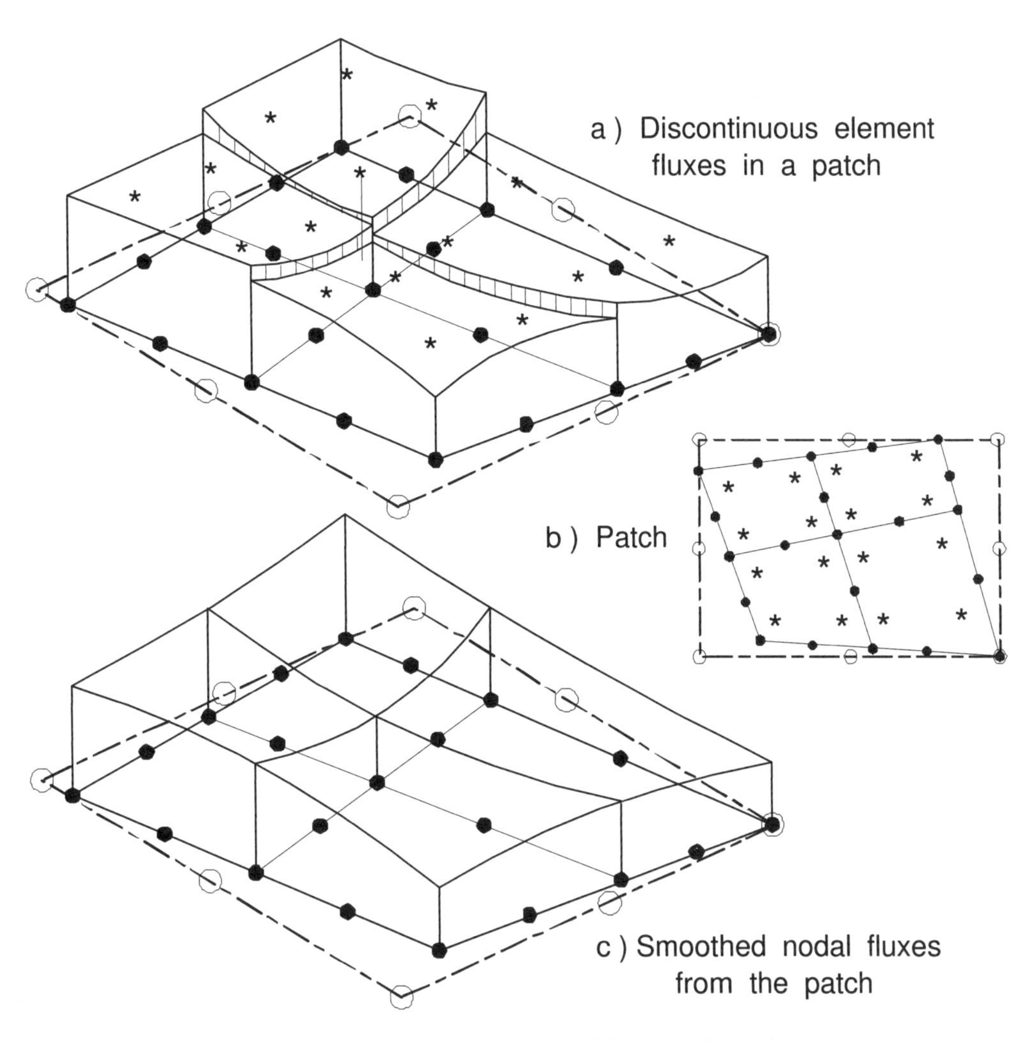

Interpolated solution: $u_h = N(x)\ U$ at node points, •
Element flux: $q_h = E\ B(x)\ U$ at Gauss points, *
Least squares patch fit of flux, F_p at patch points, ○
Interpolated node flux in patch: $q_p = N(x)\ F_p$ at nodes in patch, •
Element flux error estimate: $e_q = q_p - q_h$

Figure 2.18 *Smoothing flux values on a node based patch*

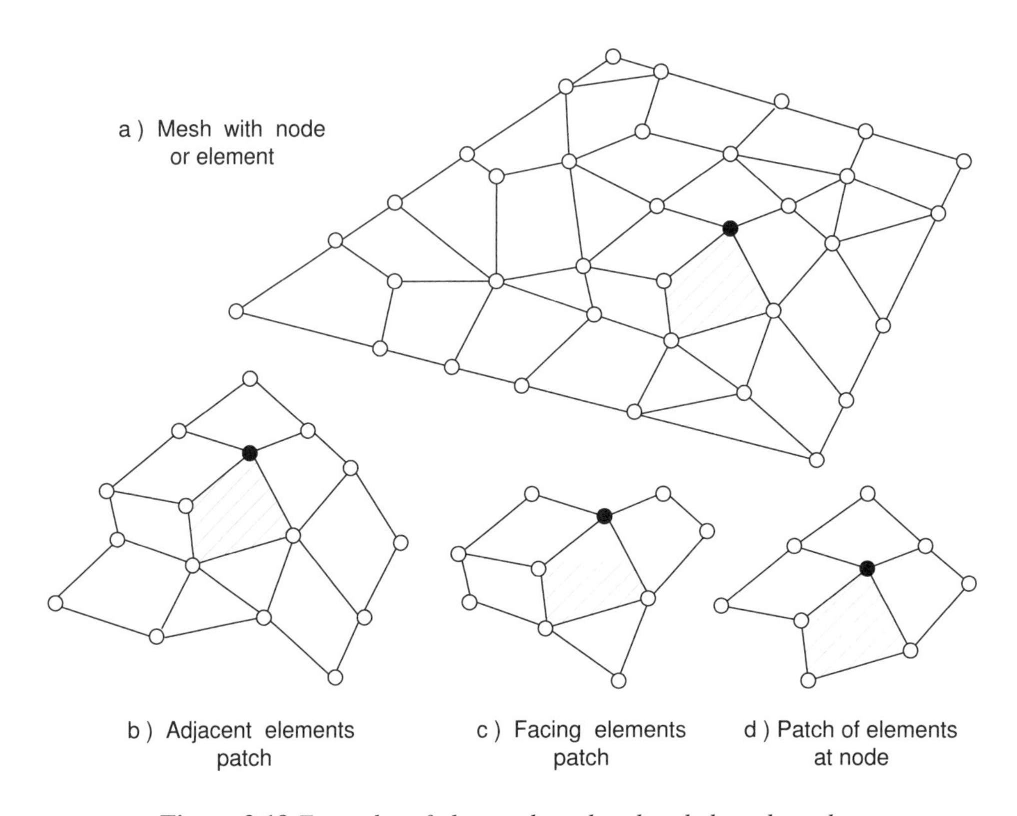

Figure 2.19 *Examples of element based and node based patches*

$$F(\mathbf{a}) = \sum_{e=1}^{n_{pe}} \sum_{j=1}^{n_q} \left[\mathbf{P}_j \, \mathbf{a} - \mathbf{E}^e \, \mathbf{B}_j^e \, \boldsymbol{\phi}^e \right]^2 \tag{2.46}$$

where n_{pe} denotes the number of elements in the patch and n_q is the number of integration points used to form $\hat{\sigma}^e$. That is, we are seeking a least squares fit through the

$$n = \sum_{e=1}^{n_{pe}} \sum_{j=1}^{n_q}$$

data points to compute the unknown coefficients, **a**, which is a rectangular matrix of flux components at each node of the patch. Note that the number rows in the least squares system will be equal to the number of nodes defining the patch 'element'. Thus the above value of n sampling points must be equal to or greater than the number of nodes on the patch 'element' (i.e., the number of coefficients in **P**). The standard least squares minimization gives the local algebraic problem $\mathbf{S}\,\mathbf{a} = \mathbf{C}$ where

$$\mathbf{S} = \sum_{e=1}^{n_{pe}} \sum_{j=1}^{n_q} \mathbf{P}^T(\phi_j,\ \eta_j)\,\mathbf{P}(\phi_j,\ \eta_j), \quad \mathbf{C} = \sum_{e=1}^{n_{pe}} \sum_{j=1}^{n_q} \mathbf{P}_j^T \, \mathbf{E}^e \, \mathbf{B}_j^e \, \boldsymbol{\Phi}^e.$$

This is solved for the coefficients **a** of the local patch fit. It is the cost of solving this small system of equations, for each patch, that we must pay in order to obtain the continuous nodal values for the fluxes. To avoid ill-conditioning common to least squares, the local patch fitting parametric space (ϕ, η) is mapped to enclose the patch of elements while using a constant Jacobian for the patch. The use of the constant Jacobian is the key to the efficient conversion of the physical stress location, $\mathbf{x}_j$, to the corresponding patch location, $\boldsymbol{\phi}_j$. Here the implementation actually employs a diagonal constant Jacobian to map the patch onto the physical domain.

Zienkiewicz and Zhu have verified numerically that the derivatives estimated in this way have an accuracy of at least order $O(h^{p+1})$, where h is the size of the element and p is the degree of the interpolation, **N**, used for the solution. There is a theorem that states that if the σ^* are super-convergent of order $O(h^{p+\alpha})$ for $\alpha > 0$, then the error estimator will be asymptotically exact. That is, the effectivity index should approach unity, $\Theta \rightarrow 1$. This means that we have the ability to accurately estimate the error and, thus, to get the maximum accuracy for a given number of degrees of freedom. There is not yet a theoretical explanation for the 'hyperconvergent' convergence (two orders higher) reported in some of the SCP numerical studies. It may be because the least square fit does not go exactly through the given Barlow points. Thus, they are really sampling nearby. In Sec. 3.10 we show that derivative sampling points for a cubic are at ± 0.577, while those for the quartic are at ± 0.707. The patch smoothing may effectively be picking up those quartic derivative estimates and jumping to a higher degree of precision.

It is also possible to make other logical choices for selecting the elements that will constitute a patch. Figure 2.19 shows two types of element based patches as well as the above node based patch. Regardless of the types of patches selected they almost always overlap with other patches which means that the mesh nodes receive several different estimates for the continuous nodal flux value. They should be quite close to each other but they need to be averaged to get the final values for the continuous nodal fluxes. It is possible to weight that averaging by the size of the contributing patch but it is simpler to just employ a straight numerical average. The implementation of the SCP recovery method will be given in full detail in the next chapter after considering other error indicator techniques.

2.12 A one-dimensional example error analysis

As a simple example of the process for recovering estimates of the continuous nodal flux values we will return to one of the one-dimensional models considered earlier. Figure 2.20 shows a five element model for a second order ODE, while the corresponding analytic, Gauss point (o), and patch averaged flux estimates are shown in Fig. 2.21. The piecewise linear flux estimate (solid line) in the latter figure was obtained by using the SCP process described above. It is the relation that will be used to describe $\sigma^*(\mathbf{x})$ for general post-processing or for use as in the stress error estimate.

For linear interpolation elements we recall that the gradient in each element is constant. The elements used two Gaussian quadrature points (in order to exactly integrate the 'mass' matrix). In Figs. 2.22 and 23 we see a zoomed view of the various flux representations near element number 2 (in a 5 element mesh). The horizontal dash-dot line through the quadrature point flux values represents the standard finite element spatial distribution, $\mathbf{B}^e \boldsymbol{\phi}^e$, of the flux in that element. Again, the solid line is the spatial form of an averaged flux from a set of patches, and the dashed line is the exact flux

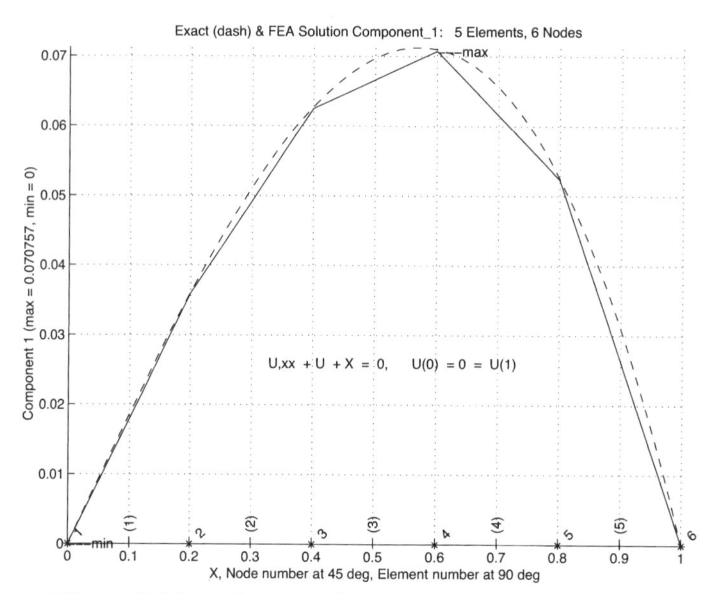

Figure 2.20 *Analytic and linear element solution results*

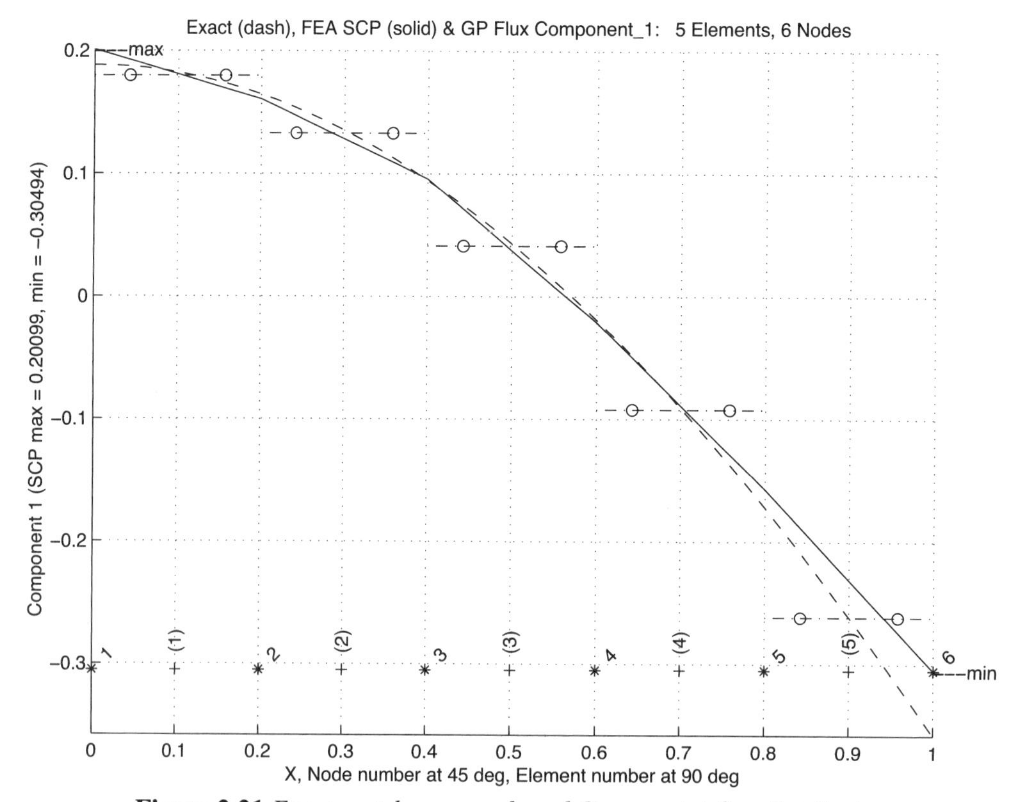

Figure 2.21 *Exact, patch averaged, and Gauss point flux distribution*

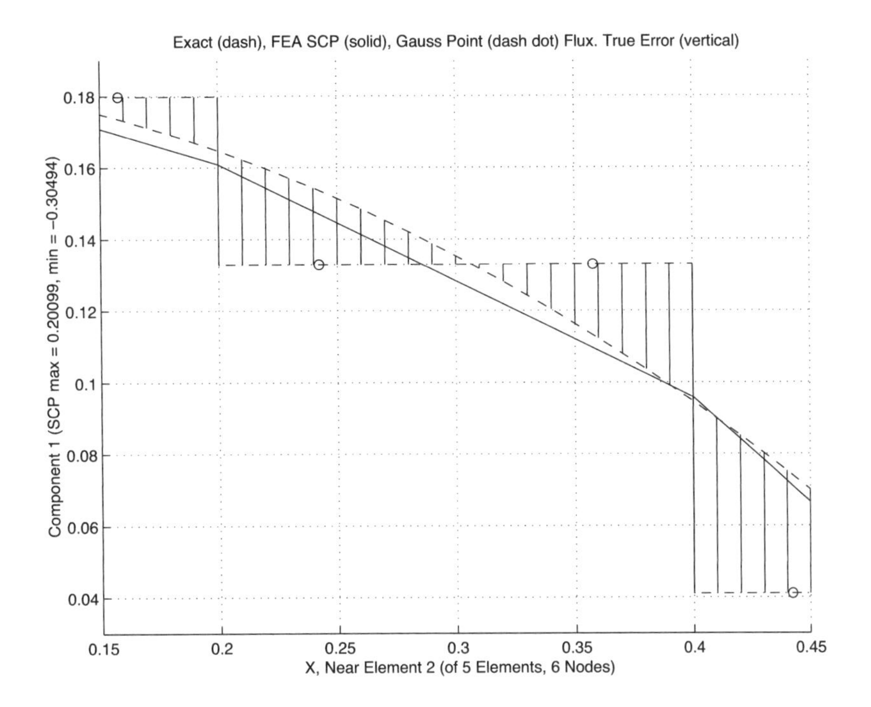

Figure 2.22 *Zoom of exact flux error near the second element*

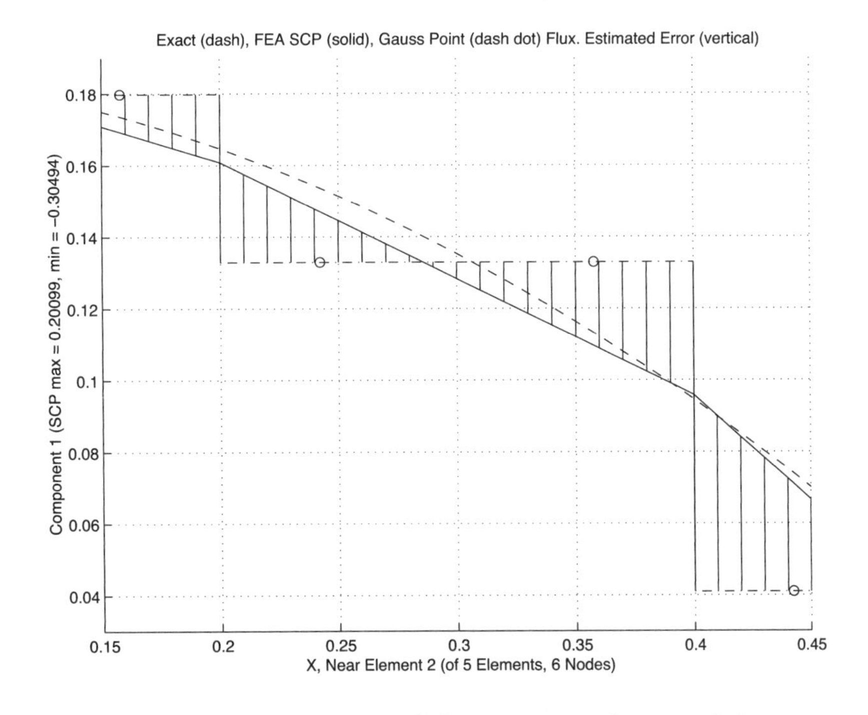

Figure 2.23 *Zoom of estimated flux error near the second element*

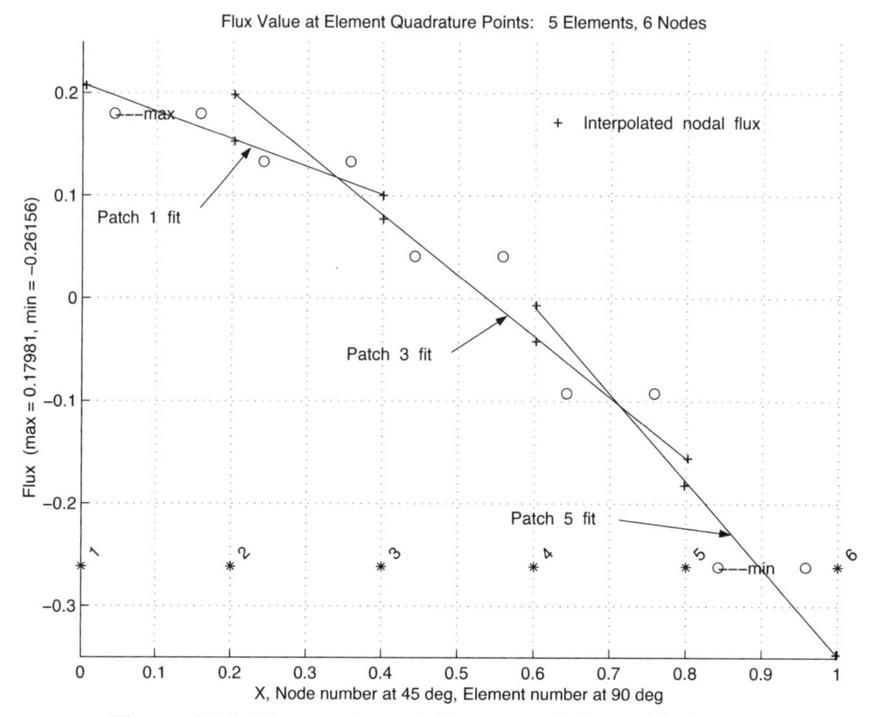

Figure 2.24 *Element flux and linear patch fits (odd elements)*

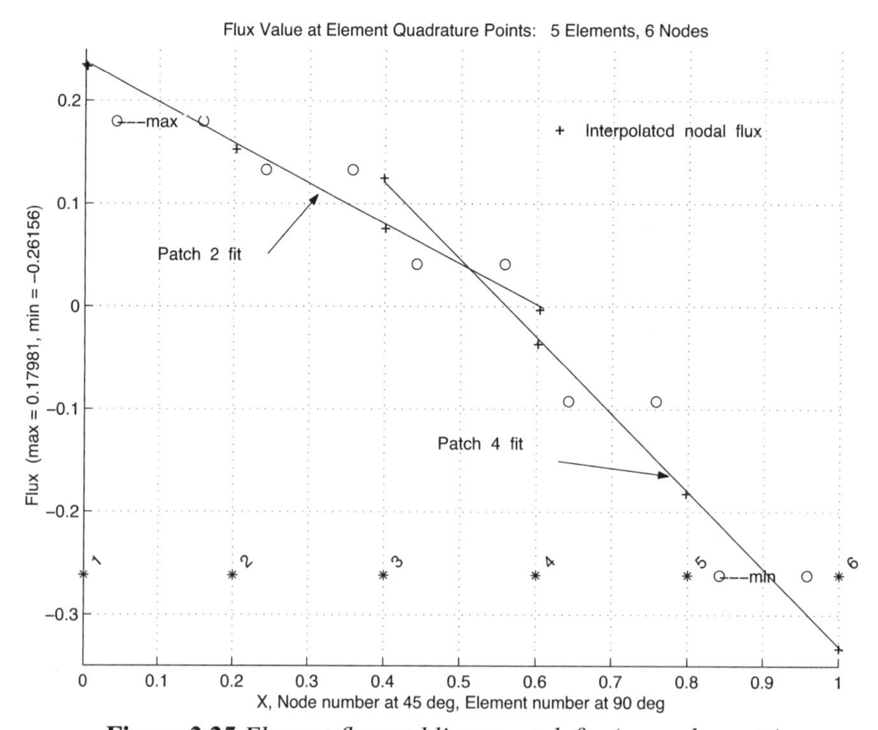

Figure 2.25 *Element flux and linear patch fits (even elements)*

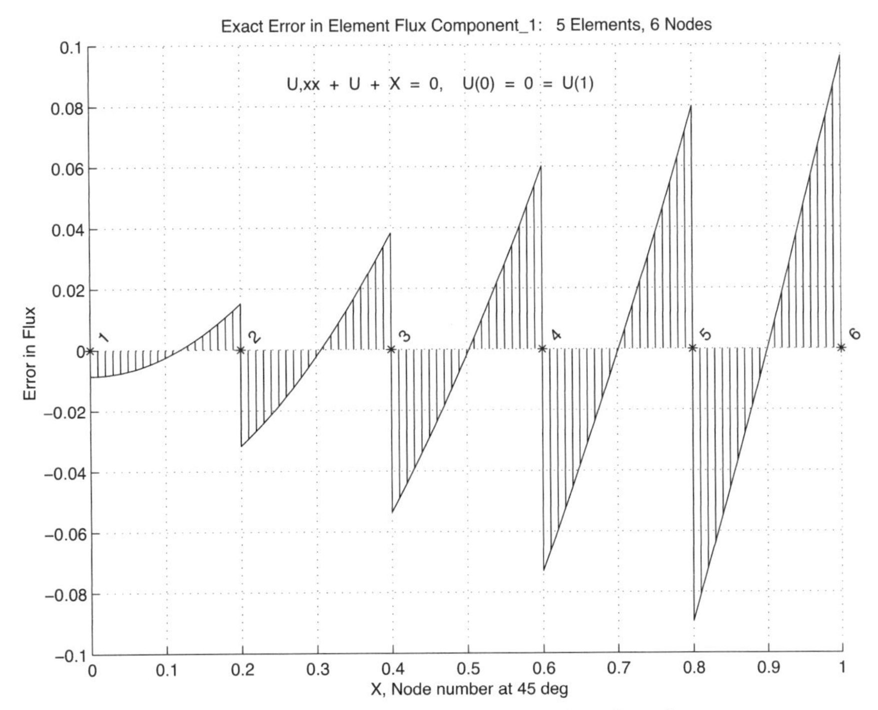

Figure 2.26 *Exact error in element flux distribution*

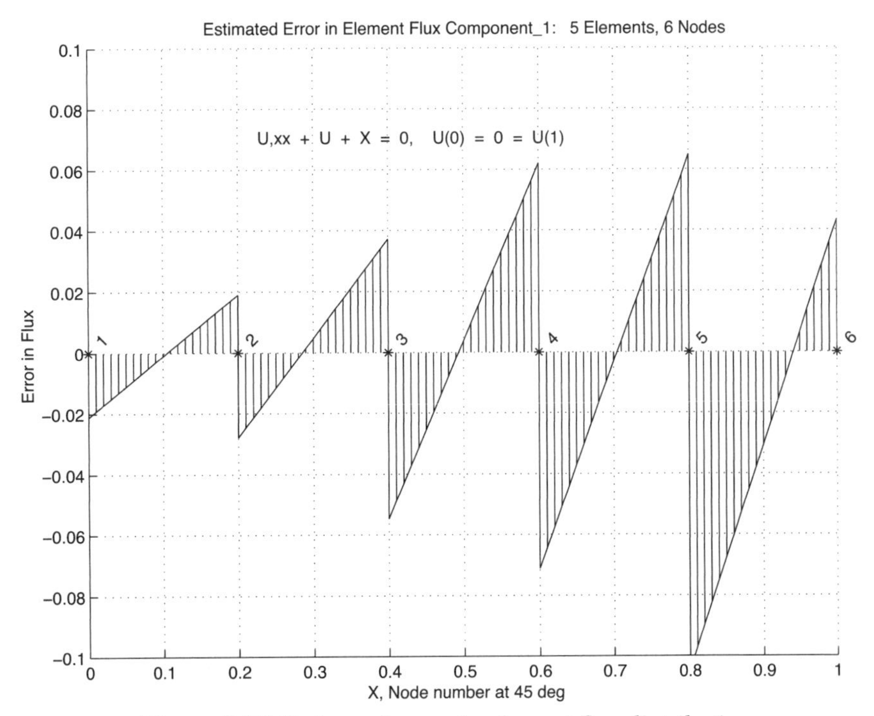

Figure 2.27 *Estimated error in element flux distribution*

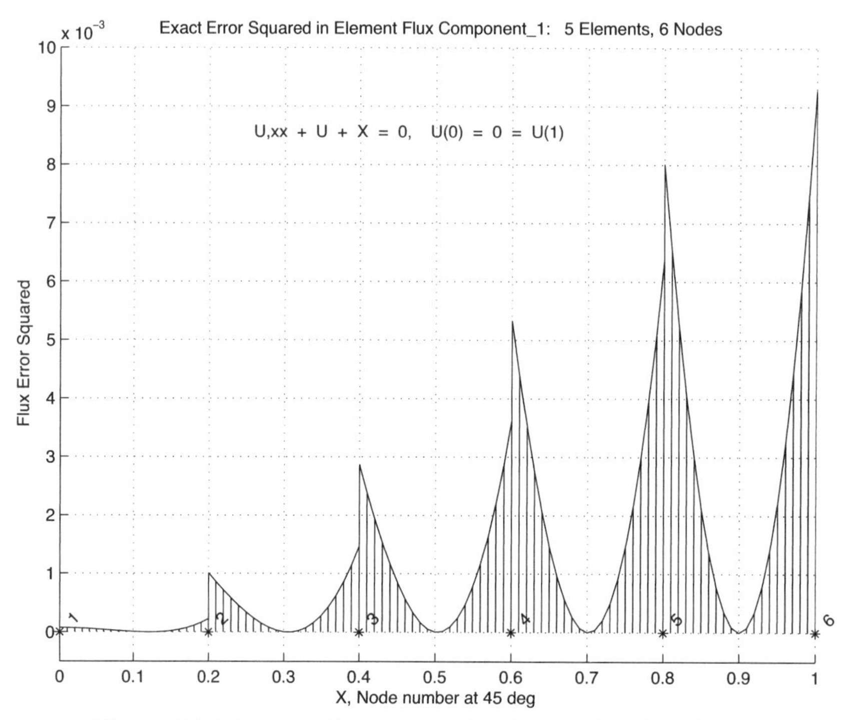

Figure 2.28 *Square of exact error in element flux distribution*

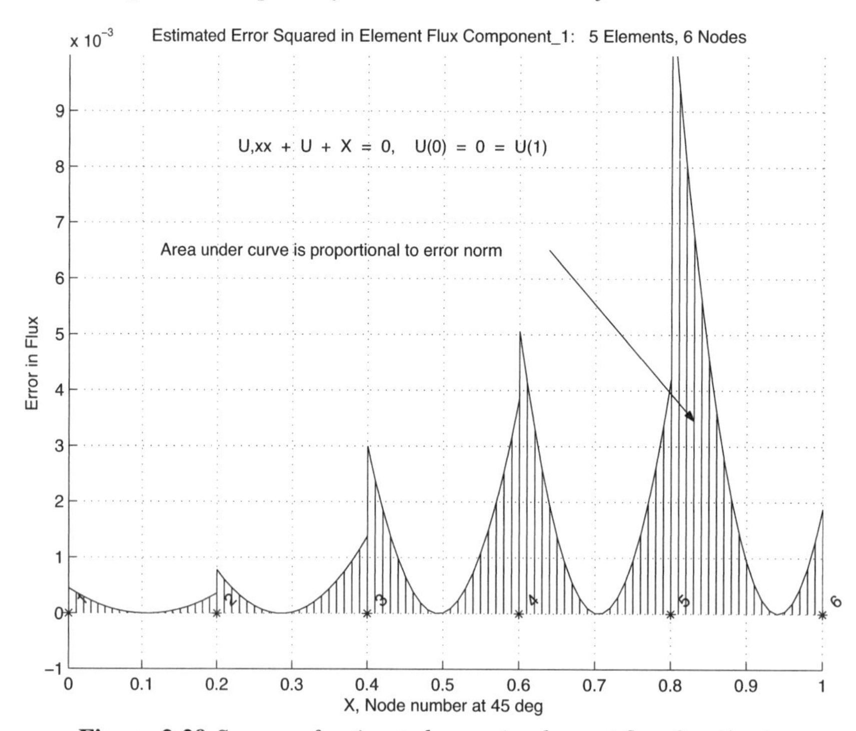

Figure 2.29 *Square of estimated error in element flux distribution*

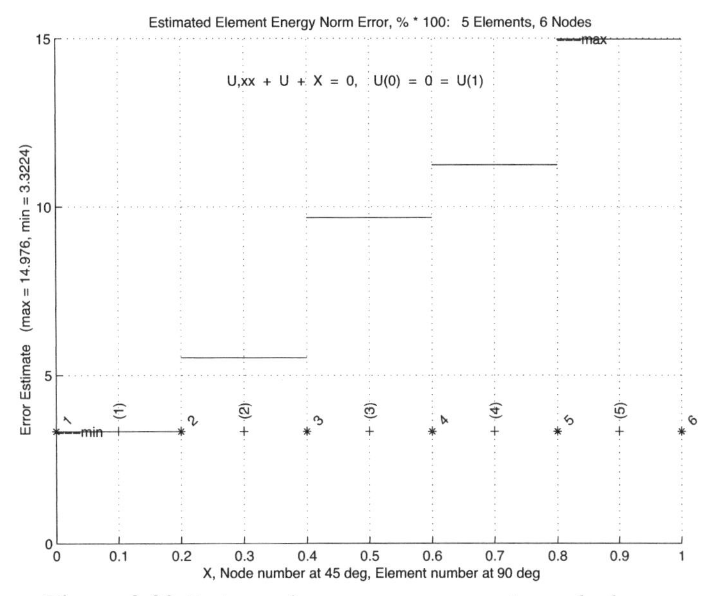

Figure 2.30 *Estimated energy norm error in each element*

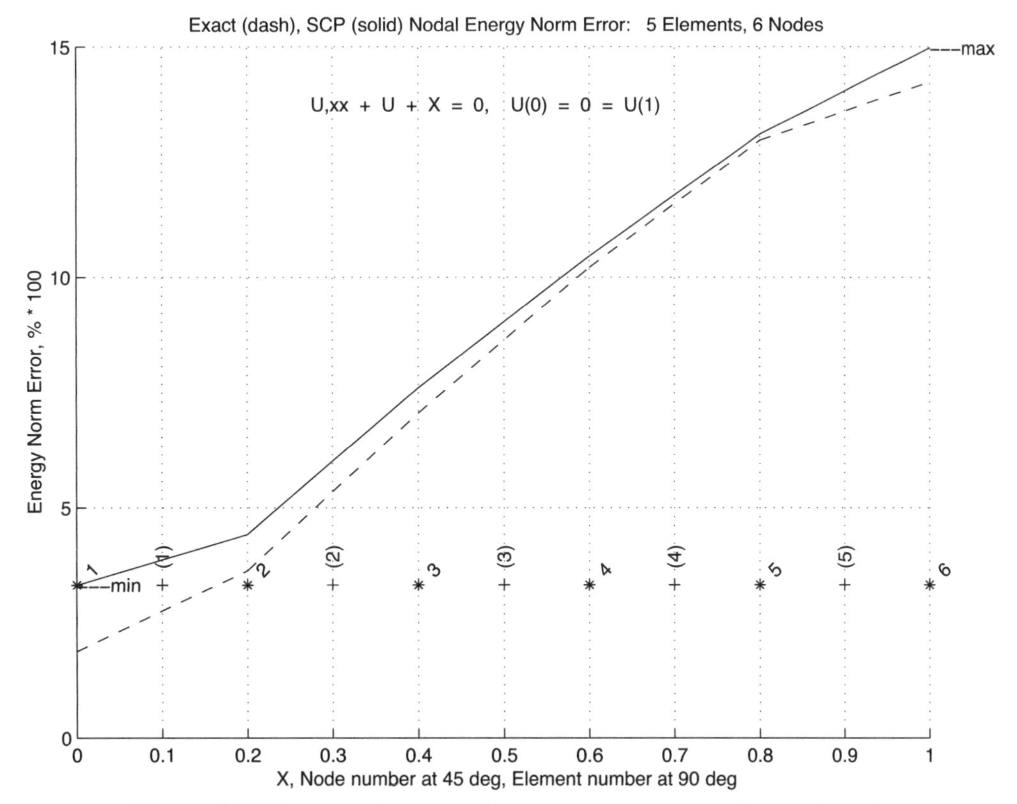

Figure 2.31 *Exact and averaged energy norm error at the mesh nodes*

distribution. In Fig. 2.22, the vertical lines show the exact error in the flux, e_σ. We will see later that it is the quantity whose norm can be used to establish an error indicator. Of course, in practice one does not know the exact flux distribution. However, even for this very crude mesh we can see that the patch averaged flux, $\boldsymbol{\sigma}^*$, is a much better estimate of the true flux,σ, than the standard element estimate, $\hat{\boldsymbol{\sigma}} = \mathbf{B}^e \boldsymbol{\phi}^e$. This is confirmed by comparing Fig. 2.23 to 2.22 where the vertical lines are the difference between the patch averaged and standard element flux values, i.e. $\boldsymbol{\sigma}^* - \hat{\boldsymbol{\sigma}}$.

The flux at each Gauss point is again plotted (as open circles) in Figs. 2.24 and 25. Also shown there are solid lines that represent the linear fit (same degree as the assumed element solution) over the elements in each patch. There were five different element patches corresponding to the five elements in the mesh. The first and last patches contained only two elements because the originating elements occurred on the boundary. The interior patches consisted of three elements each: the original element and the adjacent 'facing' element on the left and right. Once a patch fit has been obtained it is used to interpolate to the nodal values on that line (marked with a plus symbol). In that process each node in the original mesh receives multiple estimated flux values. Averaging all the estimates from each patch containing the node gives the (solid line) values shown earlier in Fig. 2.21. We will assume that the piecewise linear averaged fit for the flux in Fig. 2.21 is more accurate than the piecewise constant steps from the original element estimates.

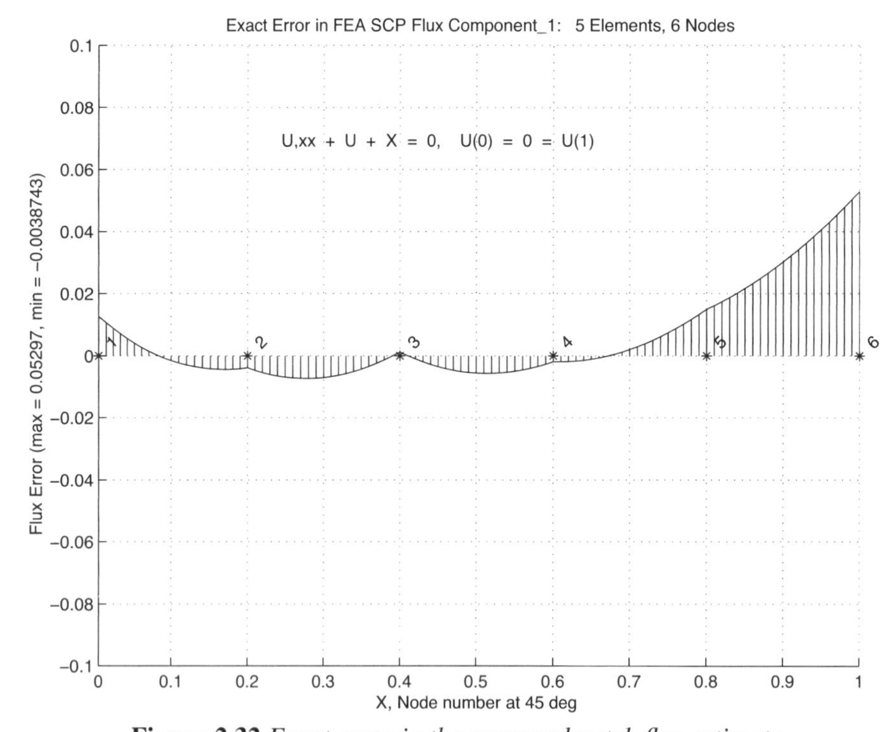

Figure 2.32 *Exact error in the averaged patch flux estimate*

The exact error in the flux, $\mathbf{e}_\sigma(\mathbf{x}) \equiv \sigma(\mathbf{x}) - \hat{\sigma}(\mathbf{x})$, is shown in Fig. 2.26. The estimated error in the flux, $\mathbf{e}_\sigma \approx \sigma^*(\mathbf{x}) - \hat{\sigma}(\mathbf{x})$, obtained from the SCP average flux data is shown to the same scale in Fig. 2.27. The comparisons are reasonably good and would improve as the mesh is refined or adapted. Both plots clearly show that the flux error from a standard element calculation, $\hat{\sigma}$, is usually largest at the nodes on the element interfaces. Note that the integral of the square of the above stress error distribution would define a stress error norm. Likewise, each element contributes to that norm and we could compare the local error to the average error to select elements for refinement or de-refinement. If we use the $\mathbf{E}^e$ matrix to scale the product of the stress error we would obtain the more common strain energy norm of the error. In either case the norm can be viewed as being directly proportional to the area under the curve defined by the stress error. Those exact and estimated error norms are shown as the hatched portions of Figs. 2.28 and 29, respectively.

Referring to Fig. 2.29, the area under the curve for each element can be used to assign a (constant) error value to each element. They are shown in Fig. 2.30. After all elements have been assigned an error value those values can be gathered at the nodes of the mesh to give a spatial approximation of the true energy norm error. Figure 2.31 illustrates such an averaging process and compares it to a similar average of the exact error. Of course, the exact error in the energy norm is a continuous function and we expect the nodally averaged values would approach the continuous values as the mesh is refined. For the crude mesh used in this example one can actually see the differences in exact and approximate plots, but for a fine mesh they usually look the same and one must rely on the numerical process to obtain useful error estimates.

Above we noted that the standard element level flux estimate, $\hat{\sigma} = \mathbf{E}^e\mathbf{B}^e\boldsymbol{\phi}^e$ is discontinuous at element interfaces and least accurate at those locations. Figure 2.32 shows the exact error in the SCP averaged flux estimates for this problem, to the same scale used to give the standard flux error in Fig. 2.26. There we see a number of improvements. The flux is continuous at the nodes. Its error is usually smallest at the element interfaces, except for nodes on the domain boundary. Usually the boundary has a significant effect and it is desirable to use smaller elements near the boundary. Special patch processes can be added to try to improve the flux estimates near boundaries but we will not consider such processes. Having illustrated the process in one-dimension we next consider a common two-dimensional test case for error estimators.

2.13 General boundary condition choices

Our discussion of the model differential equation in Eq. 2.15 has not yet led us to the need to introduce the general range of boundary condition choices that we commonly encounter in applying finite element analysis. Assume our model equation is generalized to the form

$$-\frac{d}{dx}\left(a\frac{du}{dx}\right) + b\,u + c = 0, \quad 0 < x < L. \tag{2.47}$$

Two physical examples, in one-dimension, readily come to mind where one can assign meanings to the three coefficients in this differential equation. For axial heat transfer u represents the unknown temperature, a the thermal conductivity of the material, b a convection coefficient (per unit length) on the surface, and c a heat source per unit length (and/or information on the external convection temperature). Likewise, for an axial elastic bar u is the displacement, a the material elastic modulus, b a resisting foundation stiffness, and c an axial force per unit length (like gravity). For an eigen-problem we would have $c = 0$ and b would be an unknown global constant (eigenvalue) to be computed from the finite element square matrices.

At either end of the domain one of three conditions can typically be prescribed as possible boundary conditions. The choices are known as a:

1. Dirichlet (essential) boundary condition that specifies the value of the solution on a portion of the boundary; $u(x_D) = U_D$.

2. Neumann (natural) boundary condition that specified the derivative of the solution normal to a portion of the boundary; $\pm\, a\, \partial u(x_N)/\partial n = g$. This usually represents a known flux or force, in the direction of the normal vector, defined in terms of the gradient of the solution. The sign of the a coefficient usually depends on the use of a constitutive relation, like Fourier's law, Flick's law, or Hooke's law.

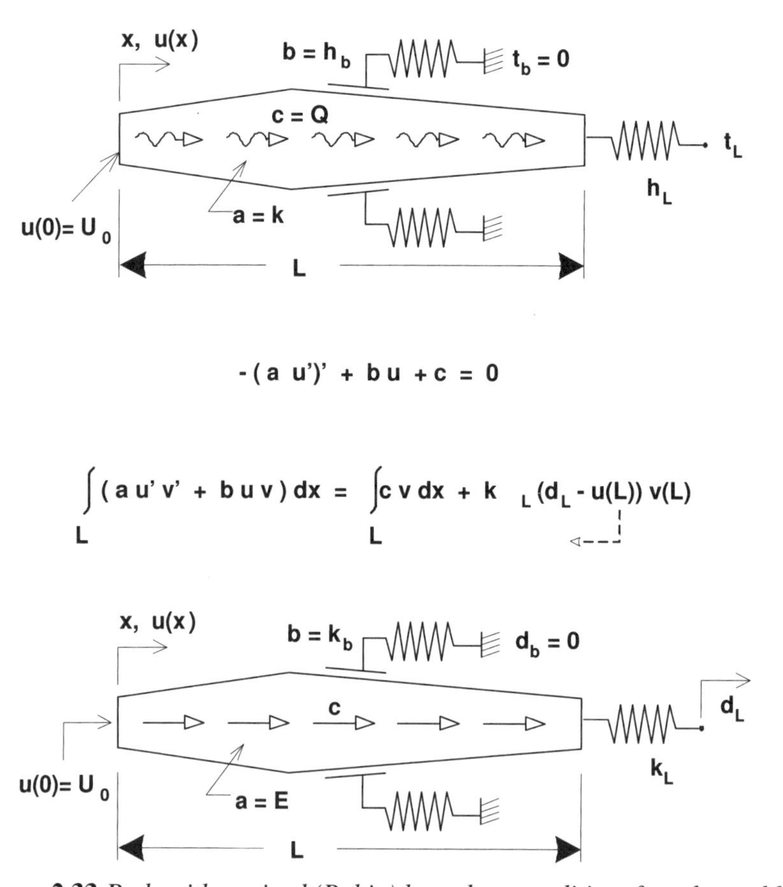

Figure 2.33 *Rods with a mixed (Robin) boundary condition, from h_L and k_L*

In heat transfer it is negative. Note that in one-dimension the direction of the outward normal reverses directions and $\partial / \partial n = \pm \partial / \partial x$.

3. Robin (mixed) boundary condition that specifies a linear combination of the normal derivative and solution value on a portion of the boundary; $\pm a\, \partial u(x_R) / \partial n + f\, u(x_R) = g$. The mixed condition occurs often in heat transfer as a convection boundary condition and sometimes as a linearized approximation to a radiation condition. In stress analysis it is less common and usually represents the effect of an elastic support foundation, with settlement. The generalized mixed condition notation here is: $a\, \partial u(x_R) / \partial n + r_1\, u(x_R) + r_2 = 0$. The most common example of this type of boundary condition is heat convection from a surface where the normal flux is $-k_n(x_R)\, \partial u(x_R) / \partial n - h(x_R)\, [u(x_R) - u_\infty(x_R)] = 0$ so that $r_1 = h$ and $r_2 = -h u_\infty$, where $u(x_R)$ is the unknown temperature on the convecting surface, $u_\infty(x_R)$ is the known surrounding fluid temperature, h is the convection coefficient, and k_n the thermal conductivity normal to the boundary.

In terms of the effects on the corresponding algebraic equation system a Dirichlet condition reduces the number of primary unknowns to be solved. That is, it reduces the size of the effective algebraic system that must be solved (see the following section). However, it also introduces an unknown nodal reaction term in the RHS that can optionally be recovered, as noted below. A Neumann condition does not change the size of the algebraic system; it simply adds known additional terms to the RHS. If that value is zero then no action is required and it is then usually called a natural boundary condition. A Robin, or mixed, condition likewise adds known additional terms to the RHS but, more importantly, it also adds a known symmetric square matrix contribution to the LHS in the rows and columns that correspond to the Robin surface nodes (usually heat convection nodes). For the one-dimensional problems considered in this chapter the Robin, or mixed condition can only occur at an end point (a single dof). For that node number the value of r_1 is added to the diagonal of the system square matrix, and the value of $-r_2$ is added to the corresponding row on the RHS (or for the common convection case add $h^b A^b$ and $h^b A^b u_\infty^b$, respectively for a convecting area A at the point). The corresponding Robin boundary contributions in 2-D will be covered in Sec. 4.3. Figure 2.33 illustrates a thermal (top) and stress problem, in 1-D, with a Dirichlet condition on the left ($x = 0$) end and a Robin condition on the right end ($x = L$).

2.14 General matrix partitions

The above small example has led to the most general form of the algebraic system that results from satisfying the required integral form: a singular matrix system with more unknowns that equations. That is because we chose to apply the essential boundary conditions last and there is not a unique solution until that is done. The algebraic system can be written in a general matrix form that more clearly defines what must be done to reduce the system to a solvable form by utilizing essential boundary condition values. Note that the system degrees of freedom, **D**, and the full equations could always be re-arranged in the following partitioned matrix form

$$\begin{bmatrix} \mathbf{S}_{uu} & \mathbf{S}_{uk} \\ \mathbf{S}_{ku} & \mathbf{S}_{kk} \end{bmatrix} \begin{Bmatrix} \mathbf{D}_u \\ \mathbf{D}_k \end{Bmatrix} = \begin{Bmatrix} \mathbf{C}_u \\ \mathbf{C}_k + \mathbf{P}_k \end{Bmatrix} \tag{2.48}$$

where $\mathbf{D}_u$ represents the unknown nodal parameters, and $\mathbf{D}_k$ represents the known essential boundary values of the other parameters. The sub-matrices $\mathbf{S}_{uu}$ and $\mathbf{S}_{kk}$ are square, whereas $\mathbf{S}_{uk}$ and $\mathbf{S}_{ku}$ are rectangular, in general. In a finite element formulation all of the coefficients in the **S** and **C** matrices are known. The $\mathbf{P}_k$ term represents that there are usually unknown generalized reactions associated with essential boundary conditions. This means that in general after the essential boundary conditions ($\mathbf{D}_k$) are prescribed the remaining unknowns are $\mathbf{D}_u$ and $\mathbf{P}_k$. Then the net number of unknowns corresponds to the number of equations, but they must be re-arranged before all the remaining unknowns can be computed.

Here, for simplicity, it has been assumed that the equations have been numbered in a manner that places the prescribed parameters (essential boundary conditions) at the end of the system equations. The above matrix relations can be rewritten as

$$\mathbf{S}_{uu}\,\mathbf{D}_u + \mathbf{S}_{uk}\,\mathbf{D}_k = \mathbf{C}_u$$

$$\mathbf{S}_{ku}\,\mathbf{D}_u + \mathbf{S}_{kk}\,\mathbf{D}_k = \mathbf{C}_k + \mathbf{P}_k$$

so that the unknown nodal parameters are obtained by inverting the non-singular square matrix $\mathbf{S}_{uu}$ in the top partitioned rows. That is,

$$\mathbf{D}_u = \mathbf{S}_{uu}^{-1}(\mathbf{C}_u - \mathbf{S}_{uk}\,\mathbf{D}_k).$$

Most books on numerical analysis assume that you have reduced the system to the non-singular form given above where the essential conditions, $\mathbf{D}_u$, have already been moved to the right hand side. Many authors use examples with null conditions ($\mathbf{D}_k$ is zero) so the solution is the simplest form, $\mathbf{D}_u = \mathbf{S}_{uu}^{-1}\,\mathbf{C}_u$. If desired, the values of the necessary reactions, $\mathbf{P}_k$, can now be determined from

$$\mathbf{P}_k = \mathbf{S}_{ku}\,\mathbf{D}_u + \mathbf{S}_{kk}\,\mathbf{D}_k - \mathbf{C}_k$$

In nonlinear and time dependent applications the reactions can be found from similar calculations. In most applications the reaction data have physical meanings that are important in their own right, or useful in validating the solution. However, this part of the calculations is optional. If one formulates a finite element model that satisfies the essential boundary conditions in advance then the second row of the partitioned system **S** matrix is usually not generated and one can not recover reaction data.

2.15 Elliptic boundary value problems

2.15.1 One-dimensional equations Since the previous model equation had an exact solution that was trigonometric our approximation with polynomials could never give an exact solution (but could reach a specified level of accuracy). Here we will consider an example where the finite element solution can be nodally exact and possibly exact everywhere. The following one-dimensional ($n_s = 1$) model problem which will serve as an analytical example:

$$k \frac{d^2 \phi}{dx^2} + Q = 0 \tag{2.49}$$

on the closed domain, $x \in]0, L[$, and is subjected to the boundary conditions $\phi(L) = \phi_L$, and $k\, d\phi/dx\,(0) = q_0$. Note that we have dropped the second term in the previous ODE. The corresponding governing integral statement to be used for the finite element model is obtained from the *Galerkin weighted residual method*, followed by integration by parts and is very similar to the first example. In this case, it states that the function, ϕ, which satisfies the essential condition, $\phi(L) = \phi_L$, and satisfies

$$I = \int_0^L \left[k\,(d\phi/dx)^2 - \phi Q \right] dx + q_0\,\phi(0) - q_L\,\phi(L) = 0, \tag{2.50}$$

also satisfies the new ODE. A typical element contribution is:

$$I^e = \mathbf{D}^{e^T}\, \mathbf{S}^e\, \mathbf{D}^e - \mathbf{D}^{e^T}\, \mathbf{C}^e, \tag{2.51}$$

where

$$\mathbf{S}^e \equiv \int_{L^e} k^e \frac{d\mathbf{H}^{e^T}}{dx} \frac{d\mathbf{H}^e}{dx}\, dx = \int_{L^e} \mathbf{B}^{e^T}\, k^e\, \mathbf{B}^e\, dx, \quad \mathbf{C}^e \equiv \int_{L^e} Q^e\, \mathbf{H}^{e^T}\, dx.$$

The system of algebraic equations from the weak form is $\mathbf{S}\,\mathbf{D} = \mathbf{C}$ where

$$\mathbf{S} = \sum_{e=1}^{n_e} \beta^{e^T}\, \mathbf{S}^e\, \beta^e, \qquad \mathbf{C} = \sum_{e=1}^{n_e} \beta^{e^T}\, \mathbf{C}^e + \sum_{b=1}^{n_b} \beta^{b^T}\, \mathbf{C}^b,$$

and where, as before, β denotes a set of symbolic bookkeeping operations. If we select a linear element with the interpolation relations given previously the element matrices are:

$$\mathbf{S}^e = \frac{k^e}{L^e}\begin{bmatrix} 1 & -1 \\ -1 & 1 \end{bmatrix}, \quad \mathbf{C}^e = \int_{L^e} \frac{Q^e}{L^e} \begin{Bmatrix} (x_2{}^e - x) \\ (x - x_1{}^e) \end{Bmatrix} dx.$$

Assuming that $Q = Q_0$, a constant, this simplifies to $\mathbf{C}^{e^T} = \lfloor 1 \quad 1 \rfloor Q_0\, L^e/2$. The exact solution of the original problem for constant $Q = Q_0$ is

$$k\,\phi(x) = k\,\phi_L + q(x - L) + Q_0\,(L^2 - x^2)/2. \tag{2.52}$$

Since for $Q_0 \neq 0$ the exact value is quadratic and the selected element is linear, our finite element model can give only an approximate solution. However, for the homogeneous problem $Q_0 = 0$, the model can (and does) give an exact solution. To compare a finite element spatial approximation with the exact one, select a two element model. Let the elements be of equal length, $L^e = L/2$. Then the element matrices are the same for both elements. The assembly process (including boundary effects) yields, $\mathbf{S}\,\mathbf{D} = \mathbf{C}$:

$$\frac{2k}{L}\begin{bmatrix} 1 & -1 & 0 \\ -1 & (1+1) & -1 \\ 0 & -1 & 1 \end{bmatrix} \begin{Bmatrix} \phi_1 \\ \phi_2 \\ \phi_3 \end{Bmatrix} = \frac{Q_0\, L}{4} \begin{Bmatrix} 1 \\ (1+1) \\ 1 \end{Bmatrix} - \begin{Bmatrix} q_0 \\ 0 \\ -q_L \end{Bmatrix}.$$

However, these equations do not yet satisfy the essential boundary condition of $\phi(L) = \phi_3 = \phi_L$. That is, $\mathbf{S}$ is singular and can not be inverted. After applying this essential boundary condition, the reduced equations are

$$\frac{2k}{L}\begin{bmatrix} 1 & -1 \\ -1 & 2 \end{bmatrix}\begin{Bmatrix} \phi_1 \\ \phi_2 \end{Bmatrix} = \frac{Q_2\, L}{4}\begin{Bmatrix} 1 \\ 2 \end{Bmatrix} - \begin{Bmatrix} q \\ 0 \end{Bmatrix} + \frac{2kt_L}{L}\begin{Bmatrix} 0 \\ 1 \end{Bmatrix},$$

or $\mathbf{S}_r\, \mathbf{D}_r = \mathbf{C}_r$. Inverting $\mathbf{S}_r$ and solving for $\mathbf{D}_r = \mathbf{S}_r^{-1}\, \mathbf{C}_r$ yields

$$\mathbf{S}_r^{-1} = \frac{L}{2k}\begin{bmatrix} 2 & 1 \\ 1 & 1 \end{bmatrix},\ \mathbf{D}_r = \begin{Bmatrix} \phi_1 \\ \phi_2 \end{Bmatrix} = \frac{Q_0\, L^2}{8k}\begin{Bmatrix} 4 \\ 3 \end{Bmatrix} - \frac{qL}{2k}\begin{Bmatrix} 2 \\ 1 \end{Bmatrix} + t_L\begin{Bmatrix} 1 \\ 1 \end{Bmatrix}.$$

These are the exact nodal values as can be verified by evaluating the solution at $x = 0$ and $x = L/2$, respectively. Thus, our finite element solution is giving an *interpolate* solution. That is, it interpolates the solution exactly at the node points, and is approximate at all other points on the interior of the element. For the homogeneous problem, $Q_0 = 0$, the finite element solution is exact at all points. These properties are common to other finite element problems. The exact and finite element solutions are illustrated in Fig. 2.34, where shaded regions show the error in the solution and its gradient. Note that the derivatives are also exact at least at one point in each element. The optimal derivative sampling points will be considered in a later section. For this differential operation, it can be shown that the center point gives a derivative estimate accurate to order $0\,(L^{e2})$, while all other points are only order $0\,(L^{e})$ accurate. For $Q = Q_0$, the center point derivatives are exact in the example.

Next, we want to utilize the last equation from the weighted residual algebraic system to recover the flux 'reaction' that is necessary to enforce the essential boundary condition at $x = L$:

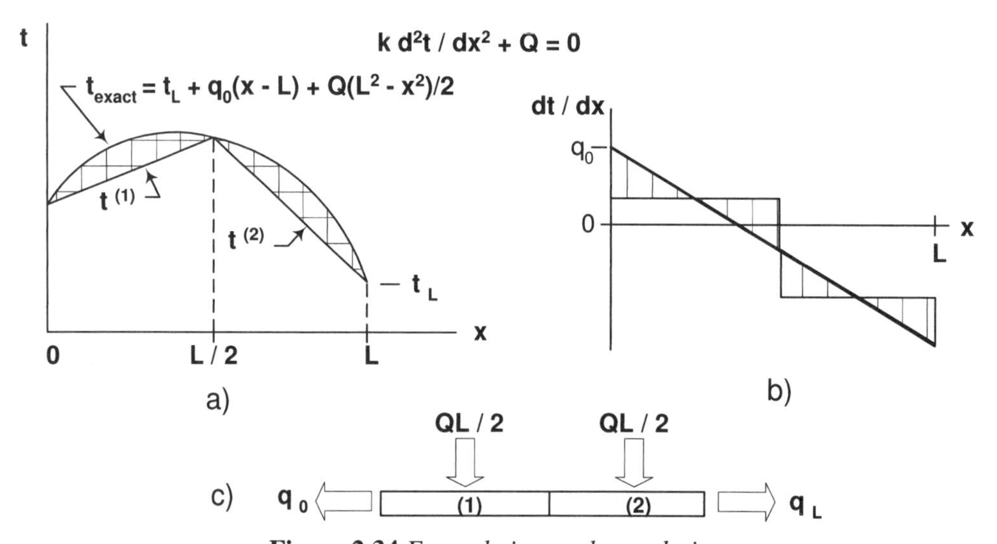

Figure 2.34 *Example interpolate solution*

$$\frac{2k}{L}\left[\,0 - 1\,\phi_2 + 1\,\phi_3\,\right] \;=\; Q_0\,\frac{L}{4} \;-\; (-\,q_L)$$

or $-Q_0 L + q_0 = q_L$, which states the flux equilibrium: that which was generated internally, $Q_0 L$, minus that which exited at $x = 0$, *must* equal that which exits at $x = L$. If one is going to save the reaction data for post-solution recovery, as illustrated above, then one could use programs like *SYSTEM_ROW_SAVE* and *GET_REACTIONS* to store and later recover the associated rows of the square matrix. It is not necessary to save these data when they are first generated since they can clearly be recomputed in a post-processing segment if desired. In the later examples we will invoke functions to save these data as they are created.

2.15.2 Two-dimensional equations

As an example of the utilization of Galerkin's method in higher dimensions, consider the following model transient linear operator:

$$L(u) \;=\; \frac{\partial}{\partial x}\left(k_x\,\frac{\partial u}{\partial x}\right) + \frac{\partial}{\partial y}\left(k_y\,\frac{\partial u}{\partial y}\right) - Q - \varsigma\,\frac{\partial u}{\partial t} \;=\; 0. \tag{2.53}$$

We get the weak form by multiplying by a test function, $w(x)$, and setting the integral to zero. The last two terms are simply

$$I_Q \;=\; \int_\Omega Q\,w\,d\,\Omega, \qquad\qquad I_\varsigma \;=\; \int_\Omega \varsigma\,w\,\frac{\partial u}{\partial t}\,d\,\Omega.$$

Our main interest is with the first two terms

$$I_k \;=\; \int_\Omega \left[\,w\,\frac{\partial}{\partial x}\left(k_x\,\frac{\partial u}{\partial x}\right) + w\,\frac{\partial}{\partial y}\left(k_y\,\frac{\partial u}{\partial y}\right)\right] d\,\Omega$$

which involve the second order partial derivatives. We can remove them by invoking Green's theorem

$$\int_\Omega \left[\,\frac{\partial \Psi}{\partial x} - \frac{\partial \gamma}{\partial y}\,\right] d\,\Omega \;=\; \int_\Gamma \left[\,\gamma\,dx + \Psi\,dy\,\right]$$

where we define $\Psi = w\,k_x\,\partial u/\partial x$, and $\gamma = -w\,k_y\,\partial u/\partial y$. We note that an alternate form of I_k is

$$I_k \;=\; \int_\Omega \left[\,\frac{\partial}{\partial x}\left(w\,k_x\,\frac{\partial u}{\partial x}\right) + \frac{\partial}{\partial y}\left(w\,k_y\,\frac{\partial u}{\partial y}\right)\right] d\,\Omega$$

$$-\int_\Omega \left[\,\frac{\partial w}{\partial x}\,k_x\,\frac{\partial u}{\partial x} + \frac{\partial w}{\partial y}\,k_y\,\frac{\partial u}{\partial y}\,\right] d\,\Omega\,.$$

The first term is re-written by Green's theorem

$$I_k = \int_\Gamma \left[-\left(w\, k_y \frac{\partial u}{\partial y} \right) dx + \left(w\, k_x \frac{\partial u}{\partial x} \right) dy \right]$$

$$- \int_\Omega \left[k_x \frac{\partial w}{\partial x} \frac{\partial u}{\partial x} + k_y \frac{\partial w}{\partial y} \frac{\partial u}{\partial y} \right] d\,\Omega .$$

For a contour integral with a unit normal vector, **n**, with components $[\, n_x \quad n_y \,]$, we note the geometric relationships are that (see Fig. 2.35) $-dx = ds \operatorname{Cos} \theta_y = ds\, n_y$, and that $dy = ds \operatorname{Cos} \theta_x = ds\, n_x$ which reduces the boundary integral to

$$\int_\Gamma w \left[k_x \frac{\partial u}{\partial x} n_x + k_y \frac{\partial u}{\partial y} n_y \right] ds = \int_\Gamma w\, k_n \frac{\partial u}{\partial n} ds$$

where $\partial u / \partial n$ is the normal gradient of u, that is $\nabla u \cdot \mathbf{n}$, and k_n is the value of the orthotropic k in the direction of the normal. The resulting Galerkin form is

$$- I = \int_\Omega \left[k_x \frac{\partial w}{\partial x} \frac{\partial u}{\partial x} + k_y \frac{\partial w}{\partial y} \frac{\partial u}{\partial y} \right] d\,\Omega + \int_\Omega Q\, w\, d\,\Omega$$

$$+ \int_\Omega \zeta\, w \frac{\partial u}{\partial t} d\,\Omega - \int_\Gamma w\, k_n \frac{\partial u}{\partial n} ds = 0 .$$

Note that Green's theorem brought in the information on the conditions of the surface. If we split this into the union of all element domains so that $\Omega = \bigcup_e \Omega^e$ and likewise the

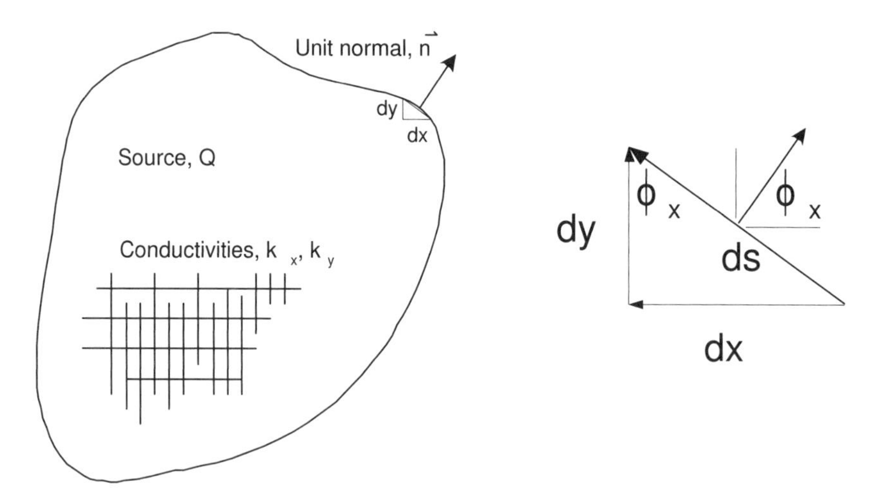

Figure 2.35 *The unit normal vector components*

boundary becomes the union of the boundary segments created by the mesh, $\Gamma = \bigcup_b \Gamma^b$, we generate four typical matrices from these terms. We interpolate, as before, with the weights $w(x, y) = \mathbf{H}^e(x, y)\,\mathbf{V}^e$, but include transient time effects by assuming that the degrees of freedom become time dependent, $\mathbf{U}^e = \mathbf{U}^e(t)$, so that $u(x, y, t) = \mathbf{H}^e(x, y)\,\mathbf{U}^e(t)$. This assumption for the time behavior makes

$$\frac{\partial u}{\partial t} = \mathbf{H}^e(x, y)\,\frac{\partial}{\partial t}\,\boldsymbol{\phi}^e(t) = \mathbf{H}^e(x, y)\,\dot{\boldsymbol{\phi}}^e. \tag{2.54}$$

These assumptions result in two square matrices

$$\mathbf{K}^e = \int_{\Omega^e}\left[\mathbf{H}^{e^T}_{,x}\,k_x\,\mathbf{H}^e_{,x} + \mathbf{H}^{e^T}_{,y}\,k_y\,\mathbf{H}^e_{,y}\right]d\Omega, \quad \mathbf{M}^e = \int_{\Omega^e}\mathbf{H}^{e^T}\,\mathbf{H}^e\,\zeta^e\,d\Omega \tag{2.55}$$

and the two source vectors

$$\mathbf{F}^e_Q = \int_{\Omega^e}\mathbf{H}^{e^T}\,Q^e\,d\Omega, \quad \mathbf{F}^b_q = \int_{\Gamma^b}\mathbf{H}^{b^T}\,k^b_n\,\frac{\partial u^b}{\partial n}\,d\Gamma = \int_{\Gamma^b}\mathbf{H}^{b^T}\,q^b_n\,d\Gamma, \tag{2.56}$$

which when assembled yield a system of ordinary differential equations in time such that $\mathbf{V}^T(\mathbf{M}\dot{\boldsymbol{\phi}} + \mathbf{K}\boldsymbol{\phi} - \mathbf{F}) = 0$, so that for arbitrary $\mathbf{V} \neq \mathbf{0}$, we have

$$\mathbf{M}\,\dot{\boldsymbol{\phi}}(t) + \mathbf{K}\,\boldsymbol{\phi}(t) = \mathbf{F}(t), \tag{2.57}$$

which is an initial value problem to be solved from the initial condition state at $\boldsymbol{\phi}(x, y, t = 0)$. Here we will note that $\mathbf{M}^e$ and $\mathbf{M}$ are usually called the *element* and *system mass* (or *capacity*) *matrices*, respectively. By making assumptions about how to approximate the time dependent vector $\dot{\boldsymbol{\phi}}(t)$ this system can be rearranged to yield a linear system $\mathbf{S}\,\boldsymbol{\phi}(t) = \mathbf{C}(t)$ to be solved at each time step for the current value of $\boldsymbol{\phi}(t)$. For the steady state case, $\partial/\partial t = 0$, this reduces to the system of algebraic equations $\mathbf{S}\,\boldsymbol{\phi} = \mathbf{C}$ considered earlier. The same procedure is easily carried out in three dimensions. Matrix $\mathbf{K}^e$ changes by having a third term involving z and for $\mathbf{F}^b_q$ we think of $d\Gamma$ as a surface area instead of a line segment.

We also note in passing that in addition to using classical finite differences in time there are two other choices for methods to treat time dependence. One is to do a separation of variables and also interpolate the element behavior over one or more time steps with $u(x, y, t) = \mathbf{H}^e(x, y)\,\mathbf{h}^e(t)\,\boldsymbol{\phi}^e$ [22]. That allows one to carry out weighted residual methods in time also. These first two choices are usually referred to as 'semi-discrete' methods for time integration. Another approach is to carry out a full space-time interpolation with $u(x, y, t) = \mathbf{H}^e(x, y, t)\,\boldsymbol{\phi}^e$ and include three-dimensional space via four-dimensional elements. The 4-D element solutions have very large memory requirements (by today's standards), and are usually restricted to a time slab (a constant time step) which simplifies the 4-D elements and the associated 4-D mesh generation. Like the above example, most 4-D models begin with the classical mechanics laws in 3-D space and then interpolate through time also. It is also possible to begin the formulation using relativity physics laws in a 4-D space-time and to form the 4-D elements therein [11]. Such an approach is not required unless the velocities involved are not small relative to the speed of light.

```
!  .............................................................  !  1
!     ***  ELEM_SQ_MATRIX PROBLEM DEPENDENT STATEMENTS FOLLOW ***  !  2
!       For required REAL (DP) :: S (LT_FREE, LT_FREE)             !  3
!       and optional REAL (DP) :: C (LT_FREE)                      !  4
!       using previous solution D (LT_FREE) with                   !  5
!    D = D_OLD + RELAXATION (D_NEW - D_OLD) via key relaxation 1.0 !  6
!    Globals TIME_METHOD, TIME_STEP, TIME along with mass matrix   !  7
!    options EL_M ((LT_FREE, LT_FREE), and DIAG_M  (LT_FREE).      !  8
!  .............................................................  !  9
!  APPLICATION DEPENDENT Galerkin FOR TRANSIENT PDE via the        ! 10
!  iterative element level assembly of semi-discrete form(s).      ! 11
!  K_e U,xx + Q - Rho_e U,t  = 0, with U(x,0) given by keyword     ! 12
!  start_value, U(0,t) and U(L,t) given by EBC constant in time    ! 13
!  Keywords time_step, time_method, scalar_source, initial_value,  ! 14
!  diagonal_mass, etc. are also available.  Stored as inefficeint  ! 15
!  source example 121, redone as efficient in example 122          ! 16
REAL(DP)  :: DL, DX_DN, Q = 0.d0 ! Length, Jacobian, source        ! 17
INTEGER   :: IQ, J               ! Loops                           ! 18
                                                                   ! 19
!   Work items for time integration                                ! 20
REAL(DP), SAVE :: Del_t = 1, K_e = 1, Rho_e = 1  ! defaults        ! 21
REAL(DP)  :: K (LT_FREE, LT_FREE), M (LT_FREE, LT_FREE)            ! 22
REAL(DP)  :: F (LT_FREE), WORK (LT_FREE, LT_FREE)                  ! 23
                                                                   ! 24
 IF ( THIS_EL == 1 ) THEN  ! first action for each iteration       ! 25
   IF ( TIME_STEP /= 1.d0 ) Del_t = TIME_STEP  ! non-default       ! 26
   TIME = START_TIME + THIS_ITER * Del_t       ! via globals       ! 27
   PRINT *, '(VIA ITERATION) TIME = ', TIME                        ! 28
 END IF                                                            ! 29
                                                                   ! 30
 K = 0 ;  M = 0 ; F = 0      ! Initialize conduction, mass, source ! 31
 IF ( EL_REAL > 1 ) THEN     ! use non-default element properties  ! 32
   K_e   = GET_REAL_LP (1) ! thermal conductivity                  ! 33
   Rho_e = GET_REAL_LP (2) ! rho*c_p                               ! 34
 END IF                                                            ! 35
 E      = K_e                                      ! Diffusion     ! 36
 DL     = COORD (LT_N, 1) - COORD (1, 1)           ! Length        ! 37
 DX_DN = DL / 2.                                   ! Jacobian      ! 38
 IF ( SCALAR_SOURCE /= 0.d0 ) Q = SCALAR_SOURCE    ! Source        ! 39
                                                                   ! 40
 CALL STORE_FLUX_POINT_COUNT ! Save LT_QP for post-processing      ! 41
 DO IQ = 1, LT_QP            ! LOOP OVER QUADRATURES               ! 42
                                                                   ! 43
!        GET INTERPOLATION FUNCTIONS, AND X-COORD                  ! 44
   H    = GET_H_AT_QP (IQ)   ! SHAPE FUNCTIONS HERE                ! 45
   XYZ  = MATMUL (H, COORD)  ! ISOPARAMETRIC                       ! 46
                                                                   ! 47
!        LOCAL AND GLOBAL DERIVATIVES, B = DGH                     ! 48
   DLH    = GET_DLH_AT_QP (IQ)         ! dH / dr                   ! 49
   DGH    = DLH / DX_DN                ! dH / dx                   ! 50
                                                                   ! 51
!       SQUARE MATRICES (STIFFNESS & MASS) & SOURCE VECTOR         ! 52
   K = K + MATMUL (TRANSPOSE(DGH), DGH)* WT (IQ) * DX_DN * K_e     ! 53
   M = M + OUTER_PRODUCT (H, H)        * WT (IQ) * DX_DN * Rho_e   ! 54
   F = F + H * Q                       * WT (IQ) * DX_DN           ! 55
                                                                   ! 56
   CALL STORE_FLUX_POINT_DATA (XYZ, E, DGH) ! for post-processing  ! 57
 END DO  ! QUADRATURE                                              ! 58
```

Figure 2.36a *Element matrices before time step changes*

```
                                                                     ! 59
!  Assemble for semi-discrete one-step time rule, via iteration:     ! 60
!  NOTE: This is very inefficient since the assembly, boundary       ! 61
!  conditions, and factorization are repeated ever "time step".      ! 62
!  (Ineffecent source example 121, see efficient example 122)        ! 63
                                                                     ! 64
 IF ( DIAGONAL_MASS ) THEN  ! use the scaled diagonal mass form      ! 65
   CALL DIAGONALIZE_SQ_MATRIX (LT_FREE, M, DIAG_M) ; M = 0.d0        ! 66
   DO J = 1, LT_FREE                                                 ! 67
     M (J, J) = DIAG_M (J)                                           ! 68
   END DO                                                            ! 69
 END IF ! diagonal instead of consistent or averaged mass matrix     ! 70
                                                                     ! 71
 SELECT CASE ( TIME_METHOD ) ! for one-step time integration rule    ! 72
   CASE ( 2 ) ! Crank-Nicolson, accuracy order O(Del_t^2)            ! 73
      WORK = M / Del_t - K / 2                                       ! 74
      C    = F + MATMUL (WORK, D)                                    ! 75
      S    = M / Del_t + K / 2                                       ! 76
   CASE ( 3 ) ! Galerkin in time, accuracy order O(Del_t^2)          ! 77
      WORK = M / Del_t - K / 3                                       ! 78
      C    = F + MATMUL (WORK, D)                                    ! 79
      S    = 2 * K / 3.d0 + M / Del_t                                ! 80
   CASE ( 4 ) ! Least Squares in time, F constant in time            ! 81
      WORK =  MATMUL (TRANSPOSE(M), M) / Del_t     &                 ! 82
           +  MATMUL (TRANSPOSE(K), M) / 2         &                 ! 83
           -  MATMUL (TRANSPOSE(M), K) / 2         &                 ! 84
           -  MATMUL (TRANSPOSE(K), K) * Del_t / 6                   ! 85
      C    = -MATMUL (TRANSPOSE(K), F) * Del_t / 2 &                 ! 86
           -  MATMUL (TRANSPOSE(M), F)             &                 ! 87
           +  MATMUL (WORK, D)                                       ! 88
      S    =  MATMUL (TRANSPOSE(K), K) * Del_t / 3 &                 ! 89
           +  MATMUL (TRANSPOSE(K), M) / 2         &                 ! 90
           +  MATMUL (TRANSPOSE(M), K) / 2         &                 ! 91
           +  MATMUL (TRANSPOSE(M), M) / Del_t                       ! 92
   ! Add generalized trapezoidal method as CASE ( 5 )                ! 93
   CASE DEFAULT ! Method 1, forward difference, order O(Del_t)       ! 94
      WORK = M / Del_t                                               ! 95
      C    = F + MATMUL (WORK, D)                                    ! 96
      S    = K + M / Del_t                                           ! 97
 END SELECT ! a one-step rule                                        ! 98
                                                                     ! 99
!    ***  END ELEM_SQ_MATRIX PROBLEM DEPENDENT STATEMENTS ***        !100
```

Figure 2.36b *New element matrices after time step changes*

If we had also included a second time derivative term, we could then see by inspection that it would simply introduce another mass matrix (with a different coefficient) times the second time derivative of the nodal degrees of freedom. This would then represent the *wave equation* whose solution in time could be accomplished by techniques to be presented later. If the time derivative terms are not present, and if the term $c\,u$ is included in the PDE and if the coefficient c is an unknown global constant, then this reduces to an *eigen-problem.* That is, we wish to determine a set of *eigenvalues*, c_i, and a corresponding set of *eigenvectors,* or *mode shapes.*

2.16 Initial value problems

Many applications, like Eq. 2.53, involve the first partial derivative with respect to time and get converted through the finite element process to a matrix system of ordinary differential equations in time, like Eq. 2.57. These are known as initial value problems, or parabolic problems, and require the additional description of the spatial value of $\boldsymbol{\phi}$ at the starting time (known as the initial condition). There are hundreds of ways to numerically integrate these matrices in time. One has to be concerned about the relative costs (storage and operation counts), stability, and accuracy of the chosen method. These topics are covered in many books on numerical methods and we will not go into them here. Instead we will simply give some insight to a few approaches.

If one assumes that Eq. 2.57 is valid at an element level for time t_s one could write

$$\mathbf{M}^e\, \dot{\boldsymbol{\phi}}^e(t_s) + \mathbf{K}\, \boldsymbol{\phi}^e(t_s) = \mathbf{F}^e\,(t_s),$$

and assume that at step number s

$$\dot{\boldsymbol{\phi}}_s = \frac{\boldsymbol{\phi}_{s+1} - \boldsymbol{\phi}_s}{\Delta t_{s+1}}$$

where Δt is the time step size, and contemplate rewriting the element relations as

$$[\mathbf{M}^e / \Delta t_{s+1}]\boldsymbol{\phi}^e_{s+1} = [\mathbf{M}^e / \Delta t_{s+1} - \mathbf{K}^e]\boldsymbol{\phi}^e_s + \mathbf{F}^e_s, \tag{2.58}$$

or as $\mathbf{S}^e(\Delta t_{s+1})\, \boldsymbol{\phi}^e_{s+1} = \mathbf{C}^e_s(\Delta t_{s+1,} t)$. Then the effective element square matrix would not change with time if the step size, Δt_{s+1} were held constant.

One could begin the time stepping with the initial condition, $\boldsymbol{\phi}_0$, and loop through the time steps, in an iterative fashion, to get the next $\boldsymbol{\phi}_{s+1}$. While this type of concept is illustrated in the element level calculations of Fig. 2.36 it is very inefficient to use this element looping (or iteration) process. It is more proper to carry out a similar process after the proper system assembly of Eq. 2.57, invoke the system initial conditions, and solve for $\boldsymbol{\phi}_{s+1}$, after applying the essential boundary conditions and/or flux conditions both of which can now have time dependent values. We will omit that level of detail here and simply note that control keyword *transient* is available in *MODEL* to carry out a transient study (see Table 2.1). It defaults to zero initial conditions and the Crank-Nicolson time integration scheme which is unconditionally stable and second order accurate in time, $O(\Delta t^2)$. The application of transient methods will be given in Chapter 11, but the main interest in this book is on spatial error estimates. The book by Huang and Usmani [8], covers two-dimensional transient error estimators in more detail than space allows here.

However, to give a brief overview of how all the transient calculations proceed we will use a simple (and slightly corrected) example from Desai [6], where he used 3 linear elements to solve the transient heat conduction problem of

$$\alpha \frac{\partial^2 \phi}{\partial x^2} = \frac{\partial \phi}{\partial t},$$

where $\alpha = k/\rho c_p$ is the thermal diffusivity, ϕ is temperature, t is time, and x is position. The bar has an initial condition of zero temperature, $\phi(x, 0) = 0$, when the two ends suddenly have different non-zero temperatures enforced and held constant with time, $\phi(0, t) = 10$, $\phi(L, t) = 20$. The element size and time step size are chosen to make the element Fourier Number, $Fo = \alpha\Delta t / \Delta x^2$, equal to unity. Here $\Delta x = L^e = 1$ (so $L = 3$)

Table 2.1 *Typical keywords for transient problems*

```
TIME_CONTROL, VALUE ! REMARKS                                      [DEFAULT]
average_mass        ! Average consistent & diagonal mass matrices    [F]
crank_nicolson      ! Select transient one-step method 2             [F]
diagonal_mass       ! Use diagonalized mass matrix                   [F]
euler_forward       ! Select transient one-step method 1             [T]
galerkin_in_time    ! Select transient one-step method 3             [F]
gen_trapezoidal     ! Select transient one-step method 5             [F]
gen_trap_next .5    ! Weight of next time step, 0 to 1             [0.5]
history_dof   12    ! DOF number for time-history graph output       [0]
history_node   3    ! Node number for time-history graph outputs     [0]
inc_save       1    ! Save after steps inc, 2*inc, 3*inc, ...        [0]
initial_value 0.    ! Initial value of transient scalar everywhere   [0]
least_sq_in_time    ! Select transient one-step method 4             [F]
save_1248           ! Save after steps 1, 2, 4, 8, 16 ...            [F]
start_time     0.   ! A time history starting time                   [0]
start_value    1.   ! Initial value of transient scalar everywhere   [0]
time_method    1    ! 1-Euler, 2-Crank-Nicolson, 3-Galerkin, etc     [1]
time_step  1.d-3    ! Time step size for time dependent solution     [1]
time_steps   128    ! Number of time steps to employ                 [5]
transient           ! Problem is first order in time                 [F]
```

and the time step is also taken as unity, $\Delta t = 1$. The thermal stiffness and mass matrix for each element are

$$\mathbf{K}^e = \frac{\alpha}{L^e}\begin{bmatrix} 1 & -1 \\ -1 & 1 \end{bmatrix}, \quad \mathbf{M}^e = \frac{L^e}{6}\begin{bmatrix} 2 & 1 \\ 1 & 2 \end{bmatrix}, \tag{2.59}$$

and the element source vector is null, $\mathbf{C}_Q^e = \mathbf{0}$. Applying the Euler (forward difference) one-step time integration rule for a uniform mesh gives system equations

$$\mathbf{K}\{\mathbf{\Phi}\}_{s+1} + \frac{1}{\Delta t}\mathbf{M}\{\mathbf{\Phi}\}_{s+1} = \{\mathbf{C}_Q\}_{s+1} + \frac{1}{\Delta t}\mathbf{M}\{\mathbf{\Phi}\}_s + \{\mathbf{C}_q\}_{s+1}.$$

Selecting the above numerical values, so $Fo = 1$, and assemblying the four equations, including the initial condition vector, $\mathbf{\Phi}_0$, but before applying the essential boundary conditions at the first and last nodes gives:

$$\frac{1}{6}\begin{bmatrix} 8 & -5 & 0 & 0 \\ -5 & 16 & -5 & 0 \\ 0 & -5 & 16 & -5 \\ 0 & 0 & -5 & 8 \end{bmatrix}\begin{Bmatrix} \Phi_1 = 10 \\ \Phi_2 \\ \Phi_3 \\ \Phi_4 = 20 \end{Bmatrix}_1 = \mathbf{C}_Q + \frac{1}{6}\begin{bmatrix} 2 & 1 & 0 & 0 \\ 1 & 4 & 1 & 0 \\ 0 & 1 & 4 & 1 \\ 0 & 0 & 1 & 2 \end{bmatrix}\begin{Bmatrix} \Phi_1 \\ \Phi_2 \\ \Phi_3 \\ \Phi_4 \end{Bmatrix}_0 + \begin{Bmatrix} -q_1 \\ 0 \\ 0 \\ q_4 \end{Bmatrix}_1$$

where the initial condition nodal vector is zero here, $\mathbf{\Phi}_0 = 0$. Applying the essential boundary conditions at this (and all) time steps (with EBC identities inserted in rows 1 and 4 after saving the originals for later reaction recovery) gives

$$\frac{1}{6}\begin{bmatrix} 6 & 0 & 0 & 0 \\ 0 & 16 & -5 & 0 \\ 0 & -5 & 16 & 0 \\ 0 & 0 & 0 & 6 \end{bmatrix}\begin{Bmatrix} \Phi_1 = 10 \\ \Phi_2 \\ \Phi_3 \\ \Phi_4 = 20 \end{Bmatrix}_1 = \begin{Bmatrix} 0 \\ 0 \\ 0 \\ 0 \end{Bmatrix} + \frac{-10}{6}\begin{Bmatrix} -6 \\ -5 \\ 0 \\ 0 \end{Bmatrix} + \frac{-20}{6}\begin{Bmatrix} 0 \\ 0 \\ -5 \\ -6 \end{Bmatrix} = \frac{1}{6}\begin{Bmatrix} 60 \\ 50 \\ 100 \\ 120 \end{Bmatrix}$$

which is solved for the first time step result of

$$\mathbf{\Phi}_1^T = [\,10.00 \quad 5.63 \quad 8.01 \quad 20.00\,].$$

For the second time step we have system equations of

$$\frac{1}{6}\begin{bmatrix} 8 & -5 & 0 & 0 \\ -5 & 16 & -5 & 0 \\ 0 & -5 & 16 & -5 \\ 0 & 0 & -5 & 8 \end{bmatrix} \begin{Bmatrix} \Phi_1 = 10 \\ \Phi_2 \\ \Phi_3 \\ \Phi_4 = 20 \end{Bmatrix}_2 = \frac{1}{6}\begin{bmatrix} 2 & 1 & 0 & 0 \\ 1 & 4 & 1 & 0 \\ 0 & 1 & 4 & 1 \\ 0 & 0 & 1 & 2 \end{bmatrix} \begin{Bmatrix} 10.00 \\ 5.63 \\ 8.01 \\ 20.00 \end{Bmatrix} = \frac{1}{6}\begin{Bmatrix} 25.63 \\ 40.53 \\ 57.67 \\ 48.01 \end{Bmatrix}.$$

Applying the essential boundary conditions at this time gives the system equations

$$\frac{1}{6}\begin{bmatrix} 6 & 0 & 0 & 0 \\ 0 & 16 & -5 & 0 \\ 0 & -5 & 16 & 0 \\ 0 & 0 & 0 & 6 \end{bmatrix} \begin{Bmatrix} \Phi_1 = 10 \\ \Phi_2 \\ \Phi_3 \\ \Phi_4 = 20 \end{Bmatrix}_2 = \frac{1}{6}\begin{Bmatrix} 60.00 \\ 90.43 \\ 157.67 \\ 120.00 \end{Bmatrix},$$

so $\mathbf{\Phi}_2^T = [\,10.00 \quad 9.68 \quad 12.88 \quad 20.00\,]$ and so on for each later step. Likewise, as done before we can recover the heat flux reactions necessary to maintain the essential boundary conditions at each step. For the first time step they can be shown to be $q_1 = 8.64$, and $q_4 = 19.99\ BTU/sec$ while at the second step they were $q_1 = 0.99$, and $q_4 = 7.93$. These eventually transition to the steady state reactions of $q_1 = -3.33$, and $q_4 = 3.33$ after about 11 steps. The corresponding steady state temperatures are linear between the two essential boundary conditions $\mathbf{\Phi}^T = [\,10.00 \quad 13.33 \quad 16.67 \quad 20.00\,]$.

To illustrate a simple one-dimensional initial value problem consider a uniform bar that is initially at a temperature of unity when suddenly the two ends are reduced to a zero temperature and we want to see the time history of the bar as it cools toward zero everywhere. The analytic solution is known for this widely used example, and is included in the *MODEL* library as *exact_case* 34. Here we use a half-symmetry model with 5 nodes and 4 linear elements. The natural BC occurs at the center point. A set of sample data for this problem are given in Fig. 2.37 and the nodal time histories are shown in Fig. 2.38 from this crude model (top) and from the exact solution (bottom, evaluated only at the nodes). The early time history of the temperature at the center (symmetry) point is given in Fig. 2.39 for a diagonal mass matrix option. The default finite element formulation employs the full consistent mass matrix. In this example we have used a diagonal form (see line 14 of Fig. 2.37). The average of the two approaches has been demonstrated to be better for some element families, like the L2 element used here. The example used the numerically integrated element matrices and thus could have used quadratic or cubic elements as well.

2.17 Eigen-problems

The square mass matrix also often occurs in eigen-problems where the assembled equations are often of the form

$$[\,\mathbf{K} - \omega_j^2 \mathbf{M}\,]\,\boldsymbol{\delta}_j = \mathbf{0}. \tag{2.60}$$

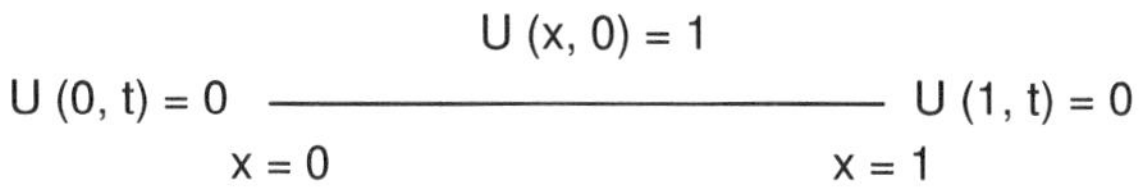

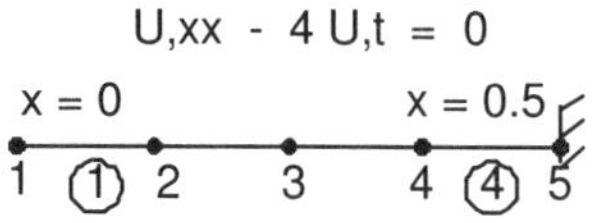

U (x, 0) = 1
U (0, t) = 0 x = 0 U (1, t) = 0 x = 1

U,xx - 4 U,t = 0
x = 0 x = 0.5
1 (1) 2 3 4 (4) 5

```
title "L2 Solution K_e U,xx - Rho_e U,t = 0, Myers values"          ! 1
example       122 ! Application source code library number          ! 2
exact_case     34 ! Analytic solution for list_exact                ! 3
list_exact        ! List given exact answers at nodes, etc          ! 4
transient         ! Problem is first order in time                  ! 5
save_pt_ans       ! Create node_results.tmp for matlab              ! 6
save_exact        ! Save exact result to exact_node_solution.tmp    ! 7
save_1248         ! Save after steps 1, 2, 4, 8, 16 ...             ! 8
time_groups   1   ! Number of groups of constant time_steps         ! 9
time_method   3   ! 1-Euler, 2-Crank-Nicolson, 3-Galerkin, 4-L Sq  !10
time_steps     64 ! Number of time steps                            !11
start_value   1.  ! Initial value of transient scalar everywhere    !12
time_step  1.d-2  ! Time step size for time dependent solution      !13
diagonal_mass     ! Use diagonalized mass matrix                    !14
# average_mass    ! Average consistent & diagonal mass matrices     !15
bar_chart         ! Include bar chart printing in output            !16
no_scp_ave        ! Do NOT get superconvergent patch averages       !17
no_error_est      ! Do NOT compute SCP element error estimates      !18
nodes         5 ! Number of nodes in the mesh                       !19
elems         4 ! Number of elements in the system                  !20
dof           1 ! Number of unknowns per node                       !21
el_nodes      2 ! Maximum number of nodes per element               !22
el_real       2 ! Number of real properties per element             !23
el_homo         ! Element properties are homogeneous                !24
space         1 ! Solution space dimension                          !25
b_rows        1 ! Number of rows in the B (operator) matrix         !26
shape         1 ! Element shape, 1=line, 2=tri, 3=quad, 4=hex       !27
gauss         2 ! Maximum number of quadrature points               !28
remarks       5 ! Number of user remarks                            !29
quit ! keyword input, remarks follow                                !30
APPLICATION DEPENDENT Galerkin FOR TRANSIENT SOLUTION               !31
 K U,xx - R U,t  = 0, with U(x,0)=1, U(0,t)=0, U(L,t)=0             !32
Myers/Akin time integration example, Fig 17.2.3 , L_e = 1/8         !33
Time integrations: 1-Euler, 2-Crank-Nicolson, 3-Galerkin in time   !34
K, R default to 1 else real el properties 1 & 2, R= 1/(16 L_e^2)   !35
 1 1 0.             ! node, bc_flag, x                              !36
 2 0 0.125          ! node, bc_flag, x                              !37
 3 0 0.25           ! node, bc_flag, x                              !38
 4 0 0.375          ! node, bc_flag, x                              !39
 5 0 0.5    ! natural BC at this center node                        !40
    1    1    2    ! elem, two nodes                                !41
    2    2    3    ! elem, two nodes                                !42
    3    3    4    ! elem, two nodes                                !43
    4    4    5    ! elem, two nodes                                !44
   1    1 0. ! node, dof, essential value                           !45
1 1. 4.0     ! el, K_e, Rho_e                                       !46
```

Figure 2.37 *Data for sudden cooling of a bar*

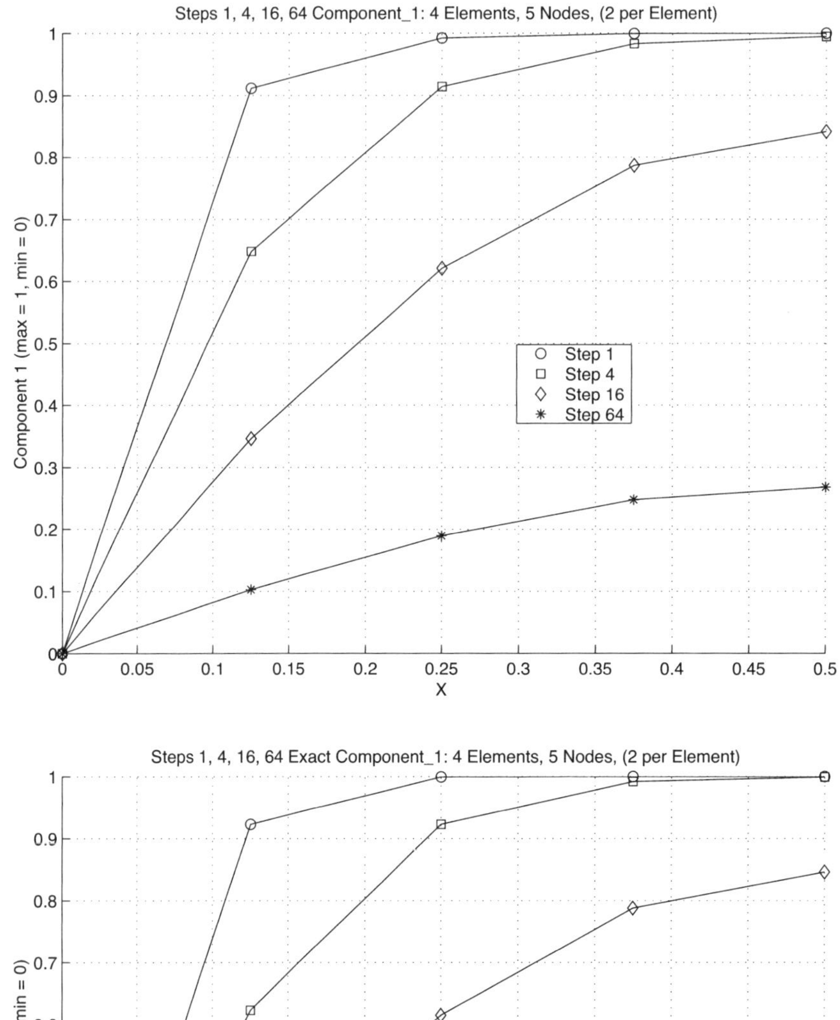

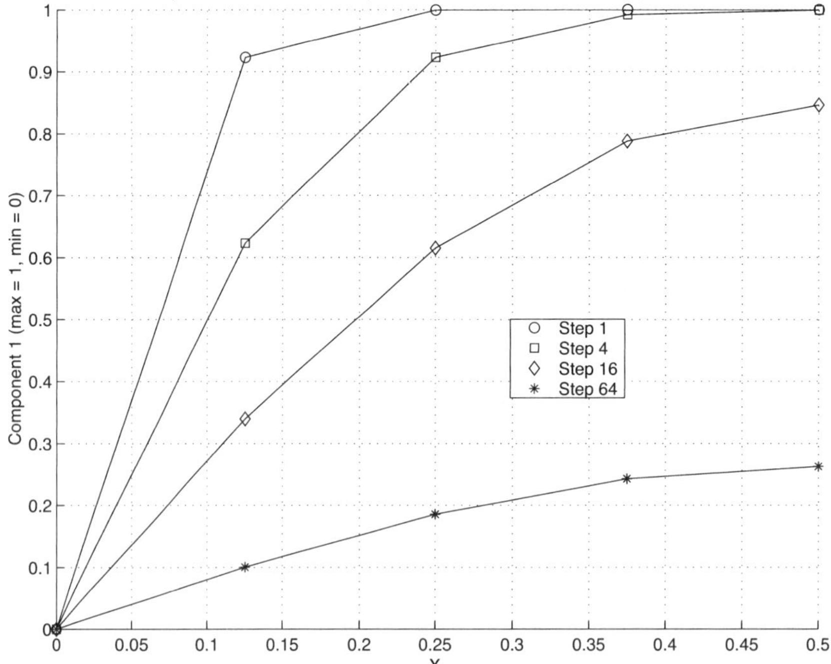

Figure 2.38 *Cooling bar finite element (top) and exact nodal time-histories*

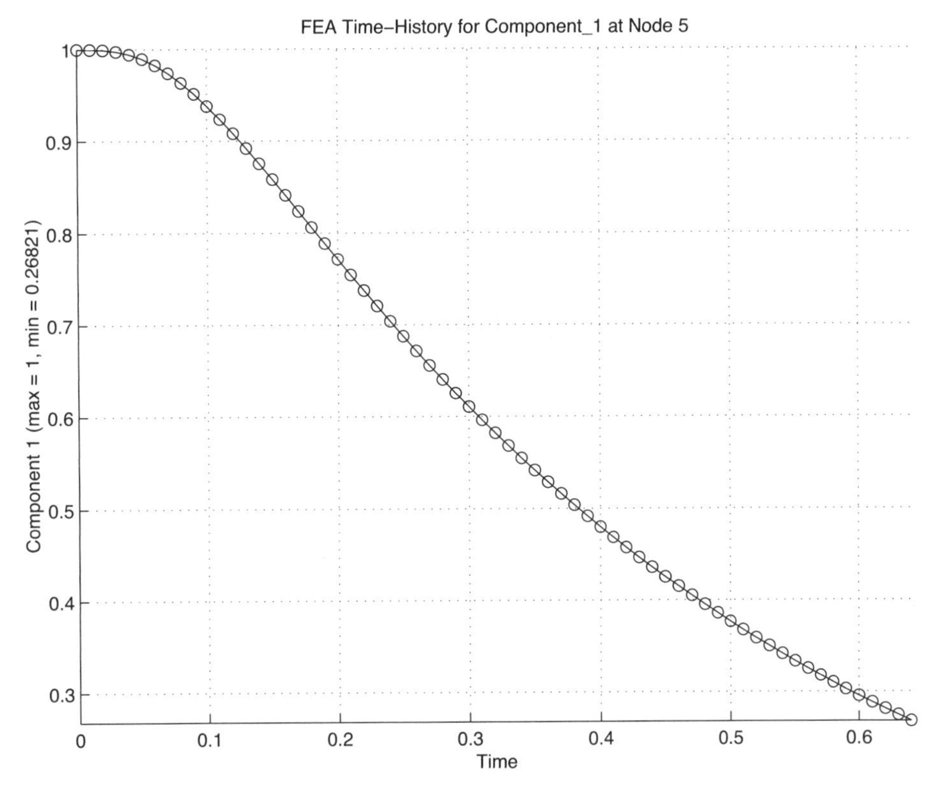

Figure 2.39 *Cooling bar center point early time history*

Such problems are discussed in detail by Bathe [4] but since we are concentrating on error estimates we will only briefly touch on such problems. Here we will use the well known Jacobi iteration process to solve small eigen-problems for the eigenvalues, ω_j, and the corresponding eigenvectors, $\boldsymbol{\delta}_j$ of the full symmetric **K** and **M**. Choosing the Jacobi method usually limits the problem size to a few hundred degrees of freedom. An eigen-problem is signaled by placing the keyword *jacobi* in the control data (see Table 2.2).

Table 2.2 *Typical keywords for eigen-problems*

```
EIGEN_CONTROL, VALUE ! REMARKS                                    [DEFAULT]
eigen_post      3 ! Number of eigenvectors to postprocess              [0]
eigen_scp       2 ! Mode to use for scp and/or averaging fluxes       [1]
eigen_show      5 ! Number of eigen values & vectors to list         [10]
jacobi            ! Use full matrix general Jacobi eigensolution     [F]
jacobi_sweeps 20  ! Max sweeps in general Jacobi eigensolution      [15]
```

2.18 Equivalent forms *

Analysis problems can be stated in different formats or forms. In finite element methods, we do not deal with the original differential equation (which is called the *strong form*). Instead, we convert to some equivalent integral form. In this optional section, we go through some mathematical manipulations in the one-dimensional case to assure the reader that they are, indeed, mathematically equivalent approaches to the solution. Here we will consider equivalent forms of certain model differential equations. For most elliptic Boundary Value Problems (*BVP*) we will find that there are three equivalent forms; the original strong form (S), the variational form (V), and the weak form, (W). The latter two integral forms will be reduced to a matrix form, (M). Consider the one-dimensional two-point boundary value problem where we use, $(\,)' = d(\,)/dx$:

$$(S)\quad \left[\begin{array}{lrcll} & -u''(x) & = & f(x) & x \in\,]0,1[\\ \text{with} & u(0) & = & 0 & \\ & u(1) & = & g. & \end{array}\right.$$

That is, given $f:\Omega \rightarrow R$ and $g \in R$ find $u:\bar{\Omega} \rightarrow R$ such that $u'' + f = 0$ on Ω and $u(0) = 0,\ u(1) = g$. The Hilbert space properties that the function, u, and its first derivative must be square integrable

$$H^1 = [\,u;\ \int_0^1 [\,u^2 + u_x^2\,]\,dx < \infty\,],$$

and introduce a linear trial solution space with the properties that it satisfies certain boundary conditions, is a Hilbert space, and produces a real number when evaluated on the closure of the solution domain: $S = [\,u; u: \bar{\Omega} \rightarrow R,\ u \in H^1,\ u(1) = g,\ u(0) = 0\,]$, and a similar linear space of weighting functions with a different set of boundary conditions: $V = [\,\omega; \omega: \bar{\Omega} \rightarrow R,\ \omega \in H^1,\ \omega(0) = 0,\ \omega(1) = 0\,]$.

The Variational Problem is: given a linear functional, F, that maps S into a real number, $F: S \rightarrow R$, given by

$$(V)\quad \left[\begin{array}{l} F(\omega) = \frac{1}{2} < \omega', \omega' > - < f, \omega > \\ \text{Find } u \in S \text{ such that } F(u) \le F(\omega) \text{ for all } \omega \in S \end{array}\right.$$

$$(W)\quad \left[\begin{array}{l} \text{The Weak Problem is find } u \in S \text{ such that for all } \omega \in V \\ \qquad < u', \omega' > \; = \; < f, \omega >. \end{array}\right.$$

For this problem, we can show that the strong, variational, and weak form imply the existence of each other: $(S) \Longleftrightarrow (V) \Longleftrightarrow (W)$. In solid mechanics applications (S) would represent the differential equations of equilibrium, (V) would denote the Principle of Minimum Total Potential Energy, and (W) would be the Principle of Virtual Work. Other physical applications have similar interpretations but the three equivalent forms often have no physical meaning. Here, we will introduce the definitions of the common symbols for the *bilinear form* $a(u, v)$ and the *linear form* (f, v):

$$a(u, v) \equiv \; < u', v' > \; = \int_0^1 u'v'\,dx, \quad (f, v) \equiv \int_0^1 f\,v\,dx.$$

Here we want to prove the relation that the strong and weak forms imply the existence of each other, $(S) \Longleftrightarrow (W)$. Let u be a solution of (S). Note that $u \in S$ since $u(1) = g$. Assume that $\omega \in V$. Now set $0 = (u'' + f)$; then we can say

$$0 = -\int_0^1 (u'' + f)\,\omega\,dx = -\int_0^1 u''\,\omega\,dx - \int_0^1 f\,\omega\,dx$$

integrating by parts

$$0 = -u'\omega\Big|_0^1 + \int_0^1 u'\omega'\,dx - \int_0^1 f\,w\,dx .$$

Note that since $\omega \in V$ we have $\omega(1) = 0 = \omega(0)$ so that

$$u'\omega\Big|_0^1 = u'(1)\,\omega(1) - u'(0)\,\omega(0) \equiv 0 ,$$

and finally

$$0 = \int_0^1 u'\omega'\,dx - \int_0^1 f\,\omega\,dx .$$

Thus, we conclude that $u(x)$ is a solution of the weak problem, (W). That is, $a(u, \omega) = (f, \omega)$ for all $\omega \in V$. Next, we want to verify that the weak solution also implies the existence of the strong form, $(W) \Leftrightarrow (S)$. Assume that $u(x) \in S$ so that $u(0) = 0$, $u(1) = g$, and assume that $\omega \in V$:

$$\int_0^1 u'\omega'\,dx = \int_0^1 f\,\omega\,dx \qquad \text{for all} \quad \omega \in V .$$

Integrating by parts

$$-\int_0^1 (u'' + f)\,\omega\,dx + u'\omega\Big|_0^1 = 0$$

but,

$$u'\omega\Big|_0^1 = u'(1)\,\omega(1) - u'(0)\,\omega(0) \equiv 0 ,$$

since $\omega \in V$, and thus

$$0 = \int_0^1 (u'' + f)\,\omega(x)\,dx, \qquad \text{for all} \quad \omega \in V .$$

Now we pick $\omega(x) \equiv \phi(x)\,(u'' + f)$ such that ϕ produces a real positive number when evaluated on the domain, $\phi\colon \overline{\Omega} \to R$, and $\phi(x) > 0$, when $x \in]0, 1[$ and $\phi(0) = \phi(1) = 0$. Is $\omega \in V$? Since $\omega(0) = 0$ and $\omega(1) = 0$, we see that it is and proceed with that substitution,

$$0 = \int_0^1 (u'' + f)^2\,\phi(x)\,dx = \int (\geq 0)(> 0)\,dx$$

which implies $(u'' + f) \equiv 0$. Thus, the weak form does imply the strong form, $(W) \Leftrightarrow (S)$ and combining the above two results we find $(S) \Leftrightarrow (W)$.

Next, we will consider the proposition that the variational form implies the existence of the weak form, $(V) \Leftrightarrow (W)$. Assume $u \in S$ is a solution to the weak form (W) and note that $v(1) = \omega(1) + u(1) = 0 + g$. Therefore, $v \in S$. Now set $\omega = v - u$ such that $v = \omega + u$, and $\omega \in V$. Given $F(v) = \frac{1}{2} < v', v' > - < f, v >$ and expanding

$$F(v) = F(u + \omega) = \tfrac{1}{2} < (u' + \omega'), (u' + \omega') > - < f, u + \omega >$$

$$= \tfrac{1}{2} < u', u' > + < u', \omega' > + \tfrac{1}{2} < \omega', \omega' > .$$

But, since u is a solution to (W) we have $< u', \omega' > - < f, \omega > \equiv 0$, and the above simplifies to $F(v) = F(u) + \frac{1}{2} < \omega', \omega' >$ so that $F(v) \geq F(u)$ for all $v \in S$ since $< \omega', \omega' > \geq 0$. This shows that u is also a solution of the variational form, (V). That is, $(W) \Longleftrightarrow (V)$, which is what we wished to show. Finally, we verify the uniqueness of the weak form, (W). Assume that there are two solutions u_1 and u_2 that are both in the space S. Then we have both $a(u_1, v) = (f, v)$, and $a(u_2, v) = (f, v)$ for all $v \in V$ and subtracting the second from the first result yields $a(u_1 - u_2, v) = 0$, so $< (u_1{}' - u_2{}'), v' > = 0$ for all $v \in V$. Consider the choice $v(x) \equiv u_1(x) - u_2(x)$. Is it in V ? We note that $v(0) = 0 - 0 = 0$ and $v(1) = g - g = 0$, so it is in V and we proceed with this choice. Thus, $v' = u_1{}' - u_2{}'$ and the inner product is

$$< (u_1{}' - u_2{}'), (u_1{}' - u_2{}') > = 0 = \int_0^1 \left[u_1{}'(x) - u_2{}'(x) \right]^2 dx.$$

This means $u_1(x) - u_2(x) = c$, and the constant, c, is evaluated from the boundary condition at $x = 0$ so $u_1(0) - u_2(0) = 0 - 0 = c$ which means that $u_1(x) = u_2(x)$, and the weak form solution, (W), is unique.

2.19 Exercises

1. The example ordinary differential equation (ODE) $d^2u/dx^2 + u + x = 0$ with $u(0) = 0 = u(1)$ has the exact solution of $u = sin(x)/sin(1) - x$. Our weighted residual approximations for a global (or single element) solution assumed a cubic polynomial

$$u^*(x) = x(1 - x)(c_1 + c_2 x) = h_1(x)c_1 + h_2(x)c_2.$$

The results were (where below + denotes a non-unique process):

Method	c_1	c_2
Collocation+	0.1935	0.1843
Least Square	0.1875	0.1695
Galerkin	0.1924	0.1707
Moments+	0.1880	0.1695
Sub-Domain+	0.1880	0.1695

a. Write a program (or spread-sheet) to plot the exact solution and the approximations on the same scale. b. Modify the above program to plot the error, $u - u^*$, for the Galerkin and Least Square approximations. c. In the future we will compare such solutions by using their norm, $\|u\|^2_{L2} = \int_L u^2 dx$ or $\|u\|^2_{H^1} = \int_L [u^2 + (du/dx)^2] dx$. Compute the $L2$ norms of the exact and Galerkin solutions. Numerical integration is acceptable. d. Compute the $L2$ norms of the error, $u - u^*$, for the Galerkin and Least Square approximation. Numerical integration is acceptable.

2. Obtain a global Galerkin approximation for the ODE $u'' + x^n = 0$, for $x \in]0, 1[$, with $u(0) = 0 = u(1)$. Assume a cubic polynomial that satisfies the two boundary conditions:

$$u^*(x) = x(1 - x)\,(\Delta_1 + x\,\Delta_2)\,.$$

Form the **S** and **C** matrices. Solve for **D** from $\mathbf{S\,D} = \mathbf{C}$. Compare to the exact solution for a) $n = 0$, b) $n = 1$, c) $n = 2$ where $u_{exact} = (x - x^{n+2})/[(n+1)(n+2)]$.

3. For the one-dimensional problems, with constant coefficients, we often need to evaluate the following four integrals:

$$a)\ \boldsymbol{C}^e = \int_{L^e} \boldsymbol{H}^T dx, \quad b)\ \boldsymbol{M}^e = \int_{L^e} \boldsymbol{H}^T \boldsymbol{H}\, dx,$$

$$c)\ \boldsymbol{S}^e = \int_{L^e} \frac{d\boldsymbol{H}^T}{dx}\frac{d\boldsymbol{H}}{dx}\, dx, \quad d)\ \boldsymbol{U}^e = \int_{L^e} \boldsymbol{H}^T \frac{d\boldsymbol{H}}{dx}\, dx$$

which are usually called the resultant source vector, the mass matrix, the stiffness matrix, and the advection matrix, respectively. Analytically integrate these four matrices for the two-noded line element using the physical coordinate 'interpolation matrix': $\boldsymbol{H}(x) = \lfloor H_1(x) \quad H_2(x) \rfloor$, where $H_1 = (x_2^e - x)/L^e$, $H_2 = (x - x_1^e)/L^e$, and $L^e = x_2^e - x_1^e$. Then repeat the four integrals for unit local coordinate interpolations: $H_1(r) = (1 - r)$, $H_2(r) = r$, where we map the coordinate from $0 \le r \le 1$ into $x(r) = x_1^e + L^e r$ using the physical element length of L^e so $dx = L^e dr$ relates the two differential measures, and the mapping yields the physical derivative as $d(\,)/dx = d(\,)/dr\; dr/dx = d(\,)/dr(1/L^e)$. (Which makes the physical units of the results the same as before.) Of course, the two approaches should yield identical algebraic matrix forms.

4. In general, we have a differential operator, $L(u)$ in Ω with essential boundary conditions $u(x) = 0$ on Γ_1 and/or flux boundary conditions $q = \partial u/\partial n = \bar{q}(x)$ on Γ_2 where Γ_1 and Γ_2 are non-overlapping parts of the total boundary $\Gamma = \Gamma_1 \bigcup \Gamma_2$. An approximate solution defines a residual error in Ω, $R(x)$. Errors in the boundary conditions may define two other boundary residual errors, such as: $R_1(x) \equiv u(x) - \bar{u}(x), x\ on\ \Gamma_1$, $R_2(x) \equiv q(x) - \bar{q}, x\ on\ \Gamma_2$. Extend our method of weighted residuals to require:

$$\int_\Omega R w\, d\Omega = \int_{\Gamma_2} (q - \bar{q}) w\, d\Gamma - \int_{\Gamma_1} (u - \bar{u}) \frac{\partial w}{\partial n}\, d\Gamma.$$

(Note that these units are consistent.) Assume L(u) is the Laplacian $\nabla^2 u$. a) Integrating by parts (using Green's Theorem) show that the above form becomes

$$\int_\Omega \frac{\partial u}{\partial x_k}\frac{\partial w}{\partial x_k}\, d\Omega = \int_{\Gamma_2} \bar{q} w\, d\Gamma + \int_{\Gamma_1} q w\, d\Gamma - \int_{\Gamma_1} \bar{u} \frac{\partial w}{\partial n}\, d\Gamma + \int_{\Gamma_1} u \frac{\partial w}{\partial n}\, d\Gamma.$$

(Hint: $\int_\Gamma f\, d\Gamma = \int_{\Gamma_1} f\, d\Gamma + \int_{\Gamma_2} f\, d\Gamma$.) This form is often used in finite element analysis. b) Integrate by parts again. Show that you obtain

$$\int_\Omega (\nabla^2 w) u\, d\Omega = -\int_{\Gamma_1} \bar{q} w\, d\Gamma - \int_{\Gamma_1} q w\, d\Gamma + \int_{\Gamma_2} u \frac{\partial w}{\partial n}\, d\Gamma + \int_{\Gamma_1} \bar{u} \frac{\partial w}{\partial n}\, d\Gamma$$

Picking $w(x)$ such that $\nabla^2 w \equiv 0$ becomes the basis for a boundary element model since only integrals over the boundary remain. We will not consider such methods further.

5. Carry out the integrations necessary to verify the matrix coefficients in **S** and **C** for the model problem in Sec. 2.5 by: a) collocation method, b) least squares, c) Galerkin method, d) subdomain method, e. method of moments.

6. Formulate the first order equation $dy/dx + Ay = F$ by a) least squares, b) Galerkin method. Use analytic integration for the linear line element (L2) to form the two element matrices. Compute a solution for $y(0) = 0$ with $A = 2, F = 10$ for 5 uniform elements over $x \leq 0.5$.

7. If the model equation Problem 1 changes to $d[a(x)\, du/dx]/dx + b(x)u + Q(x) = 0$ show that the $\mathbf{S}^e$, $\mathbf{M}^e$, and $\mathbf{C}^e$ matrices of Problem 3 change to

$$S^e = \int_{L^e} \frac{d\boldsymbol{H}^T}{dx}\, a(x)\, \frac{d\boldsymbol{H}}{dx}\, dx, \quad \boldsymbol{M}^e = \int_{L^e} \boldsymbol{H}^T\, b(x)\, \boldsymbol{H} dx, \quad \boldsymbol{C}^e = \int_{L^e} \boldsymbol{H}^T\, Q(x)\, dx.$$

8. If the model equation in Problem 1 changes to $d^2u/dx^2 + \lambda u = 0$ with $u(0) = 0 = u(1)$, where λ is an unknown global constant, how does the $\mathbf{M}^e$ matrix change? How does the classification of the assembled algebraic equations change?

9. For the global approximation examples of Sec. 2.5 plot, or accurately sketch, the weight function, $w(x)$, for each of the five methods.

10. The first two linear element Galerkin model example yielded the algebraic equation system, of Eq. 2.43, which has 3 equations with 5 unknowns. The system is singular until 2 boundary conditions are supplied to define a unique problem. Employ a pair of Dirichlet and Neumann conditions such that $u(0) = 0$ and $du(1)/dx = 0$ so that the new exact solution is $u(x) = \text{Sin}\,(x) / \text{Cos}\,(1) - x$. In other words, set $D_1 = 0$ and $q_L = 0$. a) From the algebraic system compute D_2 and D_3 and then the reaction q_0. Compare them to the exact values. b) Post-process both elements to compute the element flux, du/dx. c) Sketch the exact and approximate solution versus position, x. d) Sketch the exact and approximate flux versus position, x.

11. For the two element mesh in Fig. 2.12 use Boolean arrays and matrix multiplication and addition to assemble the given example results in Eq. 2.43.

12. Heat conduction through a layered wall is modeled by $k\, d^2u/dx^2 = 0$, where k is the thermal conductivity. With two essential boundary conditions the linear (L2) element will yield exact results, when nodes are placed on any internal material interfaces. A furnace wall has inside and outside temperatures of 1500 F and 150 F, respectively, and it is made of firebrick, insulator, and red brick having conductivities of 0.72, 0.08, and 0.5 BTU/hr ft F, respectively. The corresponding layer thicknesses are 9, 5, and 7.5 inches. Use three unequal length elements to find the internal interface temperatures and the two wall reaction heat fluxes. Post-process the elements for their gradient, and for the flux $q = -k\, du/dx$. Hint, we expect the same q value in each element.

13. Use three equal length elements to solve the problem in Fig. 2.15 ($L^e = 1/3$). Obtain the nodal values, reactions, and post process for the element gradients. Compare the nodal values to the exact solution.

14. The transverse deflection, v, of a thin beam is given by the fourth order ODE: $d^2[EI(x)d^2v/dx^2]/dx^2 = p(x)$, where E is the material modulus of elasticity, I is the moment of inertia of the cross-section, and p is a distributed load per unit length. Obtain a Galerkin integral form by integrating the first term twice by parts. In the boundary terms we usually call $\Theta = dv/dx$ the slope, $M = EI\,d^2v/dx^2$ the moment (or couple) at a point, and $F = EI\,d^3v/dx^3$ the transverse shear (or force) at the point.

15. Solve Eq 2.43 for a non-zero Neumann condition of $du/dx(1) = Cotan(1) - 1$ and $u(0) = 0$, and compare the results to the same exact solution.

16. The system $d^2u/dx^2 + Q(x) = 0$ with $u(0) = 0 = u(1)$ has a source of $Q = 1$ for $x \leq 1/2$ and 0 for $x > 1/2$. a) Explain why it is preferable for an element end to be placed at $x = 1/2$. b) If you used 3 linear (L2) elements explain why it is better to place the two smaller elements in the left half of the domain. c) Using element length ratios of 1:1:2 verify that the interior nodal values both have an exact value of 1/16. d) Using three equal length ($L^e = 1/3$) elements verify that the source vector for the second element is $\mathbf{C}^{eT} = [3 \quad 1]/24$ and that the (exact) interior nodal values are 5/72 and 1/24.

17. Obtain a Galerkin solution of $y'' - 2y'x/g + 2y/g = g$, for $g = (x^2 + 1)$, on $0 \leq x \leq 1$ with the boundary conditions $y(0) = 2$, $y(1) = 5/3$.

18. For the differential equation in Problem 2.17 if we have one essential boundary condition of $y(0) = 1$ and one Neumann flux boundary condition of $dy/dx(1) = -4/3$ the exact solution is unchanged. Obtain a Galerkin finite element solution and compare it to the exact result.

19. The Couette steady flow velocity, u, of an incompressible viscous fluid between two parallel plates, with a constant pressure gradient of dP/dx, is: $\mu d^2u/dy^2 = dP/dx$, where $u(0) = 0$ and $u(H) = U_H$ describe a lower ($y = 0$) fixed plate and an upper plate ($y = -h$) moving with a speed of U_H. One can obtain a finite element solution for the interior nodal velocities across the fluid. The associated volume flow, per unit z thickness, is $Q = \int_0^H u(y)dy$. Describe how you would post-process the elements to obtain it.

20. For the above Couette flow would you expect to obtain the exact $u(y)$ and Q values for a) one quadratic element, b) two quadratic elements, c) two linear elements? Explain why.

21. Differences between Galerkin and least squares finite element procedures are _______: a) Galerkin allows integration by parts, b) least squares always yield symmetric algebraic equations, c) least squares requires C^1 continuity for second order differential equations, d) all of the above.

22. Repeat the graphical source assembly shown in Fig. 2.12 using three elements instead of two.

23. Burnett presents detailes of a symmetric bi-material bar with non-symmetrical boundary conditions. The length is $L = 100\ cm$ with the center $20\ cm$ section being copper ($k_c = 0.92\ cal/sec - cm - C$, $\rho = 8.5\ gm/cm^3$, $c_p = 0.092\ cal/gm - C$)

while the other two ends of the bar are made of 40 cm of steel ($k_c = 0.12\ cal/sec - cm - C$, $\rho = 7.8\ gm/cm^3$, $c_p = 0.11\ cal/gm - C$). The circular bar has a radius of 2 cm. Along its full length it convects to surrounding air at $U_\infty = 20\ C$. The end area at $x = 0$ receives a flux of $q = 0.1\ cal/sec - cm^2$ and the end at $x = 100$ has a temperature of $U_{100} = 0\ C$. Obtain: a) a steady state solution with 5 equal length elements, b) a transient solution assuming an initial condition of $U(x, 0) = 20\ C$. Note that the two end 'thermal shocks' will require a finer mesh at the ends if the early time history is important. Would the steady state solution also need that mesh feature?

24. A popular one step transient integration technique is the generalized trapezoidal rule where the user provides a coefficient, $0 \le \alpha \le 1$, for the assumption that $\boldsymbol{\phi}_{s+1} = \boldsymbol{\phi}_s + [(1 - \alpha)\dot{\boldsymbol{\phi}}_s + \alpha\dot{\boldsymbol{\phi}}_{s+1}]\Delta t$. Show that this yields the assembled system: $(\mathbf{M}/\Delta t + \alpha\mathbf{K})\boldsymbol{\phi}_{s+1} = (\mathbf{M}/\Delta t + (1 - \alpha)\mathbf{K})\boldsymbol{\phi}_s + (1 - \alpha)\mathbf{P}_s + \alpha\mathbf{P}_{s+1}$. The choice of $\alpha = 1$ yields the unconditionally stable backward differences method. Verify that the above form reduces to the forms shown in Fig. 2.16b with: a) $\alpha = 0$ being the Euler method, b) $\alpha = 1/2$ yielding the Crank-Nicolson process, c) $\alpha = 2/3$ giving the Galerkin method.

25. The two matrices in Eq. 2.59 can represent the longitudinal (axial) vibration of a uniform elastic bar if $\alpha = E/\rho$ where E is its elastic modulus and ρ its mass density. Assume such a bar is fixed at $x = 0$ and free at $x = L$. Solve the eigen-problem for two assembled elements of equal length ($L^e = L/2$), and show that the first frequency differs by less than 3 percent from the exact value given by $\omega_1 = [\pi/2L]\sqrt{E/\rho}$.

26. Repeat the previous problem using three equal length elements and compare the first three natural frequences (eigenvalues) to the exact values given by $\omega_n = [(2n - 1)\pi/2L]\sqrt{E/\rho}$ where n is the mode number. Sketch the normalized mode shapes (eigenvectors) and compare them to the exact eigenfunctions of Sin $((2n - 1)\pi x/2L)$.

2.20 Bibliography

[1] Ainsworth, M. and Oden, J.T., *A Posteriori Error Estimation in Finite Element Analysis*, New York: John Wiley (2000).

[2] Axelsson, O. and Baker, V.A., *Finite Element Solution of Boundary Value Problems*, Philadelphia, PA: SIAM (2001).

[3] Babuska, I. and Strouboulis, T., *The Finite Element Method and its Reliability*, Oxford: Oxford University Press (2001).

[4] Bathe, K.J., *Finite Element Procedures*, Englewood Cliffs: Prentice Hall (1996).

[5] Ciarlet, P.G., *The Finite Element Method for Elliptical Problems*, Philadelphia, PA: SIAM (2002).

[6] Desai, C.S. and Kundu, T., *Introduction to the Finite Element Method*, Boca Raton: CRC Press (2001).

[7] Heinrich, J.C. and Pepper, D.W., *Intermediate Finite Element Method*, Philadelphia, PA: Taylor & Francis (1999).

[8] Huang, H.C. and Usmani, A.S., in *Finite Element Analysis for Heat Transfer*, London: Springer-Verlag (1994).

[9] Hughes, T.J.R., *The Finite Element Method*, Mineola: Dover Publications (2003).

[10] Liusternik, L.A. and Sobolev, V.J., *Elements of Functional Analysis*, New York: Frederick Ungar (1961).

[11] Meier, D.L., "Multi-Dimensional Astrophysical Structural and Dynamical Analysis, Development of a Nonlinear Finite Element Approach," *Astrophysics J.*, **518**, pp. 788–813 (1999).

[12] Nowinski, J.L., *Applications of Functional Analysis in Engineering*, New York: Plenum Press (1981).

[13] Oden, J.T. and Reddy, J.N., *An Introduction to the Mathematical Theory of Finite Elements*, New York: John Wiley (1976).

[14] Oden, J.T., *Applied Functional Analysis*, Englewood Cliffs: Prentice Hall (1979).

[15] Oden, J.T. and Carey, G.F., *Finite Elements: Mathematical Aspects*, Prentice Hall (1983).

[16] Oden, J.T., "The Best FEM," *Finite Elements in Analysis and Design*, **7**, pp. 103–114 (1990).

[17] Reddy, J.N. and Gartling, D.K., *The Finite Element Method in Heat Transfer and Fluid Dynamics*, Boca Raton: CRC Press (2001).

[18] Szabo, B. and Babuska, I., *Finite Element Analysis*, New York: John Wiley (1991).

[19] Whiteman, J.R., "Some Aspects of the Mathematics of Finite Elements," pp. 25–42 in *The Mathematics of Finite Elements and Applications, Vol. II*, ed. J.R. Whiteman, London: Academic Press (1976).

[20] Zhu, J.Z. and Zienkiewicz, O.C., "Superconvergence Recovery Techniques and *A Posteriori* Error Estimators," *Int. J. Num. Meth. Eng.*, **30**, pp. 1321–1339 (1990).

[21] Zienkiewicz, O.C. and Morgan, K., *Finite Elements and Approximation*, Chichester: John Wiley (1983).

[22] Zienkiewicz, O.C. and Taylor, R.L., *The Finite Element Method,* 4th Edition, New York: McGraw-Hill (1991).

[23] Zienkiewicz, O.C. and Zhu, J.Z., "Superconvergent Patch Recovery Techniques and Adaptive Finite Element Refinement," *Comp. Meth. Appl. Mech. Eng.*, **101**, pp. 207–224 (1992).

[24] Zienkiewicz, O.C. and Zhu, J.Z., "The Superconvergent Patch Recovery and *a Posteriori* Error Estimates. Part 2: Error Estimates and Adaptivity," *Int. J. Num. Meth. Eng.*, **33**, pp. 1365–1382 (1992).

[25] Zienkiewicz, O.C. and Taylor, R.L., *The Finite Element Method,* 5th Edition, London: Butterworth-Heinemann (2000).

Chapter 3

Element interpolation and local coordinates

3.1 Introduction

Up to this point we have relied on the use of a linear interpolation relation that was expressed in *global coordinates* and given by inspection. In the previous chapter we saw numerous uses of these interpolation functions. By introducing more advanced interpolation functions, **H**, we can obtain more accurate solutions. Here we will show how the common interpolation functions are derived. Then a number of expansions will be given without proof. Also, we will introduce the concept of non-dimensional *local* or element *coordinate* systems. These will help simplify the algebra and make it practical to automate some of the integration procedures.

3.2 Linear interpolation

Assume that we desire to define a quantity, u, by interpolating in space, from certain given values, **u**. The simplest interpolation would be linear and the simplest space is the line, e.g. x-axis. Thus to define $u(x)$ in terms of its values, $\mathbf{u}^e$, at selected points on an element we could choose a linear polynomial in x. That is:

$$u^e(x) \;=\; c_1^e + c_2^e x \;=\; \mathbf{P}(x)\,\mathbf{c}^e \tag{3.1}$$

where $\mathbf{P} = \lfloor 1 \quad x \rfloor$ denotes the linear polynomial behavior in space and $\mathbf{c}^{e^T} = [\,c_1^e \quad c_2^e\,]$ are undetermined constants that relate to the given values, $\mathbf{u}^e$. Referring to Fig. 3.1, we note that the element has a physical length of L^e and we have defined the nodal values such that $u^e(x_1) = u_1^e$ and $u^e(x_2) = u_2^e$. To be useful, Eq. 3.1 will be required to be valid at all points on the element, including the nodes. Evaluating Eq. 3.1 at each node of the element gives the set of identities: $u^e(x_1^e) \;=\; u_1^e \;=\; c_1^e + c_2^e\, x_1^e$, or

$$\mathbf{u}^e \;=\; \mathbf{g}^e\,\mathbf{c}^e \tag{3.2}$$

where

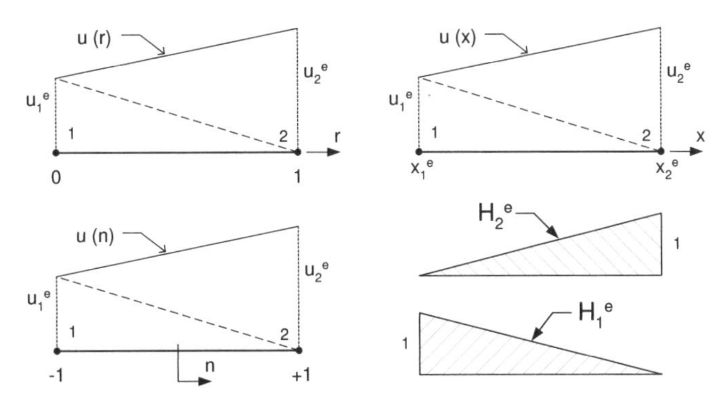

Figure 3.1 *One-dimensional linear interpolation*

$$\mathbf{g}^e = \begin{bmatrix} 1 & x_1^e \\ 1 & x_2^e \end{bmatrix}. \tag{3.3}$$

This shows that the physical constants, $\mathbf{u}^e$, are related to the polynomial constants, $\mathbf{c}^e$ by information on the geometry of the element, $\mathbf{g}^e$. Since $\mathbf{g}^e$ is a square matrix we can (usually) solve Eq. 3.2 to get the polynomial constants:

$$\mathbf{c}^e = \mathbf{g}^{e^{-1}} \mathbf{u}^e . \tag{3.4}$$

In this case the element geometry matrix can be easily inverted to give

$$\mathbf{g}^{e^{-1}} = \frac{1}{x_2^e - x_1^e} \begin{bmatrix} x_2^e & -x_1^e \\ -1 & 1 \end{bmatrix}. \tag{3.5}$$

By putting these results into our original assumption, Eq. 3.1, it is possible to write $u^e(x)$ directly in terms of the nodal values $\mathbf{u}^e$. That is,

$$u^e(x) = \mathbf{P}(x)\, \mathbf{g}^{e^{-1}} \mathbf{u}^e = \mathbf{H}^e(x)\, \mathbf{u}^e \tag{3.6}$$

or

$$u^e(x) = \lfloor 1 \quad x \rfloor \frac{1}{L^e} \begin{bmatrix} x_2^e & -x_1^e \\ -1 & 1 \end{bmatrix} \begin{Bmatrix} u_1^e \\ u_2^e \end{Bmatrix} = \left\lfloor \frac{x_2^e - x}{L^e} \quad \frac{x - x_1^e}{L^e} \right\rfloor \{\mathrm{u}^e\} \tag{3.7}$$

where H^e is called the element interpolation array. Clearly

$$\mathbf{H}^e(x) = \mathbf{P}(x)\, \mathbf{g}^{e^{-1}} . \tag{3.8}$$

From Eq. 3.6 we can see that the approximate value, $u^e(x)$ depends on the assumed behavior in space, $\mathbf{P}$, the element geometry, $\mathbf{g}^e$, and the element nodal parameters, $\mathbf{u}^e$. This is also true for two- and three-dimensional problems. Since this element interpolation has been defined in a global or physical space the geometry matrix, $\mathbf{g}^e$, and thus $\mathbf{H}^e$ will be different for every element. Of course, the algebraic form is common but the numerical terms differ from element to element. For a given type of element it is possible to make $\mathbf{H}$ unique if a local non-dimensional coordinate is utilized. This will also help reduce the amount of calculus that must be done by hand. Local coordinates are usually selected to range from 0 to 1, or from −1 to +1. These two options are also illustrated in Fig. 3.1. For example, consider the *unit coordinates* shown in Fig. 3.1 where the linear polynomial is now $\mathbf{P} = [\,1 \quad r\,]$. Repeating the previous steps yields

$$u^e(r) = \mathbf{P}(r)\,\mathbf{g}^{-1}\,\mathbf{u}^e, \quad \mathbf{g} = \begin{bmatrix} 1 & 0 \\ 1 & 1 \end{bmatrix}, \quad \mathbf{g}^{-1} = \begin{bmatrix} 1 & 0 \\ -1 & 1 \end{bmatrix} \tag{3.9}$$

so that

$$u^e(r) = \mathbf{H}(r)\,\mathbf{u}^e \tag{3.10}$$

where the unit coordinate interpolation function is

$$\mathbf{H}(r) = \lfloor (1 - r) \quad r \rfloor = \mathbf{P}\,\mathbf{g}^{-1}. \tag{3.11}$$

Expanding back to scalar form this means

$$u^e(r) = H_1(r)\,u_1^e + H_2(r)\,u_2^e = (1 - r)\,u_1^e + r u_2^e = u_1^e + r(u_2^e - u_1^e)$$

so that at $r = 0$, $u^e(0) = u_1^e$ and at $r = 1$, $u^e(1) = u_2^e$ as required.

A possible problem here is that while this simplifies $\mathbf{H}$ one may not know 'where' a given r point is located in global or physical space. In other words, what is x when r is given? One simple way to solve this problem is to note that the nodal values of the global coordinates of the nodes, x^e, are given data. Therefore, we can use the concepts in Eq. 3.10 and define $x^e(r) = \mathbf{H}(r)\,\mathbf{x}^e$, or

$$x^e(r) = (1 - r)\,x_1^e + r\,x_2^e = x_1^e + L^e\,r \tag{3.12}$$

for any r in a given element, e. If we make this popular choice for relating the local and global coordinates, we call this an *isoparametric* element. The name implies that a single (iso) set of parametric relations, $\mathbf{H}(r)$, is to be used to define the geometry, $x(r)$, as well as the primary unknowns, $u(r)$.

If we select the symmetric, or natural, local coordinates such that $-1 \le n \le +1$, then a similar set of interpolation functions are obtained. Specifically, $u^e(n) = \mathbf{H}(n)\,\mathbf{u}^e$ with $H_1(n) = (1 - n)/2$, $H_2(n) = (1 + n)/2$, or simply

$$H_i(n) = (1 + n_i\,n)/2 \tag{3.13}$$

where n_i is the local coordinate of node i. This coordinate system is often called a *natural* coordinate system. Of course, the relation to the global system is

$$x^e(n) = \mathbf{H}(n)\,\mathbf{x}^e \text{ or } x^e(r) = \mathbf{H}(r)\,\mathbf{x}^e. \tag{3.14}$$

The relationship between the unit and natural coordinates is $r = (1 + n)/2$. This will sometimes be useful in converting tabulated data in one system to the other. The above local coordinates can be used to define how an approximation changes in space. They also allow one to calculate derivatives. For example, from Eq. 3.10

$$du^e/dr = d\mathbf{H}(r)/dr\;\mathbf{u}^e \tag{3.15}$$

and similarly for other quantities of interest. Another quantity that we will find very important is the Jacobian, $J = dx/dr$. In a typical linear element, Eq. 3.12 gives

$$dx^e(r)/dr = dH_1/dr\;x_1^e + dH_2/dr\;x_2^e = -x_1^e + x_2^e$$

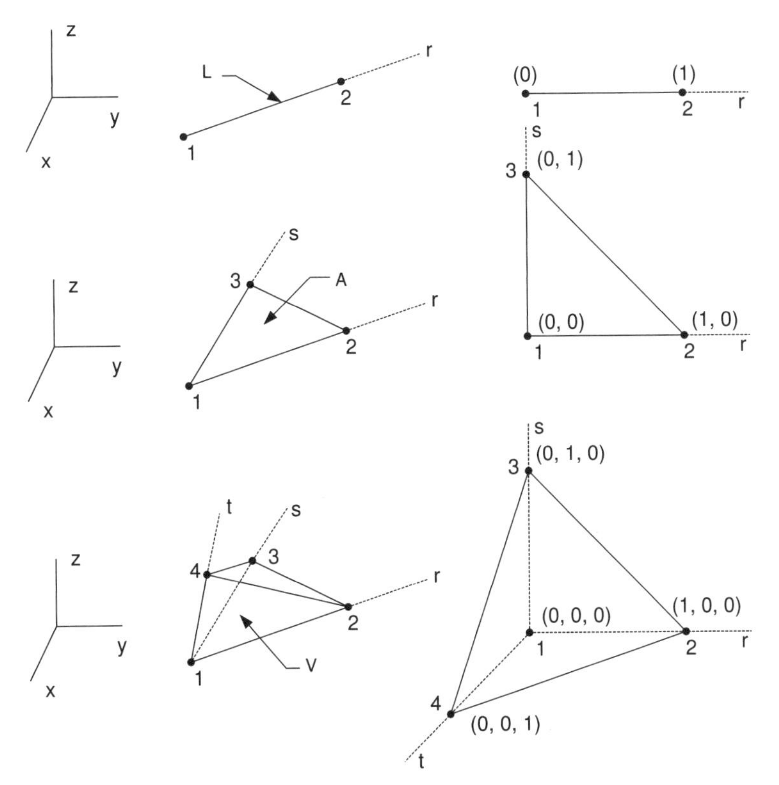

Figure 3.2 *The simplex element family in unit coordinates*

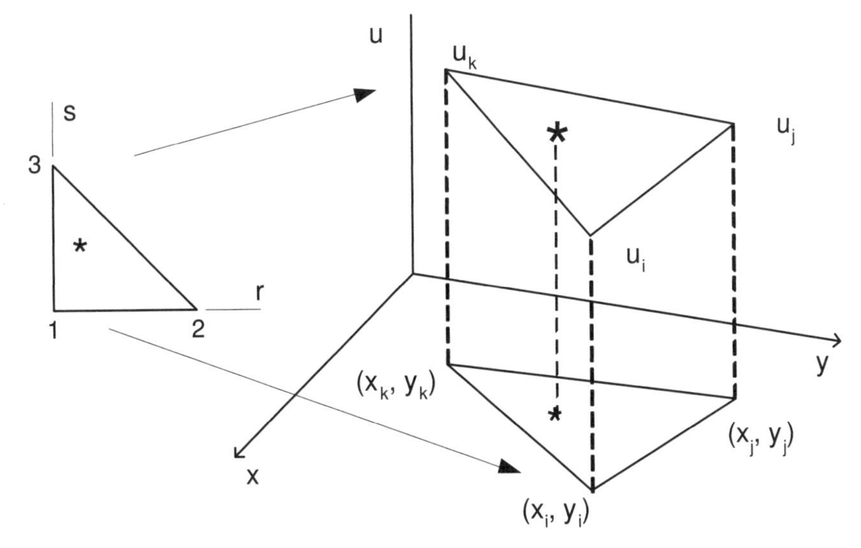

Figure 3.3 *Isoparametric interpolation on a simplex triangle*

or simply $J^e = dx^e/dr = L^e$. By way of comparison, if the natural coordinate is utilized

$$J^e = dx^e(n)/dn = L^e/2. \tag{3.16}$$

This illustrates that the choice of the local coordinates has more effect on the derivatives than it does on the interpolation itself. The use of unit coordinates is more popular with *simplex elements.* These are elements where the number of nodes is one higher than the dimension of the space. The generalization of unit coordinates for common simplex elements is illustrated in Fig. 3.2. It illustrates the general fact that for parametric element interpolation the face of a solid will degenerate to a surface element, and the edge of a volume or face element degenerates to a line element. We will prove this in Chapter 9 where we set $t = 0$ in the lower part to get the face of the middle part and there setting $s = 0$ also yields the parametric line element considered here. For simplex elements the natural coordinates becomes *area coordinates* and *volume coordinates,* which the author finds rather unnatural. Fig. 3.3 shows how the same parametric interpolations can be used for more than one purpose in an analysis. There we see that the spatial positions of points on the element are interpolated from a linear parametric triangle, and the function value is interpolated in the same way. Both unit and natural coordinates are effective for use on squares or cubes in the local space. In global space those shapes become quadrilaterals or hexahedra. The natural coordinates are more popular for those shapes.

3.3 Quadratic interpolation

The next logical spatial form to pick is that of a quadratic polynomial. Select three nodes on the line element, two at the ends and the third inside the element. In local space the third node is at the element center. Thus, the local unit coordinates are $r_1 = 0$, $r_2 = \frac{1}{2}$, and $r_3 = 1$. It is usually desirable to have x_3 also at the center of the element in global space. If we repeat the previous procedure using $u(r) = c_1 + c_2 r + c_3 r^2$, then the element interpolation functions are found to be

$$\begin{aligned} H_1(r) &= 1 - 3r + 2r^2 \\ H_2(r) &= 4r - 4r^2 \\ H_3(r) &= -r + 2r^2 \\ \hline \Sigma H_i(r) &= 1 \end{aligned} \tag{3.17}$$

These quadratic functions are completely different from the linear functions. Note that these functions have a sum that is unity at any point, r, in the element. These three functions illustrate another common feature of all C^0 Lagrangian interpolation functions. They are unity at one node and zero at all others: $H_i(r_j) = \delta_{ij}$. In natural coordinates, on $-1 \le n \le 1$, they transform to

$$H_1(n) = \frac{n(n-1)}{2}, \quad H_2(n) = 1 - n^2, \quad H_3(n) = \frac{n(n+1)}{2}. \tag{3.18}$$

$-1 < n < 1$	$0 < r < 1$
a) Linear 1 - - - - - - - - 2	
$H_1 = (1-n)/2$	$H_1 = (1-r)$
$H_2 = (1+n)/2$	$H_2 = r$
b) Quadratic 1 - - - 2 - - - 3	
$H_1 = n(n-1)/2$	$H_1 = (r-1)(2r-1)$
$H_2 = (1+n)(1-n)$	$H_2 = 4r(1-r)$
$H_3 = n(n+1)/2$	$H_3 = r(2r-1)$
c) Cubic 1 - - 2 - - 3 - - 4	
$H_1 = (1-n)(3n+1)(3n-1)/16$	$H_1 = (1-r)(2-3r)(1-3r)/2$
$H_2 = 9(1+n)(n-1)(3n-1)/16$	$H_2 = 9r(1-r)(2-3r)/2$
$H_3 = 9(1+n)(1-n)(3n+1)/16$	$H_3 = 9r(1-r)(3r-1)/2$
$H_4 = (1+n)(3n+1)(3n-1)/16$	$H_4 = r(2-3r)(1-3r)/2$

Figure 3.4 *Typical Lagrange interpolations in natural and unit coordinates*

3.4 Lagrange interpolation

Clearly this one dimensional procedure can be readily extended by adding more nodes to the interior of the element. Usually the additional nodes are equally spaced along the element. However, they can be placed in arbitrary locations. The interpolation function for such an element is known as the Lagrange interpolation polynomial. The one-dimensional m-th order *Lagrange interpolation* polynomial is the ratio of two products. For an element with $(m+1)$ nodes, r_i, $i = 1, 2, \ldots, (m+1)$, the interpolation function for the k-th node is defined in terms of the ratio of two product operators as

$$H_k^m(n) = \frac{(x-x_1)\ldots(x-x_{(k-1)})(x-x_{(k+1)})\ldots(x-x_{(m+1)})}{(x_k-x_1)\ldots(x_k-x_{(k-1)})(x_k-x_{(k+1)})\ldots(x_k-x_{(m+1)})}. \tag{3.19}$$

This is a complete m-th order polynomial in one dimension. It has the property that $H_k(n_i) = \delta_{ik}$. That is, the function for node k is unity at that node but zero at all other nodes on the element.

For local coordinates, say n, given on the domain $[-1, 1]$, a typical quadratic term $(m = 2)$ for the center node at $n = 0$ $(k = 2)$ on an element with three equally spaced nodes is given by

$$H_2(n) = \frac{(n-(-1))\,(n-1)}{(0-(-1))\,(0-1)} = (1-n^2).$$

This validates the second term in Eq. 3.18. The leftmost and middle node parametric interpolations are found in a similar way. Their algebraic sum, for any n value, is unity, as seen from Eq. 3.18. Figure 3.4 shows typical node locations and interpolation functions for members of this family of complete polynomial functions on simplex elements. Of course, the two choices for the parametric spaces in that figure are related by $n = 2r - 1$. Figure 3.5 shows the typical coding of a quadratic line element (subroutines SHAPE_3_L and DERIV_3_L).

3.5 Hermitian interpolation

All of the interpolation functions considered so far have C^0 continuity between elements. That is, the function being approximated is continuous between elements but its derivative is discontinuous. However, we know that some applications, such as a beam analysis, also require that their derivative be continuous. These C^1 functions are most easily generated by using derivatives, or slopes, as nodal degrees of freedom.

```
SUBROUTINE SHAPE_3_L (X, H)                                          ! 1
! *-* *-* *-* *-* *-* *-* *-* *-* *-* *-* *-* *-* *-*               ! 2
! CALCULATE SHAPE FUNCTIONS OF A 3 NODE LINE ELEMENT                 ! 3
!             IN NATURAL COORDINATES                                 ! 4
! *-* *-* *-* *-* *-* *-* *-* *-* *-* *-* *-* *-* *-*               ! 5
Use Precision_Module                                                 ! 6
 IMPLICIT NONE                                                       ! 7
 REAL(DP), INTENT(IN)  :: X                                          ! 8
 REAL(DP), INTENT(OUT) :: H (3)                                      ! 9
                                                                     !10
! H = ELEMENT SHAPE FUNCTIONS                                        !11
! X = LOCAL COORDINATE OF POINT,      -1 TO +1                       !12
! LOCAL NODE COORD. ARE -1,0,+1     1-----2-----3                    !13
                                                                     !14
  H (1) = 0.5d0*(X*X - X)                                            !15
  H (2) = 1.d0 - X*X                                                 !16
  H (3) = 0.5d0*(X*X + X)                                            !17
END SUBROUTINE SHAPE_3_L                                             !18
                                                                     !19
SUBROUTINE DERIV_3_L (X, DH)                                         !20
! *-* *-* *-* *-* *-* *-* *-* *-* *-* *-* *-* *-* *-*               !21
! FIND LOCAL DERIVATIVES FOR A 3 NODE LINE ELEMENT                   !22
!             IN NATURAL COORDINATES                                 !23
! *-* *-* *-* *-* *-* *-* *-* *-* *-* *-* *-* *-* *-*               !24
Use Precision_Module                                                 !25
 IMPLICIT NONE                                                       !26
 REAL(DP), INTENT(IN)  :: X                                          !27
 REAL(DP), INTENT(OUT) :: DH (3)                                     !28
                                                                     !29
! DH = LOCAL DERIVATIVES OF SHAPE FUNCTIONS (SHAPE_3_L)              !30
! X  = LOCAL COORDINATE OF POINT,      -1 TO +1                      !31
! LOCAL NODE COORD. ARE -1,0,+1     1----2----3                      !32
                                                                     !33
  DH (1) = X - 0.5d0                                                 !34
  DH (2) = - 2.d0 * X                                                !35
  DH (3) = X + 0.5d0                                                 !36
END SUBROUTINE DERIV_3_L                                             !37
```

Figure 3.5 *Coding a Lagrangian quadratic line element*

$$x = Lr \qquad (\ )' = d(\ )/dx$$

a) $C^1 : u = H_1 u_1 + H_2 u_1' + H_3 u_2 + H_4 u_2'$

$$H_1(r) = (2r^3 - 3r^2 + 1)$$
$$H_2(r) = (r^3 - 2r^2 + r) L$$
$$H_3(r) = (3r^2 - 2r^3)$$
$$H_4(r) = (r^3 - r^2) L$$

b) $C^2 : u = H_1 u_1 + H_2 u_1' + H_3 u_1'' + H_4 u_2 + H_5 u_2' + H_6 u_2''$

$$H_1 = (1 - 10r^3 + 15r^4 - 6r^5)$$
$$H_2 = (r - 6r^3 + 8r^4 - 3r^5) L$$
$$H_3 = (r^2 - 3r^3 + 3r^4 - r^5) L^2/2$$
$$H_4 = (10r^3 - 15r^4 + 6r^5)$$
$$H_5 = (7r^4 - 3r^5 - 4r^3) L$$
$$H_6 = (r^3 - 2r^4 + r^5) L^2/2$$

c) $C^3 : u = H_1 u_1 + H_2 u_1' + H_3 u_1'' + H_4 u_1'''$
$+ H_5 u_2 + H_6 u_2' + H_7 u_2'' + H_8 u_2'''$

$$H_1 = (1 - 35r^4 + 84r^5 - 70r^6 + 20r^7)$$
$$H_2 = (r - 20r^4 + 45r^5 - 36r^6 + 10r^7)/L$$
$$H_3 = (r^2 - 10r^4 + 20r^5 - 15r^6 + 4r^7) L^2/2$$
$$H_4 = (r^3 - 4r^4 + 6r^5 - 4r^6 + r^7) L^3/6$$
$$H_5 = (35r^4 - 84r^5 + 70r^6 - 20r^7)$$
$$H_6 = (10r^7 - 34r^6 + 39r^5 - 15r^4) L$$
$$H_7 = (5r^4 - 14r^5 + 13r^6 - 4r^7) L^2/2$$
$$H_8 = (r^7 - 3r^6 + 3r^5 - r^4) L^3/6$$

Figure 3.6 *C^1 to C^3 Hermitian interpolation in unit coordinates*

The simplest element in this family is the two node line element where both y and dy/dx are taken as nodal degrees of freedom. Note that a global derivative has been selected as a degree of freedom. Since there are two nodes with two dof each, the interpolation function has four constants, thus, it is a cubic polynomial. The form of this *Hermite polynomial* is well known. The element is shown in Fig. 3.7 along with the interpolation functions and their global derivatives. The latter quantities are obtained from the relation between local and global coordinates, e.g., Eq. 3.12. On rare occasions one may also need to have the second derivatives continuous between elements. Typical C^2 equations of this type are also given in Fig. 3.6 and elsewhere. Since derivatives have also been introduced as nodal parameters, the previous statement that $\Sigma H_i = 1$ is no longer true (unless i is limited to the u_i values).

```
SUBROUTINE SHAPE_C1_L (R, L, H)                                    ! 1
! *-* *-* *-* *-* *-* *-* *-* *-* *-* *-* *-* *-* *-* *-*          ! 2
! SHAPE FUNCTIONS FOR CUBIC HERMITE IN UNIT COORDINATES            ! 3
! *-* *-* *-* *-* *-* *-* *-* *-* *-* *-* *-* *-* *-* *-*          ! 4
Use Precision_Module                                               ! 5
 IMPLICIT NONE                                                     ! 6
 REAL(DP), INTENT(IN)  :: R, L                                     ! 7
 REAL(DP), INTENT(OUT) :: H (4)                                    ! 8
                                                                   ! 9
! L = PHYSICAL LENGTH OF ELEMENT      1----------2 ---> R          !10
! R = LOCAL COORDINATE OF POINT      R=0          R=1              !11
! H = SHAPE FUNCTIONS ARRAY                                        !12
! DOF ARE FUNCTION AND SLOPE, WRT X, AT EACH NODE                  !13
! D()/DX = D()/DR DR/DX = 1/L * D()/DR                             !14
                                                                   !15
  H(1) = 1.d0 - 3.0*R*R + 2.0*R*R*R                                !16
  H(2) = (R - 2.0*R*R + R*R*R)*L                                   !17
  H(3) =  3.0*R*R - 2.0*R*R*R                                      !18
  H(4) = (R*R*R - R*R)*L                                           !19
END SUBROUTINE SHAPE_C1_L                                          !20
                                                                   !21
SUBROUTINE DERIV_C1_L (R, L, DH)                                   !22
! *-* *-* *-* *-* *-* *-* *-* *-* *-* *-* *-* *-* *-* *-*          !23
! FIRST DERIVATIVES OF CUBIC HERMITE IN UNIT COORDINATES           !24
! *-* *-* *-* *-* *-* *-* *-* *-* *-* *-* *-* *-* *-* *-*          !25
Use Precision_Module                                               !26
 IMPLICIT NONE                                                     !27
 REAL(DP), INTENT(IN)  :: R, L                                     !28
 REAL(DP), INTENT(OUT) :: DH (4)                                   !29
                                                                   !30
! L = PHYSICAL LENGTH OF ELEMENT         1 -------- 2 --> R        !31
! R = LOCAL COORDINATE OF POINT          R=0          R=1          !32
! DH = FIRST PHYSICAL DERIVATTVES OF H                             !33
                                                                   !34
  DH (1) = 6.d0 * (R * R - R) / L                                  !35
  DH (2) = 1.d0 - 4.d0 * R + 3.d0 * R * R                          !36
  DH (3) = 6.d0 * (R - R * R) / L                                  !37
  DH (4) = 3.d0 * R * R - 2.d0 * R                                 !38
END SUBROUTINE DERIV_C1_L                                          !39
                                                                   !40
SUBROUTINE DERIV2_C1_L (R, L, D2H)                                 !41
! *-* *-* *-* *-* *-* *-* *-* *-* *-* *-* *-* *-* *-* *-*          !42
! 2ND DERIVATIVES OF CUBIC HERMITE IN UNIT COORDINATES             !43
! *-* *-* *-* *-* *-* *-* *-* *-* *-* *-* *-* *-* *-* *-*          !44
Use Precision_Module                                               !45
 IMPLICIT NONE                                                     !46
 REAL(DP), INTENT(IN)  :: R, L                                     !47
 REAL(DP), INTENT(OUT) :: D2H (4)                                  !48
                                                                   !49
! L = PHYSICAL LENGTH OF ELEMENT         1 -------- 2 --> R        !50
! R = LOCAL COORDINATE OF POINT          R=0          R=1          !51
! D2H = SECOND DERIVATIVES OF H                                    !52
                                                                   !53
  D2H (1) = 6.d0 * (R + R - 1.d0) / L**2                           !54
  D2H (2) = ( - 4.d0 + 6.d0 * R) / L                               !55
  D2H (3) = 6.d0 * (1.d0 - R - R) / L**2                           !56
  D2H (4) = (6.d0 * R - 2.d0) / L                                  !57
END SUBROUTINE DERIV2_C1_L                                         !58
```

Figure 3.7 *The C^1 Hermite cubic line element*

3.6 Hierarchical interpolation

Recently some alternate types of interpolation have become popular. They are called *hierarchical functions*. The unique feature of these polynomials is that the higher order polynomials contain the lower order ones. This concept is shown in Fig. 3.8. Thus, to get new functions you simply add some terms to the old functions. To illustrate this concept let us return to the linear element in local natural coordinates. In that element

$$u^e(n) = H_1(n)\, u_1^e + H_2(n)\, u_2^e \tag{3.20}$$

where the two H_i are given in Eq. 3.10. We want to generate a quadratic interpolation form that will not destroy these H_i as Eq. 3.17 did. The key to accomplishing this goal is to note that the second derivative of Eq. 3.11 is everywhere zero. Thus, if we introduce an additional degree of freedom related to the second derivative of u it will not affect the linear terms. Figure 3.8 shows the linear element where we have added a third midpoint ($n = 0$) control node to be associated with the quadratic additions. At the third node let the degree of freedom be the second local derivative, d^2u/dr^2. Upgrade the approximation by setting

$$u(n) = H_1(n)\, u_1^e + H_2(n)\, u_2^e + H_3(n)\, d^2\, u^e / dn^2 \tag{3.21}$$

where the hierarchical quadratic addition is: $H_3(n) = c_1 + c_2\, n + c_3\, n^2$. The three constants are found from the conditions that it vanishes at the two original nodes, so as not to change H_1 and H_2, and the second derivative is unity at the new midpoint node. The result is

$$H_3(n) = (n^2 - 1)/2\,. \tag{3.22}$$

The concept is extended to a cubic hierarchical element by using the new function in conjunction with the third tangential derivative at the center.

The higher order hierarchical functions are becoming increasingly popular. They utilize the higher derivatives at the center node. We introduce the notation $m \rightarrow n$ to denote the presence of consecutive tangential derivatives from order m to order n. The value of the function is implied by $m = 0$. These functions must vanish at the end nodes. Finally, we usually want the function $H_{p+1}(n),\ p \geq 2$ to have its p-th derivative take on a value of unity at the center node. The resulting functions are not unique. A common set of hierarchical functions in natural coordinates $-1 \leq n \leq 1$ are

$$H_p(n) = (n^p - b)/p!, \qquad p \geq 2 \tag{3.23}$$

where $b = 1$ if p is even, and $b = n$ if p is odd. The first six members of this family are shown in Fig. 3.9. Note that the even functions approach a rectangular shape as $p \rightarrow \infty$, but there is not much change in their form beyond the fourth order polynomial. Likewise, the odd functions approach a sawtooth as $p \rightarrow \infty$, but they change relatively little after the cubic order polynomial. These observations suggest that for the above hierarchical choice it may be better to stop at the fourth order polynomial and refine the mesh rather than adding more hierarchical degrees of freedom. However, this form might capture shape boundary layers or shocks better than other choices. These relations are zero at the ends, $n = \pm 1$. The first derivatives of these functions are

$$H'_{p+1} = [\, pn^{(p-1)} - b'\,]/p!$$

and since b'' is always zero, the second derivatives are

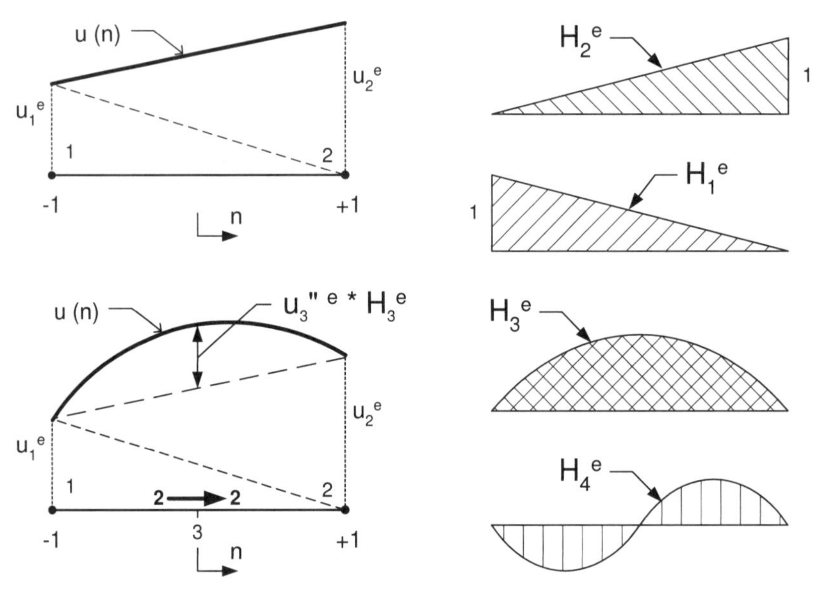

Figure 3.8 *Concept of hierarchical shape functions*

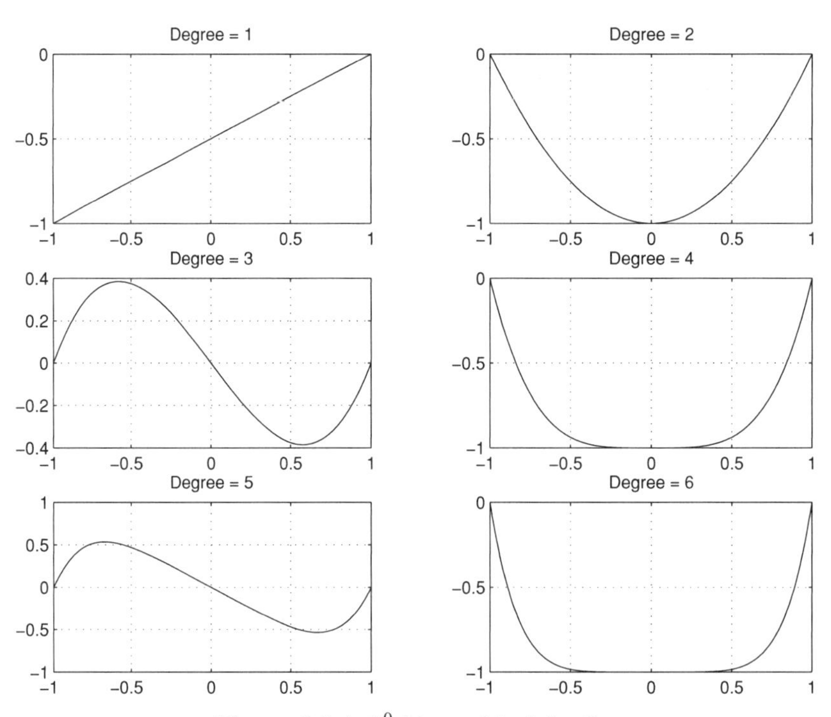

Figure 3.9 *A C^0 hierarchical family*

$$H''_{p+1} = p(p-1)\,n^{(p-2)}/p! = n^{(p-2)}/(p-2)!.$$

Proceeding in this manner it is easy to show by induction that the m-th derivative is

$$H_{p+1}^{(m)}(n) = n^{(p-m)}/(p-m)!, \quad m \geq 2. \tag{3.24}$$

At the center point, $n = 0$, the derivative has a value of

$$H_{p+1}^{(m)}(0) = \begin{cases} 0 & \text{if } m \neq p \\ 1 & \text{if } m = p. \end{cases}$$

We will see later that when hierarchical functions are utilized, the element matrices for a p-th order polynomial are partitions of the element matrices for a $(p+1)$ order polynomial. A typical cubic element would be built by using the degree 2 and 3 hierarchical functions shown in Fig. 3.9.

The element square matrix will always involve an integral of the product of the derivatives of the interpolation functions. If those derivatives were orthogonal then they would result in a diagonal square matrix. That would be very desirable. Thus, it is becoming popular to search for interpolation functions whose derivatives are close to being orthogonal. It is well known that integrals of products of *Legendre polynomials* are orthogonal. This suggests that a useful trick would be to pick interpolation functions that are integrals of Legendre polynomials so that their derivatives are Legendre polynomials. Such a trick is very useful in the so-called *p – method* and *hp – method* of adaptive finite element analysis. For future reference we will observe that the first four Legendre polynomials on the domain of $-1 \leq x \leq 1$ are [1, 11]:

$$\begin{aligned} P_0(x) &= 1 \\ P_1(x) &= x \\ P_2(x) &= (3x^2 - 1)/2 \\ P_3(x) &= (5x^3 - 3x)/2 \\ P_4(x) &= (35x^4 - 30x^2 + 3)/8 \end{aligned} \tag{3.25}$$

Legendre polynomials can be generated from the *recursion formula*:

$$(n+1)\,P_{n+1}(x) = (2n+1)\,x\,P_n(x) - nP_{n-1}(x), \qquad n \geq 1$$

and

$$n\,P'_{n+1}(x) = (2n+1)\,x\,P'_n(x) - (n+1)\,P'_{n-1}(x). \tag{3.26}$$

To avoid roundoff error and unnecessary calculations, these recursion relations should be used instead of Eq. 3.25 when computing these polynomials. They have the orthogonality property:

$$\int_{-1}^{+1} P_i(x)\,P_j(x)\,dx = \begin{cases} \dfrac{2}{2i+1} & \text{for } i = j \\ 0 & \text{for } i \neq j. \end{cases} \tag{3.27}$$

To create a family of functions for potential use as hierarchical interpolation functions we next consider the integral of the above polynomials. Define a new function

$$\gamma_j(x) = \int_{-1}^{x} P_{j-1}(t)\, dt. \tag{3.28}$$

A handbook of mathematical functions [1] shows the useful relation for Legendre polynomials that

$$(2j-1)\, P_{j-1}(t) = P'_j(t) - P'_{j-2}(t) \tag{3.29}$$

where $(\ )'$ denotes dP/dt. The integral of the derivative is evaluated by inspection so

$$\gamma_j(x) = [\, P_j(x) - P_{j-2}(x)\,]/(2j-1) \tag{3.30}$$

since the lower limit terms cancel each other because

$$P_j(-1) = \begin{cases} 1 & j \text{ even} \\ -1 & j \text{ odd}. \end{cases}$$

We may want to multiply by a constant to scale such a function in a special way. For example, to make its second derivative unity at $x = 0$. Thus, for use as interpolation functions we will consider the family of functions defined as

$$\phi_j(x) = [\, P_j(x) - P_{j-2}(x)\,] \,/\, \lambda_j \equiv \psi_j(x)/\lambda_j \tag{3.31}$$

where λ_j is a constant to be selected later. From the definition of the Legendre polynomials, we see that the first few values of $\psi_j(x)$ that are of interest are:

$$\begin{aligned} \psi_2(x) &= 3(x^2 - 1)/2 \\ \psi_3(x) &= 5(x^3 - x)/2 \\ \psi_4(x) &= 7(5x^4 - 6x^2 + 1)/8 \\ \psi_5(x) &= 9(7x^5 - 10x^3 + 3x)/8 \\ \psi_6(x) &= 11(21x^6 - 35x^4 + 15x^2 - 1)/16 \end{aligned} \tag{3.32}$$

These functions are shown in Fig. 3.10 along with a linear polynomial. Note that each function has its number of roots (zero values) equal to the order of the polynomial. The previous set had only two roots for the even order polynomials and three roots for the odd order polynomials (excluding the linear one). Thus, this is clearly a different type of function for hierarchical use. These would be more expensive to integrate numerically since there are more terms in each function. Note that the $\psi_j(x)$ have the property that they vanish at the ends of the domain: $\psi_j(\pm 1) \equiv 0,\ j \geq 2$. A popular choice for the midpoint hierarchical interpolation functions is to pick

$$H_j(x) = \phi_{j-1}(x), \qquad j \geq 3 \tag{3.33}$$

where the scaling is chosen to be

$$\lambda_j \equiv \sqrt{4j-2}\,. \tag{3.34}$$

The reader should note for future reference that if the above domain of $-1 \leq x \leq 1$ was the edge of a two-dimensional element then the above derivatives would be viewed as tangential derivatives on that edge. The same is true for edges of solid elements. Hierarchical enrichment is not just restricted to C^0 functions, but has also been used with Hermite functions as well. Earlier we saw the C^1 cubic Hermite and the C^2 fifth order Hermite polynomials. The cubic has nodal dof that are the value and slope of the solution at each end. If we desire to add a center hierarchical enrichment, then that

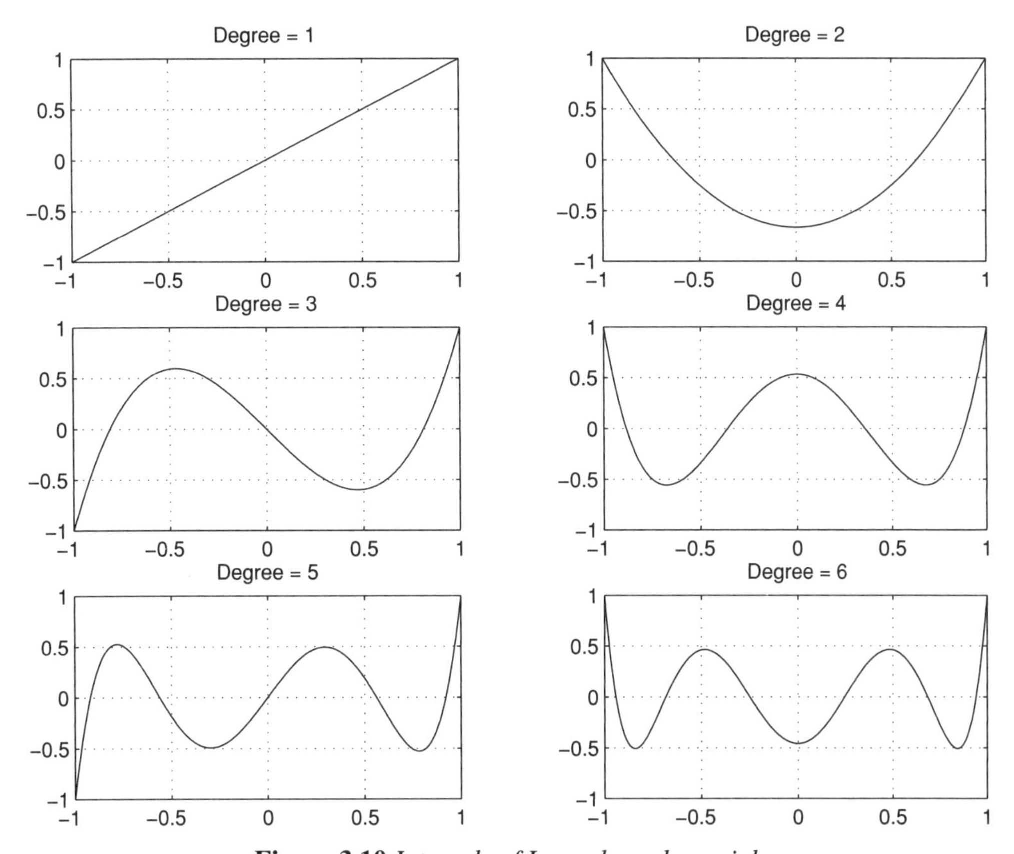

Figure 3.10 *Integrals of Legendre polynomials*

function should have a zero value and slope at each end. In addition, since the fourth derivative of the cubic polynomial is zero, we select that quantity as the first hierarchical dof. In natural coordinates $-1 \le a \le 1$, we have $p - 3$ internal functions for $p \ge 3$. One possible choice is

$$H_p^{(0)} = \frac{1}{p\,!}\left[\, a^{p/2} - 1 \,\right]^2, \qquad p \ge 4, \text{ even}$$
$$\frac{1}{p\,!}\left[\, a^{(p-1)/2} - 1 \,\right]^2, \qquad p \ge 5, \text{ odd}. \tag{3.35}$$

For example, for $p = 4$, $H_p^{(0)} = [\, a^4 - 2a^2 + 1 \,] / 24$, which is zero at both ends as is its first derivative $d\,H_p^{(0)} / d\,a = [\, 4a^3 - 4a \,] / 24$, while its fourth local derivative is unity for all a. We associate that constant dof with the center point, $a = 0$. A similar set of enhancements that have zero second derivatives at the ends can be used to enrich the C^2 Hermite family of elements.

3.7 Space-time interpolations *

Most books on finite elements limit the interpolation methods to physical space and do not cover combined space-time interpolations even though they have proved useful for more than three decades. As their name implies they are used in transient problems, like those of Section 2.16, or in wave propagation, computational fluid dynamics or structural dynamics. Very early applications of space-time elements were given by Oden for structural dynamics and by Bruch and Zyvoloski [6, 7] who described transient heat transfer. Space-time elements can be made continuous in time, like the one step semi-discrete time integration methods of Section 2.16. However, then they would generally be unstructured in time and the dimension of the problem (and mesh) formulation increases by one. That can be particularly confusing for mesh generation and for result visualization when normal transient three-dimensional problems become four-dimensional. That is not too bad for one-dimensional space, as given in [6] and as illustrated in Fig. 3.11.

There we see the parametric local forms of a linear 1-D space element with 2 nodes extended to a full unstructured 2-D triangle (simplex) with 3 nodes in space-time, or a structured rectangular element with 4 nodes. In the latter case, for simplicity, we assume that it covers a fixed interval, or 'slab', of time so local nodes 1 and 2 are at the same first time while nodes 3 and 4 are at the same later time. (That is different from a space-time quadrilateral where all four nodes could be at a different time.) View that as simply translating the space element in time and you can see that any space-time slab element (in any dimension of physical space) will simply have twice as many nodes as its spatial form. Thus, you can use the common interpolations give earlier in this chapter and later in Chapter 9 for the spatial forms. You also only have to input the usual spatial coordinates and connectivity and the program hides the doubling of the element nodes and their translation in time.

The accuracy of the time integrations are the same as the classical semi-discrete methods when the space-time slabs are made continuous with each other. However, many users of space-time slab elements employ elements discontinuous in time. Dettmer and Peric [9] have shown that such formulations have accuracy in time of order Δt^3 instead of Δt as obtained in the continuous linear interpolation in time. That happens because the time interface is treated with a weak (integral) continuity requirement. Tezduyar and his research group, see [12, 13] for example, have solved numerous very large transient, non-linear, 3-D complex flow geometries with such techniques.

3.8 Nodally exact interpolations *

The analytic solution to a differential equation is generally viewed as the sum of a homogeneous solution and a particular solution. It has been proved by Tong [14] and others [15] that if the finite element interpolation functions are the exact solution to the homogeneous differential equation ($Q = 0$), then the finite element solution of a non-homogeneous (non-zero) source term will *always* be exact at the nodes. Clearly, this also means that if the source is zero, then this type of solution would be exact everywhere. It is well known that the exact solution of the homogeneous equations for the bar on an elastic foundation (or a rod conducting and convecting heat) will generally involve

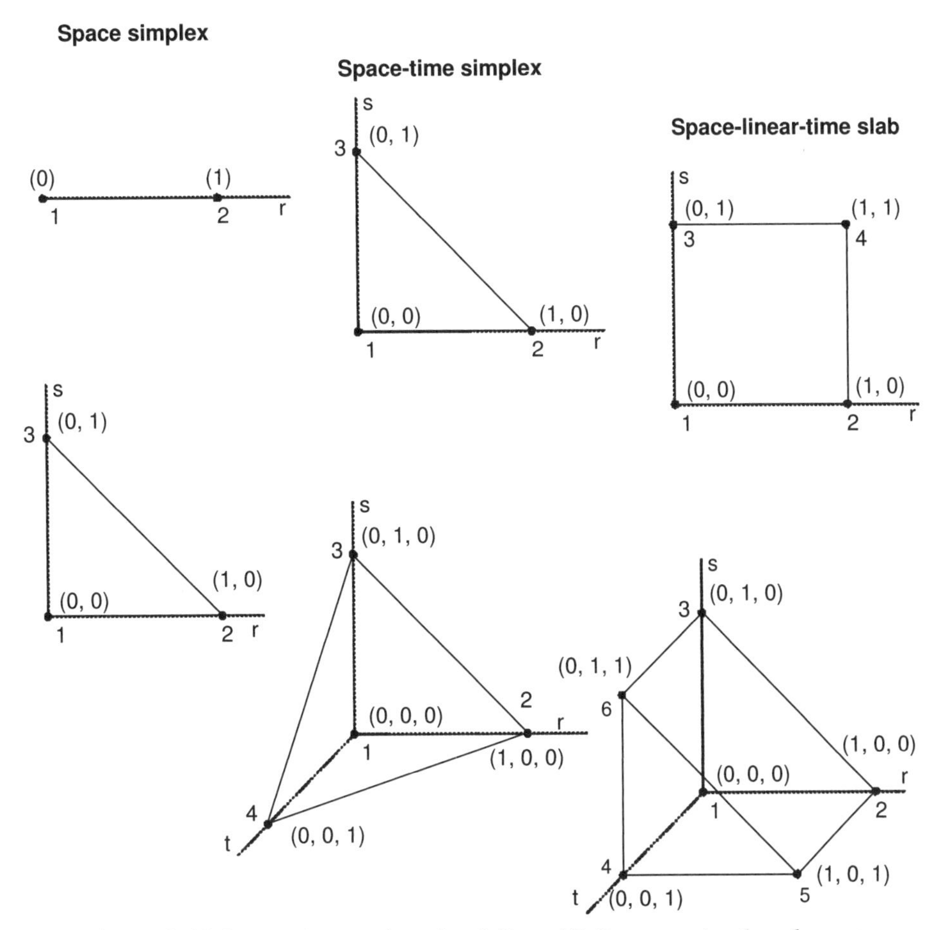

Figure 3.11 *Space-time options for 1-D and 2-D space simplex elements*

hyperbolic functions. Therefore, if we replaced the polynomial interpolations with the homogeneous hyperbolic functions we can assure ourselves of results that are at least exact at the nodes. For the problems considered here, it can be shown that the typical element 1-d matrices obtained from interpolating with the exact homogeneous solutions are summarized in Tables 3.1 and 3.2. In practice, using hyperbolic functions with large arguments can break down due to the way their values are computed.

3.9 Interpolation error*

To obtain a physical feel for the typical errors involved, we consider a one-dimensional model. A hueristic argument will be used. The Taylor's series of a function v at a point x:

Table 3.1 *Homogeneous solution interpolation for semi-infinite axial bar on a foundation*

a) PDE : $\dfrac{d}{dx}\left(EA\,\dfrac{du}{dx} \right) - k u = Q, \quad m = k/EA, \quad k > 0$

b) Homogeneous interpolation : $H_1 = e^{-mx}$

c) Stiffness matrix : $K_{11} = \dfrac{m\,EA}{2} + \dfrac{k}{2m}$

d) Force vector : $F_1 = Q/m, \qquad Q = \text{constant}$

e) Mass matrix : $M_{11} = \rho/2m$

Table 3.2 *Homogeneous solution interpolation for finite axial bar on a foundation*

a) PDE : $\dfrac{d}{dx}\left(EA\,\dfrac{du}{dx} \right) - k u = Q, \quad m = k/EA, \quad k > 0$

b) Homogeneous interpolation : $S = \sinh(mL^e), \qquad C = \cosh(mL^e)$

$$\mathbf{H} = \frac{1}{S}\left[\, \sinh[\, m(L-x)\,] \quad \sinh(mx)\,\right]$$

c) Stiffness matrix : $a = k + EA\,m^2, \qquad b = k - EA\,m^2$

$$\mathbf{K} = \frac{1}{2m\,S^2}\begin{bmatrix} (a \sinh(2\,mL^e) - b\,mL^e) & (b \sinh(2\,mL^e) - a\,S) \\ \text{symmetric} & (a \sinh(2\,mL^e) - b\,mL^e) \end{bmatrix}$$

d) Force vector : $Q = Q_1(1 - x/L^e) + Q_2\,x/L^e$

$$\mathbf{F} = \frac{1}{m}\begin{Bmatrix} \dfrac{Q_1(C-1)}{S} + \dfrac{(Q_2 - Q_1)\,(1 - mL^e/S)}{m\,L^e} \\ \dfrac{Q_1(C-1)}{S} + \dfrac{(Q_2 - Q_1)\,(mL^e \coth(mL^e) - 1)}{m\,L^e} \end{Bmatrix}$$

e) Mass matrix : $\mathbf{M} = \dfrac{\zeta}{2m\,S^2}\begin{bmatrix} (\sinh(2\,mL^e) - mL^e) & (S + \sinh(2\,mL^e)) \\ \text{symmetric} & (\sinh(2\,mL^e) - mL^e) \end{bmatrix}$

$$v(x+h) = v(x) + h\,\frac{\partial v}{\partial x}(x) + \frac{h^2}{2}\,\frac{\partial^2 v}{\partial x^2}(x) + \cdots \tag{3.36}$$

The objective here is to show that if the third term is neglected, then the relations for the linear line element are obtained. That is, the third term is a measure of the interpolation error in the linear element. For an element with nodes at i and j, we use Eq. 3.36 to estimate the function at node j when h is the length of the element:

$$v_j = v_i + h\,\frac{\partial v}{\partial x}(x_i).$$

Solving for the gradient at node i yields

$$\frac{\partial v}{\partial x}(x_i) = \frac{(v_j - v_i)}{h} = \frac{\partial v}{\partial x}(x_j)$$

which is the constant previously obtained for the derivative in the linear line element. Thus, we can think of this type of element as representing the first two terms of the Taylor series. The omitted third term is a measure of the error associated with the element. Its value is proportional to the product of the second derivative and the square of the element size.

If the exact solution is linear so that the first derivative is constant, then the second derivative, $\partial^2 v/\partial x^2$, is zero and there is no error in the element. Otherwise, the second derivative and element error do not vanish. If the user wishes to exercise control over this relative error, then the element size, h, must be varied, or we must use a higher degree interpolation for the element. If we think in terms of the bar element, then v and $\partial v/\partial x$ represent the displacement and strain, respectively. The contribution to the error represents the strain gradient (and stress gradient). Therefore, we must use our engineering judgment to make the element size, h, small in regions of large strain gradients (stress concentrations). Conversely, where the strain gradients are small, we can increase the element size, h, to reduce the computational cost. A similar argument can be stated for the heat conduction problem. Then, v is the temperature, $\partial v/\partial x$ describes the temperature gradient (heat flux), and the error is proportional to the flux gradient. If one does not wish to vary the element sizes, h, then to reduce the error, one must add higher order polynomial terms of the element interpolation functions so that the second derivative is present in the element. These two approaches to error control are known as the *h-method* and the *p-method,* respectively.

The previous comments have assumed the use of a uniform mesh, that is, h was the same for all elements in the mesh. Thus, the above error discussions have not considered the interaction of adjacent elements. The effects of adjacent element sizes have been evaluated for the case of a continuous bar subject to an axial load. An error term, in the governing differential equation, due to the finite element approximation at node j has been shown to be

$$E = -\frac{h}{3}(1-a)\,\frac{\partial^3 v}{\partial x^3}(x_j) + \frac{h^2}{12}\left(\frac{1+a^3}{1+a}\right)\frac{\partial^4 v}{\partial x^4}(x_j) + \cdots \tag{3.37}$$

where h is the size of one element and ah is the size of the adjacent element. Here it is seen that for a smooth variation ($a \doteq 1$) or a uniform mesh ($a = 1$), the error in the approximated ODE is of the order h squared. However, if the adjacent element sizes differ greatly ($a \neq 1$), then a larger error of order h is present. This suggests that it is desirable to have a gradual change in element sizes when possible. One should avoid placing a small element adjacent to one that is many times larger. Today the process of error estimation in a finite element analysis is a well established field of applied mathematics. This knowledge can be incorporated into a finite element software system. The *MODEL* code has this ability.

3.10 Gradient estimates *

In our finite element calculations we often have a need for accurate estimates of the derivatives of the primary variable. For example, in plane stress or plane strain analysis, the primary unknowns which we compute are the displacement components of the nodes. However, we often are equally concerned about the strains and stresses which are computed from the derivatives of the displacements. Likewise, when we model an ideal fluid with a velocity potential, we actually have little or no interest in the computed potential; but we are very interested in the velocity components which are the derivatives of the potential. A logical question at this point is: what location in the element will give me the most accurate estimate of derivatives? Such points are called *optimal points* or *Barlow points* [5] or *superconvergent points*. A heuristic argument for determining their location can be easily presented. Let us begin by recalling some of our previous observations. In Secs. 2.6.2 and 3.4, we found that our finite element solution example was an *interpolate* solution, that is, it was exact at the node points and approximate elsewhere. Such accuracy is rare but, in general, one finds that the computed values of the primary variable are most accurate at the node points. Thus, for the sake of simplicity we will assume that the element's nodal values are exact, or superconvergent.

We have taken our finite element approximation to be a polynomial of some specific order, say m. If the exact solution is also a polynomial of order m, then our finite element

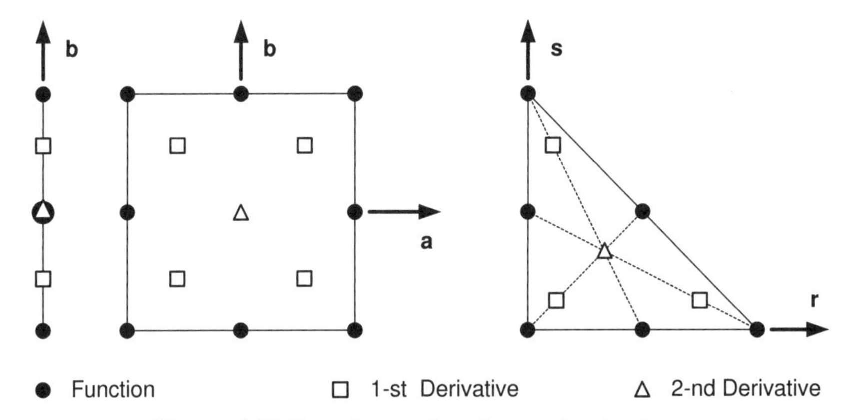

Figure 3.12 *Sampling points for quadratic elements*

solution will be exact everywhere in the element. In addition, the finite element derivative estimates will also be exact. It is rare to have such good luck. In general, we must expect our results to only be approximate. However, we can hope for the next best thing to an exact solution. That would be where the exact solution is a polynomial that is one order higher, say $n = m + 1$, than our finite element polynomial. Let the subscripts E and F denote the exact and finite element solutions, respectively. Consider a one-dimensional formulation in natural coordinates, $-1 < a < +1$. Then the exact solution could be written as

$$U_E(a) = \mathbf{P}_E(a)\,\mathbf{V}_E = \lfloor 1 \quad a \quad a^2 \cdots a^m \quad a^n \rfloor \mathbf{V}_E,$$

and our approximate finite element polynominal solution would be

$$U_F(a) = \mathbf{P}_F(a)\,\mathbf{V}_F = \lfloor 1 \quad a \quad a^2 \cdots a^m \rfloor \mathbf{V}_F$$

where $n = (m + 1)$, as assumed above. In the above $\mathbf{V}_E$ and $\mathbf{V}_F$ represent different vectors of unknown constants. In the domain of a typical element, these two forms should be almost equal. If we assume that they are equal at the nodes, then we can equate $u_E(a_j) = u_F(a_j)$ where a_j is the local coordinate of node j. Then the following identities are obtained: $\mathbf{P}_F(a_j)\,\mathbf{V}_F = \mathbf{P}_E(a_j)\,\mathbf{V}_E,\ 1 \le k \le m$, or symbolically

$$\mathbf{A}_F\,\mathbf{V}_F = \mathbf{A}_E\,\mathbf{V}_E \tag{3.38}$$

where the rectangular array $\mathbf{A}_E$ has one more column than the square matrix $\mathbf{A}_F$, but otherwise they are the same. Indeed, upon closer inspection we should observe that $\mathbf{A}_E$ can be partitioned into a square matrix that is the same as $\mathbf{A}_F$ and an additional column so that $\mathbf{A}_E = [\,\mathbf{A}_F \mid \mathbf{C}_E\,]$ where the column is $\mathbf{C}_E^T = [\,a_1^n \ a_2^n \ a_3^n \cdots a_m^n\,]$. If we solve Eq. 3.38 we can relate the finite element constants, $\mathbf{V}_F$, to the exact constants, $\mathbf{V}_E$, at the nodes of the element. Thus, multiplying by the inverse of the square matrix $\mathbf{A}_F$, Eq. 3.38 gives the relationship between the finite element nodal values and exact values as $\mathbf{V}_F = \mathbf{A}_F^{-1}\,\mathbf{A}_E\,\mathbf{V}_E = [\,\mathbf{I} \mid \mathbf{A}_F^{-1}\,\mathbf{C}_E]\,\mathbf{V}_E = [\,\mathbf{I} \mid \mathbf{E}\,]\,\mathbf{V}_E$ or simply

$$\mathbf{V}_F = \mathbf{K}\,\mathbf{V}_E \tag{3.39}$$

where $\mathbf{K} = \mathbf{A}_F^{-1}\,\mathbf{A}_E$ is a rectangular matrix with constant coefficients. Therefore, we can return to Eq. 3.38 and relate everything to $\mathbf{V}_E$. This gives $u_F(a) = \mathbf{P}_F(a)\,\mathbf{K}\mathbf{V}_E = \mathbf{P}_E(a)\,\mathbf{V}_E = u_E(a)$ so that for arbitrary $\mathbf{V}_E$, one probably has the finite element polynomial and the exact polynomial related by $\mathbf{P}_F(a)\,\mathbf{K} = \mathbf{P}_E(a)$. Likewise, the derivatives of this relation should be approximately equal.

As an example, assume a quadratic finite element in one-dimensional natural coordinates, $-1 < a < +1$. The exact solution is assumed to be cubic. Therefore,

$$\mathbf{P}_F = \lfloor 1 \quad a \quad a^2 \rfloor, \qquad \mathbf{V}_F^T = \lfloor V_1 \quad V_2 \quad V_3 \rfloor_F,$$

$$\mathbf{P}_E = \lfloor 1 \quad a \quad a^2 \mid a^3 \rfloor, \qquad \mathbf{V}_E^T = \lfloor V_1 \quad V_2 \quad V_3 \mid V_4 \rfloor_E\,.$$

Selecting the nodes at the standard positions of $a_1 = -1$, $a_2 = 0$, and $a_3 = 1$ gives:

$$\mathbf{A}_F = \begin{bmatrix} 1 & -1 & 1 \\ 1 & 0 & 0 \\ 1 & 1 & 1 \end{bmatrix}, \qquad \mathbf{A}_F^{-1} = \frac{1}{2}\begin{bmatrix} 0 & 2 & 0 \\ -1 & 0 & 1 \\ 1 & -2 & 1 \end{bmatrix},$$

$$\mathbf{A}_E = \left[\begin{array}{ccc|c} 1 & -1 & 1 & -1 \\ 1 & 0 & 0 & 0 \\ 1 & 1 & 1 & 1 \end{array}\right], \qquad \mathbf{C}_E = \begin{Bmatrix} -1 \\ 0 \\ 1 \end{Bmatrix},$$

$$\mathbf{A}_F^{-1}\,\mathbf{C}_E = \begin{Bmatrix} 0 \\ 1 \\ 0 \end{Bmatrix} \equiv \mathbf{E}, \quad \mathbf{K} = \left[\begin{array}{ccc|c} 1 & 0 & 0 & 0 \\ 0 & 1 & 0 & 1 \\ 0 & 0 & 1 & 0 \end{array}\right].$$

For an interpolate solution, the two equivalent forms are exact at the three nodes $(a = \pm 1, \ a = 0)$ and inside the element. Then the product expands to $\mathbf{P}_F\,\mathbf{K} = [\,1 \quad a \quad a^2 \quad a\,]$. Only the last polynomial term differs from $\mathbf{P}_E$. By inspection we see that term is $a\,V_4 = a^3\,V_4$ which is valid when a is evaluated at any of the three nodes. Equating the first derivatives at the optimum point a_0,

$$\lfloor\, 0 \quad 1 \quad 2a_0 \quad 1 \,\rfloor = \lfloor\, 0 \quad 1 \quad 2a_0 \quad 3a_0^2 \,\rfloor,$$

or simply $1 = 3a_0^2$ so that $a_0 = \pm 1/\sqrt{3} = \pm 0.577$. These are the usual Gauss points used in the two point integration rule. Similarly, the *optimal location*, a_s, for the second derivative is found from $\lfloor\, 0 \quad 0 \quad 2 \quad 0 \,\rfloor = \lfloor\, 0 \quad 0 \quad 2 \quad 6a_s \,\rfloor$, so that $a_s = 0$, the center of the element. The same sort of procedure can be applied to 2-D and 3-D elements. Generally, we find that derivative estimates are least accurate at the nodes. The derivative estimates are usually most accurate at the tabulated integration points. That is indeed fortunate, since it means we get a good approximation of the element square matrix. The typical sampling positions for the C^0 quadratic elements are shown in Fig. 3.12. The C^1 line elements have the same points except that the function and slope are most accurate at the end points while the best second and third derivative locations are at the marked interior points. It is easy to show that the center of the linear element is the optimum position for sampling the first derivative. Since the front of partition **K** is an identity matrix, **I**, we are really saying that an exact nodal interpolate solution implies that $\mathbf{P}_F\,(a)\,\mathbf{A}_F^{-1}\,\mathbf{C}_E = a^n$. Let the vector $\mathbf{A}_F^{-1}\,\mathbf{C}_E$ denote an extrapolation vector, say **E**. Then, the derivatives would be the same in the two systems at points where

$$\left(\frac{d^k}{da^k}\,\mathbf{P}_F\,(a)\right)\mathbf{E} = \left(\frac{d^k}{da^k}\,a^n\right), \qquad 0 \le k \le n-1. \tag{3.40}$$

For example, the above quadratic element interpolate of a cubic solution gave

$$\begin{array}{ll} k = 0, & [\,1 \quad a \quad a^2\,] \\ k = 1, & [\,0 \quad 1 \quad 2a\,] \\ k = 2, & [\,0 \quad 0 \quad 2\,] \end{array} \begin{Bmatrix} 0 \\ 1 \\ 0 \end{Bmatrix} = \begin{matrix} a^3 \\ 3\,a^2 \\ 6\,a \end{matrix}$$

which are only satisfied for

k	a_k
0	$-1,\ 0,\ +1$
1	$\pm 1/\sqrt{3}$
2	0

which are the locations shown for the line element in Fig. 3.12. That figure also illustrates that the first derivatives are usually most accurate at the quadrature points.

3.11 Exercises

1. For a quadratic element, with $J^e = dx/dr = L^e$ use the unit coordinate interpolation in Fig. 3.4 to evaluate the matrices:

$$a)\ \ \mathbf{C}^e = \int_{L^e} \mathbf{H}^T\, dx,\ \ b)\ \ \mathbf{M}^e = \int_{L^e} \mathbf{H}^T\, \mathbf{H}\, dx,$$

$$c)\ \ \mathbf{S}^e = \int_{L^e} \frac{d\mathbf{H}^T}{dx}\, \frac{d\mathbf{H}}{dx}\, dx,\ \ d)\ \ \mathbf{U}^e = \int_{L^e} \mathbf{H}^T\, \frac{d\mathbf{H}}{dx}\, dx\,.$$

   Also give the sum of all of the coefficients of each matrix.

2. Solve the above problem by using the natural coordinate version, $-1 \le n \le 1$, from Fig, 3.4.

3. Referring to Fig. 3.4 verify, in both coordinate systems, that $\sum H_j = 1$ for the a) linear, b) quadratic, c) cubic interpolations.

4. Referring to the linear interpolations in Fig. 3.4 verify that $H_j(r_i) = \delta_{ij}$ for both local coordinate systems.

5. For a quadratic (3 node) line element in parametric space assume the the solution value is constant, say c, at each node. Write and simplify the analytic interpolated value in terms of the parametric coordinate. Also obtain the local (parametric) derivative of the interpolated value.

6. Problem 2.13 involved 3 elements and 4 degrees of freedom. We could have used a single 4 node cubic element instead. If you do that the 2 internal (non-zero) node values are $u_2 = 0.055405, u_3 = 0.068052$. Use these computed values with the cubic interpolation functions in Fig. 3.4 to plot the single element solution in comparison to the exact solution. Also plot the element and exact gradient.

7. The beam element of Problem 2.14 requires C^1 continuity provided by the cubic Hermite in Fig. 3.6. The element stiffness matrix and resultant generalized load vector are

$$\mathbf{S}^e = \int_{L^e} \mathbf{B}^e(x)^T\ EI^e(x)\,\mathbf{B}^e(x)\,dx\,,\quad \mathbf{C}^e_p = \int_{L^e} \mathbf{H}^e(x)^T\ p^e(x)\,dx\,,$$

   where

$$\mathbf{B}^e = \frac{d^2\mathbf{H}}{dx^2} = \frac{1}{(L^e)^2}\frac{d^2\mathbf{H}}{dr^2} = \frac{1}{L^2}\Big[(12r-6)\;\; L(6r-4)\;\; (6-12r)\;\; L(6r-2)\Big].$$

a) Verify that the results for a cubic element are:

$$\mathbf{S}^e = \frac{EI}{L^3}\begin{bmatrix} 12 & & & \text{sym.} \\ 6L & 4L^2 & & \\ -12 & -6L & 12 & \\ 6L & 2L^2 & -6L & 4L^2 \end{bmatrix}, \quad \mathbf{C}_p^e = p^e L \begin{Bmatrix} 1/2 \\ L/12 \\ 1/2 \\ -L/12 \end{Bmatrix}$$

where L denotes the element length, and p^e is assumed constant. b) Assume $p(x)$ varies linearly from p_1^e to p_2^e at the nodes of the element and verify that

$$\mathbf{C}_p^e = \frac{L}{20}\begin{bmatrix} 7 & 3 \\ L & 2L/3 \\ 3 & 7 \\ -2L/3 & -L \end{bmatrix}\begin{Bmatrix} p_1 \\ p_2 \end{Bmatrix}^e .$$

8. Use the 3 node element interpolation of Eq. 3.17 in the geometry mapping of Eq. 3.14 to evaluate the local Jacobian $J^e(r)$. a) Show that it will not be constant except for the special case where the interior node is exactly in the middle of the element in physical space, $x_2^e = (x_1^e + x_3^e)/2$. b) Evaluate J^e if the interior node is placed at the quarter length position instead.

9. Solve Problem P2.17 using the least squares finite element method instead.

10. A bar hanging under its own weight has its axial deflection, u, governed by $EA\, d^2u/dx^2 + \gamma A = 0$ over the length, $0 \le x \le L$, Where E is the material's elastic modulus, γ its weight per unit volume, and A is the cross-sectional area. The top point is restrained, $u(0) = 0$. The stress on any cross-section is $\sigma = E\, du/dx$. The free end (at $x = L$) is stress free so $du/dx(L) = 0$. Assume a constant area, A, so that the exact deflection is $\gamma L^2/2E$. a) Analytically solve this problem with one quadratic element where the stiffness matrix and resultant force vector are:

$$\mathbf{S}^e = \frac{E^e A^e}{3\,L^e}\begin{bmatrix} 7 & -8 & 1 \\ -8 & 16 & -8 \\ 1 & -8 & 7 \end{bmatrix}, \quad \mathbf{C}^e = \frac{\gamma^e A^e L^e}{6}\begin{Bmatrix} 1 \\ 4 \\ 1 \end{Bmatrix},$$

and the local derivative, du/dr, can be obtained from Eq. 3.14. Compute the end and mid-length deflections, and the reaction force at the top (which should be equal and opposite to the total weight $W = \gamma AL$). Also recover the stress values at the two ends and the mid-length. b) Repeat the study with two linear elements, c) compare the two solutions.

3.12 Bibliography

[1] Abramowitz, M. and Stegun, I.A., *Handbook of Mathematical Functions*, National Bureau of Standards (1964).

[2] Akin, J.E., *Finite Elements for Analysis and Design*, London: Academic Press (1994).

[3] Babuska, I., Griebel, M., and Pitkaranta, J., "The Problem of Selecting Shape Functions for a *p*-Type Finite Element," *Int. J. Num. Meth. Eng.*, **28**, pp. 1891–1908 (1989).

[4] Bank, R.E., "Hierarchical Bases and the Finite Element Method," *Acta Numerica*, **5**, pp. 1–45 (1996).

[5] Barlow, J., "Optimal Stress Locations in Finite Element Models," *Int. J. Num. Meth. Eng.*, **10**, pp. 243–251 (1976).

[6] Bruch, J.C. Jr. and Zyvoloski, G., "A Finite Element Weighted Residual Solution to One-Dimensional Field Problems," *Int. J. Num. Meth. Eng.*, **6**, pp. 577–585 (1973).

[7] Bruch, J.C. Jr. and Zyvoloski, G., "Transient Two-Dimensional Heat Conduction Problems Solved by the Finite Element Method," *Int. J. Num. Meth. Eng.*, **8**, pp. 481–494 (1974).

[8] Desai, C.S. and Kundu, T., *Introduction to the Finite Element Method*, Boca Raton: CRC Press (2001).

[9] Dettmer, W. and Peric, D., "An Analysis of the Time Integration Algorithms for the Finite Element Solutions of Incompressible Navier-Stokes Equations Based on a Stabilised Formulation," *Comp. Meth. Appl. Mech. Eng.*, **192**, pp. 1177–1226 (2003).

[10] Krishnamoorthy, C.S., *Finite Element Analysis: Theory and Programming*, New York: McGraw-Hill (1994).

[11] Szabo, B. and Babuska, I., *Finite Element Analysis*, New York: John Wiley (1991).

[12] Tezduyar, T.E. and Ganjoo, D.K., "Petrov-Galerkin Formulations with Weighting Functions Dependent Upon Spatial and Temporal Discretization," *Comp. Meth. Appl. Mech. Eng.*, **59**, pp. 47–71 (1986).

[13] Tezduyar, T.E., "Stabilized Finite Element Formulations for Incompressible Flow Computations," *Advances in Applied Mechanics*, **28**, pp. 1–44 (1991).

[14] Tong, P. and Rossettos, J.N., *Finite Element Method: Basic Techniques and Implementation*, MIT Press (1977).

[15] Zienkiewicz, O.C. and Taylor, R.L., *The Finite Element Method,* 4th Edition, New York: McGraw-Hill (1991).

[16] Zienkiewicz, O.C. and Taylor, R.L., *The Finite Element Method,* 5th Edition, London: Butterworth-Heinemann (2000).

Chapter 4

One-dimensional integration

4.1 Introduction

Since the finite element method is based on integral relations it is logical to expect that one should strive to carry out the integrations as efficiently as possible. In some cases we will employ exact integration. In others we may find that the integrals can become too complicated to integrate exactly. In such cases the use of numerical integration will prove useful or essential. The important topics of local coordinate integration and Gaussian quadratures will be introduced here. They will prove useful when dealing with higher order interpolation functions in complicated element integrals.

4.2 Local coordinate Jacobian

We have previously seen that the utilization of local element coordinates can greatly reduce the algebra required to establish a set of interpolation functions. Later we will see that some 2-D elements must be formulated in local coordinates in order to meet the interelement continuity requirements. Thus, we should expect to often encounter local coordinate interpolation. However, the governing integral expressions must be evaluated with respect to a unique global or physical coordinate system. Clearly, these two coordinate systems must be related. The relationship for integration with a change of variable (change of coordinate) was defined in elementary concepts from calculus. At this point it would be useful to review these concepts from calculus. Consider a definite integral

$$I = \int_a^b f(x)\, dx, \qquad a < x < b \tag{4.1}$$

where a new variable of integration, r, is to be introduced such that $x = x(r)$. Here it is required that the function $x(r)$ be continuous and have a continuous derivative in the interval $\alpha \le r \le \beta$. The region of r directly corresponds to the region of x such that when r varies between α and β, then x varies between $a = x(\alpha)$ and $b = x(\beta)$. In that case

$$I = \int_a^b f(x)\, dx = \int_\alpha^\beta f(x(r))\, \frac{dx}{dr}\, dr \tag{4.2}$$

or
$$I = \int_{\alpha}^{\beta} f(r)\, J\, dr \tag{4.3}$$

where $J = dx/dr$ is called the *Jacobian* of the coordinate transformation.

4.3 Exact polynomial integration *

If we utilize the unit coordinates, then $\alpha = 0$ and $\beta = 1$. Then from Sec. 3.2, the Jacobian is $J = L^e$ in an element domain defined by linear interpolation. By way of comparison, if one employs natural coordinates, then $\alpha = -1$, $\beta = +1$, and from Eq. 3.16 $J = L^e/2$. Generally, we will use interpolation functions that are polynomials. Thus, the element integrals of them and/or their derivatives will also contain polynomial terms. Therefore, it will be useful to consider expressions related to typical polynomial terms. A typical polynomial term is r^m where m is an integer. Thus, from the above

$$I = \int_{x_1^e}^{x_2^e} r^m\, dx = \int_0^1 r^m\, L^e\, dr = L^e \left. \frac{r^{(1+m)}}{1+m} \right|_0^1 = \frac{L^e}{(1+m)} . \tag{4.4}$$

A similar expression can be developed for the natural coordinates. It gives

$$I = \int_{L^e} n^m\, dx = \frac{L^e}{m+1} \begin{cases} 0 & \text{if } m \text{ is odd} \\ 1 & \text{if } m \text{ is even .} \end{cases} \tag{4.5}$$

Later we will tabulate the extension of these concepts to two- and three-dimensional integrals. As an example of the use of Eq. 4.5, consider the integration of the outer product of $\mathbf{H}^e$ with itself:

$$I = \int_{L^e} \mathbf{H}^{e^T}\, \mathbf{H}^e\, dx .$$

Recall that the integral of a matrix is the matrix resulting from the integration of each of the elements of the original matrix. If linear interpolation is selected for $\mathbf{H}^e$ on a line element then typical terms will include H_1^2, $H_1\, H_2$, etc. Thus, one obtains:

$$I_{11} = \int_{L^e} H_1^2(r)\, dx = \int_{L^e} (1-r)^2\, dx = L^e(1 - 2/2 + 1/3) = L^e/3$$

$$I_{12} = \int_{L^e} H_1\, H_2\, dx = \int_{L^e} (1-r)\, r dx = L^e\,(1/2 - 1/3) = L^e/6$$

$$I_{22} = \int_{L^e} H_2^2\, dx = \int_{L^e} r^2\, dx = L^e/3$$

so that
$$I = \frac{L^e}{6} \begin{bmatrix} 2 & 1 \\ 1 & 2 \end{bmatrix} . \tag{4.6}$$

Similarly, if one employs the Lagrangian quadratic $\mathbf{H}$ in Eq. 3.17 one obtains:

$$I = \frac{L^e}{30} \begin{bmatrix} 4 & -1 & 2 \\ -1 & 4 & 2 \\ 2 & 2 & 16 \end{bmatrix} . \tag{4.7}$$

By way of comparison, if one selects the hierarchical quadratic polynomial in Eq. 3.22 the above integral becomes

$$I = \frac{L^e}{6}\left[\begin{array}{cc|c} 2 & 1 & -1/4 \\ 1 & 2 & -1/4 \\ - & - & - \\ -1/4 & -1/4 & 1/10 \end{array}\right].$$

Note that the top left portion of this equation is the same as Eq. 4.6 which was obtained from the linear polynomial. This desirable feature of hierarchical elements was mentioned in Sec. 3.6.

Of course, all of the above integrals are so simple one could note that the integral is simply the area under the curve and there are simple algebraic relations for defining the area under power law curves. Figure 4.1 reviews such relations and illustrates the terms needed in Eq. 4.6.

Before leaving the subject of simplex integrations one should give consideration to the common special case of axisymmetric geometries, with coordinates (*rho, z*). Recall from calculus that the Theorem of Pappus relates a differential volume and surface area to a differential area and length in the (*rho, z*) plane of symmetry, respectively. That is, $dV = 2\pi\rho \, dA$ and $dS = 2\pi\rho dl$, where ρ denotes the radial distance to the differential element. Thus, typical axisymmetric surface integrals reduce to

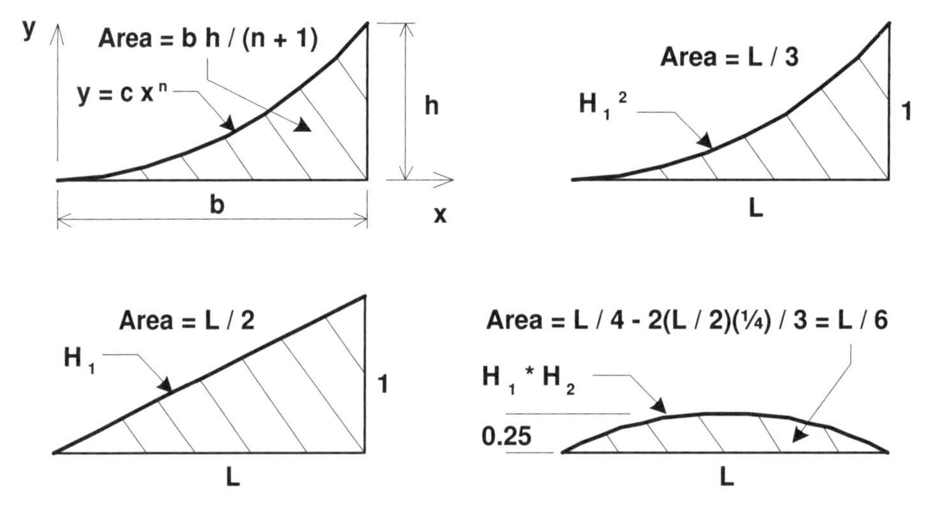

Figure 4.1 *Exact integrals by inspection*

$$\mathbf{I}_\Gamma = \int_\Gamma \mathbf{H}^T \, d\Gamma = 2\pi \int_{L^e} \mathbf{H}^T \rho \, dl = 2\pi \left(\int_{L^e} \mathbf{H}^T \mathbf{H} \, dl \right) \rho^e = \frac{2\pi L^e}{6} \begin{bmatrix} 2 & 1 \\ 1 & 2 \end{bmatrix} \rho^e ,$$

since $\rho = \mathbf{H}\rho^e$. Many workers like to omit the 2π term and work on a per-unit-radian basis so that they can more easily do both two-dimensional and axisymmetric calculations with a single program.

4.4 Numerical integration

Numerical integration is simply a procedure that approximates (usually) an integral by a summation. To review this subject we refer to Fig. 4.2. Recall that the integral

$$I = \int_a^b f(x) \, dx \tag{4.8}$$

can be viewed graphically as the area between the x-axis and the curve $y = f(x)$ in the region of the limits of integration. Thus, we can interpret numerical integration as an approximation of that area. The *trapezoidal rule* of numerical integration simply approximates the area by the sum of several equally spaced trapezoids under the curve between the limits of a and b. The height of a trapezoid is found from the integrand, $y_j = y(x_j)$, evaluated at equally spaced points, x_j and x_{j+1}. Thus, a typical contribution is $A = h(y_j + y_{j+1})/2$, where $h = x_{j+1} - x_j$ is the spacing. Thus, for n_q points (and $n_q - 1$ spaces), the well-known approximation is

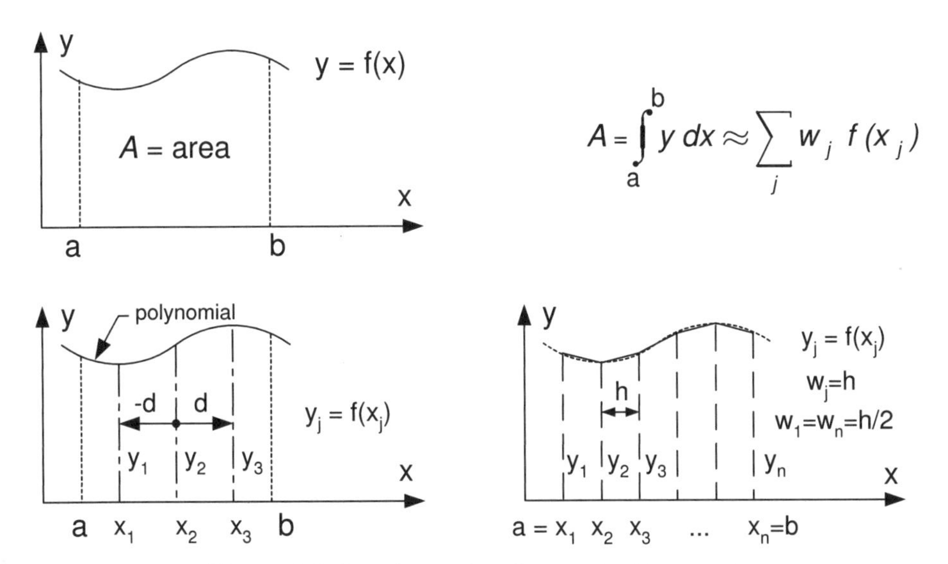

Figure 4.2 *One-dimensional numerical integration*

$$I \approx h\left(\frac{1}{2}\, y_1 + y_2 + y_3 + \cdots + y_{n-1} + \frac{1}{2}\, y_n\right), \quad I \approx \sum_{j=1}^{n} w_j\, f(x_j) \tag{4.9}$$

where $w_j = h$, except $w_1 = w_n = h/2$. A geometrical interpretation of this is that the area under curve, I, is the sum of the products of certain heights, $f(x_j)$ times some corresponding widths, w_j. In the terminology of numerical integration, the locations of the points, x_j, where the heights are computed are called *abscissae* and the widths, w_j, are called *weights*. Another well-known approximation is the *Simpson rule*, which uses parabolic segments in the area approximation. For most functions the above rules may require 20 to 40 terms in the summation to yield acceptable accuracy. We want to carry out the summation with the minimum number of terms, n_q, in order to reduce the computational cost. What is the minimum number of terms? The answer depends on the form of the integrand $f(x)$. Since the parametric geometry usually involves polynomials we will consider that common special case for $f(x)$.

The famous mathematician Gauss posed this question: What is the minimum number of points, n_q, required to exactly integrate a polynomial, and what are the corresponding abscissae and weights? If we require the summation to be exact when $f(x)$ is any one of the power functions $1, x, x^2, \ldots, x^{2n-1}$, we obtain a set of $2n$ conditions that allow us to determine the n_q abscissae, x_i, and their corresponding n_q weights, w_j. The n_q *Gaussian quadrature* points are symmetrically placed with respect to the center of the interval, and will exactly integrate a polynomial of order $(2n_q - 1)$. The center point location is included in the abscissae list when n_q is odd, but the end points are never utilized in a Gauss rule. The Gauss rule data are usually tabulated for a non-dimensional *unit coordinate* range of $0 \le t \le 1$, or for a *natural coordinate* range of $-1 \le t \le +1$. Table 4.1 presents the low-order Gauss rule data in natural coordinates, and the alternate unit coordinate data are in Table 4.2. A two-point Gauss rule can often exceed the accuracy of a 20-point trapezoidal rule. When computing norms of an exact solution, to be compared to the finite element solution, we often use the trapezoidal rule. That is because the exact solution is usually not a polynomial and the Gauss rule may not be accurate.

Sometimes it is desirable to have a numerical integration rule that specifically includes the two end points in the abscissae list when $(n \ge 2)$. The *Lobatto rule* is such an alternate choice. Its n_q points will exactly integrate a polynomial of order $(2n - 3)$ for $n_q > 2$. Its data are included in Table 4.3. It is usually less accurate than the Gauss rule but it can be useful. Mathematical handbooks give tables of Gauss or Lobatto data for much higher values of n_q. Some results of Gauss's work are outlined below. Let y denote $f(x)$ in the integral to be computed. Define a change of variable

$$x(n) = 1/2\,(b - a)\, n + 1/2\,(b + a) \tag{4.10}$$

so that the non-dimensional limits of integration of n become -1 and $+1$. The new value of $y(n)$ is

$$y = f(x) = f\,[\,1/2\,(b - a)\, n + 1/2\,(b + a)\,] = \Phi(n)\,. \tag{4.11}$$

Noting from Eq. 4.10 that $dx = 1/2\,(b - a)\, dn$, the original integral becomes

Table 4.1 *Abscissas and weights for Gaussian quadrature*

$$\int_{-1}^{+1} f(x)\,dx = \sum_{i=1}^{n_q} w_i\, f(x_i)$$

$\pm\, x_i$		w_i
0.00000 00000 00000 00000 0000	$n_q = 1$	2.00000 00000 00000 00000 000
0.57735 02691 89625 76450 9149	$n_q = 2$	1.00000 00000 00000 00000 000
0.77459 66692 41483 37703 5835	$n_q = 3$	0.55555 55555 55555 55555 556
0.00000 00000 00000 00000 0000		0.88888 88888 88888 88888 889
0.86113 63115 94052 57522 3946	$n_q = 4$	0.34785 48451 37453 85737 306
0.33998 10435 84856 26480 2666		0.65214 51548 62546 14262 694
0.90617 98459 38663 99279 7627	$n_q = 5$	0.23692 68850 56189 08751 426
0.53846 93101 05683 09103 6314		0.47862 86704 99366 46804 129
0.00000 00000 00000 00000 0000		0.56888 88888 88888 88888 889
0.93246 95142 03152 02781 2302	$n_q = 6$	0.17132 44923 79170 34504 030
0.66120 93864 66264 51366 1400		0.36076 15730 48138 60756 983
0.23861 91860 83196 90863 0502		0.46791 39345 72691 04738 987
0.94910 79123 42758 52452 6190	$n_q = 7$	0.12948 49661 68869 69327 061
0.74153 11855 99394 43986 3865		0.27970 53914 89276 66790 147
0.40584 51513 77397 16690 6607		0.38183 00505 05118 94495 037
0.00000 00000 00000 00000 0000		0.41795 91836 73469 38775 510

Table 4.2 *Unit abscissas and weights for Gaussian quadrature*

$$\int_{0}^{1} f(x)\,dx = \sum_{i=1}^{n_q} w_i\, f(x_i)$$

x_i		w_i
0.50000 00000 00000 00000 000	$n_q = 1$	1.00000 00000 00000 00000 000
0.21132 48654 05187 11774 543	$n_q = 2$	0.50000 00000 00000 00000 000
0.78867 51345 94812 88225 457		0.50000 00000 00000 00000 000
0.11270 16653 79258 31148 208	$n_q = 3$	0.27777 77777 77777 77777 778
0.50000 00000 00000 00000 000		0.44444 44444 44444 44444 444
0.88729 83346 20741 68851 792		0.27777 77777 77777 77777 778
0.06943 18442 02973 71238 803	$n_q = 4$	0.17392 74225 68726 92868 653
0.33000 94782 07571 86759 867		0.32607 25774 31273 07131 347
0.66999 05217 92428 13240 133		0.32607 25774 31273 07131 347
0.93056 81557 97026 28761 197		0.17392 74225 68726 92868 653
0.04691 00770 30668 00360 119	$n_q = 5$	0.11846 34425 28094 54375 713
0.02307 65344 94715 84544 818		0.23931 43352 49683 23402 065
0.50000 00000 00000 00000 000		0.28444 44444 44444 44444 444
0.76923 46550 52841 54551 816		0.23931 43352 49683 23402 065
0.95308 99229 69331 99639 881		0.11846 34425 28094 54375 713

Table 4.3 *Abscissas and weight factors for Lobatto integration*

$$\int_{-1}^{+1} f(x)\, dx \approx \sum_{i=1}^{n_q} w_i\, f(x_i)$$

$\pm x_i$		w_i
0.00000 00000 00000	$n_q = 1$	2.00000 00000 00000
1.00000 00000 00000	$n_q = 2$	1.00000 00000 00000
1.00000 00000 00000	$n_q = 3$	0.33333 33333 33333
0.00000 00000 00000		1.33333 33333 33333
1.00000 00000 00000	$n_q = 4$	0.16666 66666 66667
0.44721 35954 99958		0.83333 33333 33333
1.00000 00000 00000	$n_q = 5$	0.10000 00000 00000
0.65465 36707 07977		0.54444 44444 44444
0.00000 00000 00000		0.71111 11111 11111
1.00000 00000 00000	$n_q = 6$	0.06666 66666 66667
0.76505 53239 29465		0.37847 49562 97847
0.28523 15164 80645		0.55485 83770 35486

$$I = \frac{1}{2}(b-a)\int_{-1}^{1} \Phi(n)\, dn\,. \tag{4.12}$$

Gauss showed that the integral in Eq. 4.12 is given by

$$\int_{-1}^{1} \Phi(n)\, dn = \sum_{i=1}^{n_q} W_i\, \Phi(n_i)\,,$$

where W_i and n_i represent tabulated values of the *weight functions* and *abscissae* associated with the n_q points in the non-dimensional interval (−1, 1). The final result is

$$I = \frac{1}{2}(b-a)\sum_{i=1}^{n_q} W_i\, \Phi(n_i) = \sum_{i=1}^{n_q} f(x(n_i))\, W_i\,. \tag{4.13}$$

Gauss also showed that this equation will exactly integrate a polynomial of degree $(2n_q - 1)$. For a higher number of space dimensions (which range from −1 to +1), one obtains a multiple summation. Since Gaussian quadrature data are often tabulated in references for the range $-1 \le n \le +1$, it is popular to use the natural coordinates in defining element integrals. However, one can convert the tabulated data to any convenient system such as the unit coordinate system where $0 \le r \le 1$. The latter may be more useful on triangular regions. As an example of Gaussian quadratures, consider the following one-dimensional integral:

$$I = \int_1^2 \begin{bmatrix} 2 & \dfrac{2x}{(1+2x^2)} \\ 2x & \end{bmatrix} dx = \int_1^2 \mathbf{F}(x)\, dx\,.$$

If two Gauss points are selected ($n_q = 2$), then the tabulated values from Table 4.1 give $W_1 = W_2 = 1$ and $r_1 = 0.57735 = -r_2$. The change of variable gives $x(r) = (r+3)/2$, so that $x(r_1) = 1.788675$ and $x(r_2) = 1.211325$. Therefore, from Eq. 4.13

$$I = \tfrac{1}{2}(2-1)\Big[\, W_1\,\mathbf{F}(\,x(r_1)\,) + W_2\,\mathbf{F}(\,x(r_2)\,)\,\Big]$$

$$= \tfrac{1}{2}(1)\left((1)\begin{bmatrix} 2 & 2(1.788675) \\ sym. & 1+2(1.788675)^2 \end{bmatrix} + (1)\begin{bmatrix} 2 & 2(1.211325) \\ sym. & 1+2(1.211325)^2 \end{bmatrix} \right)$$

$$I = \begin{bmatrix} 2.00000 & 3.00000 \\ 3.00000 & 5.66667 \end{bmatrix},$$

which is easily shown to be in good agreement with the exact solution. As another example consider a typical term in Eq. 4.8. Specifically, from Eqs. 3.16 and 3.18

$$I_{33} = \int_{L^e} H_3^2 \, dx = \frac{L^e}{2}\int_{-1}^{+1} (1-n^2)^2 \, dn \,.$$

Since the polynomial terms to be integrated are fourth order, we should select $(2n_q - 1) = 4$, or $n_q = 3$ for an integer number of Gaussian points. Then,

$$I_{33} = \int_{L^e} H_3^2 \, dx = \frac{L^e}{2}\int_{-1}^{+1} H_3^2\,(n)\, dn$$

$$I_{33} = \frac{L^e}{2}\bigg(0.55556\Big[\, 1.00000 - (-0.77459)^2 \,\Big]^2$$

$$+ \; 0.88889\Big[\, 1.00000 - (0.0)^2 \,\Big]^2 + 0.55556\Big[\, 1.00000 - (+0.77459)^2 \,\Big]^2 \bigg)$$

$$I_{33} = \frac{L^e}{2}(0.08889 + 0.88889 + 0.08889) = 0.5333\, L^e\,,$$

which agrees well with the exact value of 16 $L^e/30$, when using 6 digits.

4.5 Variable Jacobians

When the parametric space and physical space have the same number of dimensions then the Jacobian is a square matrix. Otherwise, we need to use more calculus to evaluate the integrals. For example, we often find the need to execute integrations along a two-dimensional curve defined by our one-dimensional parametric representation. Consider a planar boundary curve in the xy-plane such as that shown in Fig. 4.3. We may need to know its length and first moments (centroid), which are defined as

$$L = \int_L ds\,, \qquad \bar{x}\, L = \int_L x(r)\, ds, \qquad \bar{y}\, L = \int_L y(r)\, ds,$$

respectively, where ds denotes the physical length of a segment, dr, of the parametric length. To evaluate these quantities we need to convert to an integral in the parametric space. For example,

$$L = \int_L ds = \int_0^1 \frac{ds}{dr}\, dr\,.$$

To relate the physical and parametric length scales, we use the planar relation that

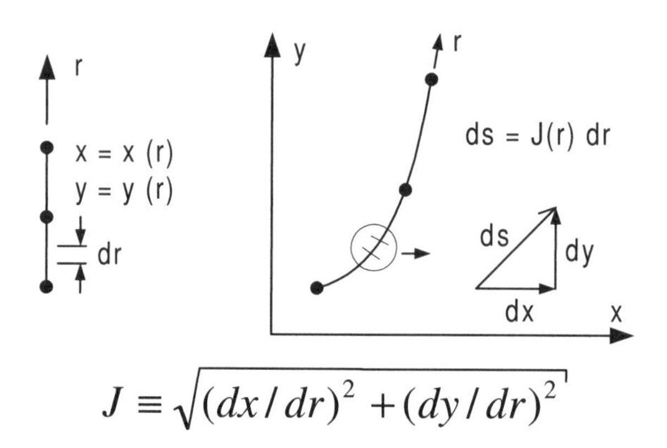

Figure 4.3 *A variable curve metric or Jacobian*

$ds^2 = dx^2 + dy^2$. Since both x and y are defined in terms of r, we can extend this identity to the needed quantity

$$\left(\frac{ds}{dr}\right)^2 = \left(\frac{dx}{dr}\right)^2 + \left(\frac{dy}{dr}\right)^2$$

where dx/dr can be found from the spatial interpolation functions, etc. for dy/dr. Thus, our physical length is defined in terms of the parametric coordinate, r, and the spatial data for the nodes defining the curve location in thc xy-planc (i.c., x^e and y^e) :

$$L = \int_0^1 \sqrt{\left(\frac{dx}{dr}\right)^2 + \left(\frac{dy}{dr}\right)^2}\, dr$$

where

$$\frac{dx(r)}{dr} = \sum_{i=1}^{n_b} \frac{d\,H_i(r)}{dr} x_i^e, \qquad \frac{dy(r)}{dr} = \sum_{i=1}^{n_b} \frac{d\,H_i(r)}{dr} y_i^e.$$

Note that this does preserve the proper units for L, and it has what we could refer to as a variable Jacobian. The preceding integral is trivial only when the planar curve is a straight line ($n_b = 2$). Then, from linear geometric interpolation $dx/dr = x_2^e - x_1^e$ and $dy/dr = y_2^e - y_1^e$ are both constant, and the result simplifies to

$$L^e = \sqrt{\left(x_2^e - x_1^e\right)^2 + \left(y_2^e - y_1^e\right)^2} \int_0^1 dr = L$$

which, by inspection, is exact. For any other curve shape, these integrals become unpleasant to evaluate, and we would consider their automation by means of numerical integration.

To complete this section we outline an algorithm to automate the calculation of the y-centroid:

1. Recover the n_b points describing the curve, $\mathbf{x}^e$, $\mathbf{y}^e$.

2. Recover the n_q-point quadrature data, r_j and w_j.

3. Zero the integrals: $L = 0$, $\bar{y}\, L = 0$.

4. Loop over all the quadrature points: $1 \le j \le n_q$. At each local quadrature point, t_j:

A. Find the length scales (i.e., Jacobian):

(i) Compute local derivatives of the n_b interpolation functions

$$\mathbf{DL\_H_j} \equiv \left. \frac{\partial \mathbf{H}}{\partial \mathrm{r}} \right|_{\mathrm{r=r_j}}$$

(ii) Get x- and y-derivatives from curve data

$$\frac{dx}{dr_j} \equiv \mathbf{DL\_H_j}\ \mathbf{x}^e = \sum_{i=1}^{n_b} \frac{\partial H_i(r_j)}{\partial r}\ x_i^e\ , \qquad \frac{dy}{dr_j} \equiv \mathbf{DL\_H_j}\ \mathbf{y}^e$$

(iii) Find length scale at point r_j

$$\frac{ds}{dr_j} = \sqrt{\left(\frac{dx}{dr}\right)_j^2 + \left(\frac{dy}{dr}\right)_j^2}$$

B. Evaluate the integrand at point r_j:

(i) Evaluate the n_b interpolation functions:

$$\mathbf{H}_j \equiv \mathbf{H}(r_j)$$

(ii) Evaluate y from curve data

$$y_j = \mathbf{H}_j\, \mathbf{y}^e = \sum_{i=1}^{N} H_i(r_j)\ y_i^e$$

C. Form products and add to previous values

$$L = L + \frac{ds}{dr_j}\, w_j, \qquad \bar{y}\, L = \bar{y}\, L + y_j\, \frac{ds}{dr_j}\, w_j .$$

5. Evaluate items computed from the completed integrals: $\bar{y} = \bar{y}\, L/L$, etc. for $\bar{x}$.

Note that to automate the integration we simply need: 1) storage of the quadrature data, r_j and w_j, 2) access to the curve data, $\mathbf{x}^e$, $\mathbf{y}^e$, 3) a subroutine to find the parametric derivative of the interpolation functions at any point, r, and 4) a function program to evaluate the integrand(s) at any point, r. This usually requires (see Step 4B) a subroutine to evaluate the interpolation functions at any point, r. The evaluations of the interpolation products, at a point r_j, can be thought of as a dot product of the data array $\mathbf{x}^e$ and the evaluated interpolation quantities $\mathbf{H}_j$ and $\mathbf{DL\_H_j}$.

4.6 Exercises

1. In Tables 4.1 and 4.3 the sum of the weights is exactly 2, but in Table 4.2 the sum is exactly 1. Explain why.

2. Note that the Gauss abscissas are always interior to the domain and symmetrically placed. The Lobatto rules for $n_q \geq 2$ always includes the two end points. Discuss an advantage or disadvantage of including the end points.

3. For a one-dimensional quadratic element, with a constant Jacobian, use Gaussian quadratures to numerically evaluate the matrices:

$$a)\ \ \mathbf{C}^e = \int_{L^e} \mathbf{H}^T \, dx, \qquad b)\ \ \mathbf{M}^e = \int_{L^e} \mathbf{H}^T \, \mathbf{H} \, dx,$$

$$c)\ \ \mathbf{S}^e = \int_{L^e} \frac{d\mathbf{H}^T}{dx} \frac{d\mathbf{H}}{dx} \, dx, \quad d)\ \ \mathbf{U}^e = \int_{L^e} \mathbf{H}^T \frac{d\mathbf{H}}{dx} \, dx\,.$$

4 Use the exact integrals for $\mathbf{S}^e$ and $\mathbf{M}^e$ from Problem 3 to hard code the equivalent of Fig. 2.11 for a three node quadratic line element. Note that such an element will also allow for $Q(x) = x^2$ as well as the original linear source term. In that case, of course, the exact solution changes.

5. Use Lobatto quadrature to evaluate the matrices in Problem 3. Note that $\mathbf{M}^e$ becomes a diagonal matrix.

4.7 Bibliography

[1] Abramowitz, M. and Stegun, I.A., *Handbook of Mathematical Functions*, National Bureau of Standards (1964).
[2] Carey, G.F. and Oden, J.T., *Finite Elements – Computational Aspects*, Englewood Cliffs: Prentice Hall (1984).
[3] Hinton, E. and Owen, D.R.J., *Finite Element Programming*, London: Academic Press (1977).
[4] Hughes, T.J.R., *The Finite Element Method*, Mineola: Dover Publications (2003).
[5] Smith, I.M. and Griffiths, D.V., *Programming the Finite Element Method,* 3rd Edition, Chichester: John Wiley (1998).
[6] Stroud, A.H. and Secrest, D., *Gaussian Quadrature Formulas*, Englewood Cliffs: Prentice Hall (1966).
[7] Zienkiewicz, O.C. and Taylor, R.L., *The Finite Element Method,* 5th Edition, London: Butterworth-Heinemann (2000).

Chapter 5

Error estimates for elliptic problems

5.1 Introduction

Having obtained a finite element solution, we would like to be able to estimate the error in that solution and, perhaps, have the analysis program correct itself. Currently, that is a practical option for an elliptic partial differential equation (PDE). Here we will outline the basic method and notation of that class of error estimation. Consider a problem posed by the PDE written as

$$\mathbf{L}\,\phi + Q \;=\; 0 \quad in\ \Omega \tag{5.1}$$

with the essential boundary condition $\phi = \phi_o$ on boundary Γ_ϕ, and a prescribed traction, or flux, $\mathbf{t} = \mathbf{t}_o$ on the boundary Γ_t with $\Gamma = \Gamma_\phi \cup \Gamma_t$. Here $\mathbf{L}$ is a linear differential operator that can usually be written in the symmetric form

$$\mathbf{L} \;\equiv\; D^T\,\mathbf{E}\,D \tag{5.2}$$

where D is a lower order operator and the symmetric constitutive array $\mathbf{E}$ contains material information. The gradient quantities of interest are denoted as

$$\boldsymbol{\varepsilon} \;\equiv\; D\,\phi \tag{5.3}$$

and the flux quantities, $\mathbf{q}$, by some constitutive relation

$$\mathbf{q} \;=\; \pm\,\mathbf{E}\,\boldsymbol{\varepsilon}. \tag{5.4}$$

On the boundary, Γ, of Ω we are often interested in a traction, $\mathbf{t}$, defined in terms of the fluxes by

$$\mathbf{t} \;=\; \mathbf{G}\,\mathbf{q} \tag{5.5}$$

where $\mathbf{G}$ is usually defined in terms of the components of the normal vector, $\mathbf{n}$.

For example, in isotropic conduction ϕ is the temperature, Q, in internal volumetric heat source, $\mathbf{E} = k\,\mathbf{I}$, where k is the thermal conductivity, $\mathbf{I}$ is the identity matrix, and D is simply the gradient operator

$$D^T = \nabla^T = \left[\frac{\partial}{\partial x} \quad \frac{\partial}{\partial y} \quad \frac{\partial}{\partial z} \right]$$

so that **L** becomes the Laplacian, $\mathbf{L} = \nabla^T k \, \mathbf{I} \, \nabla$, and for the common case of constant k:

$$\mathbf{L} = k\left(\frac{\partial^2}{\partial x^2} + \frac{\partial^2}{\partial y^2} + \frac{\partial^2}{\partial z^2} \right).$$

In this example $\boldsymbol{\varepsilon}$ is simply the gradient vector

$$\boldsymbol{\varepsilon} = \nabla \phi \, , \quad \boldsymbol{\varepsilon}^T = \left| \frac{\partial \phi}{\partial x} \quad \frac{\partial \phi}{\partial y} \quad \frac{\partial \phi}{\partial z} \right]$$

and the Fourier Law (note the negative sign) defines the heat flux vector

$$\mathbf{q} = -k \, \mathbf{I} \, \nabla \phi = -k \, \nabla \phi, \; q^T = [q_x \quad q_y \quad q_z].$$

Likewise, for $\mathbf{G} = \mathbf{n}$ the boundary traction is the normal heat flux:

$$\mathbf{t} = \mathbf{n} \cdot \mathbf{q} = q_x n_x + q_y n_y + q_z n_z = q_n = -k \, \partial \phi / \partial n.$$

For the one-dimensional case of heat conduction these all reduce to scalars with

$$D = \partial / \partial x, \; \mathbf{E} = k, \; \boldsymbol{\varepsilon} = \partial \phi / \partial x, \; n_x = \pm 1, \; \mathbf{q} = q_x = -k \, \partial \phi / \partial x, \; \mathbf{t} = \pm q_x,$$

and the governing differential equation $\mathbf{L}\phi + Q = 0$ becomes

$$\frac{\partial}{\partial x}\left(k \frac{\partial \phi}{\partial x} \right) + Q = 0$$

in Ω with $\phi = \phi_0$ on Γ_0. While on the boundary Γ_t, the traction $\mathbf{t} = q_n = -k \, n_x \, \partial \phi / \partial x$, and has an assigned value of $q_n = t_0$.

Likewise, for a problem in planar elasticity, $\boldsymbol{\phi}$ and $\boldsymbol{\varepsilon}$ become the displacement vector components $\boldsymbol{\phi} = [u \quad v]^T$ and strain matrix components $\boldsymbol{\varepsilon} = [\varepsilon_x \quad \varepsilon_y \quad \gamma]$, respectively, which are related by the differential operator

$$D = \begin{bmatrix} \frac{\partial}{\partial x} & 0 \\ 0 & \frac{\partial}{\partial y} \\ \frac{\partial}{\partial y} & \frac{\partial}{\partial x} \end{bmatrix}.$$

The corresponding fluxes or stress tensor components are $\mathbf{q} = \boldsymbol{\sigma} \equiv [\sigma_x \quad \sigma_y \quad \tau]^T$ which are related to the strains, $\boldsymbol{\varepsilon}$, by the symmetric Hooke's Law 'stress-strain' matrix, **E** (without a minus sign). The source **Q** generalizes to the body force vector $\mathbf{Q} = \mathbf{X} = [X_x \quad X_y]^T$. Finally, the surface traction vector $\mathbf{t} = \mathbf{T} = [T_x \quad T_y]$ is related to the surface stresses, $\boldsymbol{\sigma}$, and the components of the outward unit normal vector, **n**, by $\mathbf{T} = \mathbf{G} \, \boldsymbol{\sigma}$ where

$$\mathbf{G} = \begin{bmatrix} n_x & 0 & n_y \\ 0 & n_y & n_x \end{bmatrix}.$$

In a finite element method we seek a solution $\hat{\phi}$ which, in turn, yields the approximations for the gradient and flux terms, $\hat{\boldsymbol{\varepsilon}}$ and $\hat{\boldsymbol{q}}$. The standard interpolation gives

$$\phi \approx \hat{\phi} = \mathbf{N}(x)\,\mathbf{\Phi}^e \quad \mathbf{x} \textit{ in } \Omega^e \tag{5.6}$$

with a corresponding gradient estimate

$$\varepsilon \approx \hat{\varepsilon} = D\,\mathbf{N}(\mathbf{x})\,\mathbf{\Phi}^e \equiv \mathbf{B}^e(\mathbf{x})\,\mathbf{\Phi}^e \tag{5.7}$$

for $\mathbf{x}$ in Ω^e. Likewise, the flux approximation is

$$\sigma \approx \hat{\sigma} = \mathbf{E}^e\,\mathbf{B}^e(\mathbf{x})\,\mathbf{\Phi}^e. \tag{5.8}$$

In this notation the element square matrix and source vector are

$$\mathbf{S}^e = \int_{\Omega^e} \mathbf{B}^{e^T}\,\mathbf{E}^e\,\mathbf{B}^e\,d\Omega, \quad \mathbf{C}^e_Q = \int_{\Omega^e} Q^e\,\mathbf{N}^{e^T}\,d\Omega \tag{5.9}$$

and the boundary traction contribution, if any, is

$$\mathbf{C}^b_{q_n} = \int_{\Gamma^b} q^b_n\,\mathbf{N}^{b^T}\,d\Gamma. \tag{5.10}$$

When the element degrees of freedom subset, $\mathbf{\Phi}^e \subset \mathbf{\Phi}$, have been computed and gathered from substitution into Eqs. 5.6-8, the local errors in an element domain are

$$e_\phi(\mathbf{x}) \equiv \phi(\mathbf{x}) - \hat{\phi}(\mathbf{x}) \tag{5.11}$$

$$\mathbf{e}_\varepsilon(\mathbf{x}) \equiv \varepsilon(\mathbf{x}) - \hat{\varepsilon}(\mathbf{x}), \qquad \mathbf{x}\ \varepsilon\ \Omega^e \tag{5.12}$$

$$\mathbf{e}_\sigma(\mathbf{x}) \equiv \sigma(\mathbf{x}) - \hat{\sigma}(\mathbf{x}). \tag{5.13}$$

These quantities can be either positive or negative so we will mainly be interested in their absolute value or some normalized measure of them. We will employ integral *norms* for our error measures. On a linear space we can show that a norm has the properties given in Sec. 2.2. In finite elements we often use the *inner product* defined as

$$< u,\ v > \equiv \int_{\Omega} u(\mathbf{x})\,v(\mathbf{x})\,d\Omega \tag{5.14}$$

which possesses a natural norm defined as

$$\|\,\phi\,\|^2 = <\phi,\ \phi> = \int_{\Omega} \phi(\mathbf{x})\,\phi(\mathbf{x})\,d\Omega. \tag{5.15}$$

This is also called the L_2 *norm*, since it involves the integral of the square of the argument. We wish to minimize the error in the solution, e_ϕ. However, for elliptical problems it can be shown that this corresponds to minimizing the error energy norm, or other related measures. Error estimates commonly employ one of the following norms:

1. The error energy norm $\|\,e\,\|$ defined as

$$\| e \| = \left[\int_{\Omega} (\varepsilon - \hat{\varepsilon})^T \mathbf{E} (\varepsilon - \hat{\varepsilon}) \, d\Omega \right]^{\frac{1}{2}}$$

$$= \left[\int_{\Omega} (\varepsilon - \hat{\varepsilon})^T (\sigma - \hat{\sigma}) \, d\Omega \right]^{\frac{1}{2}} = \left[\int_{\Omega} \mathbf{e}_{\varepsilon}^T \mathbf{e}_{\sigma} \, d\Omega \right]^{\frac{1}{2}} \tag{5.16}$$

$$= \left[\int_{\Omega} (\sigma - \hat{\sigma})^T \mathbf{E}^{-1} (\sigma - \hat{\sigma}) \, d\Omega \right]^{\frac{1}{2}} .$$

2. The L_2 flux or stress error norm

$$\| e_{\sigma} \|_{L_2} = \left[\int_{\Omega} (\sigma - \hat{\sigma})^T (\sigma - \hat{\sigma}) \, d\Omega \right]^{\frac{1}{2}} = \left[\int_{\Omega} \mathbf{e}_{\sigma}^T \mathbf{e}_{\sigma} \, d\Omega \right]^{\frac{1}{2}} . \tag{5.17}$$

3. The root mean square stress error, $\Delta\sigma$, given by

$$\Delta\sigma = \| e_{\sigma} \|_{L_2} / \Omega^{\frac{1}{2}} . \tag{5.18}$$

4. In general, any of these norms is the sum of the corresponding element norms:

$$\| \phi \|^2 = \sum_{e} \| \phi \|_e^2, \quad \| \phi \|_e^2 = \int_{\Omega^e} \phi^2 \, d\Omega \tag{5.19}$$

   and the domain is the union of all of the element domains, $\Omega = \bigcup_{e} \Omega^e$.

A relative percentage error can be defined as $\eta = 100 \times \| e \| / \| \phi \|$ which represents a weighted root mean square percentage error in the stresses. We can compute a similar estimate relative to the L_2 norms. In most of the literature on the subject of error estimators there is a discussion of the effectivity index, Θ. It is simply the ratio of the estimated error divided by the exact error, $\Theta = \| e \|_{fea} / \| e \|_{exact}$. Usually an

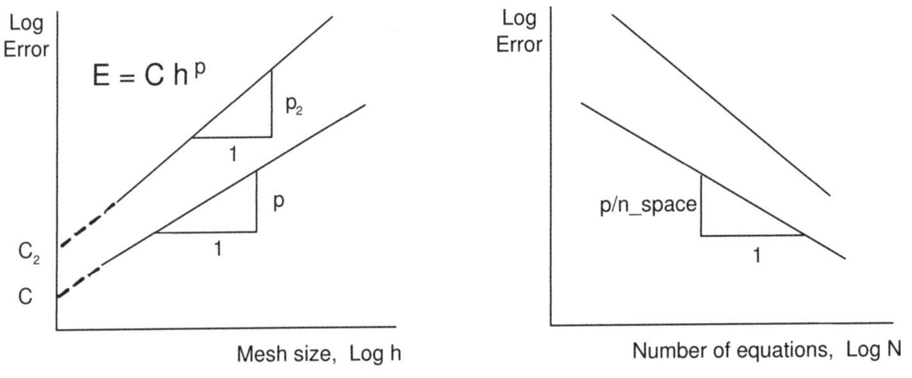

Figure 5.1 *Asymptotic convergence rates for finite elements*

analytical solution is employed to compute the exact error (and to assign the problem source, Q, and boundary conditions, ϕ_0), but sometimes very high precision numerical results are used. Clearly, one should search for methods where the effectivity index is very close to unity. Some methods employ a constant, determined by numerical experiment, to increase their effectivity index to near unity for a specific element type.

From studies in interpolation theory, the finite element approximation is known to converge in the energy norm when $\| e \| \le Ch^p$, for $p > 0$, where h is the distance between nodes on a uniform mesh (the characteristic element length), p is called the rate of convergence. The rate depends on the degree of the polynomial used to approximate ϕ, the order of the highest derivative of ϕ in the weak form, and whether there are local singularities in the domain. The constant, C, is independent of ϕ and will be influenced by the shape of the domain and whether Dirichlet or Neumann boundary conditions are employed. Typically $p = k + 1 - m > 0$ where k is the degree of the highest complete polynomial used in the interpolation and m is the order of the highest derivative of ϕ in the weak form. Remember that simplex elements always use complete polynomials by Lagrangian and Serendipity elements use incomplete polynomials. Note that the above equation for the error would be a straight line plot for a log-log plot of error versus mesh size, as shown in Fig. 5.1. In that case the slope of the line is the rate of convergence, p. Such a convergence relation can also be expressed in terms of the number of equations associated with the mesh. In a one-dimensional problem the number of equations, N, is proportional to $1/h$ while in two-dimensions it depends on $1/h^2$, etc. Then the (absolute value of) the slope of the line is the convergence rate divided by the dimension of the space.

5.2 Error estimates

In general, we do not know the exact strain, $\boldsymbol{\varepsilon}$, or stress values, $\boldsymbol{\sigma}$, in Eqs. 5.3 and 5.4. We do have piecewise continuous estimates for the element strains, $\hat{\boldsymbol{\varepsilon}}$, and stresses, $\hat{\boldsymbol{\sigma}}$, in the element interiors. However, unlike the solution ϕ, these estimates are generally discontinuous between elements. For homogeneous domains (homogeneous $\mathbf{E}$), we expect the exact $\boldsymbol{\varepsilon}$ and $\boldsymbol{\sigma}$ to be continuous. At the interface of two different homogeneous materials ($\mathbf{E}_1$ and $\mathbf{E}_2$), we expect the gradients, $\boldsymbol{\varepsilon}$, to be discontinuous and the fluxes, $\boldsymbol{\sigma}$, to usually be continuous normal to the interface of the two materials (or continuous tangent to the interface in some electromagnetic applications). In most elliptical problems, we expect the normal flux component to be continuous, but the tangential component along the interface may be discontinuous. For some electromagnetic problems the reverse is true for interface flux components. In a homogeneous domain a continuous estimate of $\boldsymbol{\varepsilon}$ and $\boldsymbol{\sigma}$ should be more accurate than would be the piecewise continuous $\hat{\boldsymbol{\varepsilon}}$ and $\hat{\boldsymbol{\sigma}}$. Denote such continuous approximations by $\boldsymbol{\varepsilon}^*$ and $\boldsymbol{\sigma}^*$, respectively. That is, $\hat{\boldsymbol{\sigma}}$ is discontinuous across element boundaries, while the $\boldsymbol{\sigma}^*$ are constructed to be continuous across those boundaries. Then, within an element, the error estimators with good accuracy are

$$\mathbf{e}_{\varepsilon} \approx \boldsymbol{\varepsilon}^*(\mathbf{x}) - \hat{\boldsymbol{\varepsilon}}(\mathbf{x}), \quad \mathbf{e}_{\sigma} \approx \boldsymbol{\sigma}^*(\mathbf{x}) - \hat{\boldsymbol{\sigma}}(\mathbf{x}). \tag{5.20}$$

There are various procedures for obtaining nodal values of the strains, ε^*, or stresses, σ^*, that will yield a continuous solution over the domain. Probably the most common early approach was simply an averaging based on the number and/or size of elements contributing to a node. The continuous nodal stresses were obtained by averaging the values from surrounding elements. However, this simple averaging process does not have any mathematical foundation relative to the original problem and can not be used as part of an effective error estimator. A precise mathematical procedure for computing the nodal values directly was given early in the development of finite element methods by Oden, et al. [8, 17]. However, that 'Conjugate Stress' approach required the assembly of element contributions and solving a system of equations equal in size to the number of nodes in the system. More recently for elliptical problems it has been shown that a Super-Convergent Patch (SCP) of elements provides a way to recover accurate continuous nodal fluxes or nodal gradients that can be used in an error estimator. Ainsworth and Oden [3] have carried out an extensive review of the most useful error estimation techniques. They consider both elliptical equations and other classes of problems such as the Navier-Stokes equations.

In Chapter 2 we showed that a patch based averaging process is one way to estimate the value of σ^*. While we will employ mainly that SCP approach some other methods have proven practical. We will look briefly at hierarchical and flux balancing methods as alternate ways of estimating the error. Then we will follow with a chapter outlining the details of the super-convergence patch averaging and error estimation.

5.3 Hierarchical error indicator

Zienkiewicz and Morgan [31] have given a detailed study of how hierarchical interpolation functions can be employed to compute an error estimate. Here we will outline this approach in one-dimension. They define the error norm as

$$\| e \|_E^2 = -\int_{\Omega} e\, r\, d\Omega \tag{5.21}$$

where the error is $e = \phi - \hat{\phi}$ and r is the residual error on the interior of the domain

$$L\hat{\phi} + q = r \neq 0. \tag{5.22}$$

Now we enrich the current approximate solution $\hat{\phi}$ to get a more accurate (higher degree) approximation by adding the next hierarchical bubble function $\phi^* = \hat{\phi} + H_b\, a_b$ where a_b is the next unknown hierarchical degree of freedom. If we take this as representing the correction solution ($\phi \approx \phi^*$), then we have $e^e = H_b^e\, a_b$ and

$$\| e^e \|_E = a_b \int_{\Omega^e} H_b^{e^T}\, r^e\, d\Omega. \tag{5.23}$$

If one can estimate the degree of freedom a_b, then we have an error indicator. If it is the only new dof, and if the hierarchical functions are orthogonal, the new system equilibrium equations are

$$\begin{bmatrix} \mathbf{S} & \mathbf{0} \\ \mathbf{0} & s_{bb} \end{bmatrix} \begin{Bmatrix} \mathbf{a} \\ a_b \end{Bmatrix} = \begin{Bmatrix} \mathbf{C} \\ c_b \end{Bmatrix}, \tag{5.24}$$

where $\mathbf{S}$ and $\mathbf{C}$ were the previous system matrices, and s_{bb} and c_b are the new element (and system) stiffness and source terms, respectively. From this diagonal system, we compute the new term $a_b = c_b / s_{bb}$, that is,

$$c_b = \int_{\Omega^e} H_b^T \, q^e \, d\Omega$$

or from the internal residual definition and the above orthogonality,

$$c_b = \int_{\Omega^e} H_b^T \,(r - L\hat{\phi})\, d\Omega = \int_{\Omega^e} H_b^T \, r^e \, d\Omega.$$

Therefore, this error indicator simplifies to $\| e^e \|_E = a_b \, c_b = c_b^2 / s_{bb}$.

In the following we will use this approach on a one-dimensional sample problem. We will see that the effectivity index is only about one-half, which is unacceptably far from the desired value of unity. While we could introduce a 'fudge factor' constant of two, it is wiser to search for a method, like the SCP recovery, that would yield an effectivity index that is always much closer to unity. Consider the Zienkiewicz and Morgan (Z-M) hierarchical error estimator for their Example 8.1 of [31] expanded to consider the local element errors and flux balances. The model problem is

$$-\frac{d^2\phi}{dx^2} + Q = 0, \qquad x \in \,]0, L[\,, \qquad \phi(0) = 0,\ \phi(L) = 0 \tag{5.25}$$

with the exact solution $\phi = Q(x - L)\, x/2$, so $\phi' = Q(2x - L)/2$. Using the Galerkin approximation:

$$\int_L \phi Q \, dx - \int_L \phi \, \phi_{,\,xx} \, dx = \int_L \phi Q \, dx - \phi \, \phi_{,\,x} \Big|_0^L + \int_L \phi_{,\,x}^2 \, dx = 0$$

or finally

$$\int_L \phi_{,\,x}^2 \, dx = -\int_L \phi Q \, dx + \phi \, \phi_{,\,x} \Big|_0^L .$$

Splitting the domain into elements and using our interpolations $\phi_h = \mathbf{H}^e \, \mathbf{u}^e$ this reduces to the matrix form:

$$\sum_e \mathbf{u}^{e^T} \mathbf{K}^e \, \mathbf{u}^e = -\sum_e \mathbf{u}^{e^T} \mathbf{F}_Q^e + u(L)\, \phi_{,\,x}(L) - u(0)\, \phi_{,\,x}(0)$$

with the typical element matrices defined (with $\mathbf{E} = \mathbf{I}$) as

$$\mathbf{K}^e = \int_{L^e} \mathbf{H}_{,\,x}^{e^T} \, \mathbf{H}_{,\,x}^e \, dx, \qquad \mathbf{F}_Q^e = \int_{L^e} \mathbf{H}^{e^T} \, Q^e \, dx.$$

For an initial linear interpolation with constant coefficients

$$\mathbf{K}^e = \frac{1}{L^e} \begin{bmatrix} 1 & -1 \\ -1 & 1 \end{bmatrix}, \qquad \mathbf{F}_Q^e = \frac{Q^e L^e}{2} \begin{Bmatrix} 1 \\ 1 \end{Bmatrix}.$$

First, consider a trivial single element solution. By inspection, $L^e = L$ so that

$$\frac{1}{L^e} \begin{bmatrix} 1 & -1 \\ -1 & 1 \end{bmatrix} \begin{Bmatrix} u_1 \\ u_2 \end{Bmatrix} = -\frac{QL^e}{2} \begin{Bmatrix} 1 \\ 1 \end{Bmatrix} + \begin{Bmatrix} -\phi_{,\,x}(0) \\ +\phi_{,\,x}(L) \end{Bmatrix}$$

but $u_1 = u_2 = 0$ from the boundary conditions. There are no unknown degrees of freedom to compute so we go directly to the flux recovery and error estimates. Solving for the flux gives $\phi_{,x}(0) = -QL^e/2$ and $\phi_{,x}(L) = QL^e/2$ as the two necessary nodal flux values. Checking we see that a useless solution has still given nodal fluxes that are exact as $L^e \equiv L$. The recovered nodal flux resultants are exact despite the fact that the single element solution is trivial, i.e., $\phi_h = \mathbf{H}^e\ \mathbf{u}^e = \mathbf{H}^e\ \mathbf{O}^e = 0$ (which is exact at nodes). The single element solution is useless in estimating the solution error. In the energy norm the error measure is

$$\| e \|^2 = -\int_L (\phi - \phi_h)(-\phi_{h,x}\ x + Q_h)\ dx = -\int_L e\ r\ dx$$

where r is the interior residual. To compute an error indicator, we add a quadratic hierarchical term to the linear element so $\overset{*}{\phi}_h = \phi_h + u_3^e\ H_3^e$ where $H_3^e(x) = x(L - x)$ in global space, or $H_3^e(r) = r(1 - r)$ in a local unit coordinate space. The Z-M error indicator is

$$\| e^e \|^2 = [\int_{L^e} H_3\ r\ dx\ ]^2 / K_{33}^e\ , \quad K_{33}^e = \int_{L^e} H_3 [\ -H_3''\ ]\ dx$$

is the new hierarchical stiffness term, and $e^e = \overset{*}{\phi}_h - \phi_h$. Here

$$K_{33}^e = \int_{L^e} r(1 - r)[\ \frac{-1}{L^{e^2}}(-2)\ ]\ dx = 1/3L^e$$

and

$$I^e = \int_{L^e} H_3^e R^e dx = \int_{L^e} r(1 - r)\ Q^e dx = Q^e\ L^e/6$$

so $\| e^e \|^2 = [Q^e L^e/6]^2/(1/3L^e) = Q^{e^2}\ L^{e^3}/12$ which happens to be exact for one element. We now repeat the solution and error indicators for two elements of equal size with $L^e = L/2$. The equilibrium equations are

$$\frac{1}{L^e}\begin{bmatrix} 1 & -1 & 0 \\ -1 & 2 & -1 \\ 0 & -1 & 1 \end{bmatrix}\begin{Bmatrix} u_1 \\ u_2 \\ u_3 \end{Bmatrix} = -\frac{QL^e}{2}\begin{Bmatrix} 1 \\ 2 \\ 1 \end{Bmatrix} + \begin{Bmatrix} -\phi_{,x}(0) \\ 0 \\ +\phi_{,x}(L) \end{Bmatrix}.$$

Setting $u_1 = 0 = u_3$ the remaining second equation yields $u_2 = -QL^{e^2}/2$, but $L^e = L/2$ so that $u_2 = -QL^2/8$ which is exact. Recovering the fluxes from equilibrium, we first check the global reactions: $\phi_{,x}(0) = -QL^e = -QL/2$, and $\phi_{,x}(L) = +QL/2$, which are both exact. Next, we find the fluxes on each element necessary for local equilibrium:

$$e = 1, \quad \frac{1}{L^e}\begin{bmatrix} 1 & -1 \\ -1 & 1 \end{bmatrix}\begin{Bmatrix} u_1 \\ u_2 \end{Bmatrix} = -\frac{QL^e}{2}\begin{Bmatrix} 1 \\ 1 \end{Bmatrix} + \begin{Bmatrix} -\phi_{,x}(x_1) \\ +\phi_{,x}(x_2) \end{Bmatrix}$$

$$-\frac{1}{L^e}\begin{Bmatrix} -QL^{e^2}/2 \\ +QL^{e^2}/2 \end{Bmatrix} + \frac{QL^e}{2}\begin{Bmatrix} 1 \\ 1 \end{Bmatrix} = QL^e\begin{Bmatrix} 1 \\ 0 \end{Bmatrix} = \begin{Bmatrix} -\phi_{,x}(x_1) \\ +\phi_{,x}(x_2) \end{Bmatrix}$$

which are exact since $L^e = L/2$. Likewise, for $e = 2$,

$$\begin{Bmatrix} -\phi_{,x}(x_2) \\ +\phi_{,x}(x_3) \end{Bmatrix} = QL^e \begin{Bmatrix} 0 \\ 1 \end{Bmatrix}.$$

The equilibrium of these global and local fluxes is sketched in Fig. 5.2. Note that the flux is zero at the symmetry point $(x = x_2)$ as expected. Since Q is a constant, the previously developed element error indicator, $\| e^e \|^2 = Q^{e^2} L^{e^3}/12$, is still valid for each element and the system error estimate is $\| e \|^2 = \sum_{e=1}^{n_e} = 2 \| e^e \|^2 = Q^{e^2} L^{e^3}/6$ and since $L^e = L/2$ we get $\| e \|^2 = Q^2 L^3/48$ compared to the exact value of $Q^2 L^3/24$. Thus, the total error is underestimated by a factor of two, but the indicator correctly shows each to have the same amount of error.

If we select two unequal elements, we still get exact values for the nodal values and fluxes. That is, if we let $L^e = L/4$ and $L^e = 3L/4$, respectively, we see the results in Fig. 5.2. There we note drastic differences in the local errors in each of the two elements. Checking our error indicators we get

$$e = 1, \quad \| e^e \|^2 = Q^{e^2} L^{e^3}/12 = Q^2 L^3/768$$

$$e = 2, \quad \| e^e \|^2 = Q^2/12\,(3L/4)^3 = 27 Q^2 L^3/768.$$

Clearly, this indicates that the error in the second element is 27 times as large as for that in the first element. Thus, the second element would be selected for refinement. Of course, the total error estimate for the two unequal elements is

$$\sum_{e=1}^{n_e} = 2 \| e^e \|^2 = 28 Q^2 L^3/768, \quad \text{and} \quad \| e \|^2_{exact} = \frac{7}{96} Q^2 L^3 = 56\, Q^2 L^3/768.$$

Refining the second mesh by placing a new node at $x = 3L/4$ gives the results in Fig. 5.2. Clearly, the first and third elements have the same indicators $\| e^e \|^2 = Q^2 L^3/768$ while

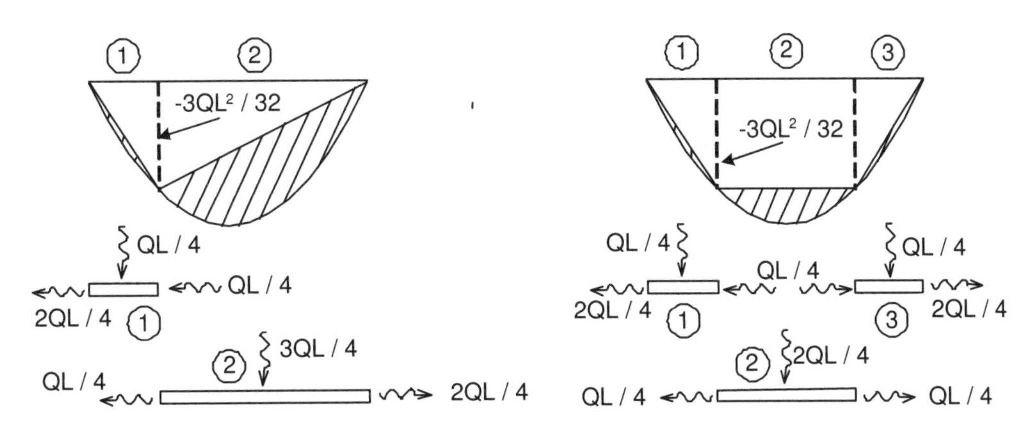

Figure 5.2 *Sample two and three element solutions, with flux values*

the middle element has a value of $\| e^e \|^2 = Q^2 L^3 / 96$. The total error estimate is

$$\| e \|^2 = \sum_{e=1}^{n_e} = 2 \, \| e^e \|^2 = Q^2 L^3/768 \, [1 + 8 + 1] = 10 \, Q^2 L^3/768.$$

Therefore, we notice that 10 percent of the error is in each of the two small elements and the remaining 80 percent is in the middle element. The exact error and effectivity measures are:

$$\| e \|^2_{exact} = \frac{10}{384} Q^2 L^2, \qquad \frac{\| e \|^2}{\| e \|^2_{exact}} = 0.5.$$

Finally, we observe the effects of four equally spaced elements on the error indicators. The system error indicator is the same for all elements and

$$\| e \|^2 = \sum_{e=1}^{n_e} = 4 \, \| e^e \|^2 = Q^{e^2} \, L^{e^3}/3 = Q^2 \, L^3/192,$$

and the exact value is $2 Q^2 \, L^3/192$, and once again we get an effectivity of only 50 percent. Since we want an error estimator with an effectivity index near unity, this method is not as desirable as the SCP recovery despite correctly giving the relative element error.

5.4 Flux balancing error estimates

Kelly *et al* [12-14, 24] conducted early studies, on simple structural and thermal models, in the use of residual equilibrium in establishing error estimates for the finite element method. That includes looking at the resultant flux (integral) at the element interfaces and assigning a percentage of it to each of the two elements so as to maintain equilibrium (i.e., balance the flux). Generalizing the functional analysis of such a procedure leads to a powerful flux balancing error estimate.

Ainsworth and Oden [1−3] have developed a local patch error estimator that is very well justified through detailed functional analysis, is robust, and economical to implement, and gives very accurate local error estimates for any order interpolation functions. That is, it usually produces an effectivity index that is very close to unity and is much more reliable than other methods known to the author. By using a dual variational formulation, they have proved that this estimator provides an *upper bound estimate* of the true error. The Ainsworth-Oden flux balancing method uses a local patch of elements for each master node. A typical patch includes all elements connected, or constrained, to the master node. The goal is to choose a linear averaging function α_{KL} between each pair of adjacent elements, K and L, such that the residual internal error, r, and inter-element gradient jumps, R, are in equilibrium; that is,

$$\int_{\Omega} r \, d\Omega + \int_{\Gamma} R \, d\Gamma = 0. \tag{5.26}$$

They provide a detailed procedure for implementing this method, including pseudo-code for the flux-splitting algorithm. The equilibrium fluxes are used to compute the local error estimator. A summary of the method is as follows:

```
for each master node in patch A do
    begin
        calculate a modified topology matrix, T
        factorize the matrix, L U ≡ T
        for every element e in the patch do
            begin
                calculate mean flux source, b^e
                calculate inter-element weight, ζ_j^e
                assemble patch source, b
            end
        solve for patch constants λ ; L U λ = b
        for every inter-element edge Γ_KL between elements K and L in patch do
            begin
                α_KL = 1/2 + ( λ_K − λ_L )/ζ_KL
            end
end
```

with the topology matrix defined as

$$T_{jk} = \begin{cases} (1 + \text{number of elements in patch}), & \text{if } j = k \\ 0, & \text{if } \Omega^j \text{ and } \Omega^k \text{ are neighbors in patch} \\ 1, & \text{otherwise.} \end{cases}$$

Letting Ψ be a piecewise linear function that is unity at the master node and zero on the patch boundary, the mean source is defined in terms of the model equation

$$-\nabla \cdot (k \nabla u) + \mathbf{b} \cdot \nabla u + cu = f \tag{5.27}$$

as

$$\mathbf{b}^e = L^e(\Psi) - B^e(\hat{u}, \Psi) + \int_{\Gamma^e \backslash \Gamma} \Psi < \mathbf{n}^e \cdot k \nabla \hat{u} >_{\frac{1}{2}} d\Gamma$$

$$L^e(\Psi) = \int_{\Omega^e} f \, \Psi \, d\Omega + \int_{\Gamma_n} k \frac{\partial u}{\partial n} \Psi \, d\Gamma$$

$$B^e(\hat{u}, \Psi) = \int_{\Omega^e} (k \nabla \hat{u} \cdot \nabla \Psi + \Psi \mathbf{b} \cdot \nabla \hat{u} + c \hat{u} \Psi) \, d\Omega$$

$$< \mathbf{n}^e \cdot k \nabla \hat{u} > = \mathbf{n}^e \cdot \frac{1}{2} \left(k^e \nabla \hat{u} \Big|_{\Omega^e} + k^j \nabla \hat{u} \Big|_{\Omega^j} \right)$$

and the inter-element weight is

$$\zeta_j^e = -\int_{\Gamma_j^e} \Psi \left(\mathbf{n}^e \cdot k^e \nabla \hat{u} \Big|_{\Omega^e} + \mathbf{n}^j \cdot k^j \nabla \hat{u} \Big|_{\Omega^j} \right) d\Gamma.$$

The actual flux-splitting function on the boundary between nodes K and L is

$$\alpha_{KL}(s) = \sum_A \alpha_{KL} \Psi(s)$$

where the sum has taken over all patches containing edge KL (and a non-zero Ψ). Once the fluxes are in equilibrium, the error, $e = u - \hat{u}$, is bounded above by the norm

$$\| e \|^2 \le \frac{1}{\beta^2} \sum_{e=1}^{n_e} \| \phi^e \|^2$$

where $\beta > 0$ is a constant depending on the norm selected ($\beta = 1$ for the standard energy norm), and ϕ is obtained by solving the element local Neumann problem

$$a^e(\phi, w) = L^e(w) - B^e(\hat{u}, w) + \tag{5.28}$$

$$+ \int_{\Gamma^e} w\, \mathbf{n}^e \cdot \left[(1 - \alpha_{KL}(s))\, k^e \nabla \hat{u} \Big|_{\Omega^e} + \alpha_{KL}(s)\, k^j \nabla \hat{u} \Big|_{\Omega^f} \right] d\Gamma.$$

The examples by Ainsworth and Oden show this procedure to be accurate and economical. The effectivity index, Θ, is usually very near unity as desired, and is usually above 0.9 for even crude initial mesh calculations. While this is also a recommended method, we choose to implement SCP recovery due to its simplicity.

5.5 Element adaptivity

Upon completing the loop over all elements we have the element norms, the element volume, the system norms,

$$\|\mathbf{e}\|^2 = \sum_e^{n_e} \|\mathbf{e}^e\|^2, \tag{5.29}$$

and the system volume. The allowed error energy is obtained from the product of the strain energy norm and the user input value of the allowed percentage error, η (keyword input value *scp_allow_error_%*). That number is used, in turn, to evaluate two allowed error densities in dividing by the square root of the number of elements and the square root of the volume to yield mean element and volumetric references, respectively. One of these reference values will be used to rank the relative error measures in each element. The system norm values are printed along with the two allowable reference values for the energy error and the system volume. For each element we will list the element error norm, its percentage of the strain energy norm, and a refinement parameter for that element. The element error energy norms are summed to get the total energy in the error to compare to the total strain energy norm and the allowed percentage of error. Here, the refinement parameter is based on the volumetric error density, so for element j the refinement parameter is

$$\xi_j = (\|\mathbf{e}_j\| / \sqrt{\Omega_j}) / (\eta \|\mathbf{e}\| / \sqrt{\Omega})$$

or

$$\xi_j = \frac{\|\mathbf{e}_j\|}{\eta \|\mathbf{e}\|} \left(\frac{\Omega}{\Omega_j} \right)^{\frac{1}{2}}. \tag{5.30}$$

Here it is informative to note that for a uniform mesh all the n_e element volumes are constant with a value of

$$\Omega_j = \Omega / n_e$$

and the refinement indicator becomes the same as originally employed by Zienkiewicz and Zhu, namely:

$$\xi_j = \frac{\| \mathbf{e}_j \|}{\eta \| \mathbf{e} \| \sqrt{n_e}} . \tag{5.31}$$

By combining such an indicator with interpolation error analysis, one can predict the desired element size or polynomial degree. For each element, i, we define the ratio ξ to indicate needed refinement when $\xi_i > 1$ and de-refinement when $\xi_i < 1$.

5.6 H-adaptivity

The refinement indicator of Eq. 5.31 is the ratio of the current estimated error in an element to that desired in the element. From interpolation theory the asymptotic convergence rates for an element in a uniform mesh is $\| e \| = Ch^p$, for $p > 0$, where h is the characteristic element length (distance between nodes), p is the degree of the polynomial, and C is a constant that depends on the shape of the domain and the boundary conditions (Dirichlet versus Neumann). In h-adaptivity we hold the polynomial degree, p, constant and seek a new element size, say h_{new}. Thus, we can also view the refinement indicator as related to the current and new sizes, namely for the j-th element:

$$\xi_j = \frac{Ch_j^p}{Ch_{new}^p} . \tag{5.32}$$

Cancelling the problem constant, C, and factoring out the polynomial order, p, the new element size should be

$$h_{new} = h_j / \xi_j^{1/p} . \tag{5.33}$$

It should be noted that some analysts like to normalize the asymptotic rate by dividing the element size by its initial size. That is, after a few iterations they employ $(h/h_0)^p$ where h_0 was the element size in the original mesh.

The relation in Eq. 5.33 is used to output a sequential list of desired element sizes to be utilized as input by an automatic mesh generator. It could be arbitrarily associated with the element centroid. Here it is output at each current node using the average value of all the elements connected to the node. Huang and Usmani provide such an automatic mesh generator for two-dimensional applications [11]. A modified version of their source code was used for several of the examples given herein.

5.7 P-adaptivity

In p-adaptivity it is less clear how to proceed and several hueristic approaches have been used. It is clear that the new polynomial degree must be an integer. The change in degree should be small (1 or 2) because higher order elements are expensive. Also numerical studies show that the nature of the error is different in even and odd order polynomials. The interior error (measured well by the SCP) is more important in even polynomials, while the interface flux jump error is more important in odd order elements.

Thus, one could simply assume the element size will be constant and note which integer increase in p would give a refinement indicator slightly smaller that the one obtained from the error estimator. In global p-refinements typically use the largest integer found in this way (but limiting the new polynomial to ≤ 6). However, that can be very expensive. There is another empirical equation for estimating the new polynomial degree. If one assumes Lagrange interpolation and views h as the distance between nodes then a real number estimate of the new degree is

$$p_{Est} = p \times \xi_j^{1/p} \tag{5.34}$$

which would have to be rounded to an integer value, p_{new}. Typical results from this estimate are illustrated below:

ξ	p_{old}	$\xi^{1/p_{old}}$	p_{Est}	Operation	p_{new}
4.00	2	2.00	4.0	Enrich	4
1.00	2	1.00	2.0	No Change	2
0.50	2	0.71	1.4	No Change	2
0.05	2	0.22	0.45	Degrade	1
4.00	3	1.59	4.76	Enrich	5
2.00	3	1.26	3.78	Enrich	4
0.50	3	0.79	2.38	Degrade	2

Clearly, arbitrary rules govern how to round the estimated p in the fourth column. The third line value of 1.4 might just as easily been thought of as a degrade to a linear polynomial ($p_{new} = 1$).

5.8 HP-adaptivity

It has been proved (see Ainsworth and Oden [3] for example) that an hp-adaptive system gives the optimal convergence (maximum accuracy for a given number of equations). However, its programming is difficult and requires careful planning of the data base structure. In an hp-adaptive solution one needs to pick which item to change first. Since p changes are relatively expensive and must be limited to integers it may be best to select p_{new} first and to restrict the change in degree, say n to 0 or + – 1. Then due to the integer choice on p some of the estimated refinement (or de-refinement) still needs to occur by also selecting a new mesh size. We can envision the refinement indicator as having two contributions, $\xi = \xi_p \times \xi_h$. If the new integer degree, $(p + n)$, was based on the current element size then the now known numerical value $\xi_p = h^p / h^{(p+n)}$ can be used to get the needed remaining spatial refinement indicator , ξ_h. Note that the product relation is

$$\xi = \frac{h^p}{h_{new}^{(p+n)}} = \frac{h^p}{h^{(p+n)}} \times \frac{h^{(p+n)}}{h_{new}^{(p+n)}} = \xi_p \times \xi_h \tag{5.35}$$

which with ξ and ξ_p known simplifies to

$$\xi_h = (\xi / \xi_p) = (h / h_{new})^{(p+n)}$$

or finally

$$h_{new} = h / \xi_h^{1/(p+n)} . \tag{5.36}$$

Even with these rough estimates of desired changes one may need other rules to assure that the mesh size and local degree do not change rapidly from one solution iteration to the next, or oscillate between large and small values.

5.9 Exercises

1. A four element model of our previous example differential equation, $u,xx + u + x = 0$, $u(0) = 0 = u(1)$ yields the exact and finite element solution as shown in Prob. 5.1a and the true flux is shown in Prob. 5.1b. The finite element flux estimates in the elements consist of the four constant steps listed above. Obtain a nodal continuous flux estimate by using element based patches (four in total). Show the estimated (eyeball) linear fit on each patch. For each patch use a unique symbol to show the interpolated nodal flux values. At each original mesh node average the nodal flux values from all patches. Plot a piecewise linear curve through those average flux values and compare it to the exact flux curve in Prob. 5.1b. Utilize those numerical results, for the two-noded linear element (L2).

2. Resolve the example in Chapter 2 for $k\, d^2\phi / dx^2 + Q = 0$ with $\phi(L) = \phi_L$, and $k\, d\phi / dx\,(0) = q_0$ using a four element model. Plot the results compared to the exact solution. Obtain the gradient estimate in each element at its centroid, and plot it against the true gradient. Obtain a nodal continuous flux estimate by using element based patches (four in total). Show the estimated (eyeball) linear fit on each

```
*** OUTPUT OF RESULTS AND EXACT VALUES IN NODAL ORDER ***        ! 1
NODE,  X-Coord,      DOF_1,       EXACT1,                        ! 2
   1  0.0000E+00  0.0000E+00  0.0000E+00                         ! 3
   2  2.5000E-01  4.3758E-02  4.4014E-02                         ! 4
   3  5.0000E-01  6.9345E-02  6.9747E-02                         ! 5
   4  7.5000E-01  5.9715E-02  6.0056E-02                         ! 6
   5  1.0000E+00  0.0000E+00 -2.2829E-10                         ! 7
                                                                 ! 8
*** FE AND EXACT FLUX COMPONENTS AT INTEGRATION POINTS ***       ! 9
ELEMENT, PT,  X-Coord,     FX_1,      EX_1,                      !10
       1  1  5.283E-02  1.750E-01  1.867E-01                     !11
       1  2  1.972E-01  1.750E-01  1.654E-01                     !12
ELEMENT, PT,  X-Coord,     FX_1,      EX_1,                      !13
       2  1  3.028E-01  1.023E-01  1.343E-01                     !14
       2  2  4.472E-01  1.023E-01  7.155E-02                     !15
ELEMENT, PT,  X-Coord,     FX_1,      EX_1,                      !16
       3  1  5.528E-01 -3.852E-02  1.137E-02                     !17
       3  2  6.972E-01 -3.852E-02 -8.890E-02                     !18
ELEMENT, PT,  X-Coord,     FX_1,      EX_1,                      !19
       4  1  8.028E-01 -2.389E-01 -1.745E-01                     !20
       4  2  9.472E-01 -2.389E-01 -3.060E-01                     !21
```

P5.1 Four linear element results

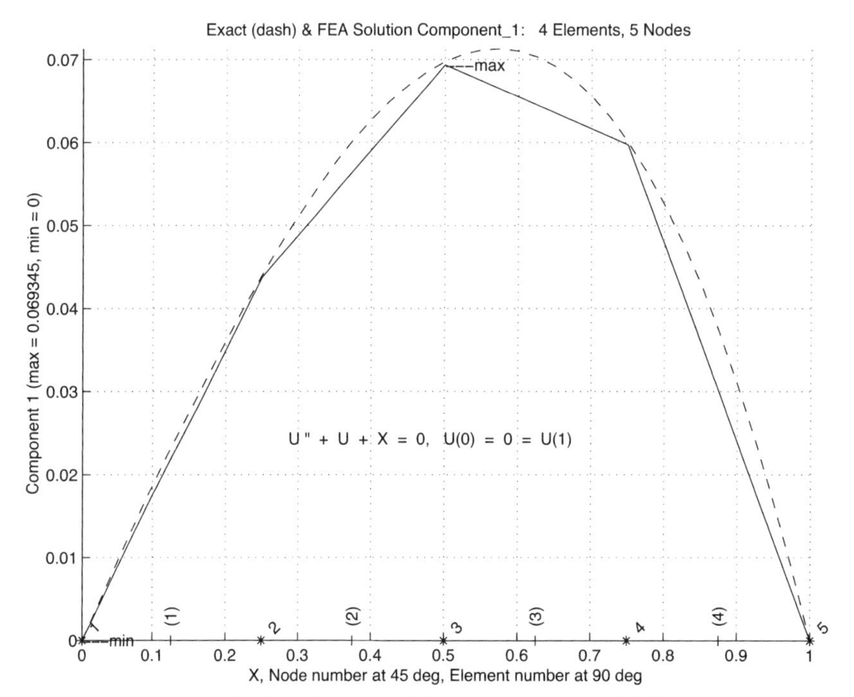

Prob. 5.1a Exact and FEA solution of ODE

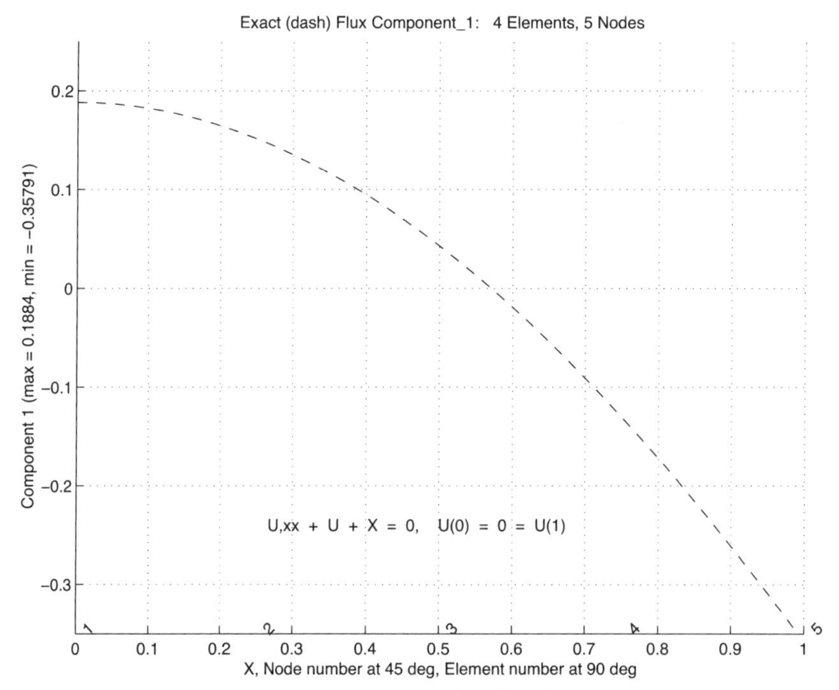

Prob. 5.1b Exact flux for ODE

patch. For each patch use a unique symbol to show the interpolated nodal flux values. At each original mesh node average the nodal flux values from all patches. Plot a piecewise linear curve through those average flux values and compare it to the exact gradient. Use a) $L = 1$, $k = 1$, $q_0 = 2$, $\phi_L = 1$, and $Q = Q_0 = 1$ so that the exact solution is given by $k\,\phi(x) = k\,\phi_L + q_0(x - L) + Q_0(L^2 - x^2)/2$, b) $L = 1$, $k = 1$, $q_0 = 1/12$, $\phi_L = 0$, and $Q = x^2$ so the exact solution is $\phi(x) = (x - x^4)/12$.

5.10 Bibliography

[1] Ainsworth, M. and Oden, J.T., "A Procedure for *a Posteriori* Error Estimation for *h-p* Finite Element Methods," *Comp. Meth. Appl. Mech. Eng.*, **101**, pp. 73–96 (1992).

[2] Ainsworth, M. and Oden, J.T., "A Unified Approach to *a Posteriori* Error Estimation Based on Element Residual Methods," *Numer. Math.*, **65**, pp. 23–50 (1993).

[3] Ainsworth, M. and Oden, J.T., *A Posteriori Error Estimation in Finite Element Analysis*, New York: John Wiley (2000).

[4] Babuska, I., Strouboulis, T., and Upadhyay, C.S., "A Model Study of the Quality of *a Posteriori* Error Estimators for Finite Element Solutions of Linear Elliptic Problems, with Particular Reference to the Behavior near the Boundary," *Int. J. Num. Meth. Eng.*, **40**, pp. 2521–2577 (1997).

[5] Babuska, I. and Strouboulis, T., *The Finite Element Method and its Reliability*, Oxford: Oxford University Press (2001).

[6] Barnhill, R.E. and Whiteman, J.R., "Error Analysis of Finite Element Methods with Triangles for Elliptic Boundary Value Problems," in *The Mathematics of Finite Elements and Applications*, ed. J.R. Whiteman, London: Academic Press (1973).

[7] Blacker, T. and Belytschko, T., "Superconvergent Patch Recovery with Equilibrium and Conjoint Interpolant Enhancements," *Int. J. Num. Meth. Eng.*, **37**, pp. 517–536 (1995).

[8] Brauchli, H.J. and Oden, J.T., "On the Calculation of Consistent Stress Distribution in Finite Element Applications," *Int. J. Num. Meth. Eng.*, **3**, pp. 317–325 (1971).

[9] Ciarlet, P.G., *The Finite Element Method for Elliptical Problems*, Philadelphia, PA: SIAM (2002).

[10] Cook, R.D., Malkus, D.S., Plesha, N.E., and Witt, R.J., *Concepts and Applications of Finite Element Analysis*, New York: John Wiley (2002).

[11] Huang, H.C. and Usmani, A.S., in *Finite Element Analysis for Heat Transfer*, London: Springer-Verlag (1994).

[12] Kelly, D.W., "The Self Equilibration of Residuals and Complementary *a Posteriori* Error Estimates in the Finite Element Method," *Int. J. Num. Meth. Eng.*, **20**, pp. 1491–1506 (1984).

[13] Kelly, D.W. and Isles, J.D., "Procedures for Residual Equilibrium and Local Error Estimation in the Finite Element Method," *Comm. in Applied Num. Methods*, **5**, pp. 497–505 (1989).

[14] Kelly, D.W. and Isles, J.D., "A Procedure for *A Posteriori* Error Analysis for the Finite Element Method which Contains a Bounding Measure," *Computers & Structures*, **31**(1), pp. 63–71 (1989).
[15] Krizek, M., Neittaanmaki, P., and Stenberg, R., *Finite Element Methods: Superconvergence, Post-Processing and a Posteriori Estimates*, New York: Marcel Dekker, Inc. (1998).
[16] Ladeveze, D. and Leguillon, D., "Error Estimate Procedure in the Finite Element Method and Applications," *SIAM J. Num. Anal.*, **20**(3), pp. 485–509 (1983).
[17] Oden, J.T., *Finite Elements of Nonlinear Continua*, New York: McGraw-Hill (1972).
[18] Oden, J.T., *Applied Functional Analysis*, Englewood Cliffs: Prentice Hall (1979).
[19] Oden, J.T., "The Best FEM," *Finite Elements in Analysis and Design*, **7**, pp. 103–114 (1990).
[20] Szabo, B. and Babuska, I., *Finite Element Analysis*, New York: John Wiley (1991).
[21] Wiberg, N.-E., Abdulwahab, F., and Ziukas, S., "Enhanced Superconvergent Patch Recovery Incorporating Equilibrium and Boundary Conditions," *Int. J. Num. Meth. Eng.*, **37**, pp. 3417–3440 (1994).
[22] Wiberg, N.-E., Abdulwahab, F., and Ziukas, S., "Improved Element Stresses for Node and Element Patches Using Superconvergent Patch Recovery," *Comm. Num. Meth. Eng.*, **11**, pp. 619–627 (1995).
[23] Wiberg, N.-E., "Superconvergent Patch Recovery – A Key to Quality Assessed FE Solutions," *Adv. Eng. Software*, **28**, pp. 85–95 (1997).
[24] Yang, J.D., Kelly, D.W., and Isles, J.D., "*A Posteriori* Point-wise Upper Bound Error Estimates in the Finite Element Method," *Intern. J. for Num. Meth. Engr.*, **36**, pp. 1279–1298 (1993).
[25] Zhang, Z. and Zhu, J.Z., "Analysis of the Superconvergent Patch Recovery Technique and *a posteriori* Error Estimator in the Finite Element Method (I)," *Computer Methods in Applied Mechanics and Engineering*, **123**, pp. 173–187 (1995).
[26] Zhang, Z., "Ultraconvergence of the Patch Recovery Technique," *Math. Comp.*, **65**, pp. 1431–1437 (1996).
[27] Zhang, Z., "Derivative Superconvergence Points in Finite Element Solutions of Poisson's Equation for the Serendipity and Intermediate Families – A Theoretical Justification," *Math. Comp.*, **67**, pp. 541–552 (1998).
[28] Zhang, Z. and Zhu, J.Z., "Analysis of the Superconvergent Patch Recovery Technique and *a posteriori* Error Estimator in the Finite Element Method (II)," *Computer Methods in Applied Mechanics and Engineering*, **163**, pp. 159–170 (1998).
[29] Zhang, Z., "Ultraconvergence of the Patch Recovery Technique II, with Graph," *Math. Comp.*, **69**, pp. 141–158 (2000).

[30] Zhu, J.Z. and Zienkiewicz, O.C., "Superconvergence Recovery Techniques and *A Posteriori* Error Estimators," *Int. J. Num. Meth. Eng.*, **30**, pp. 1321–1339 (1990).

[31] Zienkiewicz, O.C. and Morgan, K., *Finite Elements and Approximation*, Chichester: John Wiley (1983).

[32] Zienkiewicz, O.C. and Zhu, J.Z., "A Simple Error Estimator and Adaptive Procedure for Practical Engineering Analysis," *Int. J. Num. Meth. Eng.*, **24**, pp. 337–357 (1987).

[33] Zienkiewicz, O.C., Kelley, D.W., Gago, J., and Babuska, I., "Hierarchal Finite Element Approaches Error Estimates and Adaptive Refinement," pp. 313–346 in *The Mathematics of Finite Elements and Applications, VI*, ed. J.R. Whiteman, London: Academic Press (1988).

[34] Zienkiewicz, O.C., Zhu, J.Z., Craig, A.W., and Ainsworth, M., "Simple and Practical Error Estimation and Adaptivity," pp. 100–114 in *Adaptive Methods for Partial Differential Equations*, ed. J.E. Flaherty et al., SIAM (1989).

[35] Zienkiewicz, O.C. and Zhu, J.Z., "Superconvergent Patch Recovery Techniques and Adaptive Finite Element Refinement," *Comp. Meth. Appl. Mech. Eng.*, **101**, pp. 207–224 (1992).

[36] Zienkiewicz, O.C. and Zhu, J.Z., "The Superconvergent Patch Recovery and *a Posteriori* Error Estimates. Part 2: Error Estimates and Adaptivity," *Int. J. Num. Meth. Eng.*, **33**, pp. 1365–1382 (1992).

Chapter 6

Super-convergent patch recovery

6.1 Patch implementation database

Since the super-convergent patch (SCP) recovery method is relativity easy to understand and is accurate for a wide range of problems, it was selected for implementation in the educational program *MODEL*. Its implementation is designed for use with most of the numerically integrated 1-D, 2-D, and 3-D elements in the *MODEL* library. Most of the literature on the SCP recovery methods is limited to a single element type and a single patch type. The present version is somewhat more general in allowing a mixture of element shapes in the mesh and a mesh that is either linear, quadratic, or cubic in its polynomial degree.

This represents actually the third version of the SCP algorithm and is a simplification of the first two. The method given here was originally developed for a p-adaptive code where all the elements could have a different polynomial degree on each element edge. That version was then extended to an object-oriented F90 p-adaptive program that also included equilibrium error contributions as suggested by Wiberg [12] and others. Including the p-adaptive features and the object-oriented features made the data base more complicated and required more planning and programming than desirable in an introductory text such as this one. However, the version given here has shown to be robust and useful.

The SCP recovery process is clearly heuristic in nature, so some arbitrary choices need to be made in the implementation. We begin by defining a 'patch' to be a local group of elements surrounding at least one interior node or being adjacent to a boundary node. The original research in SCP recovery methods used patches sequentially built around each node in the mesh. Later it was widely recognized that one could use a patch for every element in the mesh. Therefore, three types of patches will be defined here:

1. Node-based patch: An adjacent group of elements associated with a particular node.
2. Element-based patch: All elements adjacent to a particular element.
3. Face-based patch: This subset of the element-based patch includes only the adjacent elements that share a common face with the selected element. For two-dimensional

elements this means that they share a common edge.

Those three choices for patches were shown in Fig. 2.19. Any of these three definitions of a patch requires that one have a mesh 'neighbors list'. That is, we will need a list of elements adjacent to each node, or a list of elements adjacent to each element, or the subset list of facing element neighbors. These can be expensive lists to create, but are often needed for other purposes and are sometime supplied by a mesh generation code or an equation re-ordering program.

Here several routines are included for creating the lists and printing them. Normally, since those neighbor lists are of unknown variable lengths, they would be stored in a linked list data structure. Here, for simplicity, they have been placed in rectangular arrays and padded with trailing zeros. This wastes a little storage space but keeps the general code simpler. Several of these routines are actually invoked at the mesh input stage as part of the data checking process. The neighbor lists are usually quite large and are not usually listed but can be (via keywords *list_el_to_el* or *pt_el_list*).

The primary routines for establishing the node-based patches are the subroutines COUNT_L_AT_NODE and FORM_L_ADJACENT_NODE, seen in Figs. 6.1 and 2, while the

```
SUBROUTINE COUNT_L_AT_NODES (N_ELEMS, NOD_PER_EL, MAX_NP, &          ! 1
                             NODES, L_TO_N_SUM)                      ! 2
! * * * * * * * * * * * * * * * * * * * * * * * * * * *              ! 3
! COUNT NUMBER OF ELEMENTS ADJACENT TO EACH NODE                     ! 4
!        (TO SIZE ELEM ADJACENT TO ELEM LIST)                        ! 5
! * * * * * * * * * * * * * * * * * * * * * * * * * * *              ! 6
 IMPLICIT NONE                                                       ! 7
 INTEGER, INTENT(IN)  :: N_ELEMS, NOD_PER_EL, MAX_NP                 ! 8
 INTEGER, INTENT(IN)  :: NODES (N_ELEMS, NOD_PER_EL)                 ! 9
 INTEGER, INTENT(OUT) :: L_TO_N_SUM (MAX_NP)                         !10
                                                                     !11
 INTEGER :: ELEM_NODES (NOD_PER_EL)                                  !12
 INTEGER :: IE, IN, N_TEMP                                           !13
                                                                     !14
! ELEM_NODES      = INCIDENCES ARRAY FOR A SINGLE ELEMENT            !15
! L_TO_N_SUM (I)  = NUMBER OF ELEM NEIGHBORS OF NODE I               !16
! MAX_NP          = TOTAL NUMBER OF NODES                            !17
! N_ELEMS         = TOTAL NUMBER OF ELEMENTS                         !18
! NOD_PER_EL      = NUMBER OF NODES PER ELEMENT                      !19
! NODES           = SYSTEM ARRAY OF ALL ELEMENTS INCIDENCES          !20
                                                                     !21
  L_TO_N_SUM = 0                 ! INITIALIZE                        !22
  DO IE = 1, N_ELEMS             ! LOOP OVER ALL ELEMENTS            !23
!       EXTRACT INCIDENCES LIST FOR ELEMENT IE                       !24
    CALL ELEMENT_NODES (IE, NOD_PER_EL, NODES, ELEM_NODES)           !25
    DO IN = 1, NOD_PER_EL        ! loop over each node               !26
      N_TEMP = ELEM_NODES (IN)                                       !27
      IF ( N_TEMP < 1 )  CYCLE ! to a real node                      !28
        L_TO_N_SUM (N_TEMP) = L_TO_N_SUM (N_TEMP) + 1                !29
    END DO ! over IN                                                 !30
  END DO ! over elements                                             !31
END SUBROUTINE COUNT_L_AT_NODES                                      !32
```

Figure 6.1 *Computing the neighbor array sizes*

```
SUBROUTINE FORM_L_ADJACENT_NODES (N_ELEMS, NOD_PER_EL, MAX_NP, &    ! 1
                                  NODES, NEIGH_N, L_TO_N_NEIGH)     ! 2
! * * * * * * * * * * * * * * * * * * * * * * * * * * * * * * *    ! 3
!       TABULATE ELEMENTS ADJACENT TO EACH NODE                     ! 4
!         (TO SIZE ELEM ADJACENT TO ELEM LIST)                      ! 5
! * * * * * * * * * * * * * * * * * * * * * * * * * * * * * * *    ! 6
IMPLICIT NONE                                                       ! 7
 INTEGER, INTENT(IN)  :: N_ELEMS, NOD_PER_EL, MAX_NP, NEIGH_N       ! 8
 INTEGER, INTENT(IN)  :: NODES (N_ELEMS, NOD_PER_EL)                ! 9
 INTEGER, INTENT(OUT) :: L_TO_N_NEIGH (NEIGH_N, MAX_NP)             !10
                                                                    !11
 INTEGER :: ELEM_NODES (NOD_PER_EL), COUNT (MAX_NP) ! scratch       !12
 INTEGER :: IE, IN, N_TEMP                                          !13
                                                                    !14
! ELEM_NODES     = INCIDENCES ARRAY FOR A SINGLE ELEMENT            !15
! L_TO_N_SUM (I) = NUMBER OF ELEM NEIGHBORS OF NODE I               !16
! MAX_NP         = TOTAL NUMBER OF NODES                            !17
! N_ELEMS        = TOTAL NUMBER OF ELEMENTS                         !18
! NOD_PER_EL     = NUMBER OF NODES PER ELEMENT                      !19
! NODES          = SYSTEM ARRAY OF INCIDENCES OF ALL ELEMENTS       !20
                                                                    !21
  L_TO_N_NEIGH = 0 ; COUNT = 0   ! INITIALIZE                       !22
  DO IE = 1, N_ELEMS             ! LOOP OVER ALL ELEMENTS           !23
                                                                    !24
!        EXTRACT INCIDENCES LIST FOR ELEMENT IE                     !25
    CALL ELEMENT_NODES (IE, NOD_PER_EL, NODES, ELEM_NODES)          !26
                                                                    !27
    DO IN = 1, NOD_PER_EL       ! loop over each node               !28
      N_TEMP = ELEM_NODES (IN)                                      !29
      IF ( N_TEMP < 1 )  CYCLE ! to a real node                     !30
        COUNT (N_TEMP) = COUNT (N_TEMP) + 1                         !31
        L_TO_N_NEIGH (COUNT (N_TEMP), N_TEMP) = IE                  !32
    END DO ! over IN nodes                                          !33
  END DO ! over elements                                            !34
END SUBROUTINE FORM_L_ADJACENT_NOD                                  !35
```

Figure 6.2 *Find elements at every node*

element-based patches use the two similar subroutines COUNT_ELEMS_AT_ELEM and FORM_ELEMS_AT_EL which are given in Figs. 6.3 and 4. These routines are also useful in validating meshes that have been prepared by hand. Building lists of neighbors can take a lot of processing but they are useful in plotting and post-processing.

In *MODEL* the default is to use an element-based patch. However, one can investigate other options by utilizing some of the available control keywords given in Fig. 6.5. Having selected a patch type, we should now give consideration to the kind of data that will be needed for the SCP recovery. There are two main segments in the process:

1. Averaging the patch and system nodal fluxes.
2. Using the system nodal fluxes in the calculation of an error estimate.

The whole basis of the SCP recovery is that there are special locations within an element where we can show that the derivatives are most accurate or exact for a given polynomial degree. We refer to such locations as element super-convergent points. They are sometimes called *Barlow points*. The derivation of the locations generally shows them

```
SUBROUTINE COUNT_ELEMS_AT_ELEM (N_ELEMS, NOD_PER_EL, MAX_NP, &     ! 1
           L_FIRST, L_LAST, NODES, NEEDS, L_TO_L_SUM, N_WARN)     ! 2
! * * * * * * * * * * * * * * * * * * * * * * * * * * * * * * *   ! 3
!    COUNT NUMBER OF ELEMENTS ADJACENT TO OTHER ELEMENTS          ! 4
! * * * * * * * * * * * * * * * * * * * * * * * * * * * * * * *   ! 5
IMPLICIT NONE                                                     ! 6
 INTEGER, INTENT(IN)    :: N_ELEMS, NOD_PER_EL, MAX_NP, NEEDS     ! 7
 INTEGER, INTENT(IN)    :: L_FIRST (MAX_NP), L_LAST (MAX_NP)      ! 8
 INTEGER, INTENT(IN)    :: NODES (N_ELEMS, NOD_PER_EL)            ! 9
 INTEGER, INTENT(OUT)   :: L_TO_L_SUM (N_ELEMS)                   !10
 INTEGER, INTENT(INOUT) :: N_WARN                                 !11
                                                                  !12
 INTEGER :: ELEM_NODES (NOD_PER_EL), NEIG_NODES (NOD_PER_EL)      !13
 INTEGER :: FOUND, IE, IN, L_TEST, L_START, L_STOP, N_TEST        !14
 INTEGER :: NEED, KOUNT, NULLS                                    !15
                                                                  !16
! ELEM_NODES      = INCIDENCES ARRAY FOR A SINGLE ELEMENT         !17
! KOUNT           = NUMBER OF COMMON NODES                        !18
! L_FIRST (I)     = ELEMENT WHERE NODE I FIRST APPEARS            !19
! L_LAST (I)      = ELEMENT WHERE NODE I LAST APPEARS             !20
! L_TO_L_SUM (I)  = NUMBER OF ELEM NEIGHBORS OF ELEMENT I         !21
! MAX_NP          = TOTAL NUMBER OF NODES                         !22
! NEEDS           = NUMBER OF COMMON NODES TO BE A NEIGHBOR       !23
! N_ELEMS         = TOTAL NUMBER OF ELEMENTS                      !24
! NOD_PER_EL      = NUMBER OF NODES PER ELEMENT                   !25
! NODES           = SYSTEM ARRAY OF INCIDENCES OF ALL ELEMENTS    !26
                                                                  !27
  L_TO_L_SUM = 0 ; NEED = MAX (1, NEEDS)    ! INITIALIZE          !28
                                                                  !29
  MAIN : DO IE = 1, N_ELEMS        ! LOOP OVER ALL ELEMENTS       !30
    FOUND = 0                      ! INITIALIZE                   !31
                                                                  !32
!        EXTRACT INCIDENCES LIST FOR ELEMENT IE                   !33
    CALL ELEMENT_NODES (IE, NOD_PER_EL, NODES, ELEM_NODES)        !34
                                                                  !35
!        ESTABLISH RANGE OF POSSIBLE ELEMENT NEIGHBORS            !36
    L_START = N_ELEMS ; L_STOP = 0                                !37
    DO IN = 1, NOD_PER_EL                                         !38
      N_TEST  = ELEM_NODES (IN)                                   !39
      IF ( N_TEST < 1 ) CYCLE! to a real node                     !40
        L_START = MIN (L_START, L_FIRST (N_TEST) )                !41
        L_STOP  = MAX (L_STOP,  L_LAST  (N_TEST) )                !42
    END DO                                                        !43
                                                                  !44
```

Figure 6.3a *Interface and data for elements joining element*

```
!          LOOP OVER POSSIBLE ELEMENT NEIGHBORS                         !45
    IF ( L_START <= L_STOP) THEN                                        !46
     RANGE : DO L_TEST = L_START, L_STOP                                !47
       IF ( L_TEST /= IE) THEN                                          !48
        KOUNT = 0  ! NO COMMON NODES                                    !49
                                                                        !50
!       LOOP OVER INCIDENCES OF POSSIBLE ELEMENT NEIGHBOR               !51
        CALL ELEMENT_NODES (L_TEST,NOD_PER_EL,NODES,NEIG_NODES)         !52
        LOCAL : DO IN = 1, NOD_PER_EL                                   !53
         N_TEST = NEIG_NODES (IN)                                       !54
         IF ( N_TEST < 1 .OR. N_TEST > MAX_NP )  THEN                   !55
           PRINT *, 'INVALID NODE ', N_TEST, ' AT ', L_TEST             !56
           N_WARN = N_WARN + 1 ! INCREMENT WARNING                      !57
           CYCLE LOCAL ! to a real node                                 !58
         END IF ! IMPOSSIBLE NODE                                       !59
         IF ( L_FIRST (N_TEST) > IE ) CYCLE LOCAL ! to next node        !60
         IF ( L_LAST  (N_TEST) < IE ) CYCLE LOCAL ! to next node        !61
                                                                        !62
!            COMPARE WITH INCIDENCES OF ELEMENT IE                      !63
           IF ( ANY ( ELEM_NODES == N_TEST ) ) THEN                     !64
             KOUNT = KOUNT + 1                                          !65
             IF ( KOUNT == NEED ) THEN ! IS A NEIGHBOR                  !66
               FOUND = FOUND + 1                                        !67
               EXIT LOCAL ! this L_TEST element search loop             !68
             END IF ! NUMBER NEEDED                                     !69
           END IF                                                       !70
        END DO LOCAL ! over in                                          !71
       END IF                                                           !72
     END DO RANGE ! over candidate element L_TEST                       !73
    END IF ! a possible candidate                                       !74
    L_TO_L_SUM (IE) = FOUND                                             !75
  END DO MAIN ! over all elements                                       !76
                                                                        !77
  PRINT *, 'MAXIMUM NUMBER OF ELEMENT NEIGHBORS = ', &                  !78
            MAXVAL (L_TO_L_SUM)                                         !79
  NULLS = COUNT ( L_TO_L_SUM == 0 ) !  CHECK DATA                       !80
  IF ( NULLS > 0 ) THEN                                                 !81
    PRINT *, 'WARNING, ', NULLS, ' ELEMENTS HAVE NO NEIGHBORS'          !82
    N_WARN = N_WARN + 1 ! INCREMENT WARNING                             !83
  END IF                                                                !84
END SUBROUTINE COUNT_ELEMS_AT_ELEM                                      !85
```

Figure 6.3b *Fill the neighbor array and validate*

```
SUBROUTINE FORM_ELEMS_AT_EL (N_ELEMS, NOD_PER_EL, MAX_NP,        &   !  1
                             L_FIRST, L_LAST, NODES, N_SPACE,    &   !  2
                             L_TO_L_SUM, L_TO_L_NEIGH,           &   !  3
                             NEIGH_L, NEEDS, ON_BOUNDARY)            !  4
! * * * * * * * * * * * * * * * * * * * * * * * * * * * * * * *     !  5
!    FORM LIST OF ELEMENTS ADJACENT TO OTHER ELEMENTS                 !  6
! * * * * * * * * * * * * * * * * * * * * * * * * * * * * * * *     !  7
 IMPLICIT NONE                                                       !  8
 INTEGER, INTENT(IN) :: N_ELEMS, NOD_PER_EL, MAX_NP, NEIGH_L         !  9
 INTEGER, INTENT(IN) :: L_FIRST (MAX_NP), L_LAST (MAX_NP)            ! 10
 INTEGER, INTENT(IN) :: NODES (N_ELEMS, NOD_PER_EL)                  ! 11
 INTEGER, INTENT(IN) :: L_TO_L_SUM   (N_ELEMS)                       ! 12
 INTEGER, INTENT(IN) :: N_SPACE, NEEDS ! for pt, edge, face          ! 13
 INTEGER, INTENT(OUT)   :: L_TO_L_NEIGH (NEIGH_L, N_ELEMS)           ! 14
 LOGICAL, INTENT(INOUT) :: ON_BOUNDARY  (N_ELEMS)                    ! 15
                                                                     ! 16
 INTEGER :: ELEM_NODES (NOD_PER_EL), NEIG_NODES (NOD_PER_EL)         ! 17
 INTEGER :: IE, IN, L_TEST, L_START, L_STOP, N_TEST                  ! 18
 INTEGER :: FOUND, NEXT, SUM_L_TO_L                                  ! 19
 INTEGER :: IO_1, KOUNT, NEED, N_FACES, WHERE                        ! 20
                                                                     ! 21
! ON_BOUNDARY   = TRUE IF ELEMENT HAS A FACE ON BOUNDARY              ! 22
! ELEM_NODES    = INCIDENCES ARRAY FOR A SINGLE ELEMENT               ! 23
! FOUND         = CURRENT NUMBER OF LOCAL NEIGHBORS                   ! 24
! KOUNT         = CURRENT NUMBER OF COMMON NODES                      ! 25
! L_FIRST (I)   = ELEMENT WHERE NODE I FIRST APPEARS                  ! 26
! L_LAST (I)    = ELEMENT WHERE NODE I LAST APPEARS                   ! 27
! L_TO_L_NEIGH  = ELEM NEIGHBOR J OF ELEMENT I                        ! 28
! L_TO_L_SUM    = NUMBER OF ELEM NEIGHBORS OF ELEMENT I               ! 29
! NEEDS         = NUMBER OF COMMON NODES TO BE A NEIGHBOR             ! 30
! NEIGH_L       = MAXIMUM NUMBER OF NEIGHBORS AT A ELEMENT            ! 31
! MAX_NP        = TOTAL NUMBER OF NODES                               ! 32
! N_ELEMS       = TOTAL NUMBER OF ELEMENTS                            ! 33
! NOD_PER_EL    = NUMBER OF NODES PER ELEMENT                         ! 34
! NODES         = SYSTEM ARRAY OF INCIDENCES OF ELEMENTS              ! 35
! WHERE         = LOCATION TO INSERT NEIGHBOR, <= MAX_FACES           ! 36
                                                                     ! 37
  NEED = MAX (1, NEEDS) ; L_TO_L_NEIGH = 0  ! INITIALIZE             ! 38
  MAIN : DO IE = 1, N_ELEMS                 ! ELEMENT LOOP           ! 39
                                                                     ! 40
    SUM_L_TO_L = L_TO_L_SUM (IE)              ! MAX NEIGHBORS        ! 41
    FOUND = COUNT (L_TO_L_NEIGH (:, IE) > 0) ! PREVIOUSLY FOUND      ! 42
    IF ( FOUND == SUM_L_TO_L ) CYCLE MAIN     ! ALL FOUND            ! 43
                                                                     ! 44
!         EXTRACT INCIDENCES LIST FOR ELEMENT IE                      ! 45
    CALL ELEMENT_NODES (IE, NOD_PER_EL, NODES, ELEM_NODES)           ! 46
                                                                     ! 47
!         ESTABLISH RANGE OF POSSIBLE ELEMENT NEIGHBORS               ! 48
    L_START = N_ELEMS + 1 ; L_STOP = 0                               ! 49
    DO IN = 1, NOD_PER_EL                                            ! 50
      L_START = MIN (L_START, L_FIRST (ELEM_NODES (IN)) )            ! 51
      L_STOP  = MAX (L_STOP,  L_LAST  (ELEM_NODES (IN)) )            ! 52
    END DO                                                           ! 53
    L_START = MAX (L_START, IE+1) ! SEARCH ABOVE IE ONLY             ! 54
                                                                     ! 55
```

Figure 6.4a *Interface and data for element neighbors*

```
!        LOOP OVER POSSIBLE ELEMENT NEIGHBORS                        ! 56
    IF ( L_START <= L_STOP) THEN                                     ! 57
      RANGE : DO L_TEST = L_START, L_STOP                            ! 58
        KOUNT = 0   ! NO COMMON NODES                                ! 59
                                                                     ! 60
!         EXTRACT NODES OF L_TEST                                    ! 61
        CALL ELEMENT_NODES (L_TEST,NOD_PER_EL,NODES,NEIG_NODES)      ! 62
                                                                     ! 63
!         LOOP OVER INCIDENCES OF POSSIBLE ELEMENT NEIGHBOR          ! 64
        LOCAL : DO IN = 1, NOD_PER_EL                                ! 65
         N_TEST = NEIG_NODES (IN)                                    ! 66
         IF ( N_TEST < 1 ) CYCLE ! to a real node                    ! 67
         IF (L_FIRST (N_TEST) > IE) CYCLE LOCAL ! to next node       ! 68
         IF (L_LAST  (N_TEST) < IE) CYCLE LOCAL ! to next node       ! 69
                                                                     ! 70
!           COMPARE WITH INCIDENCES OF ELEMENT IE                    ! 71
          IF ( ANY ( ELEM_NODES == N_TEST ) ) THEN                   ! 72
            KOUNT = KOUNT + 1          ! SHARED NODE COUNT           ! 73
            IF ( KOUNT == NEED ) THEN ! NEIGHBOR PAIR FOUND          ! 74
              FOUND = FOUND + 1        ! INSERT THE PAIR             ! 75
                                                                     ! 76
!             NOTE: THIS INSERT IS NOT ORDERED.                      ! 77
              WHERE = FOUND ! OR ORDER THE CURRENT FACE              ! 78
              L_TO_L_NEIGH (WHERE, IE) = L_TEST   ! 1 of 2           ! 79
                                                                     ! 80
              NEXT = COUNT ( L_TO_L_NEIGH(:, L_TEST) > 0 )           ! 81
              WHERE = NEXT+1 ! OR ORDER THE NEIGHBOR FACE            ! 82
                                                                     ! 83
              IF ( L_TO_L_SUM (L_TEST) > NEXT )  &                   ! 84
                L_TO_L_NEIGH (NEXT+1, L_TEST) = IE   ! 2 of 2        ! 85
              IF ( SUM_L_TO_L == FOUND ) CYCLE MAIN ! ALL            ! 86
              CYCLE RANGE ! this L_TEST element search loop          ! 87
            END IF ! NUMBER NEEDED                                   ! 88
          END IF ! SHARE AT LEAST ONE COMMON NODE                    ! 89
        END DO LOCAL ! over N_TEST                                   ! 90
      END DO RANGE ! over candidate element L_TEST                   ! 91
    END IF ! a possible candidate                                    ! 92
  END DO MAIN! over all elements                                     ! 93
                                                                     ! 94
  IF ( NEED >= N_SPACE ) THEN ! EDGE OR FACE NEIGHBOR DATA           ! 95
!     SAVE THE ELEMENT NUMBERS THAT FACE THE BOUNDARY                ! 96
    DO IE = 1, N_ELEMS                                               ! 97
      CALL GET_LT_FACES (IE, N_FACES)                                ! 98
      IF ( N_FACES > 0 ) THEN ! MIGHT BE ON THE BOUNDARY             ! 99
        IF ( ANY (L_TO_L_NEIGH (1:N_FACES, IE) == 0)) THEN           !100
          ON_BOUNDARY (IE) = .TRUE.                                  !101
        END IF  ! ON BOUNDARY                                        !102
      END IF ! POSSIBLE ELEMENT                                      !103
    END DO ! OVER ELEMENTS                                           !104
  END IF ! SEARCH OF FACING NEIGHBORS                                !105
END SUBROUTINE FORM_ELEMS_AT_EL                                      !106
```

Figure 6.4b *Fill the neighbor array and check boundary*

```
# SCP_WORD     TYPICAL_VALUE ! REMARKS                                  [DEFAULT]
debug_scp            ! Debug the SCP averaging process                   [F]
face_nodes         3 ! Number of shared nodes on an element face         [d]
grad_base_error      ! Base error estimates on gradients only            [F]
list_el_to_el        ! List elements adjacent to elements                [F]
no_scp_ave           ! Do NOT get superconvergent patch averages         [F]
no_error_est         ! Do NOT compute SCP element error estimates        [F]
pt_el_list           ! List all the elements at each node                [F]
scp_center_only      ! Use center node or element only in average        [T]
scp_center_no        ! Use all elements in the patch in average          [F]
scp_deg_inc        1 ! Increase patch degree by this (1 or 2)            [0]
scp_max_error 5.     ! Allowed % error in energy norm                    [1]
scp_neigh_el         ! Element based patch, all neighbors (default)      [T]
scp_neigh_face       ! Element based patch, facing neighbors             [F]
scp_neigh_pt         ! Nodal based patch, all element neighbors          [F]
scp_only_once        ! Scatter to a node only once per patch             [T]
scp_2nd_deriv        ! Recover 2nd derivatives data also                 [F]
```

Figure 6.5 *Optional SCP control keywords*

to coincide with the Gaussian quadrature points (as illustrated here in Secs. 3.8 and 6.5). Here we will assume that the minimum number of quadratic points needed to properly form the element matrices have locations that correspond to the element super-convergent points, or are reasonably close to them. Thus, as we process each element to build its square matrix, we will want to save, at each quadrature point, its physical location in space and the differential operator matrix, **B**, that will allow the accurate gradients to be computed from the local nodal solution. Looking ahead to the error estimation or other post-processing, we know that at times we will also want to have the constitutive matrix, **E**, so we will also save it. Note that we are allowing for different, but compatible, element shapes in the mesh (and patches) and they would require different numerical integration rules within each shape.

Now we should look ahead to how the above data are to be recovered in the SCP section of the code. The main observation is that, for an unstructured mesh, the element numbers for the elements adjacent to a particular node or element are totally random. While we have a straight forward way to save the above data in a sequential fashion, we need to recover the element data in a random fashion. Thus, we either need to build a database that allows random access recovery of that sequential information or we must decide to re-compute the data in each element of each patch. The author considers the latter to be too expensive, so we select the new database option. In the examples that are presented later the reader will note function calls to save these data, but they could be omitted if the user was willing to pay the cost of recomputing the data.

For the database structure to save and recover the data we could select linked lists, or a tree structure, but there is a simpler way. Fortran has always had a feature known as a "direct access" file that allows the user to randomly recover or change data. The actual data structure employed is left up to the group that writes the compiler, and is mainly hidden from the user. However, the user must declare the 'record number' of the data set to be recovered or changed. Likewise, the record number of each data set must be given as the data are saved to the random access file. This means that some logical way will be

```
SUBROUTINE POST_PROCESS_GRADS (NODES, DD, ITER)                    ! 1
! * * * * * * * * * * * * * * * * * * * * * * * * * * * *          ! 2
!     SAVE ELEMENT GRADIENTS AS SCP INPUT RECORDS                  ! 3
! * * * * * * * * * * * * * * * * * * * * * * * * * * * *          ! 4
Use System_Constants ! for L_S_TOT, N_D_FRE, NOD_PER_EL,           ! 5
    ! N_L_TYPE, N_PRT, THIS_EL, U_FLUX                             ! 6
Use Elem_Type_Data   ! for PT (LT_PARM, LT_QP), WT (LT_QP),        ! 7
    ! G (LT_GEOM, LT_QP), DLG (LT_PARM, LT_GEOM, LT_QP),           ! 8
    ! H (LT_N), DLH (LT_PARM, LT_N , LT_QP), C (LT_FREE),          ! 9
    ! S (LT_FREE, LT_FREE), ELEM_NODES (LT_N), D (LT_FREE)         !10
Use Interface_Header ! for GET_ELEM_*                              !11
 IMPLICIT NONE                                                     !12
 REAL(DP), INTENT(IN) :: DD (N_D_FRE)                              !13
 INTEGER,  INTENT(IN) :: NODES (L_S_TOT, NOD_PER_EL), ITER         !14
 INTEGER :: IE, LT    ! Loops, element type                        !15
                                                                   !16
! D          = NODAL PARAMETERS ASSOCIATED WITH AN ELEMENT         !17
! DD         = ARRAY OF SYSTEM DEGREES OF FREEDOM                  !18
! INDEX      = SYSTEM DOF NOS ASSOCIATED WITH ELEMENT              !19
! ITER       = CURRENT ITERATION NUMBER                            !20
! ELEM_NODES = THE NOD_PER_EL INCIDENCES OF THE ELEMENT            !21
! NOD_PER_EL = NUMBER OF NODES PER ELEMENT                         !22
! N_D_FRE    = TOTAL NUMBER OF SYSTEM DEGREES OF FREEDOM           !23
! N_L_TYPE   = NUMBER OF DIFFERENT ELEMENT TYPES USED              !24
! N_ELEMS    = NUMBER OF ELEMENTS IN SYSTEM                        !25
! NODES      = ELEMENT INCIDENCES OF ALL ELEMENTS                  !26
! U_FLUX     = BINARY UNIT TO STORE GRADIENTS OR FLUXES            !27
                                                                   !28
  LT = 1 ! INITIALIZE ELEMENT TYPES                                !29
  WRITE (N_PRT, "(/,'BEGIN SCP SAVE, ITER =', I4)") ITER           !30
                                                                   !31
!-->   LOOP OVER ELEMENTS                                          !32
  DO IE = 1, N_ELEMS ! for elements, boundary segments             !33
    CALL SET_THIS_ELEMENT_NUMBER (IE) ! Set THIS_EL                !34
                                                                   !35
!        VALIDATE ELEMENT TYPE                                     !36
    IF ( N_L_TYPE > 1) LT = L_TYPE (IE) ! GET TYPE                 !37
    IF ( LT /= LAST_LT ) THEN ! this is a new type                 !38
      CALL SET_ELEM_TYPE_INFO (LT) ! Set controls                  !39
    END IF ! a new element type                                    !40
                                                                   !41
!      RECOVER ELEMENT DEGREES OF FREEDOM                          !42
    ELEM_NODES = GET_ELEM_NODES (IE, LT_N, NODES)                  !43
    INDEX = GET_ELEM_INDEX (LT_N, ELEM_NODES)                      !44
    D = GET_ELEM_DOF (DD)  ! Get all nodal dof                     !45
                                                                   !46
!-->  USE DOF TO RECOVER FLUXES, LIST, SAVE FOR SCP                !47
    IF ( USE_EXACT_FLUX ) THEN                                     !48
      CALL LIST_ELEM_AND_EXACT_FLUXES (U_FLUX, IE)                 !49
    ELSE                                                           !50
      CALL LIST_ELEM_FLUXES (U_FLUX, IE)                           !51
    END IF ! an exact solution is known                            !52
  END DO ! over all elements                                       !53
END SUBROUTINE POST_PROCESS_GRADS                                  !54
```

Figure 6.6 *Preparing data for averaging or post-processing* needed to create a unique number for each record at any quadrature point in the mesh.

For a mesh with a single element type and a single integration rule, we could write a simple equation for the record number. Here we are allowing a mixture of element types and quadrature rules, so we store the record number at each quadrature point in an integer array sized for the maximum number of elements and the maximum number of quadrature points per element. The record numbers are created sequentially as the element matrices are integrated. A file structure, SCP_RECORD_NUMBER, is supplied for randomly recovering the integer record number at any integration point in any element. Like any other file used in a program, a random access file must be opened. It is opened as a *DIRECT* access file of *UNFORMATTED*, or binary, records to minimize storage. We must also declare the length of the data records. It is actually hardware-dependent, so F90 includes an intrinsic function, *INQUIRE*(*IOLENGTH*), that will compute the record length given a list of variables and/or arrays to be included in each record. The unit number assigned to the random access file holding the SCP records is given the variable name *U_SCPR*.

As mentioned above the SCP process can be used to determine the average nodal fluxes and to use them to compute the element error estimates. The general outline of the process is as follows:

1. Preliminary
   a. Build a list of element neighbors.
   b. Open the sequential file unit *U_FLUX* to receive element data related to flux calculations. Those data can also be used for optional post-processing.
   c. Compute the record length necessary to store the coordinates and flux components at a point.
   d. Open the random access file unit *U_SCPR* that will receive the quadrature point coordinates and flux components.

2. Element Matrices Generation Loop
   a. For each element save its number of integration points to file unit *U_FLUX*.
   b. Within the numerical integration loop of the element sequentially save the arrays **XYZ**, **E**, and **B** at each point so that the gradients and/or flux components can be found at the point.
   c. When all elements have been processed rewind the file *U_FLUX* to its beginning.

3. Flux Calculations and Saving Them for Averaging

   After the solution has been obtained it is possible to compute the flux (and gradient) components within each element so that they can be smoothed to nodal values. The element flux calculation is done in subroutine POST_PROCESS_GRADS, as detailed in Fig. 6.6. First, the SCP record number is set to zero. Next, each element is processed in a loop:
   a. Recover the element type;
   b. Gather the nodal degrees of freedom of the element;

```
SUBROUTINE LIST_ELEM_FLUXES (N_FILE, IE)                              ! 1
! * * * * * * * * * * * * * * * * * * * * * * * * * * * *             ! 2
! LIST ELEMENT FLUXES AT QUADRATURE POINTS, ON N_FILE                 ! 3
! * * * * * * * * * * * * * * * * * * * * * * * * * * * *             ! 4
Use System_Constants ! for N_ELEMS, N_R_B, N_SPACE,                   ! 5
    ! FLUX_NAME, XYZ_NAME, N_FILE5, IS_ELEMENT, U_PLT4,               ! 6
    ! U_SCPR, GRAD_BASE_ERROR                                         ! 7
Use Elem_Type_Data   ! for LT_FREE, D (LT_FREE)                       ! 8
 IMPLICIT NONE                                                        ! 9
 INTEGER,  INTENT(IN) :: N_FILE, IE                                   !10
 INTEGER,  SAVE :: TEST_IE, TEST_IP, J, N_IP, EOF, IO_1               !11
 REAL(DP), SAVE :: DERIV_MAX = 0.d0                                   !12
                                                                      !13
!                Automatic Arrays                                     !14
 REAL(DP) :: XYZ (N_SPACE), E (N_R_B, N_R_B),          &              !15
             B (N_R_B, LT_FREE), STRAIN (N_R_B + 2), &                !16
             STRESS (N_R_B + 2)                                       !17
                                                                      !18
! B       = GRADIENT VERSUS DOF MATRIX: (N_R_B, LT_FREE)              !19
! D       = NODAL PARAMETERS ASSOCIATED WITH AN ELEMENT               !20
! E       = ELEMENT CONSTITUTIVE MATRIX AT GAUSS POINT                !21
! LT_N    = NUMBER OF NODES PER ELEMENT                               !22
! LT_FREE = NUMBER OF DEGREES OF FREEDOM PER ELEMENT                  !23
! N_ELEMS = TOTAL NUMBER OF ELEMENTS                                  !24
! N_R_B   = NUMBER OF ROWS IN B AND E MATRICES                        !25
! N_SPACE = DIMENSION OF SPACE                                        !26
! N_FILE  = UNIT FOR POST SOLUTION MATRICES STORAGE                   !27
! STRAIN  = GENERALIZED STRAIN OR FLUX VECTOR                         !28
! STRESS  = GENERALIZED STRESS OR GRADIENT VECTOR                     !29
! XYZ     = SPACE COORDINATES AT A POINT                              !30
! U_PLT4  = UNIT TO STORE PLOT DATA, IF > 0                           !31
! U_SCPR  = BINARY UNIT FOR SUPER_CONVERGENT PATCH RECOVERY           !32
                                                                      !33
!-->      FIRST CALL: PRINT TITLES, INITIALIZE, OPEN FILE             !34
 IF ( IE == 1) THEN                          ! FIRST ELEMENT          !35
   REWIND (N_FILE) ; RECORD_NUMBER = 0   ! INITIALIZE                 !36
   WRITE (N_PRT, 5) XYZ_NAME (1:N_SPACE), FLUX_NAME (1:N_R_B)         !37
   5 FORMAT (/,                                                  &    !38
   '** FLUX COMPONENTS AT ELEMENT INTEGRATION POINTS **', &           !39
   /, 'ELEMENT, PT, ', (6A12) )                                       !40
                                                                      !41
!    OPEN FLUX PLOT FILE IF ACTIVE (BINARY FASTER)                    !42
   IF (N_FILE5 >0) OPEN (N_FILE5,FILE='el_qp_xyz_grads.tmp',&         !43
      ACTION='WRITE', STATUS='REPLACE', IOSTAT = IO_1)                !44
   IF (U_PLT4 > 0) OPEN (U_PLT4,FILE='el_qp_xyz_fluxes.tmp',&         !45
      ACTION='WRITE', STATUS='REPLACE', IOSTAT = IO_1)                !46
 END IF ! THIS IS THE FIRST ELEMENT                                   !47
                                                                      !48
```

Figure 6.7a *Interface and data to establish SCP records*

```
!   IS THIS AN ELEMENT, BOUNDARY, OR ROBIN SEGMENT ?               !49
  IF ( IS_ELEMENT ) THEN ! ELEMENT RESULTS                          !50
                                                                    !51
    READ (N_FILE, IOSTAT = EOF) N_IP  ! # INTEGRATION POINTS        !52
                                                                    !53
!-->     READ COORDS, CONSTITUTIVE, AND DERIVATIVE MATRIX           !54
    DO J = 1, N_IP  ! OVER ALL INTEGRATION POINTS                   !55
      READ (N_FILE, IOSTAT = EOF) XYZ, E, B                         !56
                                                                    !57
!        CALCULATE DERIVATIVES,  STRAIN = B * D                     !58
      STRAIN (1:N_R_B) = MATMUL (B, D)                              !59
                                                                    !60
!        FLUX FROM CONSTITUTIVE DATA                                !61
      STRESS (1:N_R_B) = MATMUL (E, STRAIN (1:N_R_B))               !62
                                                                    !63
!-->     PRINT COORDINATES AND FLUX AT THE POINT                    !64
      WRITE (N_PRT, '(I7, I3, 10(ES12.4))')    &                    !65
            IE, J, XYZ, STRESS (1:N_R_B)                            !66
                                                                    !67
!-->     STORE FLUX RESULTS TO BE PLOTTED LATER, IF USED            !68
      IF (U_PLT4  > 0) WRITE (U_PLT4,  '( (10(1PE6.5)) )') &        !69
                             XYZ, STRESS (1:N_R_B)                  !70
      IF (N_FILE5 > 0) WRITE (N_FILE5, '( (10(1PE6.5)) )') &        !71
                             XYZ, STRAIN (1:N_R_B)                  !72
                                                                    !73
!        SAVE COORDINATES & FLUX FOR SCP FLUX AVERAGING             !74
      IF ( U_SCPR > 0 ) THEN ! SCP recovery is active               !75
        RECORD_NUMBER              = RECORD_NUMBER + 1              !76
        SCP_RECORD_NUMBER (IE, J) = RECORD_NUMBER                   !77
                                                                    !78
        IF ( GRAD_BASE_ERROR ) THEN ! User override                 !79
          WRITE (U_SCPR, REC = RECORD_NUMBER) &                     !80
                XYZ, STRAIN (1:SCP_FIT)                             !81
        ELSE                        ! Usual case                    !82
          WRITE (U_SCPR, REC = RECORD_NUMBER) &                     !83
                XYZ, STRESS (1:SCP_FIT)                             !84
        END IF ! GRAD VS FLUX IS DESIRED                            !85
      END IF ! SCP RECOVERY                                         !86
                                                                    !87
    END DO  ! OVER INTEGRATION POINTS                               !88
  END IF ! ELEMENT OR BOUNDARY SEGMENT OR MIXED BOUNDARY            !89
  CALL UPDATE_SCP_STATUS !  FLAG IF SCP DATA WERE SAVED             !90
END SUBROUTINE LIST_ELEM_FLUXES                                     !91
```

Figure 6.7b *Saving element gradient or flux data to files*

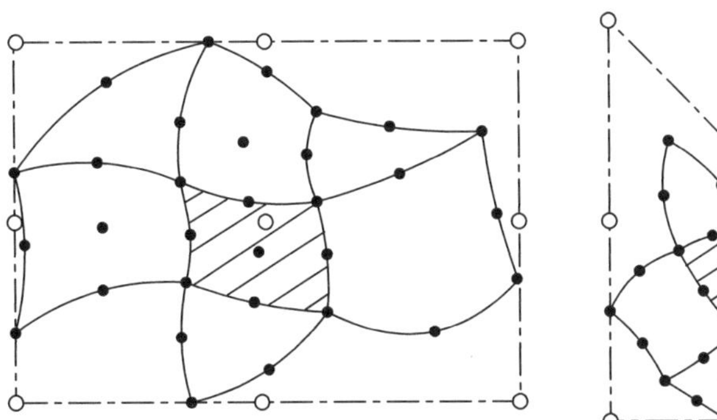

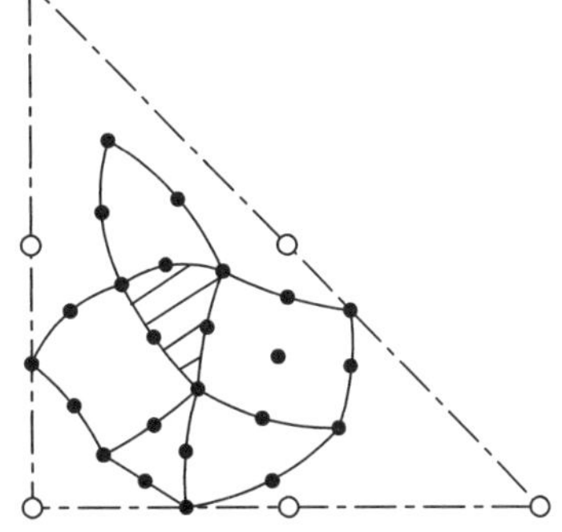

Figure 6.8 *Bounding a group of elements with a constant Jacobian patch*

c. Read the number of quadrature points in the element from *U_FLUX*;

d. Quadrature Point Loop

For each integration point, in routine *LIST_ELEM_FLUXES*, sequentially recover the **XYZ**, **E**, and **B** arrays from *U_FLUX*. Multiply **B** by the element *dof* to get the gradients or strains at the point, and then multiply those by the constitutive array, **E**, to get the fluxes, or stresses at the point. The element and quadrature point numbers are then printed along with their coordinates and flux, or stress, components. Lastly, the SCP database is updated by incrementing the record number by one, and then writing the coordinates and flux component arrays to the random access file, *U_SCPR*, as that record is to be later randomly recovered in the patch smoothing process. The above details are shown in Fig. 6.7.

6.2 SCP nodal flux averaging

Having developed the above database on unit *U_SCPR* we can now average the flux components at every node in each patch and then average them for each node in the mesh. Here we assume an element based patch system for calculating the averages. In subroutine *CALC_SCP_AVE_NODE_FLUXES* we loop over every element and carry out the least squares fit in its associated patch. Looking ahead to that process we must select a polynomial, **P**, to be used in the patch. We must make a choice for that function. We might select a complete polynomial of a given degree, or a Serendipity polynomial of a given edge degree, etc. In the current implementation the default is to select that polynomial to be exactly the same as the polynomial used to interpolate the element for which the patch is being constructed. This means that we will select a constant Jacobian patch 'element' that has its local axes parallel to the global axes and completely surrounds the standard elements that make up the patch. This is easily done by searching for the maximum and minimum components of all of the element nodes in the patch. Such a process is illustrated in Fig. 6.8 where the active element used to select the patch element type is shown crosshatched. It would also be easy to allow the user to select a patch type and degree through a keyword control input. The full details of the process are given in the source code of Fig. 6.9 and the main points are outlined below.

The least squares flux averaging process is:

a. Zero the nodal flux array and the counter for each node.

b. Loop over each element in the mesh:

1. Extract its element neighbors to define the patch.
2. Find the spatial 'box' that bounds the patch.
3. Find the number of quadrature points in the patch (i.e. sum the count in each element of the patch).
4. Determine the element type and thus the patch 'element' shape (line, triangle, hexahedron, etc.) and the corresponding patch polynomial degree.
5. Allocate storage for the least squares fit arrays.
6. Set the fit matrix row number to zero.

```
 SUBROUTINE  CALC_SCP_AVE_NODE_FLUXES (NODES, X ,L_NEIGH, &         !  0
                                       SCP_AVERAGES)                !  1
! * * * * * * * * * * * * * * * * * * * * * * * * * * * *           !  2
!  CALCULATE THE SUPER_CONVERGENCE_PATCH AVERAGE FLUXES             !  3
!     AT ALL NODES IN THE MESH VIA THE SVD METHOD                   !  4
! * * * * * * * * * * * * * * * * * * * * * * * * * * * *           !  5
 Use System_Constants ! MAX_NP, NEIGH_P, N_ELEMS, N_PATCH,          !  6
                      ! N_QP, SCP_FIT, U_SCPR, ON_BOUNDARY          !  7
 Use Elem_Type_Data   ! for LAST_LT, LT_*                           !  8
 Use SCP_Type_Data    ! for SCP_H (SCP_N), SCP_DLH                  !  9
  IMPLICIT NONE                                                     ! 10
  INTEGER,  INTENT (IN)  :: NODES   (L_S_TOT, NOD_PER_EL)           ! 11
  INTEGER,  INTENT (IN)  :: L_NEIGH (NEIGH_P, N_PATCH)              ! 12
  REAL(DP), INTENT (IN)  :: X       (MAX_NP, N_SPACE)               ! 13
  REAL(DP), INTENT (OUT) :: SCP_AVERAGES (MAX_NP, SCP_FIT)          ! 14
                                                                    ! 15
  INTEGER :: MEMBERS    (NEIGH_P+1) ! ELEMENTS IN PATCH             ! 16
  INTEGER :: SCP_COUNTS (MAX_NP)    ! PATCH HITS PER NODE           ! 17
  INTEGER :: POINTS                 ! CURRENT PATCH EQ SIZE         ! 18
  INTEGER :: L_IN_PATCH             ! NUM OF ELEMS IN PATCH         ! 19
  INTEGER :: FIT, IL, IP, IQ, LM    ! LOOPS                         ! 20
  INTEGER :: ROW                      ! IN PATCH ARRAYS             ! 21
  INTEGER :: LT, REC_LM_IQ, SCP_STAT ! MEMBER OF PATCH              ! 22
  INTEGER, PARAMETER :: ONE = 1                                     ! 23
                                                                    ! 24
  REAL(DP) :: XYZ_MIN (N_SPACE), XYZ_MAX (N_SPACE) ! BOUNDS         ! 25
  REAL(DP) :: XYZ     (N_SPACE), FLUX    (N_R_B)   ! PT, FLUX       ! 26
  REAL(DP) :: POINT   (N_SPACE)                    ! SCP POINT      ! 27
                                                                    ! 28
  REAL(DP), ALLOCATABLE :: PATCH_SQ  (:, :)  ! SCRATCH ARRAY        ! 29
  REAL(DP), ALLOCATABLE :: PATCH_DAT (:, :)  ! SCRATCH ARRAY        ! 30
  REAL(DP), ALLOCATABLE :: PATCH_P   (:, :)  ! SCRATCH ARRAY        ! 31
  REAL(DP), ALLOCATABLE :: PATCH_WRK (:)     ! SCRATCH ARRAY        ! 32
  REAL(DP), ALLOCATABLE :: PATCH_FIT (:, :)  ! ANSWERS              ! 33
                                                                    ! 34
! L_S_TOT    = TOTAL NUMBER OF ELEMENTS & THEIR SEGMENTS            ! 35
! L_NEIGH    = ELEMENTS FORMING THE PATCH                           ! 36
! L_TYPE     = ELEMENT TYPE NUMBER                                  ! 37
! LT         = ELEMENT TYPE NUMBER (IF USED)                        ! 38
! LT_QP      = NUMBER OF QUADRATURE PTS FOR ELEMENT TYPE            ! 39
! MAX_NP     = NUMBER OF SYSTEM NODES                               ! 40
! MEMBERS    = ELEMENT NUMBERS MACKING UP A SCP                     ! 41
! NEIGH_P    = MAXIMUM NUMBER OF ELEMENTS IN A PATCH                ! 42
! NOD_PER_EL = NUMBER OF NODES PER ELEMENT                          ! 43
! NODES      = NODAL INCIDENCES OF ALL ELEMENTS                     ! 44
! N_ELEMS    = NUMBER OF ELEMENTS IN SYSTEM                         ! 45
! N_PATCH    = NUMBER OF PATCHES, MAX_NP OR N_ELEMS                 ! 46
! N_QP       = MAXIMUN NUMBER OF QUADRATURE POINTS, >= LT_QP        ! 47
! N_R_B      = NUMBER OF ROWS IN B AND E MATRICES                   ! 48
! N_SPACE    = DIMENSION OF SPACE                                   ! 49
! PATCH_FIT  = LOCAL PATCH VALUES FOR FLUX AT ITS NODES             ! 50
! POINT      = LOCAL POINT IN PATCH INTERPOLATION SPACE             ! 51
! POINTS     = TOTAL NUMBER OF QUADRATURE POINTS IN PATCH           ! 52
! SCP_AVERAGES = AVERAGED FLUXES AT ALL NODES IN MESH               ! 53
! SCP_FIT    = NUMBER IF TERMS BEING FIT, OR AVERAGED               ! 54
! SCP_H      = INTERPOLATION FUNCTIONS OF PATCH, USUALLY H           ! 55
```

Figure 6.9a *Interface and data for flux averages*

```
! SCP_N         = NUMBER OF NODES PER PATCH                          !  56
! SCP_RECORD_NUMBER = SCP DIRECT ACCESS RECORD LOCATOR               !  57
! U_SCPR        = SUPER_CONVERGENT PATCH RECOVERY UNIT               !  58
! X             = COORDINATES OF SYSTEM NODES                        !  59
! XYZ           = SPACE COORDINATES AT A POINT                       !  60
! XYZ_MAX       = UPPER BOUNDS FOR SCP GEOMETRY                      !  61
! XYZ_MIN       = LOWER BOUNDS FOR SCP GEOMETRY                      !  62
                                                                     !  63
  LT=1 ; LAST_LT=0 ; SCP_COUNTS=0 ; SCP_AVERAGES=0                   !  64
                                                                     !  65
  IF ( N_PATCH == 0 ) THEN ! No data supplied                        !  66
   PRINT *,'NO PATCHS GIVEN, SKIPPING AVERAGES'; RETURN              !  67
  END IF                                                             !  68
                                                                     !  69
  DO IP = 1, N_PATCH              ! LOOP OVER EACH PATCH             !  70
    MEMBERS  = 0 ; POINTS = 0    ! INITIALIZE                        !  71
                                                                     !  72
!     ELEMENT OR NODAL CENTERED PATCH TYPE                           !  73
    IF ( .NOT. SCP_NEIGH_PT ) THEN ! ELEMENT BASED                   !  74
!      GET ELEMENT NEIGHBORS TO DEFINE THE PATCH                     !  75
      MEMBERS = (/ IP, L_NEIGH (:, IP) /)                            !  76
    ELSE ! NODAL BASED PATCH OF ELEMENTS                             !  77
!      GET ELEMENT NEIGHBORS TO DEFINE THE PATCH                     !  78
      MEMBERS     = L_NEIGH (:, IP)                                  !  79
    END IF ! PATCH BASIS                                             !  80
                                                                     !  81
    L_IN_PATCH = COUNT ( MEMBERS > 0 )                               !  82
    IF ( L_IN_PATCH <= 1 ) CYCLE ! TO AN ACTIVE PATCH                !  83
                                                                     !  84
!    FIND TYPE OF SCP NEEDED HERE, VERIFY GEOMETRY                   !  85
    CALL DETERMINE_SCP_BOUNDS (L_IN_PATCH, MEMBERS, &                !  86
                  NODES, X, XYZ_MIN, XYZ_MAX, POINTS)                !  87
                                                                     !  88
    IF ( PATCH_ALLOC_STATUS ) THEN ! DEALLOCATE ARRAYS               !  89
      DEALLOCATE (PATCH_WRK) ; DEALLOCATE (PATCH_SQ )                !  90
      DEALLOCATE (PATCH_FIT) ; DEALLOCATE (PATCH_DAT)                !  91
      DEALLOCATE (PATCH_P  ) ; PATCH_ALLOC_STATUS=.FALSE.            !  92
    END IF ! STATUS CHECK                                            !  93
                                                                     !  94
!     ALLOCATE NEXT SET OF LOCAL PATCH RELATED ARRAYS                !  95
    ALLOCATE ( PATCH_P   (POINTS, SCP_N  ) )                         !  96
    ALLOCATE ( PATCH_DAT (POINTS, SCP_FIT) )                         !  97
    ALLOCATE ( PATCH_FIT (SCP_N , N_R_B  ) )                         !  98
    ALLOCATE ( PATCH_SQ  (SCP_N , SCP_N  ) )                         !  99
    ALLOCATE ( PATCH_WRK (SCP_N )          )                         !100
    PATCH_ALLOC_STATUS = .TRUE.                                      !101
                                                                     !102
!    ZERO PATCH WORKSPACE AND RESULTS ARRYS                          !103
    PATCH_P   = 0.d0 ; PATCH_DAT = 0.d0 ; PATCH_FIT = 0.d0           !104
    PATCH_SQ = 0.d0 ; PATCH_WRK = 0.d0                               !105
                                                                     !106
```

Figure 6.9b *Establish bounds and dynamic memory for each patch*

```
!    PREPARE LEAST SQUARES FIT MATRICES                                !107
   LAST_LT = 0 ; ROW = 0            ! INITIALIZE                       !108
   DO IL = 1, L_IN_PATCH            ! PATCH MEMBER LOOP                !109
    LM = MEMBERS (IL)               ! ELEMENT IN PATCH                 !110
                                                                       !111
    ! GET ELEMENT TYPE NUMBER                                          !112
    IF (N_L_TYPE > 1) LT=L_TYPE (LM)                                   !113
    IF ( LT /= LAST_LT ) THEN          ! ELEMENT TYPE                  !114
      CALL GET_ELEM_TYPE_DATA (LT)     ! TYPE CONTROLS                 !115
      LAST_LT = LT                                                     !116
    END IF ! a new element type and scp type                           !117
                                                                       !118
    DO IQ = 1, LT_QP              ! LOOP OVER GAUSS POINTS             !119
     ROW = ROW + 1                ! UPDATE LOCATION                    !120
                                                                       !121
!      RECOVER EACH FLUX VECTOR TO FOR LEAST SQ FIX                    !122
     REC_LM_IQ = SCP_RECORD_NUMBER (LM, IQ) ! REC NUMBER               !123
                                                                       !124
!     GET GAUSS PT COORD & FLUX SAVED IN LIST_ELEM_FLUXES              !125
     READ (U_SCPR,REC=REC_LM_IQ,IOSTAT=SCP_STAT) XYZ, FLUX             !126
     SELECT CASE ( SCP_STAT ) ! for read status                        !127
       CASE (:-1)                                                      !128
         STOP    'EOR OR EOF, CALC_SCP_AVE_NODE_FLUXES'                !129
       CASE (1:)                                                       !130
         PRINT *,'MEMBER ELEMENT = ', LM, ' AT POINT ', IQ             !131
         PRINT *,'REC_LM_IQ      = ', REC_LM_IQ                        !132
         STOP    'BAD SCP_STAT, CALC_SCP_AVE_NODE_FLUXES'              !133
       CASE DEFAULT ! NO READ ERROR                                    !134
     END SELECT ! RANDOM ACCESS READ ERROR                             !135
                                                                       !136
!       CONVERT IQ XYZ TO LOCAL PATCH POINT                            !137
     POINT = GET_SCP_PT_AT_XYZ (XYZ, XYZ_MIN, XYZ_MAX)                 !138
                                                                       !139
!        EVALUATE PATCH INTERPOLATION AT LOCAL POINT                   !140
     IF ( .NOT. SCP_SCAL_ALLOC ) CALL    &                             !141
               ALLOCATE_SCP_INTERPOLATIONS                             !142
     CALL GEN_ELEM_SHAPE (POINT,SCP_H,SCP_N,N_SPACE,ONE)               !143
                                                                       !144
!      INSERT FLUX & INTERPOLATIONS INTO PATCH MATRICES                !145
     PATCH_DAT (ROW, 1:N_R_B) = FLUX  (:)                              !146
     PATCH_P   (ROW, :)       = SCP_H (:)                              !147
    END DO ! FOR EACH IQ FLUX VECTOR                                   !148
                                                                       !149
   END DO ! FOR EACH IL PATCH MEMBER                                   !150
! ASSEMBLY OF PATCH COMPLETED                                          !151
                                                                       !152
!     VALIDATE CURRENT PATCH                                           !153
   IF ( POINTS < SCP_N ) THEN                                          !154
     PRINT *,'WARNING: SKIPPING PATCH ', &                             !155
              IP, ' WITH ONLY ', POINTS, ' EQUATIONS'                  !156
     CYCLE ! TO NEXT PATCH                                             !157
   END IF ! INSUFFICIENT DATA                                          !158
```

Figure 6.9c *Build the least squares fit in each patch*

```
!      USE SINGULAR VALUE DECOMPOSITION SOLUTION METHOD            !159
   CALL SVDC_FACTOR (PATCH_P,POINTS,SCP_N,PATCH_WRK,PATCH_SQ)      !160
   WHERE ( PATCH_WRK < EPSILON(1.d0) ) PATCH_WRK = 0.D0            !161
                                                                   !162
   DO FIT = 1, N_R_B   ! LOOP FOR EACH FLUX COMPONENT              !163
     CALL SVDC_BACK_SUBST (PATCH_P, PATCH_WRK, PATCH_SQ, &         !164
                       POINTS, SCP_N, PATCH_DAT (:, FIT),&         !165
                       PATCH_FIT (:, FIT))                         !166
   END DO ! FOR FLUXES   COMPONENTS                                !167
                                                                   !168
!  INTERPOLATE AVERAGES TO ALL NODES IN THE PATCH. SCATTER         !169
!  PATCH NODAL AVERAGES TO SYSTEM NODES, INCREMENT COUNTS          !170
                                                                   !171
   IF (DEBUG_SCP .AND. IP==1) PRINT *,'AVERAGING FLUXES'           !172
   CALL EVAL_SCP_FIT_AT_PATCH_NODES (IP, NODES, X,        &        !173
        L_IN_PATCH,MEMBERS, XYZ_MIN, XYZ_MAX, PATCH_FIT, &         !174
        SCP_AVERAGES, SCP_COUNTS)                                  !175
                                                                   !176
! NOTE: use Loubignac iteration here for new solution              !177
                                                                   !178
!       DEALLOCATE LOCAL PATCH RELATED ARRAYS                      !179
   IF (SCP_SCAL_ALLOC) CALL DEALLOCATE_SCP_INTERPOLATIONS          !180
   DEALLOCATE (PATCH_WRK) ; DEALLOCATE (PATCH_SQ )                 !181
   DEALLOCATE (PATCH_FIT) ; DEALLOCATE (PATCH_DAT)                 !182
   DEALLOCATE (PATCH_P  ) ; PATCH_ALLOC_STATUS=.FALSE.             !183
                                                                   !184
  END DO ! FOR EACH IP PATCH IN MESH                               !185
                                                                   !186
!   FINALLY, AVERAGE FLUXES FOR EACH NODAL HIT COUNT               !187
  DO FIT = 1, MAX_NP ! FOR ALL NODES                               !188
    IF ( SCP_COUNTS (FIT) /= 0 ) THEN ! ACTIVE NODE                !189
      SCP_AVERAGES (FIT, :) = SCP_AVERAGES (FIT, :) &              !190
                            / SCP_COUNTS (FIT)                     !191
    ELSE  ! could skip since initialized                           !192
      SCP_AVERAGES (FIT, :) = 0.D0                                 !193
    END IF                                                         !194
  END DO ! FOR (AN UNWEIGHTED) AVERAGE                             !195
                                                                   !196
!    REPORT AVERAGED MAX & MIN VALUES AND LOCATIONS                !197
  CALL MAX_AND_MIN_SCP_AVE_F90 (SCP_AVERAGES)                      !198
END SUBROUTINE CALC_SCP_AVE_NODE_FLUXES                            !199
```

Figure 6.9d *Factor each patch then average at all nodes*

7. For each element in the patch loop over the following steps:
   A. Find its type and quadrature rule
   B. Loop over each of its quadrature points
      1) Increment the row number by one.
      2) Use the element number and quadrature point number pair as subscripts in the *SCP_RECORD_NUMBER* function to recover the random record number for that point.
      3) Read the physical coordinates and flux components from random access file *U_SCPR* by using that record number.
      4) Use the constant Jacobian of the patch to convert the physical

location to the corresponding non-dimensional coordinates in the patch. Note that this helps reduce the numerical ill-conditioning that is common in a least squares fit process.

5) Evaluate the patch interpolation polynomial at the local point (by utilizing the standard element interpolation library). Insert it into the left hand side of this row of the coefficient matrix.
6) Substitute the flux components into the right hand side data matrix, in the same row. Of course the number of columns on the right hand side is the same as the number of flux components. This is also the size of the patch result matrix, **a**, to be computed.

8. Having completed the loop over all the elements in this patch we now have the rectangular arrays cited in Eq. 2.45 but we have not computed their actual matrix products as shown in Eq. 2.46. While that equation is the standard way to describe a least squares fit we do not actually use that process. Instead, we try to avoid possible numerical ill-conditioning by using an equivalent but more powerful process called the singular value decomposition algorithm [10]. That process first factors the associated patch square matrix (in subroutine *SVDC_FACTOR*) and then recovers the rectangular array of local continuous patch nodal flux values (with subroutine *SVDC_BACK_SUBST*). However, we want smooth flux values at the actual modes of the elements, not values at the patch nodes. Thus, for the elements in question we need to interpolate the patch results back to the system nodes contained within the current patch.
9. Loop over nodes in this patch:
   a) Use the constant patch Jacobian to convert the node coordinates to non-dimensional coordinates of the patch
   b) Evaluate the patch interpolation matrix, ***SCP_H***, at each node point. It is usually the same as the core element interpolation functions **H**.
   c) Compute the flux components at the node by the matrix product of ***SCP_H*** and the continuous gradients at the patch nodes, **a**.
   d) Increment the nodal counter for patch contributions by one and scatter the node flux components to the rectangular system flux nodal array.

C. Optional Improvement of the Solution

At this stage in the SCP process one can use the least square smoothed gradients in this patch to get a locally improved solution value estimate at all of the patch's interior nodes. However, one may want to just do so for the single parent element about which the patch was constructed. The algorithm [9] is a form of the Loubignac iterative process:

1) Read or reform element matrices, $\mathbf{S}_e$ and $\mathbf{C}_e$, here.
2) Use a higher order quadrature rule to form the equilibrating vector

$$\mathbf{V}_e = \int_{\Omega^e} \mathbf{B}^* (\sigma^* - \hat{\sigma})\, d\Omega.$$

3) Assemble the elements in the patch into a local linear system:

$$\mathbf{S}^* \boldsymbol{\phi}_{\text{new}} = \mathbf{C} - \mathbf{V}.$$

4) Apply the previous solution as essential boundary conditions at all nodes on the patch boundary.
5) Solve for the new interior node values (always a small system but especially small for a patch of elements around a single node as in original ZZ patch paper). Call them $\boldsymbol{\phi}_e^*$.
6) Compute norm of $\boldsymbol{\phi}_e - \boldsymbol{\phi}_e^*$ to use as an additional term in the final error estimator.

D. Final nodal flux average.

As shown in Fig. 6.10, most nodes are associated with more than one patch. Having processed every patch for the mesh each node has now received as many nodal flux estimates as there were patches that contained that node. We finalize the nodal flux values by simply dividing each node's flux component sums by that integer counter, print the result, and re-save them in the rectangular array SCP_AVERAGES for use by the element error estimator or other user-defined post-processing.

6.3 Computing the SCP element error estimates

For a homogeneous domain, or sub-domain, the above nodal averaging process provides a continuous flux approximation that should be much closer to the true solution than the element discontinuous fluxes. Thus, it is reasonable to base the element error estimator on the average nodal fluxes from the SCP process. Basically, we will want to integrate the difference between the spatial distributions of the two flux estimates so that we can calculate the error norms of interest in each element. Then we will sum those

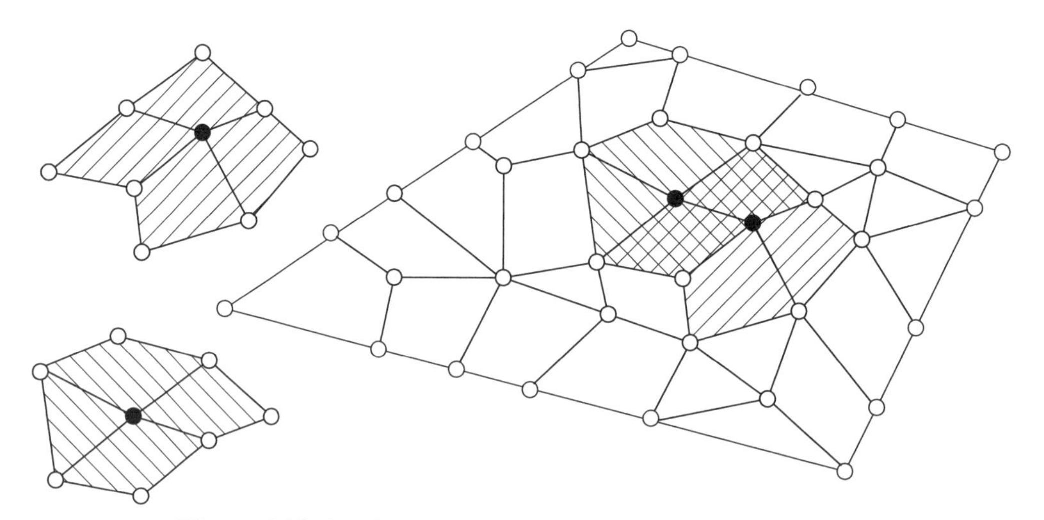

Figure 6.10 *Overlapping patches give multiple node estimates*

scalar values over all elements so that we can establish the relative errors and how they compare to the allowed value specified by the user. Of course, we will evaluate the element norms by numerical integration. This will require a higher order quadrature rule than the one needed to evaluate the element square matrix because the interpolation function, **P** (which is usually **H**), is of higher polynomial degree than the **B** matrix (which contains the derivatives of **H**) used in forming the element square matrix.

Subroutine *SCP_ERROR_ESTIMATES* implements the major steps outlined below.

1. Preliminary Setup

a. Initialize all of the norms to zero.

b. If the mesh has a constant constitutive matrix, **E**, then recover it and invert it once for later use in calculating the energy norm.

2. Loop over all elements in the mesh:

a. Recover the element type (shape, number of nodes, quadrature rule for **B**, etc.)

b. Determine the quadrature rule to integrate the **P** array (here the **H** array), allocate storage for that array to be pre-computed at each quadrature point, and then fill those arrays.

c. Extract the element's node numbers, coordinates, and *dof.*

d. At each node on the element gather the continuous nodal flux components from the system SCP averages, $\mathbf{a}^e \subset \mathbf{a}$.

e. Numerical integration loop over the element to form element norms and increment system norms:

1. Recover the **H** array at the point and its local derivatives.
2. Obtain the physical coordinates, Jacobian and its inverse.
3. If the L_2 norm of the solution is desired, interpolate for the value at the point. Increment the L_2 norm integral. If the exact solution has been provided, then compute its norm also.
4. Compute the physical gradients of the original finite element solution. Form the **B** matrix at the point for the current application, $\hat{\varepsilon} = \mathbf{B}\,\phi$. If the constitutive array, **E**, is smoothly varying, then we could evaluate it at this point (and compute its inverse). Otherwise, we employ the **E** matrix saved in the preliminary setup. Now we recover the standard element flux estimate by matrix multiplication, $\hat{\sigma} = \mathbf{E}\,\hat{\varepsilon}$. We can also increment the L_2 norm of this term if desired.
5. Now we are ready to recover the continuous flux values and approximate the stress error. We simply carry out the matrix product of the interpolation functions, **H**, and the nodal fluxes, **a**, at the quadrature point.

$$\sigma^* = \mathbf{H}\,\mathbf{a} \tag{6.1}$$

The difference between these components and those from the previous step are formed to define the stress error

$$\mathbf{e}_\sigma = \sigma^* - \hat{\sigma}. \tag{6.2}$$

If desired the square of this term (its dot product with itself) is obtained for its increment to the L_2 stress norm.

6. In this implementation we almost always use the flux error to compute the error energy norm, so at this stage we form the related triple matrix product, $\mathbf{e}_\sigma^T \mathbf{E}^{-1} \mathbf{e}_\sigma$, and increment the quadrature point contribution to the element norm

$$||\mathbf{e}^e||^2 = \sum_q^{n_q} (\sigma^* - \hat{\sigma})_q^T \mathbf{E}_q^{-1} (\sigma^* - \hat{\sigma})_q . \tag{6.3}$$

If the user has supplied an expression for the exact flux components, then they are evaluated at the physical coordinates, and the corresponding L_2 and error energy norms are updated for later comparisons and to find the effectivity index.

Having incremented all of the element norms, they are complete at the end of this quadrature loop for the current element. The active element norm values are then added to the current values of the corresponding system norms. At times we also want to use the element and system volume measures so that we can get some norm volumetric averages. Thus, the determinant of the Jacobian at the above points are also used to obtain the element volumes so that they are available for these optional calculations.

Upon completing the loop over all elements we have the element norms, the element volume, the system norms,

$$||\mathbf{e}||^2 = \sum_e^{n_e} ||\mathbf{e}^e||^2, \tag{6.4}$$

and the system volume. At this point we can then carry out the element adaptivity processes outlined at the end of the previous chapter. The coding details associated with these steps are in the single subroutine but are broken out into major segments in Fig. 6.11.

6.4 Hessian matrix *

There are times when one is also interested in the estimates of the second derivatives of the solution with respect to the spatial coordinates. Examples include the application of the Streamline Upwind Petrov Galerkin (SUPG) method for advection-diffusion problems and the inclusion of 'stabilization' (or governing PDE residual) terms in the solution of the Navier-Stokes equations. The matrix of second-order partial derivatives of a function is called the Hessian matrix. If the function is C^2, (that is, has continuous second derivatives) the Hessian is symmetric due to the equality of the mixed partial derivatives. If one is employing high-order interpolation elements, one could proceed with direct estimates of the second derivatives at the element level. Of course, we would expect a decrease in accuracy compared to the element gradient estimates. Even with higher order elements the first and second derivatives are not continuous at the boundaries of surrounding elements (that is, elements that would make up a patch). Thus, a Hessian matrix based on a patch calculation will usually not be symmetric. Assuming the use of parametric elements, we need to employ the Jacobian. The Jacobian defines the mapping from the parametric space to the physical space. In two dimensions:

```
SUBROUTINE SCP_ERROR_ESTIMATES (NODES, X, SCP_AVERAGES, DOF_SYS,& !  1
                                ELEM_ERROR_ENERGY,              & !  2
                                ELEM_REFINEMENT, ERR_MAX)         !  3
! * * * * * * * * * * * * * * * * * * * * * * * * * * * * * * *   !  4
! USE INTEGRAL OF DIFFERENCE BETWEEN THE RECOVERED AVERAGE NODAL  !  5
! FLUXES AND ORIGINAL ELEMENT FLUXES TO ESTIMATE ELEMENT ERROR    !  6
!                   IN THE ENERGY NORM                            !  7
! * * * * * * * * * * * * * * * * * * * * * * * * * * * * * * *   !  8
! Note: Debug options, 2nd derivative options, saving to          !  9
!       plotter files are not shown here to save space.           ! 10
!       See source library for full details.                      ! 11
 Use System_Constants ! for MAX_NP, NEIGH_L, N_ELEMS, N_QP,       ! 12
                      ! SCP_FIT, U_SCPR, SKIP_ERROR               ! 13
 Use Select_Source                                                ! 14
 Use Elem_Type_Data  ! for:                                       ! 15
   ! LT_FREE, LT_GEOM, LT_N, LT_PARM, LT_QP, ELEM_NODES (LT_N),&  ! 16
   ! COORD (LT_N, N_SPACE), GEOMETRY (LT_GEOM, N_SPACE),       &  ! 17
   ! C (LT_FREE), D (LT_FREE), S (LT_FREE, LT_FREE),           &  ! 18
   ! DLG (LT_PARM, LT_GEOM), DLG_QP (LT_PARM, LT_GEOM, LT_QP)  &  ! 19
   ! DLH (LT_PARM, LT_N),    DLH_QP (LT_PARM, LT_N,    LT_QP), &  ! 20
   ! DLV (LT_PARM, LT_FREE), DLV_QP (LT_PARM, LT_FREE, LT_QP), &  ! 21
   ! G (LT_GEOM), G_QP (LT_GEOM, LT_QP), H_QP (LT_N,   LT_QP), &  ! 22
   ! V (LT_FREE), V_QP (LT_FREE, LT_QP), H (LT_N),             &  ! 23
   ! PT (LT_PARM, LT_QP), WT (LT_QP), D2LH (N_2_DER, LT_N)        ! 24
 Use SCP_Type_Data                                                ! 25
 Use Interface_Header                                             ! 26
 Use Geometric_Properties                                         ! 27
  IMPLICIT NONE                                                   ! 28
  INTEGER,  INTENT (IN)  :: NODES          (L_S_TOT, NOD_PER_EL)  ! 29
  REAL(DP), INTENT (IN)  :: X              (MAX_NP,  N_SPACE)     ! 30
  REAL(DP), INTENT (IN)  :: SCP_AVERAGES (MAX_NP,  SCP_FIT)       ! 31
  REAL(DP), INTENT (IN)  :: DOF_SYS                (N_D_FRE)      ! 32
  REAL(DP), INTENT (OUT) :: ELEM_ERROR_ENERGY (N_ELEMS)           ! 33
  REAL(DP), INTENT (OUT) :: ELEM_REFINEMENT   (N_ELEMS)           ! 34
  REAL(DP), INTENT (OUT) :: ERR_MAX                               ! 35
  REAL (DP), PARAMETER :: ZERO = 0.d0                             ! 36
                                                                  ! 37
  INTEGER   :: IE, IN, IQ, LT, QP_LT, LOC_MAX (1)                 ! 38
  INTEGER   :: I_ERROR ! /= 0 IFF INVERSION OF E FAILS            ! 39
  REAL (DP) :: GLOBAL_ERROR_ENERGY,  NEAR_ZERO                    ! 40
  REAL (DP) :: GLOBAL_FLUX_NORM,     GLOBAL_FLUX_ERROR            ! 41
  REAL (DP) :: GLOBAL_FLUX_RMS,      GLOBAL_SOLUTION_L2           ! 42
  REAL (DP) :: GLOBAL_SOLUTION_ERR,  VOL                          ! 43
  REAL (DP) :: GLOBAL_STRAIN_ENERGY, STRAIN_ENERGY_NORM           ! 44
  REAL (DP) :: ELEM_STRAIN_ENERGY,   ALLOWED_ERROR                ! 45
  REAL (DP) :: ALLOWED_ERR_DENSITY,  ALLOWED_ERR_PER_EL           ! 46
  REAL (DP) :: ELEM_FLUX_NORM,       ELEM_FLUX_ERROR              ! 47
  REAL (DP) :: ELEM_FLUX_RMS,        ELEM_SOLUTION_L2             ! 48
  REAL (DP) :: ELEM_SOLUTION_ERR,    EL_ERR_ENERGY                ! 49
  REAL (DP) :: GLOBAL_H1_NORM,       GLOBAL_H2_NORM               ! 50
  REAL (DP) :: GLOBAL_H1_ERROR,      GLOBAL_H2_ERROR              ! 51
  REAL (DP) :: ELEM_H1_NORM,         ELEM_H2_NORM                 ! 52
  REAL (DP) :: ELEM_H1_ERROR,        ELEM_H2_ERROR                ! 53
  REAL (DP) :: EXACT_H1_NORM,        EXACT_H2_NORM                ! 54
  REAL (DP) :: EXACT_H1_ERROR,       EXACT_H2_ERROR               ! 55
```

Figure 6.11a *Interface and data for error estimates*

```
  REAL (DP) :: EXACT_SOL_L2,          EX_ERR_ENERGY              ! 56
  REAL (DP) :: EXACT_FLUX_ERROR,      EXACT_FLUX_NORM            ! 57
  REAL (DP) :: EXACT_STRAIN_ENERGY,   EXACT_ERR_ENERGY           ! 58
  REAL (DP) :: DET, DET_WT, TEMP, TEST, SCP_VOLUME               ! 59
  REAL (DP), ALLOCATABLE :: DOF_EL (:, :) ! D RESHAPE            ! 60
                                                                 ! 61
!                 Automatic Arrays                               ! 62
  REAL (DP) :: AJ        (N_SPACE, N_SPACE)    ! JACOBIAN          ! 63
  REAL (DP) :: AJ_INV    (N_SPACE, N_SPACE)    ! JACOBIAN INVERSE  ! 64
  REAL (DP) :: B         (N_R_B,   N_EL_FRE)   ! DIFFERENTIAL OP   ! 65
  REAL (DP) :: E         (N_R_B,   N_R_B)      ! CONSTITUTIVE      ! 66
  REAL (DP) :: E_INVERSE (N_R_B,   N_R_B)      ! INVERSE           ! 67
  REAL (DP) :: DGH       (N_SPACE, NOD_PER_EL) ! GRADIENT OF H     ! 68
  REAL (DP) :: FLUX_LT   (N_R_B,   NOD_PER_EL) ! NODAL FLUXES      ! 69
  REAL (DP) :: XYZ       (N_SPACE)             ! POINT IN SPACE    ! 70
  REAL (DP) :: SIGMA_SCP (N_R_B)               ! SCP FLUX          ! 71
  REAL (DP) :: SIGMA_HAT (N_R_B)               ! FEA FLUX          ! 72
  REAL (DP) :: DIFF      (N_R_B)               ! FLUX DIFFERENCE   ! 73
  REAL (DP) :: MEASURE   (N_ELEMS)             ! ELEMENT MEASURE   ! 74
  REAL (DP) :: ELEM_ERROR_DENSITY (N_ELEMS)    ! Per UNIT VOLUME   ! 75
  REAL (DP) :: EXAC_ERROR_ENERGY  (N_ELEMS)    ! EXACT ERR ENERGY  ! 76
  REAL (DP) :: SOLUTION (N_G_DOF), EXACT_SOL (N_G_DOF) ! AT PT     ! 77
  REAL (DP) :: SOLUTION_ERR (N_G_DOF)          ! PT ERROR RESULT   ! 78
                                                                   ! 79
! B          = GRADIENT VERSUS DOF MATRIX                          ! 80
! DGH        = GLOBAL DERIVS OF SCALAR FUNCTIONS H                 ! 81
! DOF_SYS    = DEGREES OF FREEDOM OF THE SYSTEM                    ! 82
! E          = CONSTITUTIVE MATRIX, INVERSE IS E_INVERSE           ! 83
! ELEM_ERROR_ENERGY  = ESTIMATED ELEMENT ERROR, % OF ENERGY NORM   ! 84
! ELEM_ERROR_DENSITY = ESTIMATED ELEMENT ERROR / SQRT(MEASURE)     ! 85
! ELEM_NODES         = THE NOD_PER_EL INCIDENCES OF THE ELEMENT    ! 86
! ELEM_REFINEMENT    = INDICATOR, >1 REFINE, <1 DE-REFINE          ! 87
! EXAC_ERROR_ENERGY  = ENERGY IN ERROR FROM EXACT SOLUTION         ! 88
! INDEX      = SYSTEM DOF NUMBERS ASSOCIATED WITH ELEMENT          ! 89
! L_HOMO     = 1, IF ELEMENT PROPERTIES ARE HOMOGENEOUS            ! 90
! L_S_TOT    = TOTAL NUMBER OF ELEMENTS & THEIR SEGMENTS           ! 91
! L_TYPE     = ELEMENT TYPE NUMBER LIST                            ! 92
! LT         = ELEMENT TYPE NUMBER (IF USED)                       ! 93
! LT_QP      = NUMBER OF QUADRATURE PTS FOR ELEMENT TYPE           ! 94
! MAX_NP     = NUMBER OF SYSTEM NODES                              ! 95
! MEASURE    = ELEMENT MEASURE (GENERALIZED VOLUME)                ! 96
! NOD_PER_EL = MAXIMUM NUMBER OF NODES PER ELEMENT                 ! 97
! NODES      = NODAL INCIDENCES OF ALL ELEMENTS                    ! 98
! N_QP       = MAXIMUM NUMBER OF QUADRATURE POINTS, >= LT_QP       ! 99
! N_R_B      = NUMBER OF ROWS IN B AND E MATRICES                  !100
! N_SPACE    = DIMENSION OF SPACE                                  !101
! SCP_AVERAGES = AVERAGED FLUXES AT ALL NODES IN MESH              !102
! SCP_FIT    = NUMBER IF TERMS BEING FIT, OR AVERAGED              !103
! SCP_VOLUME = SCP VOLUME USED IN GETTING RMS VALUES               !104
! SIGMA_HAT  = FLUX COMPONENTS AT PT FROM ORIGINAL ELEMENT         !105
! SIGMA_SCP  = FLUX COMPONENTS AT PT FROM SMOOTHED SCP             !106
! SOLUTION_ERR = DIFFERENCE BETWEEN SOLUTION AND EXACT_SOL         !107
! EXACT_SOL  = VALUE FROM USER SUPPLIED ROUTINE                    !108
! X          = COORDINATES OF SYSTEM NODES                         !109
! XYZ        = SPACE COORDINATES AT A POINT                        !110
```

Figure 6.11b *Automatic arrays and local variables*

```
                                                                    !111
  WRITE (N_PRT, "(/, '** BEGINNING ELEMENT ERROR ESTIMATES **')")  !112
  ELEM_ERROR_ENERGY = 0.d0 ; ERR_MAX = 0.d0          ! INITIAL     !113
  ELEM_REFINEMENT   = 0.d0 ; EXAC_ERROR_ENERGY = 0.d0 ! INITIAL    !114
  NEAR_ZERO         = TINY (1.d0)                     ! CONSTANT   !115
                                                                    !116
  GLOBAL_ERROR_ENERGY = 0.d0 ! INITIALIZE                           !117
  GLOBAL_FLUX_NORM    = 0.d0 ; GLOBAL_FLUX_ERROR     = 0.d0         !118
  GLOBAL_FLUX_RMS     = 0.d0 ; GLOBAL_SOLUTION_L2    = 0.d0         !119
  SCP_VOLUME          = 0.d0 ; GLOBAL_SOLUTION_ERR   = 0.d0         !120
  MEASURE             = 0.d0 ; GLOBAL_STRAIN_ENERGY = 0.d0          !121
  EXACT_FLUX_ERROR    = 0.d0 ; EXACT_FLUX_NORM       = 0.d0         !122
  EXACT_STRAIN_ENERGY = 0.d0 ; EXACT_ERR_ENERGY      = 0.d0         !123
  EXACT_SOL           = 0.d0 ; EXACT_SOL_L2          = 0.d0         !124
  EXAC_ERROR_ENERGY   = 0.d0 ; DIFF                  = 0.d0         !125
  GLOBAL_H1_NORM      = 0.d0 ; GLOBAL_H2_NORM        = 0.d0         !126
  GLOBAL_H1_ERROR     = 0.d0 ; GLOBAL_H2_ERROR       = 0.d0         !127
  ELEM_H1_NORM        = 0.d0 ; ELEM_H2_NORM          = 0.d0         !128
  ELEM_H1_ERROR       = 0.d0 ; ELEM_H2_ERROR         = 0.d0         !129
  EXACT_H1_NORM       = 0.d0 ; EXACT_H2_NORM         = 0.d0         !130
  EXACT_H1_ERROR      = 0.d0 ; EXACT_H2_ERROR        = 0.d0         !131
  XYZ                 = 0.d0 ; IE = 1                               !132
                                                                    !133
!      CHECK FOR POSSIBLE CONSTANT CONSTITUTIVE MATRIX              !134
  IF ( L_HOMO == 1 ) THEN ! HOMOGENEOUS                             !135
      XYZ = 0.d0 ; IE = 1 ! DUMMY ARGUMENTS                         !136
      IF ( .NOT. GRAD_BASE_ERROR ) CALL              &              !137
            SELECT_APPLICATION_E_MATRIX (IE, XYZ, E)                !138
      CALL INV_SMALL_MAT (N_R_B, E, E_INVERSE, I_ERROR)             !139
  END IF ! HOMOGENEOUS MATERIAL                                     !140
                                                                    !141
  LT = 1 ; LAST_LT = 0                                              !142
  DO  IE = 1, N_ELEMS  ! LOOP OVER ALL STANDARD ELEMENTS            !143
!-->     GET ELEMENT TYPE NUMBER                                    !144
    IF ( N_L_TYPE > 1) LT = L_TYPE (IE)  ! SAME AS LAST TYPE ?      !145
    IF ( LT /= LAST_LT ) THEN            ! this is a new type       !146
      CALL GET_ELEM_TYPE_DATA (LT)       ! CONTROLS FOR THIS TYPE   !147
      LAST_LT = LT                                                  !148
                                                                    !149
!-->     GET UPGRADED QUADRATURE RULE FOR PATCH "ELEMENT" TYPE      !150
      CALL GET_PATCH_QUADRATURE_ORDER (LT_SHAP, LT_QP, SCP_QP)      !151
                                                                    !152
!       SINCE SCP_QP >= LT_QP MUST REALLOCATE SOME ARRAYS           !153
      QP_LT = LT_QP    ! Copy to prevent overwrite                  !154
      LT_QP = SCP_QP   ! Return to original value below             !155
      IF ( TYPE_APLY_ALLOC ) CALL DEALLOCATE_TYPE_APPLICATION       !156
      CALL ALLOCATE_TYPE_APPLICATION                                !157
      IF ( TYPE_NTRP_ALLOC ) CALL DEALLOCATE_TYPE_INTERPOLATIONS    !158
      CALL ALLOCATE_TYPE_INTERPOLATIONS                             !159
      IF ( ALLOCATED (DOF_EL) ) DEALLOCATE (DOF_EL)                 !160
      ALLOCATE (DOF_EL (N_G_DOF, LT_N))                             !161
                                                                    !162
      IF ( LT_QP > 0 ) THEN ! GET QUADRATURE FOR PATCH "ELEMENT"    !163
         CALL GET_ELEM_QUADRATURES                                  !164
         CALL FILL_TYPE_INTERPOLATIONS ; END IF                     !165
    END IF ! a new element type                                     !166
                                                                    !167
```

Figure 6.11c *Set patch type and quadrature data*

```
!-->    GET ELEMENT NODE NUMBERS, COORD, DOF                              !168
   ELEM_NODES = GET_ELEM_NODES (IE, LT_N, NODES)                          !169
   CALL ELEM_COORD (LT_N, N_SPACE, X, COORD, ELEM_NODES)                  !170
   INDEX  = GET_ELEM_INDEX (LT_N, ELEM_NODES)                             !171
   D      = GET_ELEM_DOF (DOF_SYS)                                        !172
   DOF_EL = RESHAPE (D, (/ N_G_DOF, LT_N /))                              !173
                                                                          !174
!-->    EXTRACT SCP NODAL FLUXES (NOW GATHER_LT_SCP_AVERAGES)             !175
   DO IN = 1, LT_N  ! OVER NODES OF ELEMENT                               !176
     IF ( ELEM_NODES (IN) < 1 ) CYCLE ! TO VALID NODE                     !177
       FLUX_LT (1:N_R_B, IN) = SCP_AVERAGES (                  &          !178
                               ELEM_NODES (IN), 1:N_R_B)                  !179
   END DO ! FOR NODES ON ELEMENT                                          !180
                                                                          !181
!              INITIALIZE NORMS                                           !182
   ELEM_FLUX_NORM      = 0.d0 ; ELEM_FLUX_ERROR     = 0.d0                !183
   ELEM_FLUX_RMS       = 0.d0 ; ELEM_SOLUTION_L2    = 0.d0                !184
   ELEM_SOLUTION_ERR   = 0.d0 ; ELEM_STRAIN_ENERGY = 0.d0                 !185
   EL_ERR_ENERGY       = 0.d0 ; EX_ERR_ENERGY       = 0.d0                !186
   VOL                 = 0.d0 ; ELEM_ERROR_ENERGY (IE) = 0.d0             !187
                                                                          !188
   DO IQ = 1, LT_QP   ! LOOP OVER QUADRATURE POINTS                       !189
     H = GET_H_AT_QP (IQ)     ! INTERPOLATION FUNCTIONS                   !190
     XYZ = MATMUL (H, COORD) ! COORDINATES OF PT                          !191
     DLH = GET_DLH_AT_QP (IQ) ! LOCAL DERIVATIVES                         !192
                                                                          !193
!        FIND JACOBIAN AT THE PT, INVERSE AND DETERMINANT                 !194
     AJ = MATMUL (DLH (1:N_SPACE, :), COORD)                              !195
     CALL INVERT_JACOBIAN (AJ, AJ_INV, DET, N_SPACE)                      !196
     IF ( DET <= ZERO ) STOP 'BAD DET, SCP_ERROR_ESTIMATES'               !197
     DET_WT = DET * WT(IQ)                                                !198
                                                                          !199
     IF ( AXISYMMETRIC ) DET_WT = DET_WT * XYZ (1) * TWO_PI               !200
     VOL = VOL + DET_WT   ! UPDATE ELEMENT VOLUME                         !201
                                                                          !202
!      EVALUATE SOLUTION L2 NORM (ASSUMING C_0N_G_DOF)                    !203
     SOLUTION          = MATMUL (DOF_EL, H)                               !204
     ELEM_SOLUTION_L2 = ELEM_SOLUTION_L2 + DET_WT        &                !205
                      * DOT_PRODUCT (SOLUTION, SOLUTION)                  !206
                                                                          !207
!       EVALUATE APPLICATION EXACT VALUE & ERROR HERE                     !208
     IF ( USE_EXACT )  THEN                                               !209
       CALL SELECT_EXACT_SOLUTION (XYZ, EXACT_SOL)                        !210
       EXACT_SOL_L2       = EXACT_SOL_L2 + DET_WT                &        !211
                          * DOT_PRODUCT (EXACT_SOL, EXACT_SOL)            !212
       SOLUTION_ERR       = EXACT_SOL - SOLUTION                          !213
       ELEM_SOLUTION_ERR = ELEM_SOLUTION_ERR + DET_WT &                   !214
         * DOT_PRODUCT (SOLUTION_ERR, SOLUTION_ERR)                       !215
     END IF ! EXACT SOLUTION GIVEN                                        !216
                                                                          !217
     DGH  = MATMUL (AJ_INV, DLH)  ! GLOBAL DERIVATIVES                    !218
     CALL SELECT_APPLICATION_B_MATRIX (DGH, XYZ, B (:,1:LT_FREE))         !219
                                                                          !220
```

Figure 6.11d *Gather continuous flux and integrate error*

```
!        GET LOCAL STRAINS (STORE IN SIGMA_HAT)                        !221
    SIGMA_HAT (1:N_R_B) = MATMUL(B(:,1:LT_FREE),D(1:LT_FREE))         !222
                                                                       !223
!         APPLY CONSTITUTIVE RELATION (TO STRAINS IN SIGMA_SCP)        !224
    SIGMA_HAT (1:N_R_B) = MATMUL (E, SIGMA_HAT (1:N_R_B))             !225
                                                                       !226
!          GET SCP FLUX ESTIMATES & DIFFERENCE                         !227
    SIGMA_SCP (:) = MATMUL (H, TRANSPOSE(FLUX_LT (:,1:LT_N)))          !228
    DIFF          = SIGMA_SCP - SIGMA_HAT    ! SIGMA ERROR EST         !229
    ELEM_FLUX_NORM  = ELEM_FLUX_NORM                    &              !230
                  + DET_WT * DOT_PRODUCT (SIGMA_SCP,SIGMA_SCP)         !231
    ELEM_FLUX_ERROR = ELEM_FLUX_ERROR &                                !232
                  + DET_WT * DOT_PRODUCT (DIFF, DIFF)                  !233
                                                                       !234
!         INCREMENT STRAIN ENERGY & ENERGY IN THE ERROR                !235
    TEST = DOT_PRODUCT (SIGMA_SCP,MATMUL(E_INVERSE,SIGMA_SCP))         !236
    ELEM_STRAIN_ENERGY = ELEM_STRAIN_ENERGY + DET_WT * TEST            !237
                                                                       !238
    TEST = DOT_PRODUCT (DIFF, MATMUL (E_INVERSE, DIFF))                !239
    EL_ERR_ENERGY = EL_ERR_ENERGY + DET_WT * TEST                      !240
    IF (EL_ERR_ENERGY < NEAR_ZERO ) EL_ERR_ENERGY = 0.d0               !241
                                                                       !242
    IF ( USE_EXACT_FLUX )  THEN  ! GET EXACT VALUES                    !243
      CALL SELECT_EXACT_FLUX (XYZ, SIGMA_SCP (1:N_R_B))                !244
      DIFF               = SIGMA_SCP - SIGMA_HAT ! EXACT ERROR         !245
      EXACT_FLUX_NORM  = EXACT_FLUX_NORM              &                !246
        + DET_WT * DOT_PRODUCT (SIGMA_SCP, SIGMA_SCP)                  !247
      EXACT_FLUX_ERROR = EXACT_FLUX_ERROR &                            !248
        + DET_WT * DOT_PRODUCT (DIFF, DIFF)                            !249
      TEST = DOT_PRODUCT(SIGMA_SCP,MATMUL(E_INVERSE,SIGMA_SCP))        !250
      EXACT_STRAIN_ENERGY = EXACT_STRAIN_ENERGY + DET_WT*TEST          !251
      TEST = DOT_PRODUCT (DIFF, MATMUL (E_INVERSE, DIFF))              !252
      EX_ERR_ENERGY      = EX_ERR_ENERGY     + DET_WT * TEST           !253
      EXACT_ERR_ENERGY = EXACT_ERR_ENERGY + DET_WT * TEST              !254
    END IF ! EXACT FLUXES GIVEN                                        !255
  END DO ! OVER ERROR EST QP                                           !256
                                                                       !257
  EXAC_ERROR_ENERGY (IE) = EX_ERR_ENERGY + NEAR_ZERO                   !258
  ELEM_ERROR_ENERGY (IE) = EL_ERR_ENERGY + NEAR_ZERO                   !259
  MEASURE (IE)           = VOL                                         !260
  SCP_VOLUME             = SCP_VOLUME + VOL                            !261
                                                                       !262
!         COMBINE AND NORMALIZE ERROR TERMS                            !263
  GLOBAL_STRAIN_ENERGY=GLOBAL_STRAIN_ENERGY+ELEM_STRAIN_ENERGY         !264
  GLOBAL_ERROR_ENERGY = GLOBAL_ERROR_ENERGY + EL_ERR_ENERGY            !265
  GLOBAL_FLUX_NORM    = GLOBAL_FLUX_NORM    + ELEM_FLUX_NORM           !266
  GLOBAL_FLUX_ERROR   = GLOBAL_FLUX_ERROR   + ELEM_FLUX_ERROR          !267
  GLOBAL_SOLUTION_L2  = GLOBAL_SOLUTION_L2  + ELEM_SOLUTION_L2         !268
  GLOBAL_SOLUTION_ERR = GLOBAL_SOLUTION_ERR +ELEM_SOLUTION_ERR         !269
                                                                       !270
  GLOBAL_H2_NORM      = GLOBAL_H2_NORM  + ELEM_H2_NORM                 !271
  GLOBAL_H2_ERROR     = GLOBAL_H2_ERROR + ELEM_H2_ERROR                !272
                                                                       !273
  ELEM_ERROR_ENERGY (IE) = SQRT (ELEM_ERROR_ENERGY (IE))               !274
  EXAC_ERROR_ENERGY (IE) = SQRT (EXAC_ERROR_ENERGY (IE))               !275
 END DO ! OVER ALL ELEMENTS                                            !276
 LT_QP = QP_LT  ! RESET LT_QP TO ITS TRUE VALUE                        !277
                                                                       !278
```

Figure 6.11e *Update various error measure choices*

```
!           FINAL GLOBAL COMBINATIONS                                    !279
  GLOBAL_STRAIN_ENERGY=GLOBAL_STRAIN_ENERGY+GLOBAL_ERROR_ENERGY          !280
                                                                         !281
  EXACT_H2_NORM  = EXACT_H2_NORM  + EXACT_FLUX_ERROR &                   !282
                 + EXACT_SOL_L2                                          !283
  GLOBAL_H2_NORM = GLOBAL_H2_NORM + GLOBAL_FLUX_NORM &                   !284
                 + GLOBAL_SOLUTION_L2                                    !285
                                                                         !286
  STRAIN_ENERGY_NORM = SQRT (GLOBAL_STRAIN_ENERGY)                       !287
  ALLOWED_ERROR      = STRAIN_ENERGY_NORM*(PERCENT_ERR_MAX/100)          !288
  ALLOWED_ERR_DENSITY= ALLOWED_ERROR / SQRT (SCP_VOLUME)                 !289
  ALLOWED_ERR_PER_EL = ALLOWED_ERROR / SQRT (FLOAT(N_ELEMS))             !290
                                                                         !291
!       AVOID DIVISION BY ZERO IF THE ERROR IS ZERO                      !292
  ALLOWED_ERROR         = ALLOWED_ERROR       + NEAR_ZERO                !293
  ALLOWED_ERR_DENSITY   = ALLOWED_ERR_DENSITY + NEAR_ZERO                !294
  ALLOWED_ERR_PER_EL    = ALLOWED_ERR_PER_EL  + NEAR_ZERO                !295
  ERR_MAX               = ALLOWED_ERROR                                  !296
  EXACT_FLUX_ERROR      = EXACT_FLUX_ERROR    + NEAR_ZERO                !297
  EXACT_H1_ERROR        = EXACT_H1_ERROR      + NEAR_ZERO                !298
  GLOBAL_ERROR_ENERGY   = GLOBAL_ERROR_ENERGY + NEAR_ZERO                !299
  GLOBAL_FLUX_ERROR     = GLOBAL_FLUX_ERROR   + NEAR_ZERO                !300
  GLOBAL_H1_ERROR       = GLOBAL_H1_ERROR     + NEAR_ZERO                !301
  GLOBAL_SOLUTION_ERR   = GLOBAL_SOLUTION_ERR + NEAR_ZERO                !302
                                                                         !303
  GLOBAL_ERROR_ENERGY = SQRT (GLOBAL_ERROR_ENERGY)                       !304
  GLOBAL_FLUX_NORM    = SQRT (GLOBAL_FLUX_NORM)                          !305
  GLOBAL_SOLUTION_L2  = SQRT (GLOBAL_SOLUTION_L2)                        !306
  EXACT_SOL_L2        = SQRT (EXACT_SOL_L2)                              !307
  GLOBAL_SOLUTION_ERR = SQRT (GLOBAL_SOLUTION_ERR)                       !308
  IF ( SCP_VOLUME > 0.d0 ) GLOBAL_FLUX_RMS = &                           !309
       SQRT (GLOBAL_FLUX_ERROR / SCP_VOLUME)                             !310
  GLOBAL_FLUX_ERROR   = SQRT (GLOBAL_FLUX_ERROR)                         !311
                                                                         !312
!        GET EXACT VALUES, WHEN AVAILABLE                                !313
  EXACT_STRAIN_ENERGY = EXACT_STRAIN_ENERGY+EXACT_ERR_ENERGY             !314
  EXACT_STRAIN_ENERGY = SQRT (EXACT_STRAIN_ENERGY)                       !315
  EXACT_FLUX_NORM     = SQRT (EXACT_FLUX_NORM)                           !316
  EXACT_FLUX_ERROR    = SQRT (EXACT_FLUX_ERROR)                          !317
                                                                         !318
  EXACT_H2_NORM       = SQRT (EXACT_H2_NORM)                             !319
  GLOBAL_H2_NORM      = SQRT (GLOBAL_H2_NORM)                            !320
  EXACT_H2_ERROR      = SQRT (EXACT_H2_ERROR)                            !321
  GLOBAL_H2_ERROR     = SQRT (GLOBAL_H2_ERROR)                           !322
                                                                         !323
```

Figure 6.11f *Update global error measures*

```
  PRINT *,"** S_C_P  ENERGY  NORM  ERROR  ESTIMATE  DATA **"          !324
  PRINT *," "                                                          !325
  PRINT *, "DOMAIN MEASURE ............", SCP_VOLUME                   !326
  PRINT *, "AVERAGE ELEMENT MEASURE ...", SCP_VOLUME / N_ELEMS         !327
  PRINT *, "GLOBAL_SOLUTION_L2 ........", GLOBAL_SOLUTION_L2           !328
  IF ( USE_EXACT ) THEN                                                !329
    PRINT *, "EXACT_SOLUTION_L2 .........", EXACT_SOL_L2               !330
    PRINT *, "GLOBAL_SOLUTION_ERR........", GLOBAL_SOLUTION_ERR        !331
  END IF ! EXACT SOLUTION GIVEN                                        !332
  PRINT *, " "                                                         !333
  PRINT *, "STRAIN_ENERGY_NORM ........", STRAIN_ENERGY_NORM           !334
  IF ( USE_EXACT_FLUX ) PRINT *,                                  &    !335
           "EXACT_STRAIN_ENERGY_NORM ..", EXACT_STRAIN_ENERGY          !336
  PRINT *, "ALLOWED_PER_CENT_ERROR ....", PERCENT_ERR_MAX              !337
  PRINT *, "ALLOWED_GLOBAL_ERROR ......", ALLOWED_ERROR                !338
  PRINT *, "ALLOWED_ERROR_DENSITY .....", ALLOWED_ERR_DENSITY          !339
  PRINT *, "ALLOWED_ERROR_PER_ELEM ....", ALLOWED_ERR_PER_EL           !340
  PRINT *, " "                                                         !341
  PRINT *, "GLOBAL_ERROR_ENERGY .......", GLOBAL_ERROR_ENERGY          !342
  PRINT *, "GLOBAL_ERROR_PARAMETER ....",      &                       !343
            GLOBAL_ERROR_ENERGY / ALLOWED_ERROR                        !344
  PRINT *, " "                                                         !345
  PRINT *, "GLOBAL_FLUX_ERROR .........", GLOBAL_FLUX_ERROR            !346
  IF ( USE_EXACT_FLUX ) PRINT *,                               &       !347
           "EXACT_FLUX_ERROR ..........", EXACT_FLUX_ERROR             !348
  PRINT *, "GLOBAL_FLUX_NORM ..........", GLOBAL_FLUX_NORM             !349
  IF ( USE_EXACT_FLUX ) PRINT *,                              &        !350
           "EXACT_FLUX_NORM ...........", EXACT_FLUX_NORM              !351
  PRINT *, "GLOBAL_FLUX_RMS ...........", GLOBAL_FLUX_RMS              !352
                                                                       !353
!     CONVERT TO ELEMENT ERROR DENSITY                                 !354
  WHERE ( MEASURE > 0.d0 )                                             !355
    ELEM_ERROR_DENSITY = ELEM_ERROR_ENERGY / SQRT (MEASURE)            !356
  ELSEWHERE                                                            !357
    ELEM_ERROR_DENSITY = 0.d0                                          !358
  END WHERE                                                            !359
                                                                       !360
!       LIST AVERAGE TOTAL ERROR                                       !361
  TEMP = STRAIN_ENERGY_NORM / 100.d0                                   !362
  TEST = SUM ( ELEM_ERROR_ENERGY ) / TEMP                              !363
  WRITE(N_PRT,'("TOTAL % ERROR IN ENERGY NORM =",1PE8.2)') TEST       !364
                                                                       !365
!        LIST MAXIMUMS                                                 !366
  TEST    = MAXVAL ( ELEM_ERROR_ENERGY )                               !367
  LOC_MAX = MAXLOC ( ELEM_ERROR_ENERGY )                               !368
  PRINT *, " "                                                         !369
  WRITE (N_PRT, '("MAX ELEMENT ENERGY ERROR OF ",1PE8.2)') TEST       !370
  WRITE (N_PRT, '("OCCURS IN ELEMENT ", I6)') LOC_MAX (1)             !371
  TEST    = MAXVAL ( ELEM_ERROR_DENSITY )                              !372
  LOC_MAX = MAXLOC ( ELEM_ERROR_DENSITY )                              !373
  WRITE (N_PRT, '("MAX ENERGY ERROR DENSITY OF ",1PE8.2)') TEST       !374
  WRITE (N_PRT, '("OCCURS IN ELEMENT ", I6)') LOC_MAX (1)             !375
                                                                       !376
```

Figure 6.11g *List global error measures*

```
!    FINALLY, CONVERT REFINEMENT TO TRUE REFINEMENT PARAMETER     !377
  ELEM_REFINEMENT =  ELEM_ERROR_DENSITY * SQRT ( SCP_VOLUME ) &   !378
                  / ALLOWED_ERROR                                 !379
  WRITE (N_PRT, '("WITH REFINEMENT PARAMETER OF ", 1PE8.2)')  &   !380
         TEST / ALLOWED_ERR_DENSITY                               !381
  PRINT *, "------------------------------------------------"     !382
  PRINT *, "          ERROR IN      % ERROR IN    REFINEMENT"     !383
  PRINT *, "ELEMENT,  ENERGY_NORM,  ENERGY_NORM,  PARAMETER"      !384
  PRINT *, "--------------------------------------------------"   !385
                                                                  !386
  DO IE = 1, N_ELEMS ! LOOP OVER ALL ELEMENTS                     !387
    IF ( ELEM_ERROR_ENERGY  (IE) > ALLOWED_ERR_PER_EL .OR.  &     !388
         ELEM_ERROR_DENSITY (IE) > ALLOWED_ERR_DENSITY ) THEN     !389
       WRITE (N_PRT,"(I8, 3(1PE16.4),6X,A)") IE,            &     !390
         ELEM_ERROR_ENERGY (IE), ELEM_ERROR_ENERGY (IE) / &       !391
         TEMP, ELEM_REFINEMENT (IE), "Refine"                     !392
    ELSE                                                          !393
       WRITE (N_PRT,"(I8, 3(1PE16.4),6X,A)") IE,            &     !394
         ELEM_ERROR_ENERGY (IE), ELEM_ERROR_ENERGY (IE) / &       !395
         TEMP, ELEM_REFINEMENT (IE), "Refine" ; END IF            !396
  END DO ! FOR ALL ELEMENTS                                       !397
                                                                  !398
!     CONVERT TO % ERROR IN ENERGY NORM * 100                     !399
  ELEM_ERROR_ENERGY = ELEM_ERROR_ENERGY / TEMP                    !400
  IF ( TYPE_APLY_ALLOC ) CALL DEALLOCATE_TYPE_APPLICATION         !401
  IF ( TYPE_NTRP_ALLOC ) CALL DEALLOCATE_TYPE_INTERPOLATIONS      !402
END SUBROUTINE SCP_ERROR_ESTIMATES                                !403
```

Figure 6.11h *List element error and error density*

$$\begin{Bmatrix} \dfrac{\partial}{\partial r} \\[2ex] \dfrac{\partial}{\partial s} \end{Bmatrix} = \begin{bmatrix} \dfrac{\partial x}{\partial r} & \dfrac{\partial y}{\partial r} \\[2ex] \dfrac{\partial x}{\partial s} & \dfrac{\partial y}{\partial s} \end{bmatrix} \begin{Bmatrix} \dfrac{\partial}{\partial x} \\[2ex] \dfrac{\partial}{\partial y} \end{Bmatrix}. \tag{6.5}$$

Continuing this process to relate second parametric derivatives to second physical derivatives involves products and derivatives of the Jacobian array. For a two-dimensional mapping: (6.6)

$$\begin{Bmatrix} \dfrac{\partial^2}{\partial r^2} \\[2ex] \dfrac{\partial^2}{\partial s^2} \\[2ex] \dfrac{\partial^2}{\partial r\,\partial s} \end{Bmatrix} = \begin{bmatrix} \dfrac{\partial^2 x}{\partial r^2} & \dfrac{\partial^2 y}{\partial r^2} \\[2ex] \dfrac{\partial^2 x}{\partial s^2} & \dfrac{\partial^2 y}{\partial s^2} \\[2ex] \dfrac{\partial^2 x}{\partial r\,\partial s} & \dfrac{\partial^2 y}{\partial r\,\partial s} \end{bmatrix} \begin{Bmatrix} \dfrac{\partial}{\partial x} \\[2ex] \dfrac{\partial}{\partial y} \end{Bmatrix} + \begin{bmatrix} \left(\dfrac{\partial x}{\partial r}\right)^2 & \left(\dfrac{\partial y}{\partial r}\right)^2 & 2\dfrac{\partial x}{\partial r}\dfrac{\partial y}{\partial r} \\[2ex] \left(\dfrac{\partial x}{\partial s}\right)^2 & \left(\dfrac{\partial y}{\partial s}\right)^2 & 2\dfrac{\partial x}{\partial s}\dfrac{\partial y}{\partial s} \\[2ex] \dfrac{\partial x}{\partial r}\dfrac{\partial x}{\partial s} & \dfrac{\partial y}{\partial r}\dfrac{\partial y}{\partial s} & \dfrac{\partial x}{\partial s}\dfrac{\partial y}{\partial r} + \dfrac{\partial x}{\partial r}\dfrac{\partial y}{\partial s} \end{bmatrix} \begin{Bmatrix} \dfrac{\partial^2}{\partial x^2} \\[2ex] \dfrac{\partial^2}{\partial y^2} \\[2ex] \dfrac{\partial^2}{\partial x\,\partial y} \end{Bmatrix}.$$

For a constant Jacobian the first rectangular matrix on the right is zero. Otherwise, the second derivatives are clearly more sensitive to a variable Jacobian. In each case, we must invert the square matrices to obtain the first and second physical derivatives. If the Jacobian is not constant, then the effect of a distorted element would be amplified by the product terms in the square matrix of Eq. 6.6, as well as by the second derivatives in the rectangular array that multiplies the physical gradient term. Using this analytic form for the second derivative of an approximate finite element solution is certainly questionable. Clearly, for an element with a constant Jacobian, the result would be identically zero. Thus, we will actually only use this form as a tool to estimate regions of high second-derivative error, as compared to the values obtained from a patch recovery technique.

In the present software the second derivative calculations and associated output are activated by a global logical constant SCP_2ND_DERIV which is set to true by including an input keyword control of *scp_2nd_deriv* in the data file. If the norm of the estimated second derivative error is required, then the square matrix of Jacobian product terms is computed in subroutine JACOBIAN_PRODUCTS and is given the name P_AJ. Its associated inverse matrix is called P_AJ_INV. The second parametric (local) derivatives of the interpolation functions, **H**, are denoted as **D2LH** and the corresponding second physical (global) derivatives, from Eq. 6.6, are denoted by **D2GH**. Each has N_2_DER rows and a column for each interpolation function. For one-, two- and three-dimensional problems, N_2_DER has a value of 1, 3, or 6, respectively.

Since the analytic estimate will usually be compared to a smoothed patch estimate, the second derivative terms are also computed in EVAL_SCP_FIT_AT_PATCH_NODES if SCP_2ND_DERIV is true. After the element fluxes have been processed to give continuous nodal values on a patch (as described earlier), the physical gradients of the patch interpolations, SCP_DGH, are invoked to evaluate the gradients of the fluxes (i.e., the second derivatives) at all of the mesh nodes in the patch. They are then also scattered to the system array, SCP_AVERAGES, for later averaging over all patch contributions. Sometimes one may want to bias estimates on the boundary of the domain before scattering its contribution to the system averages.

After the flux components and their gradients (second derivatives) have been averaged, they are listed at the nodes, and/or saved for plotting and then passed to the routine SCP_ERROR_ESTIMATES. There the second derivative values are only used to calculate various measures or norms for them and their estimated error. While the corresponding cross derivates like $\partial^2/\partial x \partial y$ and $\partial^2/\partial y \partial x$ should be equal, they generally will differ due to various numerical approximations. All cross derivatives are computed, but only average values are used in estimating errors in the second derivatives. The larger full set of second derivatives at the nodes of an element are gathered and placed in an array called DERIV2_LT. If cross derivative estimates exist, they are averaged and placed in a smaller array called DERIV2_AVE that has the N_2_DER second derivatives at each element node. They will be interpolated to give the average second derivatives at the quadrature points used to evaluate the norms. By analogy to the first derivative norms, the N_2_DER interpolated second derivatives at a point are called DERIV2_SCP. The values computed directly from Eq. 6.6 are called DERIV2_HAT. Their values and differences are used to define norms and error estimates at the element level (ELEM_H2_NORM and ELEM_H2_ERROR) and at the system level (GLOBAL_H2_NORM and GLOBAL_H2_ERROR). If an exact solution is available for comparison, called FLUX_GRAD, it is used to compute corresponding exact values of the second spatial derivatives (EXACT_H2_NORM and EXACT_H2_ERROR). The various norm and error measures are listed, as are the SCP averaged and exact second derivatives at the system nodes. For plotting or other use, the last two items are saved to external files called pt_ave_grad_flux.tmp and pt_ex_grad_flux.tmp, respectively.

Some analysts prefer to assure a unique value for each cross-derivative. If one has solved a local patch for the continuous nodal gradients (as described above) one could use a Taylor expansion to get the second derivatives of the intermost node, or nodes of the intermost element. Let i be an intermost node of interest, ϕ_i its solution value, and $\nabla\phi_i$ its continuous gradient estimate. Then for any other node in the patch, $j \neq i$ we can seek the second derivatives at i such that

$$\phi_j = \phi_i + \left.\frac{\partial\phi}{\partial x}\right|_i (x_j - x_i) + \left.\frac{\partial\phi}{\partial y}\right|_i (y_j - y_i) + \frac{1}{2}\left.\frac{\partial^2\phi}{\partial x^2}\right|_i (x_j - x_i)^2 \tag{6.7}$$

$$+ \left.\frac{\partial^2\phi}{\partial x\partial y}\right|_i (x_j - x_i)(y_j - y_i) + \frac{1}{2}\left.\frac{\partial^2\phi}{\partial y^2}\right|_i (y_j - y_i)^2.$$

This can be used to solve for the three second derivative terms (in 2-D) by the SVD algorithm used for getting the nodal gradients.

6.5 Exercises

1. Develop a function, say *GET_SCP_PT_AT_XYZ*, to find where an element quadrature point or node is located within an element.
2. Develop a routine, say *EVAL_PT_FLUX_IN_SCP_PATCH*, to evaluate the flux components at any global coordinate point in a patch.
3. Employ the preceding patch routine to create a related subroutine, say *EVAL_SCP_FIT_AT_PATCH_NODES*, that will interpolate patch fluxes back to all mesh nodes inside the patch.
4. Develop a routine, say *FORM_NEW_EL_SIZES*, that can implement the estimated new element sizes for an h-adaptivity based on Eqs. 5.30 and 33.

6.6 Bibliography

[1] Ainsworth, M. and Oden, J.T., *A Posteriori Error Estimation in Finite Element Analysis*, New York: John Wiley (2000).

[2] Akin, J.E. and Maddox, J.R., "An RP-Adaptive Scheme for the Finite Element Analysis of Linear Elliptic Problems," pp. 427–438 in *The Mathematics of Finite Elements and Applications*, ed. J.R. Whiteman, London: Academic Press (1996).

[3] Akin, J.E. and Singh, M., "Object-Oriented Fortran 90 P-Adaptive Finite Element System," *Advances in Engineering Software*, **33**, pp. 461–468 (2002).

[4] Blacker, T. and Belytschko, T., "Superconvergent Patch Recovery with Equilibrium and Conjoint Interpolant Enhancements," *Int. J. Num. Meth. Eng.*, **37**, pp. 517–536 (1995).

[5] Cook, R.D., Malkus, D.S., Plesha, N.E., and Witt, R.J., *Concepts and Applications of Finite Element Analysis*, New York: John Wiley (2002).

[6] Huang, H.C. and Usmani, A.S., in *Finite Element Analysis for Heat Transfer*, London: Springer-Verlag (1994).

[7] Krizek, M., Neittaanmaki, P., and Stenberg, R., *Finite Element Methods: Superconvergence, Post-Processing and a Posteriori Estimates*, New York: Marcel Dekker, Inc. (1998).

[8] Lakhany, A.M. and Whiteman, J.R., "Superconvergent Recovery Operators," pp. 195–215 in *Finite Element Methods: Superconvergence, Post-Processing and a Posteriori Estimates*, New York: Marcel Dekker, Inc. (1998).

[9] Loubignac, G., Cantin, G., and Touzot, G., "Continuous Stress Fields in Finite Element Analysis," *AIAA Journal*, **15**(11), pp. 239–241 (1977).

[10] Press, W.H., Teukolsky, S.A., Vettering, W.T., and Flannery, B.P., *Numerical Recipes in Fortran 90*, New York: Cambridge University Press (1996).

[11] Szabo, B. and Babuska, I., *Finite Element Analysis*, New York: John Wiley (1991).

[12] Wiberg, N.-E., Abdulwahab, F., and Ziukas, S., "Enhanced Superconvergent Patch Recovery Incorporating Equilibrium and Boundary Conditions," *Int. J. Num. Meth. Eng.*, **37**, pp. 3417–3440 (1994).

[13] Wiberg, N.-E., Abdulwahab, F., and Ziukas, S., "Improved Element Stresses for Node and Element Patches Using Superconvergent Patch Recovery," *Comm. Num. Meth. Eng.*, **11**, pp. 619–627 (1995).

[14] Wiberg, N.-E., "Superconvergent Patch Recovery – A Key to Quality Assessed FE Solutions," *Adv. Eng. Software*, **28**, pp. 85–95 (1997).

[15] Zhu, J.Z. and Zienkiewicz, O.C., "Superconvergence Recovery Techniques and *A Posteriori* Error Estimators," *Int. J. Num. Meth. Eng.*, **30**, pp. 1321–1339 (1990).

[16] Zienkiewicz, O.C. and Zhu, J.Z., "A Simple Error Estimator and Adaptive Procedure for Practical Engineering Analysis," *Int. J. Num. Meth. Eng.*, **24**, pp. 337–357 (1987).

[17] Zienkiewicz, O.C. and Zhu, J.Z., "Superconvergent Patch Recovery Techniques and Adaptive Finite Element Refinement," *Comp. Meth. Appl. Mech. Eng.*, **101**, pp. 207–224 (1992).

[18] Zienkiewicz, O.C. and Zhu, J.Z., "The Superconvergent Patch Recovery and *a Posteriori* Error Estimates. Part 2: Error Estimates and Adaptivity," *Int. J. Num. Meth. Eng.*, **33**, pp. 1365–1382 (1992).

[19] Zienkiewicz, O.C. and Taylor, R.L., *The Finite Element Method,* 5th Edition, London: Butterworth-Heinemann (2000).

Chapter 7

Variational methods

7.1 Introduction

The Galerkin method given earlier can be shown to produce element matrix integral definitions that would be identical to those obtained from an Euler variational form, if one exists. Most nonlinear problems do not have a variational form, yet the Galerkin method and other weighted residual methods can still be used. Thus, one might ask, 'Why consider variational methods?' There are several reasons for using them. One is that if the variational integral form is known, one does not have to derive the corresponding differential equation. Also, most of the important variational statements for problems in engineering and physics have been known for over 200 years. Another important feature of variational methods is that often dual principles exist that allow one to establish both an upper bound estimate and a lower bound estimate for an approximate solution. These can be very helpful in establishing accurate error estimates for adaptive solutions. Thus, the variational methods still deserve serious study, especially the energy methods of solid mechanics.

We have seen that the weighted residual methods provide several approaches for generating approximate (or exact) solutions based on equivalent integral formulations of the original partial differential equations. *Variational Methods*, or the *Calculus of Variations* have given us another widely used set of tools for equivalent integral formulations. They were developed by the famous mathematician Euler in the mid-1700s. Since that time the variational forms of most elliptic partial differential equations have been known. It has been proved that a variational form and the Galerkin method yield the same integral formulations when a governing variational principle exists. Variational methods have thus been used to solve problems in elasticity, heat transfer, electricity, magnetism, ideal fluids, etc. Thus, it is logical to expect that numerical approximations based on these methods should be very fruitful. They have been very widely employed in elasticity and structural mechanics so we begin with that topic. Then we will introduce the finite element techniques as logical extensions of the various classical integral formulations.

7.2 Structural mechanics

Modern structural analysis relies extensively on the finite element method. Its most popular integral formulation, based on the variational calculus of Euler, is the *Principle of Minimum Total Potential Energy*. (This is also known as the principle of virtual work.) Basically, it states that the displacement field that satisfies the essential displacement boundary conditions and minimizes the total potential energy is the one that corresponds to the state of static equilibrium. This implies that displacements are our primary unknowns. They will be interpolated in space as will their derivatives, the strains. The total potential energy, Π, is the strain energy, U, of the structure minus the mechanical work, W, done by the applied forces. From introductory mechanics that the mechanical work, W, done by a force is the scalar dot product of the force vector, $\mathbf{F}$, and the displacement vector, $\mathbf{u}$, at its point of application.

To illustrate the concept of energy formulations we will review the equilibrium of the well-known linear spring. Figure 7.1 shows a typical spring of stiffness k that has an applied force, F, at the free end. That end undergoes a displacement of Δ. The work done by the single force is

$$W = \vec{\Delta} \cdot \vec{F} = \Delta F . \tag{7.1}$$

The spring stores potential energy due to its deformation. Here we call that strain energy. That energy is given by

$$U = \frac{1}{2} k \Delta^2 . \tag{7.2}$$

Therefore, the total potential energy for the loaded spring is

$$\Pi(\Delta) = U - W = \frac{1}{2} k \Delta^2 - \Delta F . \tag{7.3}$$

The equation of equilibrium is obtained by minimizing Π with respect to the displacement; that is, $\partial\Pi / \partial\Delta = 0$. This simplifies to the single scalar equation $k \Delta = F$, which is the well-known equilibrium equation for a linear spring. This example was slightly simplified, since we started with the condition that the left end of the spring had

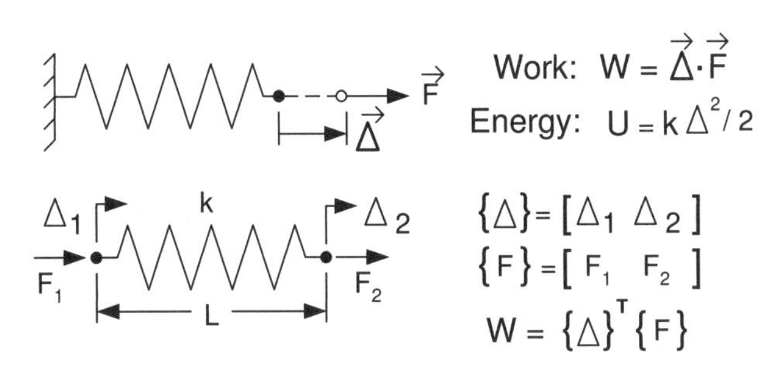

Figure 7.1 *A simple linear spring*

no displacement (an essential boundary condition). Next we will consider a spring where either end can be fixed or free to move.

The elastic bar is often modeled as a linear spring. In introductory mechanics of materials the axial stiffness of a bar is defined as $k = EA/L$ where it has a length of L, an area A, and an elastic modulus of E. Our spring model of the bar (see Fig. 7.1) has two end displacements, Δ_1 and Δ_2, and two associated axial forces, F_1 and F_2. The net deformation of the bar is $\mathbf{\Delta} = \Delta_2 - \Delta_1$. We denote the total vector of displacements as $\mathbf{D}^T = [\, \Delta_1 \quad \Delta_2 \,]$ and the associated vector of forces as $\mathbf{F}^T = [\, F_1 \quad F_2 \,]$. Then the work done on the bar is

$$W = \mathbf{D}^T \mathbf{F} = \Delta_1 F_1 + \Delta_2 F_2 .$$

The net displacement will be expressed in matrix form here to compare with the later mathematical formulations. It is $\Delta = [\, -1 \quad 1 \,]\, \mathbf{D}$. Then the spring's strain energy is

$$U = \frac{1}{2} k \Delta^2 = \frac{EA}{2L} \mathbf{D}^T \begin{Bmatrix} -1 \\ 1 \end{Bmatrix} [\, -1 \quad 1 \,]\, \mathbf{D} = \frac{1}{2} \mathbf{D}^T \mathbf{K}\, \mathbf{D}$$

where the bar stiffness is

$$\mathbf{K} = \frac{EA}{L} \begin{bmatrix} 1 & -1 \\ -1 & 1 \end{bmatrix} .$$

The total potential energy, Π, depends on all the displacements, $\mathbf{D}$:

$$\Pi(\mathbf{D}) = \frac{1}{2} \mathbf{D}^T \mathbf{K}\, \mathbf{D} - \mathbf{D}^T \mathbf{F} \tag{7.4}$$

and the equation of equilibrium comes from the minimization

$$\partial \Pi / \partial \mathbf{D} = \mathbf{0}, \quad \text{or} \quad \mathbf{K}\, \mathbf{D} = \mathbf{F} \tag{7.5}$$

represents the system of algebraic equations of equilibrium for the elastic system. These two equations do not yet reflect the presence of an essential boundary condition, and $\det(\mathbf{K}) \equiv 0$ and the system is singular. These relations were developed on physical arguments and did not involve any finite element theory. Next we will see that a one-dimensional FEA yields the same forms.

7.3 Finite element analysis

Up to this point we have considered equivalent integral forms in the classical sense and not invoked their enhancement by finite element methods. We have seen that the resulting algebraic equation systems based on a global approximate solution are fully coupled. That is, the coefficient matrix is not *sparse* (not highly populated with zeros) so that the solution or inversion cost would be high. The finite element method lets us employ an integral form to obtain a set of sparse equations to be solved for the coefficients, **D**, that yield the best approximation.

While we have looked so far mainly at one-dimensional problems in the general case we should be able to see that the residual error will involve volume integrals as well as surface integrals over part of the surface of the volume. For complicated shapes encountered in solving practical problems it is almost impossible to assume a global solution that would satisfy the boundary conditions. Even if we could do that the computational expense would probably prevent us from obtaining a solution. Both of these important practical limitations can be overcome if we utilize a *piecewise approximation* that has only local support in space. That is part of what *finite element analysis* offers us.

The basic concept is that we split the actual solution domain into a series of sub-domains, or *finite elements*, that are interconnected in such a way that we can split the required integrals into a summation of integrals over all the element domains. If we restrict the approximation to a function that exists only within the element domain then the algebraic system becomes sparse because an element only directly interacts with those elements that are connected to it. By restricting the element to a single shape, or to a small library of shapes, we can do the required integrals over that shape and use the results repeatedly to build up the integral contributions over the entire solution domain. The main additional piece of work that results is the requirement that we do some bookkeeping to keep up with the contribution of each element. We refer to this as the equation *assembly*. That topic was illustrated in Fig. 1.3, and will be discussed in more detail below. In today's terminology the assembly procedure and the post-processing procedures are a series of *gather* and *scatter* operations. In many finite element problems those concepts can be expressed symbolically as a scalar quantity, I:

$$I(\mathbf{D}) = \frac{1}{2}\,\mathbf{D}^T\mathbf{K}\,\mathbf{D} + \mathbf{D}^T\mathbf{C} \rightarrow \min \tag{7.6}$$

where $\mathbf{D}$ is a vector containing the unknown nodal parameters associated with the problem, and $\mathbf{K}$ and $\mathbf{C}$ are matrices defined in terms of the element properties and geometry. The above quantity is known as a *quadratic form*. If one uses a variational formulation then the solution of the finite element problem is usually required to satisfy the following system equations: $\partial I / \partial \mathbf{D} = \mathbf{0}$. In the finite element analysis one assumes that the (scalar) value of I is given by the sum of the element contributions. That is, one assumes

$$I(\mathbf{D}) = \sum_{e=1}^{n_e} I^e(\mathbf{D})$$

where I^e is the contribution of element number 'e'. One can (but does not in practice) define I^e in terms of $\mathbf{D}$ such that

$$I^e(\mathbf{D}) = \frac{1}{2}\,\mathbf{D}^T\mathbf{K}^e\mathbf{D} + \mathbf{D}^T\mathbf{C}^e \tag{7.7}$$

where the $\mathbf{K}^e$ are the same size as $\mathbf{K}$, but very sparse. Therefore, Eq. 7.7 is

$$I(\mathbf{D}) = \frac{1}{2}\,\mathbf{D}^T\left(\sum_{e=1}^{n_e}\mathbf{K}^e\right)\mathbf{D} + \mathbf{D}^T\left(\sum_{e=1}^{n_e}\mathbf{C}^e\right) \tag{7.8}$$

and comparing this with Eq. 7.8 one can identify the relations

$$\mathbf{K} = \sum_{e=1}^{n_e}\mathbf{K}^e, \qquad \mathbf{C} = \sum_{e=1}^{n_e}\mathbf{C}^e. \tag{7.9}$$

If n_d represents the total number of unknowns in the system, then the size of these matrices are $n_d \times n_d$ and $n_d \times 1$, respectively.

As a result of Eq. 7.9 one often sees the statement, 'the system matrices are simply the sum of the corresponding element matrices.' This is true, and indeed the symbolic operations depicted in the last equation are simple but one should ask (while preparing for the ensuing handwaving), 'in practice, how are the element matrices obtained and how does one carry out the summations?' Before attempting to answer this question, it will be useful to backtrack a little. First, it has been assumed that an element's behavior, and thus its contribution to the problem, depends *only* on those nodal parameters that are associated with the element. In practice, the number of parameters associated with a single element usually lies between a minimum of 2 and a maximum of 96; with the most common range in the past being from three to eight. (However, for hierarchical elements in three-dimensions it is possible for it to be 27!) By way of comparison, n_d can easily reach a value of several thousand, or several hundred thousand. Consider an example of a system where $n_d = 5000$. Let this system consist of one-dimensional elements with two parameters per element. A typical matrix $\mathbf{C}^e$ will contain 5000 terms and all but two of these terms will be identically zero since only those two terms of $\mathbf{D}$, 5000×1, associated with element 'e' are of any significance to element 'e'. In a similar manner one concludes that, for the present example, only four of the 25,000,000 terms of $\mathbf{K}^e$ would not be identically zero. Therefore, it becomes obvious that the symbolic procedure introduced here is not numerically efficient and would not be used in practice. There are some educational uses of the symbolic procedure that justify pursuing it a little further. Recalling that it is assumed that the element behavior depends only on those parameters, say $\boldsymbol{\phi}^e$, that are associated with element 'e', it is logical to assume that

$$I^e = \frac{1}{2} \boldsymbol{\phi}^{e^T} \mathbf{k}^e \boldsymbol{\phi}^e + \boldsymbol{\phi}^{e^T} \mathbf{c}^e . \tag{7.10}$$

If n_i represents the number of degrees of freedom associated with the element then the element vectors $\boldsymbol{\phi}^e$ and $\mathbf{c}^e$ are $n_i \times 1$ in size and the size of the square matrix $\mathbf{k}^e$ is $n_i \times n_i$. Note that in practice n_i is usually much less than n_d, but they can be equal. The matrices $\mathbf{k}^e$ and $\mathbf{c}^e$ are commonly known as *the* element matrices. For the one-dimensional element discussed in the previous example, $\mathbf{k}^e$ and $\mathbf{c}^e$ would be 2×2 and 2×1 in size, respectively, and would represent the only coefficients in $\mathbf{K}^e$ and $\mathbf{C}^e$ that are not identically zero.

All that remains is to relate $\mathbf{k}^e$ to $\mathbf{K}^e$ and $\mathbf{c}^e$ to $\mathbf{C}^e$. Obviously Eqs. 7.7 and 7.10 are equal and are the key to the desired relations. In order to utilize these equations, it is necessary to relate the degrees of freedom of the element $\boldsymbol{\phi}^e$, to the degrees of freedom of the total system $\mathbf{D}$. This is done symbolically by introducing a $n_i \times n_d$ bookkeeping matrix, $\boldsymbol{\beta}^e$, such that the following identity is satisfied:

$$\boldsymbol{\phi}^e \equiv \boldsymbol{\beta}^e \mathbf{D} . \tag{7.11}$$

Substituting this identity, Eq. 7.10 is expressed in terms of the system level unknowns as $I^e(\mathbf{D}) = \frac{1}{2} \mathbf{D}^T (\boldsymbol{\beta}^{e^T} \mathbf{k}^e \boldsymbol{\beta}^{e)} \mathbf{D} + \mathbf{D}^T \boldsymbol{\beta}^{e^T} \mathbf{c}^e$. Comparing this relation with Eq. 7.7, one can establish the symbolic relationships $\mathbf{K}^e = \boldsymbol{\beta}^{e^T} \mathbf{k}^e \boldsymbol{\beta}^e$, $\mathbf{C}^e = \boldsymbol{\beta}^{e^T} \mathbf{c}^e$ and denote the assembly process by the symbol $\underset{e}{\mathbf{A}}$ so

$$\mathbf{K} = \sum_{e=1}^{n_e} \beta^{e^T} \mathbf{k}^e \beta^e = \mathop{\mathbf{A}}_{e} \mathbf{K}^e, \quad \mathbf{C} = \sum_{e=1}^{n_e} \beta^{e^T} \mathbf{c}^e = \mathop{\mathbf{A}}_{e} \mathbf{C}^e. \tag{7.12}$$

Equation 7.12 can be considered as the symbolic definitions of the *assembly operator* and its procedures relating *the* element matrices, $\mathbf{k}^e$ and $\mathbf{c}^e$, to the total system matrices, **K** and **C**. Note that these relations involve the element connectivity (topology), β^e, as well as the element behavior, $\mathbf{k}^e$ and $\mathbf{c}^e$. Although some programs do use this procedure, it is very inefficient and thus very expensive.

For the sake of completeness, the β^e matrix will be briefly considered. To simplify the discussion, it will be assumed that each nodal point has only a single unknown scalar nodal parameter (degree of freedom). Define a mesh consisting of four triangular elements. Figure 7.2 shows both the system and element degree of freedom numbers. The system degrees of freedom are defined as $\mathbf{D}^T = [\ \Delta_1 \ \ \Delta_2 \ \ \Delta_3 \ \ \Delta_4 \ \ \Delta_5 \ \ \Delta_6\ ]$ and the degrees of freedom of element '*e*' are $\phi^{e^T} = [\ \phi_1 \ \ \phi_2 \ \ \phi_3\ ]^e$. The connectivity or topology data supplied for these elements are also shown in that figure. Thus, for element number four (e = 4), these quantities are related by

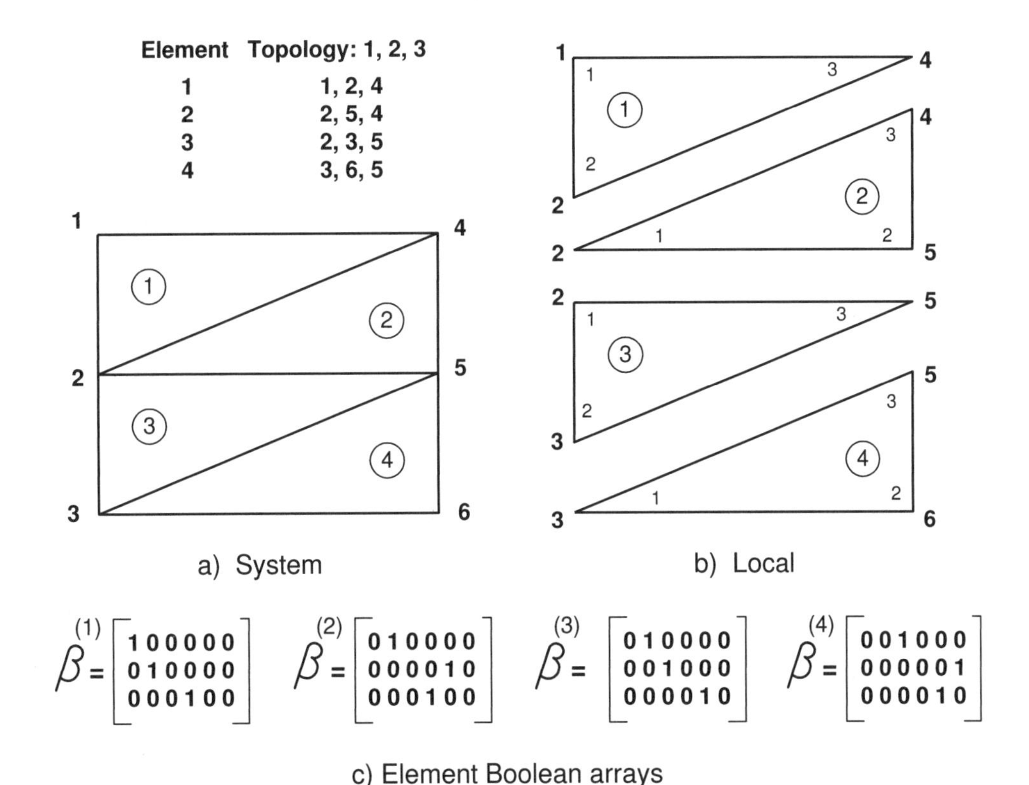

$$\beta^{(1)} = \begin{bmatrix} 1&0&0&0&0&0 \\ 0&1&0&0&0&0 \\ 0&0&0&1&0&0 \end{bmatrix} \quad \beta^{(2)} = \begin{bmatrix} 0&1&0&0&0&0 \\ 0&0&0&0&1&0 \\ 0&0&0&1&0&0 \end{bmatrix} \quad \beta^{(3)} = \begin{bmatrix} 0&1&0&0&0&0 \\ 0&0&1&0&0&0 \\ 0&0&0&0&1&0 \end{bmatrix} \quad \beta^{(4)} = \begin{bmatrix} 0&0&1&0&0&0 \\ 0&0&0&0&0&1 \\ 0&0&0&0&1&0 \end{bmatrix}$$

Figure 7.2 *Relationship between system and element degrees of freedom*

$$\boldsymbol{\phi}^{(4)^T} = \lfloor \phi_1 \quad \phi_2 \quad \phi_3 \rfloor^{(4)} = \lfloor \Delta_3 \quad \Delta_6 \quad \Delta_5 \rfloor$$

which can be expressed as the gather operation $\boldsymbol{\phi}^{(4)} \equiv \boldsymbol{\beta}^{(4)} \Delta$ where

$$\boldsymbol{\beta}^{(4)} \equiv \begin{bmatrix} 0 & 0 & 1 & 0 & 0 & 0 \\ 0 & 0 & 0 & 0 & 0 & 1 \\ 0 & 0 & 0 & 0 & 1 & 0 \end{bmatrix}.$$

The matrices $\boldsymbol{\beta}^{(1)}$, $\boldsymbol{\beta}^{(2)}$, and $\boldsymbol{\beta}^{(3)}$ can be defined in a similar manner, and are given in the figure. Since the matrix $\boldsymbol{\beta}^e$ contains only ones and zeros, it is called the element Boolean or binary matrix. Note that there is a single unity term in each row and all other coefficients are zero. Therefore, this $n_i \times n_d$ array will contain only n_i non-zero (unity) terms, and since $n_i << n_d$, the matrix multiplications of any Boolean gather or scatter operation are numerically very inefficient. There is a common shorthand method for writing any Boolean matrix to save space. It is written as a vector with the column number that contains the unity term on any row. In that form we would write $\boldsymbol{\beta}^{(4)}$ as

$$\boldsymbol{\beta}^{(4)^T} \leftrightarrow \lfloor 3 \quad 6 \quad 5 \rfloor$$

which you should note is the same as the element topology list because there is only one parameter per node. If we had formed the example with two parameters per node ($n_g = 2$) then the Boolean array would be

$$\boldsymbol{\beta}^{(4)^T} \leftrightarrow \lfloor 5 \quad 6 \quad 11 \quad 12 \quad 9 \quad 10 \rfloor .$$

This more compact vector mode was used in the assembly figures in Chapter 1. There it was given the array name *INDEX*. The transpose of a $\boldsymbol{\beta}$ matrix can be used to scatter the element terms into the system vector. For element four we see that $\boldsymbol{\beta}^{(4)^T} \mathbf{c}^{(4)} = \mathbf{C}^{(4)}$ gives

$$\begin{bmatrix} 0 & 0 & 0 \\ 0 & 0 & 0 \\ 1 & 0 & 0 \\ 0 & 0 & 0 \\ 0 & 0 & 1 \\ 0 & 1 & 0 \end{bmatrix} \begin{Bmatrix} c_1 \\ c_2 \\ c_3 \end{Bmatrix}^{(4)} = \begin{Bmatrix} 0 \\ 0 \\ C_1 \\ 0 \\ C_2 \\ C_3 \end{Bmatrix}^{(4)} .$$

This helps us to see how the scatter and sum operation for $\mathbf{C}$ in Eq. 7.12 actually works. The Boolean arrays, $\boldsymbol{\beta}$, have other properties that are useful in understanding certain other element level operations. For future reference note that $\boldsymbol{\beta}_e \boldsymbol{\beta}_e^T = \mathbf{I}$, and that if elements i and j are not connected $\boldsymbol{\beta}_i \boldsymbol{\beta}_j^T = \mathbf{0} = \boldsymbol{\beta}_j \boldsymbol{\beta}_i^T$, and if they are connected then $\boldsymbol{\beta}_i \boldsymbol{\beta}_j^T = \mathbf{X}_{ij}$ where $\mathbf{X}_{ij}$ indicates the *dof* that are common to both. That is, the Boolean array $\mathbf{X}$ is zero except for those *dof* common to both element i and j.

Although these symbolic relations have certain educational uses their gross inefficiency for practical computer calculations led to the development of the equivalent programming procedure of the 'direct method' of assembly that was discussed earlier, and illustrated in Figs. 1.3 and 4. It is useful at times to note that the identity Eq. 7.11 leads to the relation

$$\frac{\partial(\ )^e}{\partial \mathbf{D}} = \boldsymbol{\beta}^{e^T} \frac{\partial(\ )^e}{\partial \boldsymbol{\phi}^e}, \tag{7.13}$$

where $(\ )^e$ is some quantity associated with element 'e'. At this point we will begin to illustrate finite element domains and their piecewise local polynomial approximations to variational approximations by applying them to an elastic rod.

Before leaving the assembly relations for a while one should consider their extension to the case where there are more than one unknown per node. This was illustrated in Fig. 1.6 for three line elements with two nodes each and two dof per node. There the connectivity data and corresponding equation numbers are also given and we note that the connections between local and global equation numbers occur in pairs. Now we are inserting groups of two by two submatrices into the larger system matrix. It is not unusual for six dof to occur at each node. Then we assemble six by six blocks.

7.4 Continuous elastic bar

Consider an axisymmetric rod shown in Fig. 7.3. The cross-sectional area, $A(x)$, the perimeter, $p(x)$, the material modulus of elasticity, $E(x)$, and axial loading conditions would, in general, depend on the axial coordinate, x. The loading conditions could include surface tractions (shear) per unit area, $T(x)$, body forces per unit volume, $X(x)$, and concentrated point loads, P_i at point i. The axial displacement at a point will be denoted by $u(x)$, and its value at point i is u_i. The work done by a force is the product of the force, the displacement at the point of application of the force, and the cosine of the angle between the force and the displacement. Here the forces are all parallel so the cosine is either plus or minus one. Evaluating the mechanical work

$$W = \int_0^L u(x)\, X(x)\, A(x)\, dx + \int_0^L u(x)\, T(x)\, p(x)\, dx + \sum_i u_i\, P_i. \tag{7.14}$$

As mentioned earlier the total potential energy, Π, includes the *strain energy*, U, and work of the externally applied forces, W. That is, $\Pi = U - W$. In a mechanics of materials course it is shown that the strain energy per unit volume is half the product of the stress and strain. The axial strain and stress will be denoted by $\varepsilon(x)$ and $\sigma(x)$,

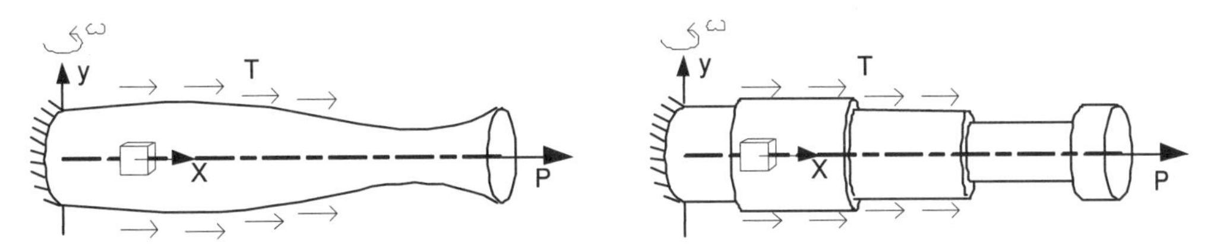

Figure 7.3 *A typical axially loaded bar*

respectively. Thus, the strain energy is

$$U = \frac{1}{2} \int_0^L \sigma(x)\ \varepsilon(x)\ A(x)\, dx\,. \tag{7.15}$$

The latter two equations have used $dV = A\,dx$ and $dS = p\,dx$ where dS is an exterior surface area. The work is clearly defined in terms of the displacement, u, since the loads would be given quantities. For example, the body force could be gravity, $X = \rho(x)\,g$, or a centrifugal load due to rotation about the y-axis, $X = \rho(x)\,x\omega^2$. Surface tractions are less common in 1-D but it could be due to a very viscous fluid flowing over the outer surface and in the x-direction.

Our goal is to develop a displacement formulation. Thus, we also need to relate both the stress and strain to the displacement, u. We begin with the *strain-displacement relation* $\varepsilon(x) = du(x)/dx$ which relates the strain to the derivative of the displacement. The stress at a point is directly proportional to the strain at the point. Thus, it is also dependent on the *displacement gradient.* The relation between stress and strain is a *constitutive relation* known as *Hooke's Law* $\sigma(x) = E(x)\ \varepsilon(x)$. Therefore, we now see that the total potential energy depends on the unknown displacements and displacement gradients. We are searching for the displacement configuration that minimizes the total potential energy since that configuration corresponds to the state of stable equilibrium. As we have suggested above, a finite element model can be introduced to approximate the displacements and their derivatives. Here, we begin by selecting the simplest model possible. That is, the two node line element with assumed linear displacement variation with x. As suggested, we now assume that each element has homogeneous properties and all the integrals can be represented as the sum of the corresponding element integrals (and the intersection of those elements with the boundary of the solution domain):

$$\Pi = \sum_{e=1}^{n_e} \Pi^e + \sum_{b=1}^{n_b} \Pi^b - \sum_i P_i\, u_i \tag{7.16}$$

where Π^e is the typical element domain contribution, and Π^b is the contribution from a typical boundary segment domain. This one-dimensional example is somewhat misleading since in general a surface traction, T, acts only on a small portion of the exterior boundary. Thus, the number of boundary domains, n_b, is usually much less than the number of elements, n_e. Here, however, $n_b = n_e$ and the distinction between the two may not become clear until two-dimensional problems are considered. Substituting Eqs. 7.14 and 7.15 into Eq. 7.3 and equating to Eq. 7.16 yields

$$\Pi^e = \tfrac{1}{2} \int_{L^e} \sigma^e\ \varepsilon^e\ A^e\, dx - \int_{L^e} u^e\ X^e\ A^e\, dx, \qquad \Pi^b = -\int_{L^b} u^b\ T^b\ p^b\, dx$$

where $u^e(x)$ and $u^b(x)$ denote the approximated displacements in the element and on the boundary surface, respectively. In this special example, $u^b = u^e$ but that is not usually true. Symbolically we interpolate such that $u^e(x) = \mathbf{H}^e(x)\ \mathbf{u}^e = \mathbf{u}^{e^T}\ \mathbf{H}^{e^T}(x)$ and likewise if we degenerate this interpolation to a portion (or sub-set) of the boundary of the element $u^b(x) = \mathbf{H}^b(x)\ \mathbf{u}^b = \mathbf{u}^{b^T}\ \mathbf{H}^{b^T}(x)$. In the example we have the unusual case that $\mathbf{u}^b = \mathbf{u}^e$ and $\mathbf{H}^b = \mathbf{H}^e$. Generally $\mathbf{u}^b$ is a sub-set of $\mathbf{u}^e$ (i.e., $\mathbf{u}^b sub-set \mathbf{u}^e$) and $\mathbf{H}^b$ is a sub-set of $\mathbf{H}^e$. This interpolation relationship gives the strain approximation in an element:

$$\varepsilon^e(x) = \frac{du^e}{dx} = \frac{d\mathbf{H}^e(x)}{dx}\mathbf{u}^e = \mathbf{u}^{e^T}\frac{d\mathbf{H}^{e^T}(x)}{dx}$$

or in more common notation

$$\boldsymbol{\varepsilon}^e = \mathbf{B}^e(x)\mathbf{u}^e \qquad (7.17)$$

where here $\mathbf{B}^e$ is called the 'strain-dislacement matrix' since it determines the mechanical strain from the element's nodal displacements. Likewise, the one-dimensional stress, $\boldsymbol{\sigma}$, is defined by 'Hooke's Law':

$$\boldsymbol{\sigma}^e(x) = \mathbf{E}^e(x)\,\boldsymbol{\varepsilon}^e(x), \qquad (7.18)$$

where $\mathbf{E}$ is the modulus of elasticity (a material property) as: Substituting into the definition of the Total Potential Energy from the elements and boundary terms:

$$\Pi^e = \frac{1}{2}\mathbf{u}^{e^T}\int_{L^e}\mathbf{B}^{e^T}E^e\,\mathbf{B}^e A^e\,\mathbf{u}^e\,dx - \mathbf{u}^{e^T}\int_{L^e}\mathbf{H}^{e^T}X^e A^e\,dx$$

$$\Pi^b = -\mathbf{u}^{b^T}\int_{L^b}\mathbf{H}^{b^T}T^b\,p^b\,dx.$$

The latter two relations can be written symbolically as

$$\Pi^e = \frac{1}{2}\mathbf{u}^{e^T}\mathbf{S}^e\,\mathbf{u}^e - \mathbf{u}^{e^T}\mathbf{C}_x^e, \qquad \Pi^b = -\mathbf{u}^{b^T}\mathbf{C}_T^b \qquad (7.19)$$

where the element stiffness matrix is

$$\mathbf{S}^e = \int_{L^e}\mathbf{B}^{e^T}E^e\,\mathbf{B}^e A^e\,dx, \qquad (7.20)$$

The vast majority of finite element problems have at least one square matrix of this form, that involves the matrix product $\mathbf{B}^{e^T}\mathbf{EB}$. We will see later that calculation of the element error estimator also requires the use of the $\mathbf{E}^e$ and $\mathbf{B}^e$ arrays. Thus, even if the $\mathbf{S}^e$ matrix is simple enough to write in closed form there are other reasons why we may want to form the $\mathbf{E}^e$ and $\mathbf{B}^e$ arrays at the same time. The element body force vector is

$$\mathbf{C}_x^e = \int_{L^e}\mathbf{H}^{e^T}X^e A^e\,dx, \qquad (7.21)$$

and the boundary segment traction vector is

$$\mathbf{C}_T^b = \int_{L^b}\mathbf{H}^{b^T}T^b\,p^b\,dx. \qquad (7.22)$$

The Total Potential Energy of the system is

$$\Pi = \frac{1}{2}\sum_e\mathbf{u}^{e^T}\mathbf{S}^e\,\mathbf{u}^e - \sum_e\mathbf{u}^{e^T}\mathbf{C}_x^e - \sum_b\mathbf{u}^{b^T}\mathbf{C}_T^b - \mathbf{u}^T\mathbf{P} \qquad (7.23)$$

where $\mathbf{u}$ is the vector of all of the unknown nodal displacements. Here we have assumed that the external point loads are applied at node points only. The last term represents the scalar, or dot, product of the nodal displacement and nodal forces. That is,

$$\mathbf{u}^T\mathbf{P} = \mathbf{P}^T\mathbf{u} = \sum_i u_i\,P_i\,.$$

Of course, in practice most of the P_i are zero. By again applying the direct assembly procedure, or from the Boolean assembly operations, the Total Potential Energy is $\Pi(\mathbf{u}) = \frac{1}{2}\,\mathbf{u}^T\,\mathbf{S}\mathbf{u} - \mathbf{u}^T\mathbf{C}$ and minimizing with respect to all the unknown displacements, $\mathbf{u}$, gives the algebraic equilibrium equations for the entire structure $\mathbf{S}\mathbf{u} = \mathbf{C}$. Therefore, we see that our variational principle has led to a very general and powerful formulation for this class of structures. It automatically includes features such as variable material properties, variable loads, etc. These were difficult to treat when relying solely on physical intuition. Although we will utilize the simple linear element none of our equations are restricted to that definition of $\mathbf{H}$ and the above symbolic formulation is valid for any linear elastic solid of any shape. If we substitute $\mathbf{H}^e$ for the linear element

$$\mathbf{H}^e(x) = \left\lfloor \frac{(x_2{}^e - x)}{L^e} \quad \frac{(x - x_1{}^e)}{L^e} \right\rfloor,$$

then

$$\mathbf{B}^e = \frac{d\mathbf{H}^e}{dx} = \left\lfloor -\frac{1}{L^e} \quad \frac{1}{L^e} \right\rfloor.$$

and assume constant properties $(\mathbf{E}^e, \mathbf{A}^e)$, then the element and boundary matrices are simple to integrate. The results are

$$\mathbf{S}^e = \frac{E^e\,A^E}{L^e}\begin{bmatrix} 1 & -1 \\ -1 & 1 \end{bmatrix}$$

$$\mathbf{C}_x^e = \frac{X^e\,A^e\,L^e}{2}\begin{Bmatrix} 1 \\ 1 \end{Bmatrix}, \qquad \mathbf{C}_T^b = \frac{T^e\,P^e\,L^e}{2}\begin{Bmatrix} 1 \\ 1 \end{Bmatrix}.$$

Also in this case one obtains the strain-displacement relation

$$\varepsilon^e = \frac{1}{L^e}\lfloor -1 \quad 1 \rfloor \begin{Bmatrix} u_1^e \\ u_2^e \end{Bmatrix} \tag{7.24}$$

which means that the strain is constant in the element but the displacement approximation is linear. It is common to refer to this element as the constant strain line element, CSL. The above stiffness matrix is the same as that obtained in Sec. 7.1. The load vectors take the resultant element, or boundary, force and place half at each node. That logical result does not carry over to more complicated load conditions, or interpolation functions and it then becomes necessary to rely on the mathematics of Eqs. 7.21 and 7.22. The implementation of this element will be given after thermal loading is defined.

As an example of a slightly more difficult loading condition consider a case where the body force varies linearly with x. This could include the case of centrifugal loading mentioned earlier. For simplicity assume a constant area A and let us define the value of the body force at each node of the element. To define the body force at any point in the element we again utilize the interpolation function and set

$$X^e(x) = \mathbf{H}^e(x)\,\mathbf{X}^e \tag{7.25}$$

where $\mathbf{X}^e$ are the defined nodal values of the body force. For these assumptions the body force vector becomes

$$\mathbf{C}_x^e = A^e \int_{L^e} \mathbf{H}^{e^T} \mathbf{H}^e \, dx \, \mathbf{X}^e .$$

For the linear element the integration reduces to

$$\mathbf{C}_x^e = \frac{A^e L^e}{6} \begin{bmatrix} 2 & 1 \\ 1 & 2 \end{bmatrix} \begin{Bmatrix} X_1^e \\ X_2^e \end{Bmatrix}.$$

This agrees with our previous result for constant loads since if $X_1^e = X_2^e = X^e$, then

$$\mathbf{C}_x^e = \frac{A^e L^e X^e}{6} \begin{Bmatrix} 2+1 \\ 1+2 \end{Bmatrix} = \frac{A^e L^e X^e}{2} \begin{Bmatrix} 1 \\ 1 \end{Bmatrix}.$$

A more common problem is that illustrated in Fig. 7.3 where the area of the member varies along the length. To approximate that case, with constant properties, one could interpolate for the area at any point as $A^e(x) = \mathbf{H}^e(x)\, \mathbf{A}^e$, then the stiffness in Eq. 7.20 becomes

$$\mathbf{S}^e = \frac{E^e}{(L^e)^2} \begin{bmatrix} 1 & -1 \\ -1 & 1 \end{bmatrix} V^e$$

where

$$V^e = \int_{L^e} A^e \, dx = \int_{L^e} \mathbf{H}^e(x) \, dx \, \mathbf{A}^e = \frac{L^e}{2} [1 \quad 1] \begin{Bmatrix} A_1^e \\ A_2^e \end{Bmatrix} = \frac{L^e (A_1^e + A_2^e)}{2}$$

is the average volume of the element. Using that volume value the body force vector is

$$\mathbf{C}_x^e = \frac{X^e}{2} \begin{Bmatrix} 1 \\ 1 \end{Bmatrix} V^e = \frac{X^e L^e (A_1^e + A_2^e)}{4} \begin{Bmatrix} 1 \\ 1 \end{Bmatrix},$$

but if we assume a constant body force per unit volume and interpolate for the local area we get

$$\mathbf{C}_x^e = X^e \int_{L^e} \mathbf{H}^{e^T} \mathbf{H}^e \, dx \, \mathbf{A}^e = \frac{X^e L^e}{6} \begin{bmatrix} 2 & 1 \\ 1 & 2 \end{bmatrix} \begin{Bmatrix} A_1^e \\ A_2^e \end{Bmatrix}.$$

The above approximations should be reasonably accurate. However, for a cylindrical bar the area is related to the radius by $A = \pi r^2$. Thus, it would be slightly more accurate to describe the radius at each end and interpolate $r^e(x) = \mathbf{H}^e(x)\, \mathbf{r}^e$ so that

$$V^e = \int_{L^e} A^e \, dx = \pi \int_{L^e} r^e(x)^2 \, dx = \mathbf{r}^{e^T} \pi \int_{L^e} \mathbf{H}^{e^T} \mathbf{H}^e \, d\mathbf{x} \, \mathbf{r}^e$$

$$= \mathbf{r}^{e^T} \frac{\pi L^e}{6} \begin{bmatrix} 2 & 1 \\ 1 & 2 \end{bmatrix} \mathbf{r}^e = \pi L^e (r_1^2 + r_1 r_2 + r_2^2)/3.$$

Of course, for a bar of constant radius (and area) all three approaches give identical resultant body force components at the nodes.

Clearly, as one utilizes more advanced interpolation functions, the integrals involved in Eqs. 7.20 to 7.22 become more difficult to evaluate. An example to illustrate the use of these element matrices and to introduce the benefits of post-solution calculations follows. Consider a prismatic bar of steel rigidly fixed to a bar of brass and subjected to a vertical load of $P = 10,000$ lb, as shown in Fig. 7.4. The structure is supported at the top point and is also subjected to a gravity (body force) load. We wish to determine the deflections, reactions, and stresses for the properties:

Element	L^e	A^e	E^e	X^e	*Topology*	
1	420"	10 sq. in.	30×10^6 psi	0.283 lb/in^3	1	2
2	240"	8 sq. in.	13×10^6 psi	0.300 lb/in^3	2	3

The first element has an axial stiffness constant of $EA/L = 0.7143 \times 10^6$ lb/in and the body force is $XAL = 1,188.6$ lb. while for the second element the corresponding terms are 0.4333×10^6 lb/in and 576 lb, respectively. The system nodal force vector is $P^T = [\,R \quad 0 \quad 10,000\,]^{lb}$. Where R is the unknown reaction at node 1. Assemblying the equations gives

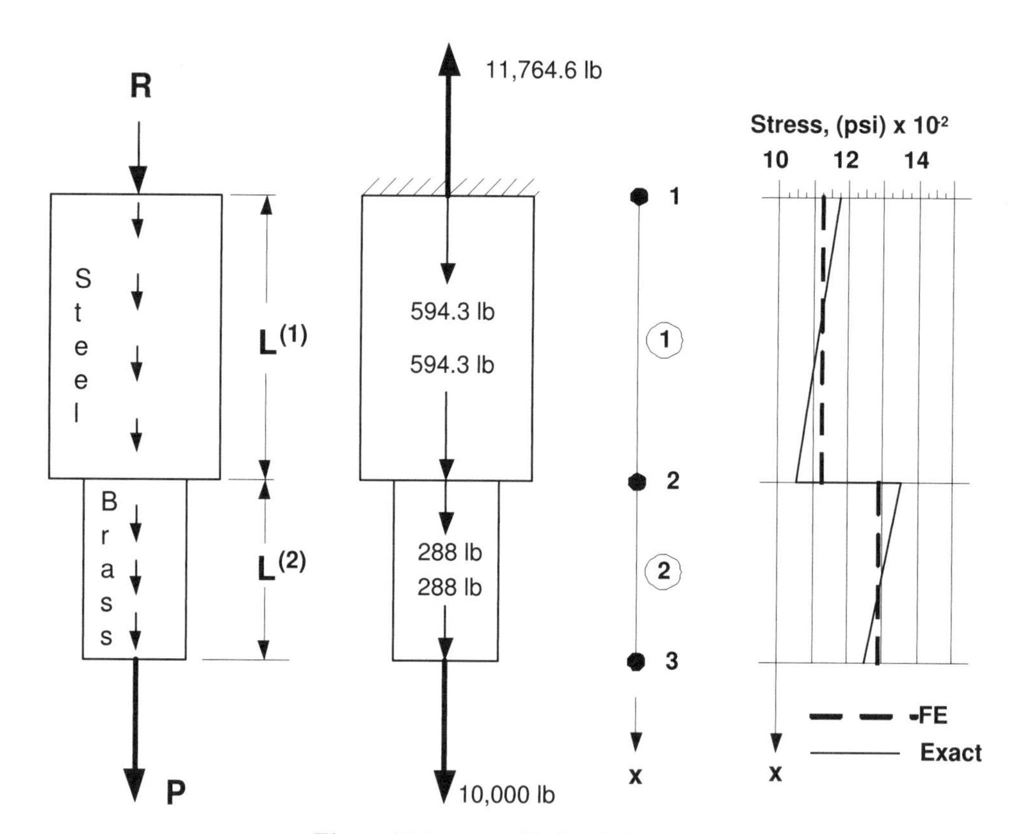

Figure 7.4 *An axially loaded system*

```
title "Steel-Brass gravity and end load" ! begin keywords  ! 1
nodes      3 ! Number of nodes in the mesh                 ! 2
elems      2 ! Number of elements in the system            ! 3
dof        1 ! Number of unknowns per node                 ! 4
el_nodes   2 ! Maximum number of nodes per element         ! 5
space      1 ! Solution space dimension                    ! 6
b_rows     1 ! Number of rows in the B (operator) matrix   ! 7
example 113 ! Application source file number               ! 8
remarks    2 ! Number of user remarks                      ! 9
el_real    5 ! Number of real properties per element       !10
loads        ! External source supplied                    !11
el_react     ! Compute & list element reactions            !12
post_el      ! Require post-processing, create n_file1     !13
quit ! keyword input, remarks follow                       !14
Nodal displacements are exact, stress exact only at center !15
Properties: A, E, DT, ALPHA, GAMMA                         !16
  1    1 0.00       ! begin node, bc_flag, x               !17
  2    0 420.                                              !18
  3    0 660.                                              !19
     1    1    2    ! begin element, connectivity          !20
     2    2    3                                           !21
 1 1 0.             ! essential bc: node, dof, value       !22
 1 10. 30.e6 0. 0. 0.283 ! el, A, E, DT, ALPHA, GAMMA      !23
 2  8. 13.e6 0. 0. 0.300 ! el, A, E, DT, ALPHA, GAMMA      !24
 3 1 1.e4 ! node, direction, load (stop with last)         !25
```

Figure 7.5 *The steel-brass bar example typical input*

```
*** REACTION RECOVERY ***                                      ! 1
NODE, PARAMETER,      REACTION, EQUATION                       ! 2
   1,  DOF_1,      -1.1765E+04      1                          ! 3
                                                               ! 4
***  OUTPUT OF RESULTS IN NODAL ORDER  ***                     ! 5
NODE,   X-Coord,      DOF_1,                                   ! 6
   1  0.0000E+00  0.0000E+00                                   ! 7
   2  4.2000E+02  1.5638E-02                                   ! 8
   3  6.6000E+02  3.9380E-02                                   ! 9
                                                               !10
** ELEMENT REACTION, INTERNAL SOURCES AND SUMMATIONS **        !11
                  ELEMENT  1                                   !12
 NODE  DOF   REACTION        ELEM_SOURCE    SUMS               !13
   1    1   -1.17646E+04     5.94300E+02                       !14
   2    1    1.05760E+04     5.94300E+02                       !15
SUM:    1   -1.18860E+03     1.18860E+03    0.00000E+00        !16
                  ELEMENT  2                                   !17
 NODE  DOF   REACTION        ELEM_SOURCE    SUMS               !18
   2    1   -1.05760E+04     2.88000E+02                       !19
   3    1    1.00000E+04     2.88000E+02                       !20
SUM:    1   -5.76000E+02     5.76000E+02    0.00000E+00        !21
                                                               !22
        E L E M E N T   S T R E S S E S                        !23
ELEMENT        STRESS    MECH. STRAIN    THERMAL STRAIN        !24
   1     1.11703E+03     3.72343E-05     0.00000E+00           !25
   2     1.28600E+03     9.89231E-05     0.00000E+00           !26
```

Figure 7.6 *The steel-brass bar example selected output*

$$10^5 \begin{bmatrix} 7.143 & -7.143 & 0 \\ -7.143 & 7.143 + 4.333 & -4.333 \\ 0 & -4.333 & 4.333 \end{bmatrix} \begin{Bmatrix} u_1 \\ u_2 \\ u_3 \end{Bmatrix}$$

$$= \begin{Bmatrix} R \\ 0 \\ 10,000 \end{Bmatrix} + \frac{1}{2} \begin{Bmatrix} 1,188.6 \\ 1,188.6 + 576. \\ 576. \end{Bmatrix} = \begin{Bmatrix} R + 594.3 \\ 882.3 \\ 10,288 \end{Bmatrix}.$$

Applying the essential condition that $u_1 = 0$

$$10^5 \begin{bmatrix} 11.476 & -4.333 \\ -4.333 & 4.333 \end{bmatrix} \begin{Bmatrix} u_2 \\ u_3 \end{Bmatrix} = \begin{Bmatrix} 882.3 \\ 10,288 \end{Bmatrix}$$

so that $u_2 = 1.5638 \times 10^{-2}$ in, $u_3 = 3.9381 \times 10^{-2}$ in, and determining the reaction from the first system equation: $R = -11,764.6$ lb. This reaction is compared with the applied loads in Fig. 7.4. Now that all the displacements are known we can *post-process* the results to determine the other quantities of interest. Substituting into the element strain-displacement relation, Eq. 7.24 gives

$$\varepsilon^{(1)} = \frac{1}{420} \lfloor -1 \quad 1 \rfloor \begin{Bmatrix} 0.0 \\ 0.01564 \end{Bmatrix} = 3.724 \times 10^{-5} \text{ in/in}$$

$$\varepsilon^{(2)} = \frac{1}{240} \lfloor -1 \quad 1 \rfloor \begin{Bmatrix} 0.01564 \\ 0.03938 \end{Bmatrix} = 9.892 \times 10^{-5} \text{ in/in}$$

and from Eq. 7.18 the element stresses are

$$\sigma^{(1)} = \mathbf{E}^{(1)} \varepsilon^{(1)} = 30 \times 10^6 (7.724 \times 10^{-5}) = 1,117 \text{ lb/in}^2$$

$$\sigma^{(2)} = \mathbf{E}^{(2)} \varepsilon^{(2)} = 13 \times 10^6 (9.892 \times 10^{-5}) = 1,286 \text{ lb/in}^2 .$$

These approximate stresses are compared with the exact stresses in Fig. 7.4. This suggests that if accurate stresses are important then more elements are required to get good estimates from the piecewise constant element stress approximations. Note that the element stresses are exact if they are considered to act only at the element center. The input data and selected results for this example are given in Figs. 7.5 and 6, respectively.

7.5 Thermal loads on a bar *

Before leaving the bar element it may be useful to note that another common loading condition can be included, that is the loading due to an initial thermal strain, so $\varepsilon = \sigma / E + \varepsilon_t$. The thermal strain, ε_t, due to a temperature rise of Δt is $\varepsilon_t = \alpha \, \Delta t$ where α is the coefficient of thermal expansion. The work term in Eq. 7.14 is extended to include this effect by adding an *initial strain* contribution

$$W_t = \int_0^L \sigma^T \, \varepsilon_t \, A(x) \, dx .$$

This defines an element thermal force vector

$$\mathbf{C}_t^e = \int_{L^e} \mathbf{B}^{e^T}(x)\, E^e(x)\, \alpha^e(x)\, \Delta t(x)\, A^e(x)\, dx \tag{7.26}$$

or for constant properties and uniform temperature rise

$$\mathbf{C}_t^{eT} = E^e\, \alpha^e\, \Delta t^e\, A^e \lfloor -1 \quad +1 \rfloor .$$

There is a corresponding change in the constitutive law such that $\sigma = \mathbf{E}\,(\,\varepsilon - \varepsilon_t\,)$.

The coding requires a few new interface items listed in Fig. 7.7. The source code for implementing the linear elastic bar element is given in Fig. 7.8. There the last action, in line 36, is to optionally save the modulus of elasticity and the strain data for later post-processing. The data keyword *post_el* activates the necessary sequential storage unit (n_file1). After the unknowns have been computed we gather typical data back to the element for use in post-processing secondary items in the element. This too requires a few interface items to the MODEL program and they are listed in Fig. 7.9. The stress recovery coding is given in Fig. 7.10 and it is invoked by the presence of the same *post_el* data keyword. Since strain, initial strain, and stress are tensor quantities that have several components, stored in subscripted arrays, a unit subscript is required to remind us that we are dealing with the one-dimensional form. Up to this point we have dealt with scalars. The mechanical strain is found in line 25 and the generalized Hooke's Law is employed, at line 28, to recover the stress.

As a numerical example of this loading consider the previous model with the statically indeterminate supports in Fig. 7.11. The left support is fixed but the right support is displaced to the left 0.001 in. and the system is cooled by 35° F. Find the stress in each member if $\alpha_1 = 6.7 \times 10^{-6}$ and $\alpha_2 = 12.5 \times 10^{-6}$ in/in F. The assembled equations are

$$10^5 \begin{bmatrix} 7.143 & -7.143 & 0 \\ -7.143 & 11.476 & -4.333 \\ 0 & -4.333 & 4.333 \end{bmatrix} \begin{Bmatrix} u_1 \\ u_2 \\ u_3 \end{Bmatrix} = 10^4 \begin{Bmatrix} -7.035 \\ 7.035 - 4.550 \\ +4.550 \end{Bmatrix} + \begin{Bmatrix} P_1 \\ 0 \\ P_2 \end{Bmatrix}$$

applying the boundary conditions that $u_1 = 0$, and $u_3 = -0.001$ in. and solving for u_2 yields $u_2 = -0.02203$ in. The reactions at points 1 and 3 are $P_1 = -54,613$ lb, and $P_3 = +54,613$ lb. Substituting into the element strain-displacement matrices yields element mechanical strains of -5.245×10^{-5} in/in, and $+8.763 \times 10^{-5}$ in/in, respectively. But, the initial thermal strains were -2.345×10^{-4}, and -4.375×10^{-4}, respectively so together they result in net tensile stresses in the elements of 5.46×10^3 and 6.83×10^3 psi, respectively. That is, the tension due to the temperature reduction exceeds the compression due to the support movement. Sample input and selected outputs for the above thermal loading example are given in Figs. 7.12 and 13, respectively. Note that the outputs for the last two examples have included the element reactions, also called the element level flux balances. These are often physically important so we will summarize how they are obtained in the next section.

```
            Interface from MODEL to ELEM_SQ_MATRIX, 2
Type    Status Name    Remarks                          (keyword)
INTEGER (IN) IE       Current element number
INTEGER (IN) LT_FREE Number of element type unknowns
INTEGER (IN) LT_N     Number of element type solution nodes
INTEGER (IN) N_R_B    Number of rows in B and E arrays (b_rows)
INTEGER (IN) N_FILE1 Optional user sequential unit (post_el)
REAL(DP) (OUT) B (N_R_B, LT_FREE)  Gradient-dof transformation
REAL(DP) (OUT) DLH (LT_PARM, LT_N) Local parametric derivatives of H
REAL(DP) (OUT) E (N_R_B, N_R_B)     Constitutive array
REAL(DP) (OUT) H_INTG (LT_N)        Integral of H array
REAL(DP) automatic XYZ (N_SPACE)    Coordinates of a point
GET_H_AT_QP          Form H array at quadrature point
GET_REAL_LP (k)      Gather real property k for current element
                     1 <= k <= (el_real)
```

Figure 7.7 *User interface to ELEM_SQ_MATRIX (part 2)*

```
! ..........................................................          ! 1
! ***  ELEM_SQ_MATRIX PROBLEM DEPENDENT STATEMENTS FOLLOW ***          ! 2
! ..........................................................          ! 3
!        (Stored as application source example 113)                   ! 4
                                                                      ! 5
!  AN AXIAL BAR BY DIRECT ENERGY APPROACH                              ! 6
!  ELEMENT REAL PROPERTIES: (1) = AREA, (2) = ELASTIC MODULUS          ! 7
!  (3) = TEMP RISE, (4) = COEFF EXPANSION, (5) = WEIGHT DENSITY        ! 8
                                                                      ! 9
  REAL(DP) :: BAR_L                    ! length                        !10
  REAL(DP) :: DELTA_T, ALPHA           ! temp rise, expansion          !11
  REAL(DP) :: AREA, GAMMA              ! area, wt. density             !12
  REAL(DP) :: M_E, THERMAL             ! modulus, thermal strain       !13
                                                                      !14
! Get properties for this element, IE                                 !15
    AREA    = GET_REAL_LP (1); M_E   = GET_REAL_LP (2)                 !16
    DELTA_T = GET_REAL_LP (3); ALPHA = GET_REAL_LP (4)                 !17
    GAMMA   = GET_REAL_LP (5)                                          !18
                                                                      !19
! Find bar length and direction cosines                               !20
    BAR_L = COORD (2, 1) - COORD (1, 1)   ! length                     !21
                                                                      !22
! Form global strain-displacement matrix                              !23
    B (1, :) = (/ -1,   1 /) / BAR_L                                   !24
                                                                      !25
! Form global stiffness, S = B' EAL B                                 !26
    S = M_E * AREA * BAR_L * MATMUL ( TRANSPOSE (B), B )               !27
                                                                      !28
! Initial (thermal) strain loading                                    !29
    THERMAL = ALPHA * DELTA_T                          ! strain        !30
    C       = B (1, :) * M_E * THERMAL * AREA * BAR_L ! force          !31
                                                                      !32
! Weight load, in positive X-direction (wt density * volume)          !33
    C = C + (/ 0.5d0, 0.5d0 /) * GAMMA * AREA * BAR_L !  weight        !34
! Save for stress post-processing (set post_el in keywords)           !35
    IF ( N_FILE1 > 0 ) WRITE (N_FILE1) M_E, B, THERMAL                 !36
!  End of application dependent code                                  !37
```

Figure 7.8 *Implementation of an elastic linear bar*

```
            Interface from MODEL to POST_PROCESS_ELEM, 1
Type   Status Name   Remarks                           (keyword)
INTEGER (IN) IE      Current element number
INTEGER (IN) LT_FREE Number of element type unknowns
INTEGER (IN) LT_N    Number of element type solution nodes
INTEGER (IN) N_R_B   Number of rows in B and E arrays (b_rows)
INTEGER (IN) N_SPACE Physical space dimension of problem (space)
INTEGER (IN) N_FILE1 Optional user sequential unit (post_el)
INTEGER (IN) N_FILE2 Optional user sequential unit (post_2)

REAL(DP) (IN)  D (LT_FREE)          Gathered element dof

REAL(DP) (OUT) B (N_R_B, LT_FREE)   Gradient-dof transformation
REAL(DP) (OUT) DGH (N_SPACE, LT_N)  Physical derivatives of H
REAL(DP) (OUT) E (N_R_B, N_R_B)     Constitutive array
REAL(DP) (OUT) H_INTG (LT_N)        Integral of H array

REAL(DP) automatic XYZ      (N_SPACE)    Coordinates of a point
REAL(DP) automatic STRAIN_0 (N_R_B)      Initial strains
REAL(DP) automatic STRAIN   (N_R_B + 2)  Mechanical strains
REAL(DP) automatic STRESS   (N_R_B + 2)  Mechanical stresses
```

Figure 7.9 *User interface to POST_PROCESS_ELEM (part 1)*

```
! .............................................................. ! 1
!  *** POST_PROCESS_ELEM PROBLEM DEPENDENT STATEMENTS FOLLOW ***  ! 2
! .............................................................. ! 3
!            (Stored as application source example 113)          ! 4
                                                                  ! 5
! AN AXIAL BAR BY DIRECT ENERGY APPROACH                          ! 6
! ELEMENT REAL PROPERTIES: (1) = AREA, (2) = ELASTIC MODULUS      ! 7
! (3) = TEMP RISE, (4) = COEFF EXPANSION, (5) = WEIGHT DENSITY    ! 8
                                                                  ! 9
!    STRESS = M_E * (MECHANICAL STRAIN - INITIAL STRAIN)          !10
                                                                  !11
  REAL(DP) :: THERMAL, M_E  ! initial strain, modulus             !12
  LOGICAL, SAVE :: FIRST = .TRUE.  ! printing                     !13
                                                                  !14
   IF ( FIRST ) THEN                  ! first call                !15
     FIRST = .FALSE. ; WRITE (6, 5)   ! print headings            !16
     5 FORMAT ('        E L E M E N T    S T R E S S E S', /, &   !17
     & ' ELEMENT      STRESS    MECH. STRAIN    THERMAL STRAIN')  !18
   END IF ! first call                                            !19
                                                                  !20
!-->  Read stress strain data from N_FILE1 (set by post_el)       !21
   READ (N_FILE1) M_E, B, STRAIN_0 (1) ! THERMAL = STRAIN_0       !22
                                                                  !23
!--> Calculate mechanical strain, STRAIN = B * D                  !24
   STRAIN (1) = DOT_PRODUCT ( B(1, :), D )                        !25
                                                                  !26
!-->  Generalized Hooke's Law                                     !27
   STRESS (1) = M_E * (STRAIN (1) - STRAIN_0 (1))                 !28
                                                                  !29
   WRITE (6, 1) IE, STRESS (1), STRAIN (1), STRAIN_0 (1)          !30
   1 FORMAT (I5, 3ES15.5)                                         !31
!  *** END POST_PROCESS_ELEM PROBLEM DEPENDENT STATEMENTS ***     !32
```

Figure 7.10 *Stress recovery for the elastic bar*

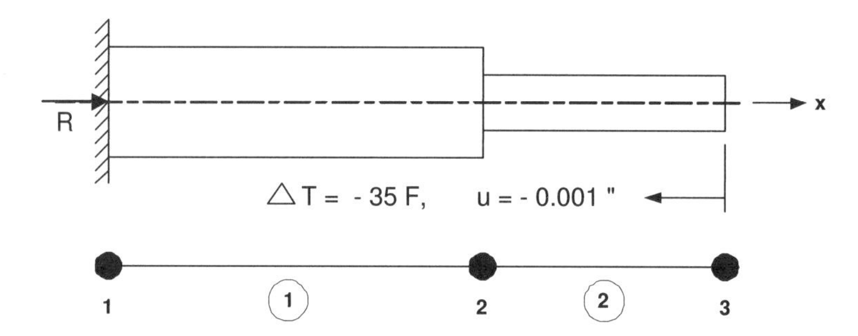

Figure 7.11 *A thermally loaded elastic bar*

7.6 Reaction flux recovery for an element*

Regardless of whether we use variational methods or weighted residual methods we are often interested in post-processing to get the *flux recovery* data for some or all of the elements in the system. Once the assembled system has been solved for the primary nodal unknowns ϕ, we are often interested in also computing the nodal forces (or fluxes) that act on each individual element. For linear structural equilibrium, or thermal equilibrium, or a general Galerkin statement the algebraic equations are of the form

$$\mathbf{S}\,\mathbf{D} = \mathbf{C} + \mathbf{P} \tag{7.27}$$

where the square matrix **S** is the assembly of the element square matrices $\mathbf{S}^e$ and the column matrix **C** is the sum of the consistent element force (or flux) matrices, $\mathbf{C}^e$, due to spatially distributed forces (or fluxes). Finally, **P** is the vector of externally applied concentrated point forces (or fluxes) that are often called element reactions. The vector **P** can also be thought of as an assembly of point sources on the elements, $\mathbf{P}^e$. This is always done if one is employing an element wavefront equation solving system. Most of the $\mathbf{P}^e$ are identically zero. When P_j is applied to the j-th node of the system we simply find the element, e, where that node makes its first appearance in the data. Then, P_j is inserted in $\mathbf{P}^e$ for that element and no entries are made in any other elements. If degree of freedom ϕ_j is given then P_j is an unknown reaction. To recover the concentrated 'external' nodal forces or fluxes associated with a specific element we make the *assumption* that a similar expression holds for the element. That is,

$$\mathbf{S}^e\,\mathbf{D}^e = \mathbf{C}^e + \mathbf{P}^e \tag{7.28}$$

This is clearly exact if the system has only one element. Otherwise, it is a reasonable approximation. When we use an energy method to require equilibrium of an assembled system, we do not exactly enforce equilibrium in every element that makes up that system. Solving the reasonable approximation gives

$$\mathbf{P}^e = \mathbf{S}^e\mathbf{D}^e - \mathbf{C}^e \tag{7.29}$$

where everything on the right hand side is known, since $\mathbf{D}^e\,sub - set\mathbf{D}$ can be recovered as a gather operation. To illustrate these calculations consider the one-dimensional

```
title "Steel-Brass cooled and deformed" ! begin keywords   ! 1
nodes      3 ! Number of nodes in the mesh                 ! 2
elems      2 ! Number of elements in the system            ! 3
dof        1 ! Number of unknowns per node                 ! 4
el_nodes   2 ! Maximum number of nodes per element         ! 5
space      1 ! Solution space dimension                    ! 6
b_rows     1 ! Number of rows in the B (operator) matrix   ! 7
example 113 ! Application library source file number       ! 8
remarks    2 ! Number of user remarks                      ! 9
el_real    5 ! Number of real properties per element       !10
el_react     ! Compute & list element reactions            !11
post_el      ! Require post-processing, create n_file1     !12
quit ! keyword input, remarks follow                       !13
Nodal displacements are exact, stresses too                !14
Properties: A, E, DT, ALPHA, GAMMA (no gravity)            !15
 1    1 0.00       ! begin node, bc_flag, x                !16
 2    0 420.                                               !17
 3    1 660.                                               !18
    1    1    2    ! begin element, connectivity           !19
    2    2    3                                            !20
1 1  0.            ! essential bc                          !21
3 1 -0.001         ! essential bc                          !22
1 10. 30.e6 -35.  6.7e-6 0. ! el, A, E, DT, ALPHA, GAMMA   !23
2  8. 13.e6 -35. 12.5e-6 0. ! el, A, E, DT, ALPHA, GAMMA   !24
```

Figure 7.12 *Data for a thermally loaded deformed bar*

```
*** INPUT SOURCE RESULTANTS ***                                        ! 1
ITEM       SUM            POSITIVE       NEGATIVE                      ! 2
   1    0.0000E+00     7.0350E+04   -7.0350E+04                        ! 3
                                                                       ! 4
*** REACTION RECOVERY ***                                              ! 5
NODE, PARAMETER,      REACTION, EQUATION                               ! 6
   1,  DOF_1,       -5.4613E+04     1                                  ! 7
   3,  DOF_1,        5.4613E+04     3                                  ! 8
                                                                       ! 9
***  OUTPUT OF RESULTS IN NODAL ORDER  ***                             !10
 NODE,  X-Coord,     DOF_1,                                            !11
    1  0.0000E+00  0.0000E+00                                          !12
    2  4.2000E+02 -2.2031E-02                                          !13
    3  6.6000E+02 -1.0000E-03                                          !14
                                                                       !15
** ELEMENT REACTION, INTERNAL SOURCES AND SUMMATIONS **                !16
                  ELEMENT  1                                           !17
 NODE  DOF   REACTION        ELEM_SOURCE    SUMS                       !18
   1    1   -5.46135E+04    7.03500E+04                                !19
   2    1    5.46135E+04   -7.03500E+04                                !20
SUM:    1    0.00000E+00    0.00000E+00    0.00000E+00                 !21
                  ELEMENT  2                                           !22
 NODE  DOF   REACTION        ELEM_SOURCE    SUMS                       !23
   2    1   -5.46135E+04    4.55000E+04                                !24
   3    1    5.46135E+04   -4.55000E+04                                !25
SUM:    1    0.00000E+00    0.00000E+00    0.00000E+00                 !26
                                                                       !27
        E L E M E N T   S T R E S S E S                                !28
ELEMENT        STRESS    MECH. STRAIN    THERMAL STRAIN                !29
   1     5.46135E+03   -5.24550E-05    -2.34500E-04                    !30
   2     6.82669E+03    8.76297E-05    -4.37500E-04                    !31
```

Figure 7.13 *Results for a thermally loaded deformed bar* stepped Steel-Brass bar system given above in Fig. 7.4 where

$$\mathbf{S} = \frac{E^e A^e}{L^e} \begin{bmatrix} 1 & -1 \\ -1 & 1 \end{bmatrix}, \quad \mathbf{C}^e = \frac{\mathbf{X}^e A^e L^e}{2} \begin{Bmatrix} 1 \\ 1 \end{Bmatrix}.$$

Now that all of the $\mathbf{D}$ are known the $\mathbf{D}^e$ can be extracted and substituted into Eq. 7.29 for each of the elements. For the first element Eq. 7.29 gives

$$\mathbf{P}^e = 7.143 \times 10^5 \begin{bmatrix} 1 & -1 \\ -1 & 1 \end{bmatrix} \begin{Bmatrix} 0 \\ 1.5638 \times 10^{-2} \end{Bmatrix} - \begin{Bmatrix} 594.3 \\ 594.3 \end{Bmatrix}$$

$$= \begin{Bmatrix} -11,170.3 \\ 11,170.3 \end{Bmatrix} - \begin{Bmatrix} 594.3 \\ 594.3 \end{Bmatrix} = \begin{Bmatrix} -11,764.6 \\ 10,576.0 \end{Bmatrix} \text{lb.} \tag{7.30}$$

Likewise for the second element

$$\mathbf{P}^e = 4.333 \times 10^5 \begin{bmatrix} 1 & -1 \\ -1 & 1 \end{bmatrix} \begin{Bmatrix} 1.5638 \times 10^{-2} \\ 3.9381 \times 10^{-2} \end{Bmatrix} - \begin{Bmatrix} 288.0 \\ 288.0 \end{Bmatrix}$$

$$= \begin{Bmatrix} -10,576.0 \\ 10,000.0 \end{Bmatrix} \text{lb.} \tag{7.31}$$

Note that if we choose to assemble these element $\mathbf{P}^e$ values we obtain the system reactions $\mathbf{P}$. That is because element contributions at all unloaded nodes are equal and opposite (Newton's Third Law) and cancel when assembled. Figure 7.14. shows these 'external' element forces when viewed on each element, as well as their assembly which matches the original system. This series of matrix operations is available in MODEL and is turned on only when the keyword *el_react* is present in the control data.

If we do the same reaction recovery for the second case of a thermal load and an enforced end displacement we get the values in Fig. 7.15. There is a subtle difference between theses two cases. In the first case the gravity load creates a net external force source. In the second case the thermal loading creates equal and opposite internal loads that cancel for no net external source. That difference can be noted in the output listings of Figs. 7.6 and 13 where the 'SUM' row of the 'ELEM_SOURCE' column is non-zero in the first (gravity) case, but zero in the second (thermal) case.

The reader is warned to remember that these calculations in Eq. 7.29 have been carried out in the global coordinate systems. In more advanced structural applications it is often desirable to transform the $\mathbf{P}^e$ back to element local coordinate system. For example, with a general truss member we are more interested in the force along the line of action of the bar rather than its x and y components. Sometimes we list both results and the user selects which is most useful. The necessary coordinate transformation is

$$\mathbf{P}^e_L = \mathbf{T}^e \, \mathbf{P}^e_g \tag{7.32}$$

where the square rotation matrix, $\mathbf{T}$, contains the direction cosines between the global axis, g, and the element local axis, L.

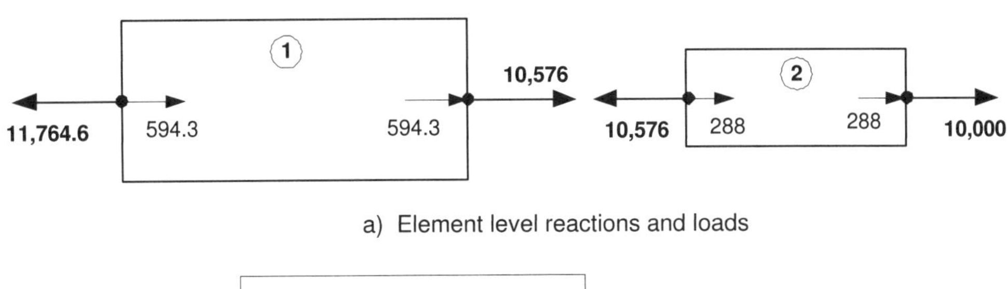

a) Element level reactions and loads

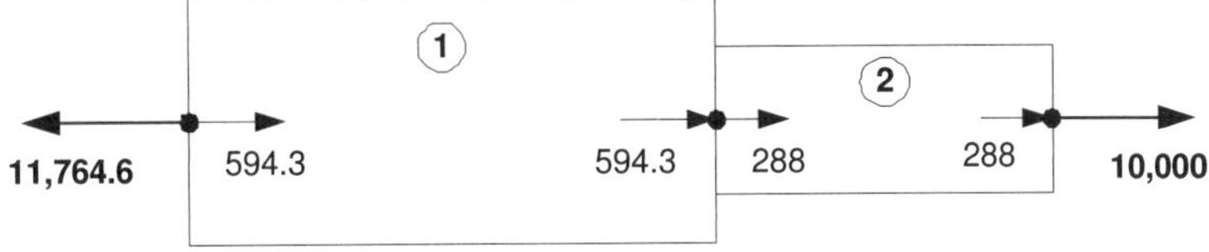

b) System reactions and loads

Figure 7.14 *Element equilibrium reaction recovery, case 1*

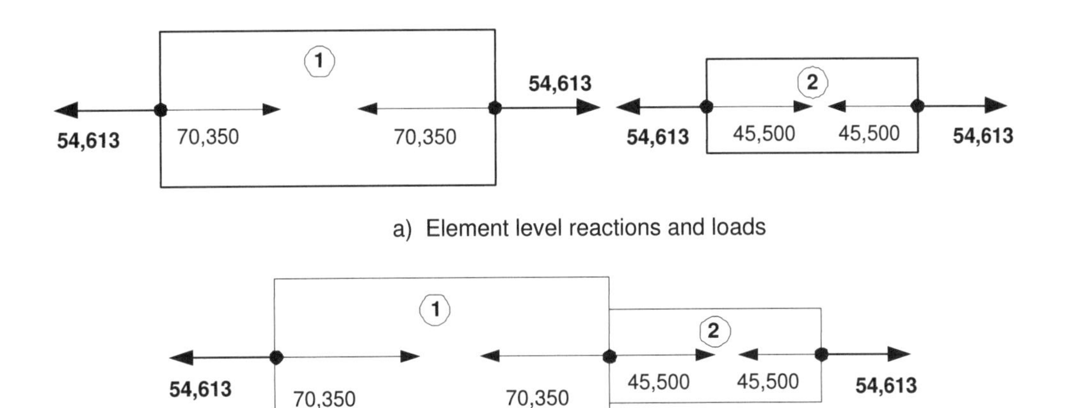

Figure 7.15 *Element equilibrium reaction recovery, case 2*

7.7 Heat transfer in a rod

A problem closely related to the previous problem is that of steady state heat transfer. Consider the heat transfer in a slender rod that has a specified temperature, θ_0, at $x = 0$ and is insulated at the other end, $x = L$. The rod has cross-sectional area, A, with a thermal conductivity of K. Thus, the rod conducts heat along its length. The rod is also surrounded by a convecting medium with a uniform temperature of θ_∞. Thus, the rod also convects heat on its outer surface area. Let the convective transfer coefficient be h and the outer perimeter of the rod be P. The governing differential equation for the temperature, $\theta(x)$, is given by Myers [6] as

$$KA \frac{d^2\theta}{dx^2} - hP(\theta - \theta_\infty) = 0, \quad 0 < x < L \tag{7.33}$$

with the essential condition $\theta(0) = \theta_0$ and the natural boundary condition $d\theta/dx\,(L) = 0$, which corresponds to an insulated right end. The exact solution can be shown to be $\theta(x) = \theta_\infty + (\theta_0 - \theta_\infty)\cosh[\,m(L - x)\,]/\cosh(mL)$ where $m^2 = hP/KA$ is a non-dimensional measure of the relative importance of convection (hP) and conduction (KA). This problem can be identified as the Euler equation of a variational principle. This principle will lead to system equations that are structured differently from our previous example with the bar. In that case, the boundary integral contributions (tractions) defined a column matrix and thus went on the right hand side of the system equations. Here we will see that the boundary contributions (convection) will also define a square matrix. Thus, they will go into the system coefficients on the left hand side of the system equations.

Generally a variational formulation of steady state head transfer involves volume integrals containing conduction terms and surface integrals with boundary heat flux, e.g., convection, terms. In our one-dimensional example both the volume and surface definitions involve an integral along the length of the rod. Thus, the distinction between volume and surface terms is less clear and the governing functional given by Myers is simply stated as a line integral. Specifically, one must render stationary the functional

$$I(\theta) = \frac{1}{2}\int_0^L [K\,A\,(d\theta/dx)^2 + h\,P\theta^2]\,dx \tag{7.34}$$

$$- \int_0^L h\,P\theta\,\theta_\infty\,dx + q_0\,\theta(0) - q_L\,\theta(L)$$

subject to the essential boundary condition(s). Divide the rod into a number of nodes and elements and introduce a finite element model where we assume

$$I = \sum_{e=1}^{n_e} I^e + \sum_{b=1}^{n_b} I^b + I^r$$

where we have defined a typical element volume contribution of

$$I^e = \frac{1}{2}\int_{L^e} K^e\,A^e\,(d\theta^e/dx)^2\,dx \tag{7.35}$$

and typical boundary contribution is

$$I^b = \frac{1}{2}\int_{L^b} h^b\,P^b\,\theta^{b2}\,dx - \int_{L^b} h^b\,P^b\,\theta^b\,\theta_\infty\,dx, \tag{7.36}$$

and I^r denotes any point flux sources or non-zero reaction contributions (here q_0 and/or q_L) that may be present. Most authors move known point sources or sinks into I^b and leave only the one or two unknown reaction terms in I^r. Using our interpolation relations as before $\theta^e(x) = \mathbf{H}^e(x)\,\mathbf{D}^e = \mathbf{D}^{e^T}\,\mathbf{H}^{e^T}$, and again in this special case $\theta^b(x) = \theta^e(x)$. Thus, these can be written symbolically as $I^e = \frac{1}{2}\,\mathbf{D}^{e^T}\,\mathbf{S}^e\,\mathbf{D}^e$, and $I^b = \frac{1}{2}\,\mathbf{D}^{b^T}\,\mathbf{S}^b\,\mathbf{D}^b - \mathbf{D}^{b^T}\,\mathbf{C}^b$. Assuming constant properties and the linear interpolation in Eq. 7.34 the element and boundary matrices reduce to

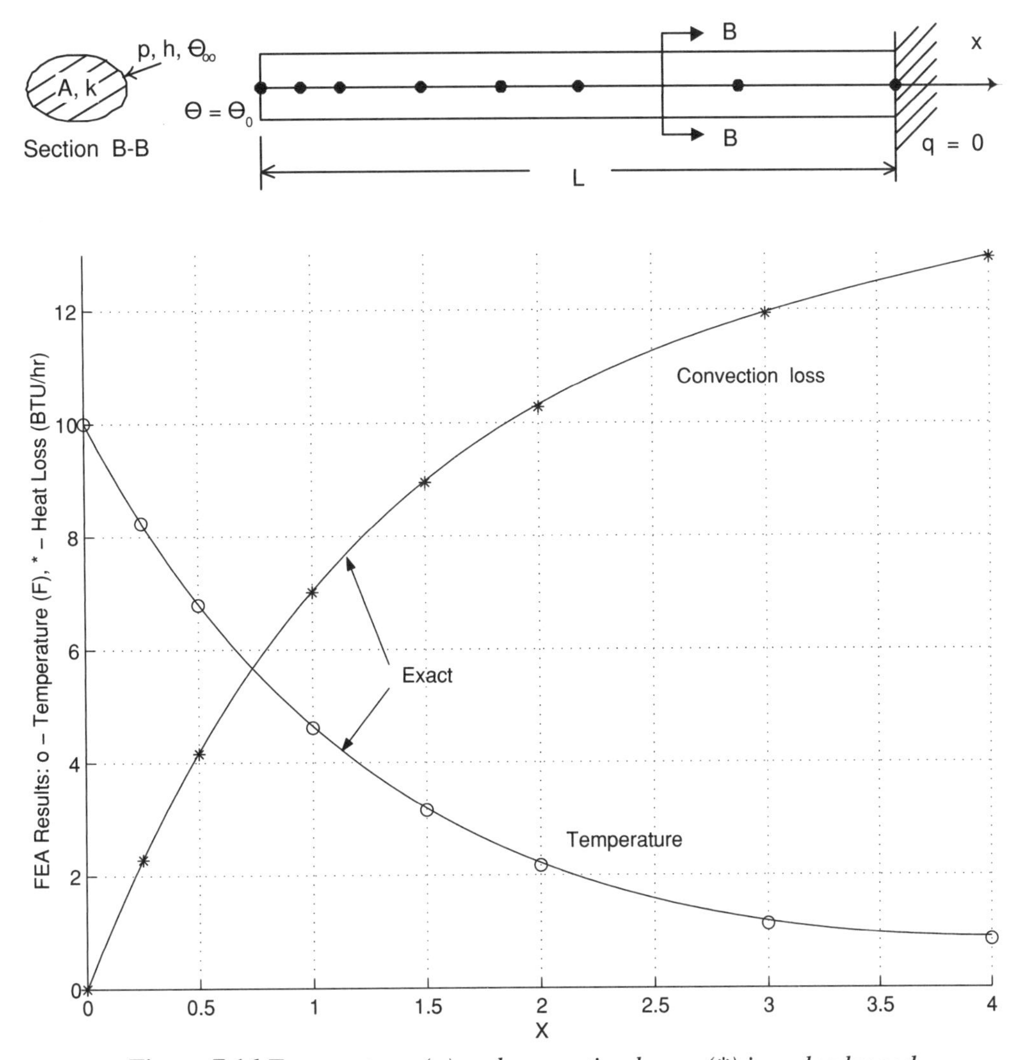

Figure 7.16 *Temperatures (o) and convection losses (*) in a slender rod*

$$\mathbf{S}^e = \frac{K^e A^e}{L^e}\begin{bmatrix} 1 & -1 \\ -1 & 1 \end{bmatrix}, \quad \mathbf{S}^b = \frac{h^b P^b L^b}{6}\begin{bmatrix} 2 & 1 \\ 1 & 2 \end{bmatrix}, \quad \mathbf{C}^b = \frac{\theta_\infty h^b P^b L^b}{2}\begin{Bmatrix} 1 \\ 1 \end{Bmatrix}.$$

Note that the conduction effect is inversely proportional to the material length while the convection effect is directly proportional to the length. If the problem is normalized so $\theta_\infty = 0$ then there is no column matrix defined and the equations will be homogeneous. As before, the assembled system equations are $\mathbf{ST} = \mathbf{C}$ where $\mathbf{S}$ is the direct assembly of $\mathbf{S}^e$ and $\mathbf{S}^b$ and $\mathbf{C}$ is assembled from the $\mathbf{C}^b$.

Another aspect of interest here is how to *post-process* the results so as to determine the convective heat loss. It should be equal and opposite to the sum of any external heat flux reactions necessary to maintain essential boundary conditions. The convection heat loss, at any point, is $dq = hP(\theta - \theta_\infty)\,dx$ where $(\theta - \theta_\infty)$ is the surface temperature difference. On a typical boundary segment this simplifies (for constant boundary data) to

$$Q^b = \int_{L^b} dq = \int_{L^b} h^b P^b [\theta^b(x) - \theta_\infty]\,dx = h^b P^b \int_{L^b} \mathbf{H}^b(x)\,dx\,[\mathbf{D}^b - \boldsymbol{\theta}_\infty] \quad (7.37)$$

where $\boldsymbol{\theta}_\infty$ is a vector with the boundary nodal values of θ_∞. For a linear element interpolation, as above, it is

$$Q^b = \tfrac{1}{2} h^b P^b L^b [1 \quad 1] (\mathbf{D}^b - \boldsymbol{\theta}_\infty) = \mathbf{P}^b (\mathbf{D}^b - \boldsymbol{\theta}_\infty). \quad (7.38)$$

Thus, if the constant array $\mathbf{P}^b$ is computed and stored for each segment then once all the temperatures are computed the boundary sub-set $\mathbf{D}^b$ can be gathered along with $\mathbf{P}^b$ to compute the loss Q^b. This is the first of several applications where we see that sometimes the post-processing will require the spatial integral of the solution and not just its gradient. Summing on the total boundary (all elements in this special case) gives the total heat loss. That value would, of course, equal the heat entering at the end $x = 0$. As a specific numerical example let $L = 4$ ft., $A = 0.01389$ ft^2, $h = 2$ BTU/hr – ft^2 F, $K = 120$ BTU/hr – ft F, $P = 0.5$ ft., and $t_0 = 10$ F. The mesh selected for this analysis are shown in Fig. 7.16 along with the results of the finite element analysis. A general implementation of this model via numerical integration is given in Fig. 7.17. It is valid for any member of the element library (currently linear through cubic interpolation). The input data for the application are given in Fig. 7.18 and selected corresponding output sets are in Fig. 7.19. The typical post-processing for recovering the integral of the product of the constant convection data and the interpolation function is shown in Fig. 7.20.

7.8 Element validation *

The successful application of finite element analysis should always include a validation of the element to be used and its implementation in a specific computer program. Usually, the elements utilized in most problem classes are very well understood and tested. However, some applications can be difficult to model, and the elements used for their analysis may be more prone to numerical difficulties. Therefore, one should subject elements to be used to a series of element validation tests. Two of the most common and important tests are the *patch test* introduced by Irons [3–5] and the *single-element tests* proposed by Robinson [8]. The single-element tests generally show the effects of element geometrical parameters such as convexity, aspect ratio, skewness, taper, and out-of-plane warping. It is most commonly utilized to test for a sensitivity to element aspect ratio. The single-element test usually consists of taking a single element in rectangular, triangular, or line form, considering it as a complete domain, and then investigating its behavior for various load or boundary conditions as a geometrical parameter is varied. An analytical solution is usually available for such a test.

The patch test has been proven to be a valid convergence test. It was developed from physical intuition and later written in mathematical forms. The basic concept is fairly simple. Imagine what happens as one introduces a very large, almost infinite,

```
!    ***  ELEM_SQ_MATRIX PROBLEM DEPENDENT STATEMENTS FOLLOW ***    ! 1
! ..........................................................        ! 2
!           (Stored as application source example 101.)             ! 3
! Combined heat conduction through, convection from a bar:          ! 4
!    K*A*U,XX - h*P*(U-U_ext) = 0, U(0)=U_0, dU/dx(L)=0             ! 5
! For globally constant data the analytic solution is:              ! 6
! U(x) = U_ext - (U_0-U_ext) * cosh [m*(L-x)]/ cosh [mL]            ! 7
! where  m^2 = h_e*P_e/(K_e*A_e), dimensionless.                    ! 8
!   Real element properties are:                                    ! 9
!   1) K_e = conductivity, BTU/ hr ft F                             !10
!   2) A_e = area of bar, ft^2                                      !11
!   3) h_e = convection, BTU/ hr ft^2 F                             !12
!   4) P_e = perimeter of area A_e, ft                              !13
!   Miscellaneous real FE data:                                     !14
!   1) U_ext = external reference temperature, F                    !15
!   Miscellaneous real data used ONLY for analytic solution:        !16
!   2) L     = exact length, ft                                     !17
!   3) U_0   = essential bc at x = 0, F                             !18
                                                                    !19
  REAL(DP) :: DL, DX_DR                    ! Length, Jacobian       !20
  REAL(DP) :: K_e, A_e, h_e, P_e, U_ext    ! properties             !21
  INTEGER  :: IQ                           ! Loops                  !22
                                                                    !23
  DL    = COORD (LT_N, 1) - COORD (1, 1) ! LENGTH                   !24
  DX_DR = DL / 2.                        ! CONSTANT JACOBIAN        !25
                                                                    !26
  U_ext = GET_REAL_MISC (1) ! external temperature                  !27
  K_e   = GET_REAL_LP (1)   ! thermal conductivity                  !28
  A_e   = GET_REAL_LP (2)   ! area of bar                           !29
  h_e   = GET_REAL_LP (3)   ! convection coefficient on perimeter   !30
  P_e   = GET_REAL_LP (4)   ! perimeter of area A_e                 !31
                                                                    !32
  E = K_e * A_e              ! constitutive array                   !33
                                                                    !34
  CALL STORE_FLUX_POINT_COUNT ! Save LT_QP, for post & error        !35
                              ! S, C, H_INTG already zeroed         !36
  DO IQ = 1, LT_QP            ! LOOP OVER QUADRATURES               !37
                                                                    !38
!        GET INTERPOLATION FUNCTIONS, AND X-COORD                   !39
    H    = GET_H_AT_QP (IQ)                                         !40
    XYZ  = MATMUL (H, COORD)  ! ISOPARAMETRIC                       !41
                                                                    !42
!        LOCAL AND GLOBAL DERIVATIVES                               !43
    DLH   = GET_DLH_AT_QP (IQ)      ! local                         !44
    DGH   = DLH / DX_DR             ! global                        !45
                                                                    !46
!          CONVECTION SOURCE                                        !47
    C = C + h_e * P_e * U_ext * H * WT (IQ) * DX_DR                 !48
                                                                    !49
!          SQUARE MATRIX, CONDUCTION & CONVECTION                   !50
    S = S + ( K_e * A_e * MATMUL (TRANSPOSE(DGH), DGH)            & !51
            + h_e * P_e * OUTER_PRODUCT (H, H) ) * WT (IQ) * DX_DR  !52
                                                                    !53
!      INTEGRATING FOR CONVECTION LOSS, FOR POST PROCESSING         !54
    H_INTG = H_INTG + h_e * P_e * H * WT (IQ) * DX_DR               !55
                                                                    !56
!      SAVE FOR FLUX AVERAGING OR POST PROCESSING, B == DGH         !57
    CALL STORE_FLUX_POINT_DATA (XYZ, E, DGH) ! for error est too    !58
  END DO   ! QUADRATURE                                             !59
  IF ( N_FILE1 > 0) WRITE (N_FILE1) H_INTG, U_ext ! if "post_el"    !60
```

Figure 7.17 *A numerically integrated thermal element*

```
title "Myer's 1-D heat transfer example, 7 L2 " ! keywords  ! 1
exact_case 1 ! Analytic solution for list_exact, etc        ! 2
list_exact   ! List given exact answers at nodes, etc       ! 3
nodes      8 ! Number of nodes in the mesh                  ! 4
elems      7 ! Number of elements in the system             ! 5
dof        1 ! Number of unknowns per node                  ! 6
el_nodes   2 ! Maximum number of nodes per element          ! 7
space      1 ! Solution space dimension                     ! 8
b_rows     1 ! Number of rows in the B (operator) matrix    ! 9
example  101 ! Application source code library number       !10
el_react     ! Compute & list element reactions             !11
remarks   21 ! Number of user remarks                       !12
gauss      2 ! Maximum number of quadrature point           !13
el_real    4 ! Number of real properties per element        !14
reals      3 ! Number of miscellaneous real properties      !15
el_homo      ! Element properties are homogeneous           !16
post_el      ! Require post-processing, create n_file1      !17
no_error_est ! Do NOT compute SCP element error estimates   !18
quit ! keyword input, remarks follow                        !19
Combined heat conduction through, convection from a bar:    !20
   K*A*U,XX - h*P*(U-U_ext) = 0, U(0)=U_0, dU/dx(L)=0       !21
For globally constant data the analytic solution is:        !22
U(x) = U_ext - (U_0-U_ext) * cosh [m*(L-x)] / cosh [mL]     !23
where  m^2 = h_e*P_e/(K_e*A_e), dimensionless.              !24
  Real element properties are:                              !25
  1) K_e = conductivity, BTU/ hr ft F                       !26
  2) A_e = area of bar, ft^2                                !27
  3) h_e = convection, BTU/ hr ft^2 F                       !28
  4) P_e = perimeter of area A_e, ft                        !29
  Miscellaneous real FE data:                               !30
  1) U_ext = external reference temperature, F              !31
  Miscellaneous real data used ONLY for analytic solution:  !32
  2) L     = exact length, ft                               !33
  3) U_0   = essential bc at x = 0, F                       !34
Element convection loss is difference in elem reactions     !35
e.g.: e=1          node  1            node  2               !36
           12.92 BTU --->*______(1)_______* ---> 10.64 BTU  !37
     Elem convection loss:  2.28 BTU --->  (physical check) !38
Note: System reaction, 12.92 BTU (12.86 exact), offsets the !39
sum of the element convection losses, 12.92 BTU (phys chk)  !40
    1    1 0.00 ! begin nodes: node, bc flag, x             !41
    2    0 0.25                                             !42
    3    0 0.50                                             !43
    4    0 1.00                                             !44
    5    0 1.50                                             !45
    6    0 2.00                                             !46
    7    0 3.00                                             !47
    8    0 4.00                                             !48
    1    1    2   ! begin elements                          !49
    2    2    3                                             !50
    3    3    4                                             !51
    4    4    5                                             !52
    5    5    6                                             !53
    6    6    7                                             !54
    7    7    8                                             !55
1 1 10.                   ! essential bc: node, dof, value  !56
1 120. 0.01389 2.0 0.5 ! el, K_e A_e h_e P_e (homogeneous)  !57
0.0 4.0 10.0              ! Miscellaneous: U_ext  L U_0     !58
```

Figure 7.18 *Example convection data*

```
*** REACTION RECOVERY ***                                              ! 1
NODE, PARAMETER,      REACTION, EQUATION                               ! 2
   1,  DOF_1,       1.2921E+01     1                                   ! 3
                                                                       ! 4
*** OUTPUT OF RESULTS AND EXACT VALUES IN NODAL ORDER ***              ! 5
NODE,  X-Coord,     DOF_1,       EXACT1,                               ! 6
   1  0.0000E+00  1.0000E+01  1.0000E+01                               ! 7
   2  2.5000E-01  8.2385E+00  8.2475E+00                               ! 8
   3  5.0000E-01  6.7878E+00  6.8051E+00                               ! 9
   4  1.0000E+00  4.6138E+00  4.6438E+00                               !10
   5  1.5000E+00  3.1496E+00  3.1877E+00                               !11
   6  2.0000E+00  2.1700E+00  2.2157E+00                               !12
   7  3.0000E+00  1.1322E+00  1.1847E+00                               !13
   8  4.0000E+00  8.4916E-01  9.0072E-01                               !14
                                                                       !15
** ELEMENT REACTION, INTERNAL SOURCES AND SUMMATIONS **                !16
                ELEMENT  1                                             !17
 NODE DOF    REACTION        ELEM_SOURCE   SUMS                        !18
   1   1    1.29210E+01     0.00000E+00                                !19
   2   1   -1.06412E+01     0.00000E+00                                !20
SUM:   1    2.27981E+00     0.00000E+00   2.27981E+00 Note             !21
                ELEMENT  2                                             !22
 NODE DOF    REACTION        ELEM_SOURCE   SUMS                        !23
   2   1    1.06412E+01     0.00000E+00                                !24
   3   1   -8.76294E+00     0.00000E+00                                !25
SUM:   1    1.87829E+00     0.00000E+00   1.87829E+00                  !26
                ELEMENT  3                                             !27
 NODE DOF    REACTION        ELEM_SOURCE   SUMS                        !28
   3   1    8.76294E+00     0.00000E+00                                !29
   4   1   -5.91252E+00     0.00000E+00                                !30
SUM:   1    2.85041E+00     0.00000E+00   2.85041E+00                  !31
                ELEMENT  4                                             !32
 NODE DOF    REACTION        ELEM_SOURCE   SUMS                        !33
   4   1    5.91252E+00     0.00000E+00                                !34
   5   1   -3.97165E+00     0.00000E+00                                !35
SUM:   1    1.94087E+00     0.00000E+00   1.94087E+00                  !36
                ELEMENT  5                                             !37
 NODE DOF    REACTION        ELEM_SOURCE   SUMS                        !38
   5   1    3.97165E+00     0.00000E+00                                !39
   6   1   -2.64175E+00     0.00000E+00                                !40
SUM:   1    1.32990E+00     0.00000E+00   1.32990E+00                  !41
                ELEMENT  6                                             !42
 NODE DOF    REACTION        ELEM_SOURCE   SUMS                        !43
   6   1    2.64175E+00     0.00000E+00                                !44
   7   1   -9.90679E-01     0.00000E+00                                !45
SUM:   1    1.65107E+00     0.00000E+00   1.65107E+00                  !46
                ELEMENT  7                                             !47
 NODE DOF    REACTION        ELEM_SOURCE   SUMS                        !48
   7   1    9.90679E-01     0.00000E+00                                !49
   8   1    0.00000E+00     0.00000E+00                                !50
SUM:   1    9.90679E-01     0.00000E+00   9.90679E-01                  !51
** ELEMENT      CONVECTION HEAT LOSS   **                              !52
   1            2.27980    <- Note reaction above, etc.                !53
   2            1.87828                                                !54
   3            2.85041                                                !55
   4            1.94087                                                !56
   5            1.32989                                                !57
   6            1.65107                                                !58
   7            0.99067                                                !59
TOTAL HEAT LOSS = 12.92103     (Exact = 12.858)                        !60
```

Figure 7.19 *Selected conduction-convection output*

```
! .............................................................  ! 1
!  *** POST_PROCESS_ELEM PROBLEM DEPENDENT STATEMENTS FOLLOW ***  ! 2
! .............................................................  ! 3
!           (Stored as application source example 101.)          ! 4
! H_INTG (LT_N) Integral of interpolation functions, H, available ! 5
                                                                  ! 6
! Linear line element face convection heat loss recover           ! 7
  REAL(DP) :: U_ext                    ! external temperature     ! 8
  REAL(DP), SAVE :: Q_LOSS, TOTAL      ! Face and total heat loss ! 9
  LOGICAL,  SAVE :: FIRST = .TRUE.     ! printing                 !10
                                                                  !11
   IF ( FIRST ) THEN                    ! first call              !12
     FIRST = .FALSE. ; WRITE (6, 5)     ! print headings          !13
     5 FORMAT ('*** CONVECTION HEAT LOSS ***', /, &               !14
     &         'ELEMENT    HEAT_LOST')                            !15
     TOTAL = 0.d0                       ! initialize              !16
   END IF ! first call                                            !17
                                                                  !18
! Get previously integrated interpolation function, times         !19
! the convection properties, h_e * P_e, now stored in H_INTG;     !20
! and the surrounding gas temperature,  U_ext, that were          !21
! saved in ELEM_SQ_MATRIX. (Indicated by keyword post_el.)        !22
! U_ext = GET_REAL_MISC (1) ! external temperature                !23
! h_e   = GET_REAL_LP (3)   ! perimeter convection coefficient    !24
! P_e   = GET_REAL_LP (4)   ! perimeter length of area            !25
                                                                  !26
   IF ( N_FILE1 > 0) READ (N_FILE1) H_INTG, U_ext ! if "post_el"  !27
                                                                  !28
!   HEAT LOST : Integral over bar length of hp * (T - T_inf)      !29
   D (1:LT_N) = D(1:LT_N) - U_ext      ! Temp difference at nodes !30
   Q_LOSS = DOT_PRODUCT (H_INTG, D)    ! Face loss integral       !31
   TOTAL  = TOTAL + Q_LOSS             ! Running total            !32
                                                                  !33
   PRINT '(I6, ES15.5)', IE, Q_LOSS                               !34
   IF ( IE == N_ELEMS ) PRINT *, 'TOTAL = ', TOTAL                !35
!  *** END POST_PROCESS_ELEM PROBLEM DEPENDENT STATEMENTS ***     !36
```

Figure 7.20 *Element convection heat loss recovery*

number of elements. Clearly, they would become very small in size. If we think of the quantities being integrated to form the element matrices, we can make an observation about how the solution would behave in this limit. The integrand, such as the strain energy, contains derivative terms that would become constant as the element size shrinks toward zero. Thus, to be valid in the limit, the element formulation must, at least, be able to yield the correct results in that state. That is, to be assured of convergence one must be able to exactly satisfy the state where the derivatives, in the governing integral statement, take on constant or zero values. This condition can be stated as a mathematical test or as a simple numerical test. The latter option is what we want here. The patch test provides a simple numerical way for a user to test an element, or complete computer program, to verify that it behaves as it should.

We define a patch of elements to be a mesh where at least one node is completely surrounded by elements. Any node of this type is referred to as an *interior node*. The other nodes are referred to as *exterior* or *perimeter nodes*. We will compute the dependent variable at all interior nodes. The derivatives of the dependent variable will be computed in each element. The perimeter nodes are utilized to introduce the essential boundary conditions and/or loads required by the test. Assume that the governing integral statement has derivatives of order n. We would like to find boundary conditions that would make those derivatives constant. This can be done by selecting an arbitrary n-th order polynomial function of the global coordinates to describe the dependent variable in the global space that is covered by the patch mesh. Clearly, the n-th order derivatives of such a function would be constant as desired. The assumed polynomial is used to define the essential boundary conditions on the perimeter nodes of the patch mesh. This is done by substituting the input coordinates at the perimeter nodes into the assumed function and computing the required value of the dependent variable at each such node. Once all of the perimeter boundary conditions are known, the solution can be numerically executed. The resulting values of the dependent variable are computed at each interior node. To pass the patch test, these computed internal values must agree with the value found when their internal nodal coordinates are substituted into the assumed global polynomial. However, the real test is that when each element is checked, the calculated n-th order derivatives must agree with the arbitrary assumed values used to generate the global function. If an element does not satisfy this test, it should not be used. The patch test can also be used for other purposes. For example, the analyst may wish to distort the element shape and/or change the numerical integration rule to see what effect that has on the numerical accuracy of the patch test.

As a simple elementary example of an analytic solution of the patch test, consider the bar element. The smallest possible patch is one with two line elements. Such a patch has two exterior nodes and one interior node. For simplicity, let the lengths of the two elements be equal and have a value of L. The governing integral statement contains only the first derivative of u. An arbitrary linear function can be selected for the patch test, since it would have a constant first derivative. Therefore, select $u(x) = a + bx$ for $0 \le x \le 2L$, where a and b are arbitrary constants. Assembling a two-element patch:

$$\frac{AE}{L}\begin{bmatrix} 1 & -1 & 0 \\ -1 & (1+1) & -1 \\ 0 & -1 & 1 \end{bmatrix}\begin{Bmatrix} u_1 \\ u_2 \\ u_3 \end{Bmatrix} = \begin{Bmatrix} P_1 \\ 0 \\ P_3 \end{Bmatrix}$$

where P_1 and P_3 are the unknown reactions associated with the prescribed external displacements. These two exterior patch boundary conditions are obtained by substituting their nodal coordinates into the assumed patch solution:

$$u_1 = u(x_1) = a + b(0) = a, \quad u_3 = u(x_2) = a + b(2L) = a + 2bL.$$

Modifying the assembled equations to include the patch boundary conditions gives

$$\frac{AE}{L}\begin{bmatrix} 0 & -1 & 0 \\ 0 & 2 & 0 \\ 0 & -1 & 0 \end{bmatrix}\begin{Bmatrix} a \\ u_2 \\ a + 2bL \end{Bmatrix} = \begin{Bmatrix} P_1 \\ 0 \\ P_3 \end{Bmatrix} - \frac{aAE}{L}\begin{Bmatrix} 1 \\ -1 \\ 0 \end{Bmatrix} - \frac{(a + bL)\,AE}{L}\begin{Bmatrix} 0 \\ -1 \\ 1 \end{Bmatrix}.$$

Retaining the independent second equation gives the displacement relation

$$\frac{2AE}{L}u_2 = 0 + \frac{aAE}{L} + \frac{(a + 2bL)\,AE}{L}.$$

Thus, the internal patch displacement is $u_2 = (2a + 2bL)/2 = (a + bL)$. The value required by the patch test is $u(x_2) = (a + bx_2) = (a + bL)$. This agrees with the computed solution, as required by a valid element. The element strains are

$$e = 1: \quad \varepsilon = \frac{(u_2 - u_1)}{L} = \frac{[\,(a + bL) - a\,]}{L} = b$$

$$e = 2: \quad \varepsilon = \frac{(u_3 - u_2)}{L} = \frac{[\,(a + 2bL) - (a + bL)\,]}{L} = b.$$

Thus, all element derivatives are constant. However, these constants must agree with the constant assumed in the patch. That value is $\varepsilon = du/dx = d(a + bx)/dx = b$. Therefore, the patch test is completely satisfied. At times one also wishes to compute the reactions, i.e., P_1 and P_3. To check for possible rank deficiency in the element formulation, one should repeat the test with only enough displacements prescribed to prevent rigid body motion. (That is, to render the square matrix non-singular.) Then, the other outer perimeter nodes are loaded with the reactions found in the precious patch test. In the above example, substituting u_1 and u_2 into the previously discarded first equation yields the reaction $P_1 = -bAE$. Likewise, the third equation gives $P_3 = -P_1$, as expected. Thus, the above test could be repeated by prescribing u_1 and P_3, or P_1 and u_3. The same results should be obtained in each case. A major advantage of the patch test is that it can be carried out numerically. In the above case, the constants a and b could have been assigned arbitrary numerical values. Inputting the required numerical values of A, E and L would give a complete numerical description that could be tested in a standard program. Such a procedure also verifies that the computer program satisfies certain minimum requirements. A problem with some elements is that they can pass the patch test for a uniform mesh, but fail when an arbitrary irregular mesh is employed. Thus, as a general rule, one should try to avoid conducting the test with a regular mesh, such as that given in the above example. It would have been wiser to use unequal element lengths such as L and αL, where α is an arbitrary constant. The linear bar element should pass the test for any scaling ratio, α. However, for α near zero, numerical ill-conditioning begins to affect the answers. Data and results for the patch test of a linear bar element are shown in Figs. 7.21 and 22.

7.9 Euler's equations of variational calculus *

Euler's Theorem of Variational Calculus was given in Eqs. 1.6 - 8. When the value of ϕ is given on a portion of the boundary, Γ, we call that an essential boundary condition, or a Dirichlet type condition. When q is present in the boundary condition (of Eq. 1.8), but a is zero that case is called a Neumann type condition, or flux condition. The most common Neumann type is the 'natural condition', $q = 0$, since then the boundary integral in Eq. 1.6 does not have to be evaluated (it is satisfied naturally). The most general form of Eq. 1.8 is called a 'mixed boundary condition', which is also known as a 'Robin type' condition. Note that a Robin condition involves a non-zero a value in Eq. 1.6 and thus it will contribute a new term to the square matrix (since u^2 leads to a quadratic form in $\mathbf{H}$). Likewise, one could reduce this to a one-dimensional form

```
title "Elastic linear bar patch test" ! begin keywords     ! 1
nodes      3 ! Number of nodes in the mesh                  ! 2
elems      2 ! Number of elements in the system             ! 3
dof        1 ! Number of unknowns per node                  ! 4
el_nodes   2 ! Maximum number of nodes per element          ! 5
space      1 ! Solution space dimension                     ! 6
b_rows     1 ! Number of rows in the B (operator) matrix    ! 7
example 113 ! Source library example number                 ! 8
remarks    3 ! Number of user remarks                       ! 9
el_real    5 ! Number of real properties per element        !10
el_homo      ! Element properties are homogeneous           !11
el_list      ! List results at each node of each element    !12
el_react     ! Compute & list element reactions             !13
post_el      ! Require post-processing, create n_file1      !14
quit ! keyword input, remarks follow                        !15
Assume u = a + bx. Let a = 5, b= -4 so that u(0) = u_1 = 5 !16
and u(2) = u_3 = -3 then all mechanical strains = -4        !17
Properties: A, E, DT, ALPHA, GAMMA (any homogeneous)        !18
 1    1 0.0        ! node, bc flag, x                       !19
 2    0 1.0        ! answer here must match 'a + bx'        !20
 3    1 2.0        ! node, bc flag, x                       !21
    1    1    2    ! begin elements                         !22
    2    2    3                                             !23
1 1  5.      ! node, dof, essential bc: u(0) = u_1 = 5      !24
3 1 -3.      ! node, dof, essential bc: u(2) = u_3 = -3     !25
1 1. 10. 0. 0. 0. ! el, A, E, DT, ALPHA, GAMMA (any A,E)    !26
```

Figure 7.21 *Example patch test data for a bar*

```
*** REACTION RECOVERY ***                                        ! 1
NODE, PARAMETER,      REACTION, EQUATION                         ! 2
   1,  DOF_1,       4.0000E+01      1                            ! 3
   3,  DOF_1,      -4.0000E+01      3                            ! 4
                                                                 ! 5
*** EXTREME VALUES OF THE NODAL PARAMETERS ***                   ! 6
PARAMETER     MAXIMUM,    NODE      MINIMUM,     NODE            ! 7
 DOF_1,     5.0000E+00,      1  -3.0000E+00,       3             ! 8
                                                                 ! 9
***  OUTPUT OF RESULTS IN NODAL ORDER  ***                       !10
NODE,  X-Coord,     DOF_1,                                       !11
   1  0.0000E+00  5.0000E+00                                     !12
   2  1.0000E+00  1.0000E+00   ! passes patch test               !13
   3  2.0000E+00 -3.0000E+00                                     !14
                                                                 !15
** ELEMENT REACTION, AND INTERNAL SOURCES **                     !16
                ELEMENT  1                                       !17
 NODE  DOF   REACTION        ELEM_SOURCE                         !18
   1    1    4.00000E+01     0.00000E+00                         !19
   2    1   -4.00000E+01     0.00000E+00                         !20
                ELEMENT  2                                       !21
 NODE  DOF   REACTION        ELEM_SOURCE                         !22
   2    1    4.00000E+01     0.00000E+00                         !23
   3    1   -4.00000E+01     0.00000E+00                         !24
                                                                 !25
        E L E M E N T    S T R E S S E S  ! pass                 !26
ELEMENT        STRESS    MECH. STRAIN    THERMAL STRAIN          !27
   1    -4.00000E+01   -4.00000E+00     0.00000E+00              !28
   2    -4.00000E+01   -4.00000E+00     0.00000E+00              !29
```

Figure 7.22 *Patch test results for a bar*

$$I = \int_a^b f(x, \phi, \frac{d\phi}{dx})\, dx + (q + a\phi)\Big|_{x=b} - (q + a\phi)\Big|_{x=a} \tag{7.45}$$

so the ordinary differential equation (ODE) is

$$\frac{\partial f}{\partial \phi} - \frac{d}{dx} \frac{\partial f}{\partial (d\phi/dx)} = 0 \tag{7.46}$$

and the natural boundary condition if ϕ is not prescribed is

$$n_x \frac{\partial f}{\partial (d\phi/dx)} + q + a\,\phi = 0. \tag{7.47}$$

Fourth order equations are formulated in the same way. For example,

$$I(\phi) = \int_a^b f\left(x, \phi, \frac{d\phi}{dx}, \frac{d^2\phi}{dx^2} \right) dx, \tag{7.48}$$

a functional involving the second derivative of ϕ, has the Euler equation

$$\frac{\partial f}{\partial \phi} - \frac{d}{dx} \frac{\partial f}{\partial (d\phi/dx)} + \frac{d^2}{dx^2} \frac{\partial f}{\partial (d^2\phi/dx^2)} = 0, \tag{7.49}$$

and the natural boundary condition of

$$\frac{\partial f}{\partial (d\phi/dx)} - \frac{d}{dx} \frac{\partial f}{\partial (d^2\phi/dx^2)} = 0 \tag{7.50}$$

if $d\phi/dx$ is specified (and ϕ unknown), and the natural condition of

$$\frac{\partial f}{\partial (d^2\phi/dx^2)} = 0 \tag{7.51}$$

if ϕ is specified (and $d\phi/dx$ is unknown).

7.10 Exercises

1. Consider heat transfer through a planar wall with a given temperature on one side and convection on the other. Employ a single linear element model where ϕ_1 is the inside temperature at left node 1, and where right node ϕ_2 is the other surface temperature adjacent to the convection fluid. Heat is convected away there at the rate of $q_h = h_a A_a (\phi_2 - \theta_a)$, where h_a is the convection coefficient over the convection area of A_a which is adjacent to the convecting fluid at temperature ϕ_a. The constant conduction matrix was given earlier and, in general, the convection matrices on boundary segment b are $\boldsymbol{S}_h^b = \int_{\Gamma}^{b} \boldsymbol{H}^{b^T} \boldsymbol{H}^b h^b d\Gamma$, and $\boldsymbol{C}_h^b = \int_{\Gamma}^{b} \boldsymbol{H}^{b^T} h^b \phi_a^b d\Gamma$. At a point surface approximation (Γ^b) the $\boldsymbol{H}^b$ is constant (unity) and the convection data (h_a^b, θ_a^b) are also constant. Create the (scalar) boundary terms, assemble, and solve for ϕ_2. Verify the result is $\phi_2 = \dfrac{A_a h_a \theta_a + (k/L)\phi_1 A^e}{(k\ A^e/L + h\ A_a)}$. Simplify for the present case where the conducting and

convecting areas are the same, $A^e = A_a = A^b$. Consider the two special cases of $h = 0$ and $h = \infty$ and discuss what they then represent as boundary conditions (Dirichlet, or Neumann).

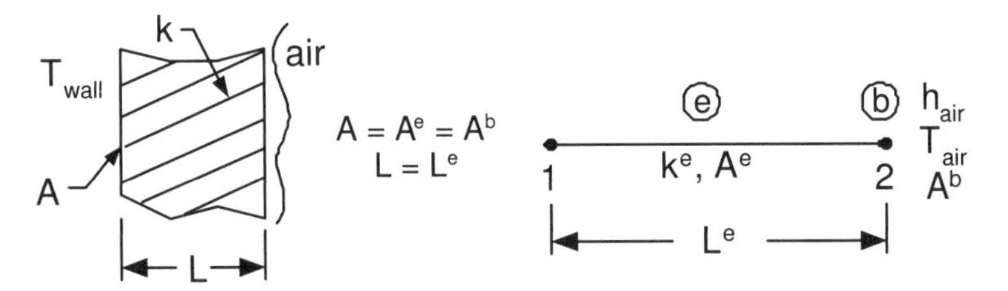

Problem P7.1 Wall conduction-convection

2. Consider a two-dimensional functional

$$I = \frac{1}{2}\int_\Omega \left[K_x\left(\frac{\partial u}{\partial x}\right)^2 + K_y\left(\frac{\partial u}{\partial y}\right)^2 - 2Qu \right] d\Omega + \int_\Gamma \left(qu + au^2/2 \right) d\Gamma.$$

Show that the partial differential equation from Euler's theorem is

$$-Q - \frac{\partial}{\partial x}\left(K_x \frac{\partial u}{\partial x} \right) - \frac{\partial}{\partial y}\left(K_y \frac{\partial u}{\partial y} \right) = 0 \qquad \varepsilon\ \Omega,$$

and that the corresponding natural boundary condition is

$$n_x K_x \frac{\partial u}{\partial x} + n_y K_y \frac{\partial u}{\partial y} + q + au = 0 \qquad \varepsilon\ \Gamma$$

or simply

$$K_n \frac{\partial u}{\partial n} + q + au = 0 \qquad \varepsilon\ \Gamma,$$

where K_n is the coefficient in the direction of the unit normal, **n**. If $K_x = K_y = K$, a constant, show this reduces to the Poisson equation, and Laplaces equation if $Q = 0$.

3. Consider the functional

$$I(u) = \int_0^L [EI\left(\frac{d^2u}{dx^2}\right)^2 - 2uQ]\,dx.$$

Show that it yields the ordinary differential equation

$$\frac{d^2}{dx^2}\left(EI \frac{d^2u}{dx^2} \right) = Q,$$

and the natural boundary condition is that

$$-\frac{d}{dx}\left(EI \frac{d^2u}{dx^2} \right) = 0.$$

This is the fourth order equation for the deflection, u, of a beam subject to a transverse load per unit length of Q. Here EI denotes the bending stiffness of the member. Explain why this equation requires four boundary conditions. Note that at any boundary condition point, one may prescribe u and/or du/dx. If u is not specified, the natural condition is that $d^2u/dx^2 = 0$. They correspond to conditions on the shear and moment, respectively.

4. A one-dimensional Poisson equation is

$$\frac{\partial^2 u}{\partial x^2} = -Q = \frac{105}{2}x^2 - \frac{15}{2}, \quad -1 < x < 1.$$

The exact solution is $u = (35x^4 - 30x^2 + 3)/8$ when the boundary conditions are

$$u(-1) = 1, \quad \partial u / \partial x(1) = 10.$$

Obtain a finite element solution and sketch it along with the exact solution.

5. A one-dimensional Poisson equation is

$$\frac{\partial^2 u}{\partial x^2} = Q = -6x - \frac{2}{\alpha^2}\left[1 - 2\left(\frac{x-\beta}{\alpha}\right)^2\right]\exp\left[-\left(\frac{x-\beta}{\alpha}\right)^2\right], \quad 0 \le x \le 1$$

where β is the interior center position of a local high gradient region, $0 < \beta < 1$, and α is a parameter that governs the amplitude of u in the region, say $\alpha = 0.05$ and $\beta = 0.5$. Obtain a finite element solution and sketch it along with the exact solution for boundary conditions

$$u(0) = \exp\left(-\frac{\beta^2}{\alpha^2}\right), \quad \frac{\partial u}{\partial x}(1) = -3 - 2\left(\frac{1-\beta}{\alpha^2}\right)\exp\left[-\left(\frac{1-\beta}{\alpha}\right)^2\right],$$

so the exact solution is given by $u = -x^3 + \exp\left[-\left(\frac{x-\beta}{\alpha}\right)^2\right]$. Note that the source term, Q, may require more integration points than does the evaluation of the square matrix.

6. Buchanan shows that an elastic cable with constant tension, T, resting on an elastic foundation of stiffness k and subjected to a vertical load per unit length of f has a vertical displacement given by the differential equation

$$T\frac{d^2 v}{dx^2} - k\,v(x) = -f.$$

a) Develop the element matrices for a linear line element, assuming constant T, k and f. b) Obtain a finite element solution where $T = 600$ lb, $k = 0.5$ lb/in^2, $f = 2$ lb/in and where $L = 120$ in is the length of the string where $v = 0$ at each end. [answer: $v_{max} = 2.54$ in] c) Replace the model in b) with a half symmetry model where $dv/dx = 0$ at the center ($x = L/2$).

7. Consider a variable coefficient problem given by

$$-[(1 + x)u'(x)]' = -1, \quad 0 \le x \le 1$$

with the boundary conditions $u(0) = 0 = u(1)$. Obtain a finite element solution and compare it to the exact result $u(x) = x - ln(1 + x)/\ln(2)$.

8. For the differential equation in Problem 2.17 if we have one essential boundary condition of $y(0) = 1$ and one Robin or mixed boundary condition of $y'(1) + y(1) = 0$ the exact solution is $y(x) = (x^4 - 3x^2 - x + 6)/6$ (which is *exact_case* 17 in the source library). Obtain a Galerkin finite element solution and compare it to the exact result. Note that this requires mixed element matrices for the point 'element' at $x = 1$. (Hint: think about what the 'normal' direction is for each end of a one-dimensional domain.)
9. For the differential equation in Problem 2.17 if we have two Robin or mixed boundary conditions of $y'(0) + y(0) = 0$ and $y'(1) - y(1) = 3$ the exact solution is $y(x) = x^4/6 + 3x^2/2 + x - 1$ (which is *exact_case* 18 in the source library). Obtain a Galerkin finite element solution and compare it to the exact result. (Hint: think about what the 'normal' direction is for each end of a one-dimensional domain.)
10. Resolve the heat transfer problem in Fig. 7.16 with the external reference temperature increased from 0 to 20 F ($\theta_\infty = 20$). Employ 3 equal length L2 linear elements, and 5 (not 4) nodes with nodes 2 and 3 both at $x = 4/3$. Connect element 1 to nodes 1 and 2 and (incorrectly) connect element 2 to nodes 3 and 4. (a) Solve the incorrect model, sketch the FEA and true solutions, and discuss why the left and right ($x \leq 4/3$ and $x \geq 4/3$) domain FEA solutions behave as they do. (Hint: Think about essential and natural boundary conditions.) (b) Enforce the correct solution by imposing a 'multiple point constraint' (MPC) that requires $1 \times t_2 - 1 \times t_3 = 0$. That is, the MPC requires the solution to be continuous at nodes 2 and 3.
11. Implement the exact integral matrices for the linear line element version of Eq. 7.34 and the matrix for recovering the convection heat loss, Eq. 7.38, in the post processing. Apply a small model to the problem in Fig. 7.16, but replace the Dirichlet boundary condition at $x = 0$ with the corresponding exact flux $q(0) = 12.86$. Compare the temperature solution with that in Fig. 7.16. Why do we still get a solution when we no longer have a Dirichlet boundary condition? (Review Problem 7.1.) [7−11]

7.11 Bibliography

[1] Bathe, K.J., *Finite Element Procedures*, Englewood Cliffs: Prentice Hall (1996).

[2] Buchanan, G.R., *Finite Element Analysis*, New York: McGraw-Hill (1995).

[3] Irons, B.M. and Razzaque, A., "Engineering Applications of Numerical Integration in the Stiffness Method," *AIAA Journal*, **4**, pp. 2035–2057 (1966).

[4] Irons, B.M. and Razzaque, A., "Experience with the Patch Test for Convergence of the Finite Element Method," pp. 557–587 in *Mathematical Foundation of the Finite Element Method*, ed. A.R. Aziz, New York: Academic Press (1972).

[5] Irons, B.M. and Ahmad, S., *Techniques of Finite Elements*, New York: John Wiley (1980).

[6] Myers, G.E., *Analytical Methods in Conduction Heat Transfer*, New York: McGraw-Hill (1971).

[7] Reddy, J.N. and Gartling, D.K., *The Finite Element Method in Heat Transfer and Fluid Dynamics*, Boca Raton: CRC Press (2001).

[8] Robinson, J., "A Single Element Test," *Comp. Meth. Appl. Mech. Engng.*, **7**, pp. 191–200 (1976).

[9] Shames, I.H. and Dym, C.L., *Energy and Finite Element Methods in Structural Mechanics*, Pittsburg: Taylor and Francis (1995).

[10] Wait, R. and Mitchell, A.R., *Finite Element Analysis and Applications*, New York: John Wiley (1985).

[11] Weaver, W.F., Jr. and Johnston, P.R., *Finite Elements for Structural Analysis*, Englewood Cliffs: Prentice Hall (1984).

[12] Zienkiewicz, O.C. and Taylor, R.L., *The Finite Element Method,* 5th Edition, London: Butterworth-Heinemann (2000).

Chapter 8

Cylindrical analysis problems

8.1 Introduction

We will slightly increase the prior one-dimensional examples by extending them to problems formulated in cylindrical coordinates. There are many problems that can accurately be modeled as being revolved about an axis. Many of these can be analyzed by employing a radial coordinate, r, and an axial coordinate, z. Solids of revolution can be formulated in terms of the two-dimensional area that is revolved about the axis. Numerous objects of that type are also very long in the axial direction and can be treated as segments of an infinite cylinder. This reduces the analysis to a one-dimensional study in the radial direction. We will begin with that common special case. We will find that changing to these *cylindrical coordinates* will make small changes in the governing differential equations and the corresponding integral theorems that govern the finite element formulation. Also the volume and surface integrals take on special forms. These use the *Theorems of Pappus*. The first states that the surface area of a revolved arc is the product of the arc length and the distance traveled by the centroid of the arc. The second states that the volume of revolution of the generating area is the product of the area and the distance traveled by its centroid. In both cases the distance traveled by the centroid, in full revolution, is $2\pi\bar{r}$ where $\bar{r}$ is the centroid radial coordinate of the arc or area. If we consider differential arcs or areas, then the corresponding differential surface or volume of revolution are $dA = 2\pi r\, dL$ and $dV = 2\pi r\, dr\, dz$.

8.2 Heat conduction in a cylinder

The previous one-dimensional heat transfer model becomes slightly more complicated here. When we consider a point on a radial line we must remember that it is a cross-section of a ring of material around the hoop of the cylinder. Thus as heat is conducted outward in the radial direction it passes through an ever increasing amount of material. The resulting differential equation for thermal equilibrium is well known:

$$\frac{1}{r}\frac{d}{dr}\left(r\,k\,\frac{d\theta}{dr}\right) + Q = 0 \tag{8.1}$$

where r is the radial distance from the axis of revolution, k is the thermal conductivity, θ is the temperature, and Q is the internal heat generation per unit volume. One can have essential boundary conditions where θ is given or as a surface flux condition

$$-\,r\,k\,\frac{d\theta}{dr} = q \tag{8.2}$$

where q is the flux normal to the surface (i.e., radially). If we multiply Eq. 8.1 by r, it would look like our previous one-dimensional form:

$$\frac{d}{dr}\left(k\,*\,\frac{d\theta}{dr}\right) + Q\,* = 0$$

where $k\,* = r\,k$ and $Q\,* = r\,Q$ could be viewed as variable coefficients. This lets us find the required integral (variational) form by inspection. It is

$$I = 2\pi\Delta z \int_L \tfrac{1}{2}\left[\,k\,*\,(d\theta/dr)^2 - Q\,*\,T\,\right] r\,dr \quad \rightarrow \min \tag{8.3}$$

where the integration limits are the inner and outer radii of the cylindrical segment under study. The typical length in the axial direction, Δz, is usually defaulted to unity. The corresponding element square conduction matrix is

$$\mathbf{S}^e = 2\pi \int_{L^e} k^e\,\frac{d\mathbf{H}^{e^T}}{dr}\,\frac{d\mathbf{H}^e}{dr}\,r\,dr \tag{8.4}$$

and the source vector (if any) is

$$\mathbf{C}_Q^e = 2\pi \int_{L^e} \mathbf{H}^{e^T}\,Q^e\,r\,dr. \tag{8.5}$$

If we consider a two node (linear) line element in the radial direction we can use our previous results to write these matrices by inspection. Noting that $L^e = (r_2 - r_1)^e$ and assuming a constant material property, k, in the element gives

$$\mathbf{S}^e = 2\pi\,\frac{k^e}{(L^e)^2}\begin{bmatrix} 1 & -1 \\ -1 & 1 \end{bmatrix}\int_{r_1}^{r_2} r\,dr$$

$$\mathbf{S}^e = 2\pi\,\frac{k^e(r_2^2 - r_1^2)^e}{2(L^e)^2}\begin{bmatrix} 1 & -1 \\ -1 & 1 \end{bmatrix} = \pi\,\frac{k^e(r_2 + r_1)^e}{(r_2 - r_1)^e}\begin{bmatrix} 1 & -1 \\ -1 & 1 \end{bmatrix}. \tag{8.6}$$

Thus, unlike the original one-dimensional case the conduction matrix depends on where the element is located, that is, it depends on how much material it includes (per unit length in the axial direction).

```
! .......................................................        ! 1
!  **  ELEM_SQ_MATRIX PROBLEM DEPENDENT STATEMENTS FOLLOW **     ! 2
! .......................................................        ! 3
!       CYLINDRICAL HEAT CONDUCTION, (see Section 8.2)           ! 4
!         (Stored as application source example 109)             ! 5
 REAL(DP), PARAMETER :: TWO_PI = 6.2831853072d0                  ! 6
 REAL(DP)            :: CONST, DET                               ! 7
 REAL(DP)            :: K_RR, SOURCE                             ! 8
 INTEGER             :: IP                                       ! 9
                                                                 !10
! 1/R * d[R K_RR dT/dR]/dR + Q = 0, Example 109                  !11
                                                                 !12
! PROP_1 = CONDUCTIVITY K_RR                                     !13
! PROP_2 = SOURCE PER UNIT VOLUME, Q                             !14
                                                                 !15
!-->  DEFINE ELEMENT PROPERTIES                                  !16
  K_RR = GET_REAL_LP (1) ; SOURCE = GET_REAL_LP (2)              !17
  E (1, 1) = K_RR ! CONSTITUTIVE                                 !18
                                                                 !19
!       STORE NUMBER OF POINTS FOR FLUX CALCULATIONS             !20
  CALL STORE_FLUX_POINT_COUNT ! Save LT_QP                       !21
                                                                 !22
!-->   NUMERICAL INTEGRATION LOOP                                !23
   DO IP = 1, LT_QP                                              !24
     H   = GET_H_AT_QP (IP)    ! INTERPOLATION FUNCTIONS         !25
     XYZ = MATMUL (H, COORD)   ! FIND RADIUS (R)                 !26
     DLH = GET_DLH_AT_QP (IP) ! FIND LOCAL DERIVATIVES           !27
     AJ = MATMUL (DLH, COORD) ! FIND JACOBIAN AT THE PT          !28
!          FORM INVERSE AND DETERMINATE OF JACOBIAN              !29
     CALL INVERT_JACOBIAN (AJ, AJ_INV, DET, N_SPACE)             !30
     CONST  = TWO_PI * DET * WT(IP) * XYZ (1) ! 2 PI*|J|*w*R     !31
                                                                 !32
!          EVALUATE GLOBAL DERIVATIVES, DGH == B                 !33
     DGH = MATMUL (AJ_INV, DLH)                                  !34
     B   = COPY_DGH_INTO_B_MATRIX (DGH) ! B = DGH                !35
                                                                 !36
     C = C + CONST * SOURCE * H ! VOLUMETRIC SOURCE OPTION       !37
                                                                 !38
!       CONDUCTION SQUARE MATRIX                                 !39
     S = S + CONST * MATMUL ((MATMUL (TRANSPOSE (B), E)), B)     !40
                                                                 !41
!-->  SAVE COORDS, E AND B MATRIX, FOR POST PROCESSING           !42
     CALL STORE_FLUX_POINT_DATA (XYZ, E, B)                      !43
   END DO                                                        !44
! ** END ELEM_SQ_MATRIX PROBLEM DEPENDENT STATEMENTS **          !45
```

Figure 8.1 *Numerically integrated cylindrical heat transfer*

Next, we assume a constant source term so the source vector becomes

$$\mathbf{C}_Q^e = 2\pi Q^e \int_{L^e} \mathbf{H}^{e^T} r\, dr.$$

But the varying radial position, r, must also be accounted for in the integration. One approach to this integration is to again use our isoparametric interpolation and let $r = \mathbf{H}^e \mathbf{R}^e$, where $\mathbf{R}^e$ notes the radial position of each node (i.e., input data). Then

$$\mathbf{C}_Q^e = 2\pi Q^e \int_{L^e} \mathbf{H}^{e^T} \mathbf{H}^e \, dr \, \mathbf{R}^e.$$

For a general element type we would have to evaluate Eq. 8.7 by numerical integration. Then the summation is

$$\mathbf{C}_Q^e = 2\pi \sum_q^{n_q} \mathbf{H}_q^{e^T} \mathbf{H}_q^e \, r_q \, Q_q \, |\mathbf{J}_q^e| w_q$$

where we have interpolated for the position, $r_q = \mathbf{H}_q^e \mathbf{R}^e$ and allowed for a spatially varying source input at the nodes with $Q_q = \mathbf{H}_q^e \mathbf{Q}^e$. A similar expression would evaluate the square matrix in Eq. 8.4. Returning to the linear line element, we have previously exactly evaluated the matrix product integral (mass matrix) and can write the resultant constant source term as

$$\mathbf{C}_Q^e = \frac{2\pi Q^e L^e}{6} \begin{bmatrix} 2 & 1 \\ 1 & 2 \end{bmatrix} \begin{Bmatrix} r_1 \\ r_2 \end{Bmatrix}^e = \frac{2\pi Q^e L^e}{6} \begin{bmatrix} 2r_1 + r_2 \\ r_1 + 2r_2 \end{bmatrix}^e. \tag{8.7}$$

Note that the result depends on the element location because the source is being created in a larger volume of material as the radius increases.

As a simple numerical example consider a cylinder with constant properties, no internal heat generation, an inner radius temperature of $\theta = 100$ at $r = 1$, and an outer radius temperature of $\theta = 10$ at $r = 2$. Select a model with four equal length elements and five nodes. Numbering the nodes radially we have essential boundary conditions of $\theta_1 = 100$ and $\theta_5 = 10$. Considering the form in Eq. 8.6, we note that the element values of $(r_2 + r_1)^e / L^e$ are 9, 11, 13, and 15, respectively. Therefore, we can write the assembled system equations as

$$\pi k^e \begin{bmatrix} 9 & -9 & 0 & 0 & 0 \\ -9 & (9+11) & -11 & 0 & 0 \\ 0 & -11 & (11+13) & -13 & 0 \\ 0 & 0 & -13 & (13+15) & -15 \\ 0 & 0 & 0 & -15 & 15 \end{bmatrix} \begin{Bmatrix} \theta_1 \\ \theta_2 \\ \theta_3 \\ \theta_4 \\ \theta_5 \end{Bmatrix} = \begin{Bmatrix} q_1 \\ 0 \\ 0 \\ 0 \\ -q_5 \end{Bmatrix}.$$

Applying the essential boundary conditions, and dividing both sides by the leading constant gives the reduced set

$$\begin{bmatrix} 20 & -11 & 0 \\ -11 & 24 & -13 \\ 0 & -13 & 28 \end{bmatrix} \begin{Bmatrix} \theta_2 \\ \theta_3 \\ \theta_4 \end{Bmatrix} = 100 \begin{Bmatrix} 9 \\ 0 \\ 0 \end{Bmatrix} + 10 \begin{Bmatrix} 0 \\ 0 \\ 15 \end{Bmatrix}.$$

Solving yields the internal temperature distribution of $\theta_2 = 71.06$, $\theta_3 = 47.39$, and $\theta_4 = 27.34$. Comparing with the exact solution of $\theta = [\theta_1 \ln(r_5/r) + \theta_5 \ln(r/r_1)] / \ln(r_5/r_1)$ shows that our approximation is accurate to at least three significant figures. Also note that both the exact and approximate temperature distributions are independent of the thermal conductivity k. This is true only because the internal heat generation Q was zero. Of course, k does have some effect on the two external heat fluxes (thermal

```
title 'Cylindrical Heat Transfer, Sec. 8.2'                  ! 1
axisymmetric   ! Problem is axisymmetric, x radius           ! 2
example    109 ! Source library example number               ! 3
data_set    01 ! Data set for example (this file)            ! 4
nodes        5 ! Number of nodes in the mesh                 ! 5
elems        2 ! Number of elements in the system            ! 6
dof          1 ! Number of unknowns per node                 ! 7
el_nodes     3 ! Maximum number of nodes per element         ! 8
space        1 ! Solution space dimension                    ! 9
b_rows       1 ! Number of rows in the B matrix              !10
shape        1 ! Element shape, 1=line, 2=tri, 3=quad        !11
remarks      7 ! Number of user remarks                      !12
gauss        3 ! Maximum number of quadrature point          !13
el_real      2 ! Number of real properties per element       !14
el_homo        ! Element properties are homogeneous          !15
no_error_est   ! No SCP element error estimates              !16
quit ! keyword input, remarks follow                         !17
1 1/r * d[r K_rr dT/dr]/dr + Q = 0, Example 109              !18
2 Real FE problem properties are:                            !19
3 K_rr = GET_REAL_LP (1) conductivity                        !20
4 Q    = GET_REAL_LP (2) source per unit length              !21
5 Mesh T=100, r=1 *----*----*----*----* r= 2, T=10           !22
6 Nodes, (Elem)  1    2(1) 3    4(2) 5, K = 1, Q = 0         !23
7 T = [T_1*ln(r_5/r) + T_5*ln(r/r_1)]/ln(r_5/r_1)           !24
1    1   1.     ! begin nodes flag x                         !25
2    0   1.25                                                !26
3    0   1.50   ! note exact T_3=47.35                       !27
4    0   1.75                                                !28
5    1   2.0                                                 !29
 1    1    2    3    ! begin elements                        !30
 2    3    4    5                                            !31
1 1 100.             ! essential bc                          !32
5 1 10.                                                      !33
 1 1. 0.             ! Elem K Q                              !34
```

Figure 8.2 *Typical data for cylindrical conduction*

```
T = [T_1*ln(r_5/r) + T_5*ln(r/r_1)]/ln(r_5/r_1)       ! 1
                                                      ! 2
*** REACTION RECOVERY ***                             ! 3
   NODE, PARAMETER,      REACTION, EQUATION           ! 4
      1,  DOF_1,       8.1591E+02      1              ! 5
      5,  DOF_1,      -8.1591E+02      5              ! 6
                                                      ! 7
***  OUTPUT OF RESULTS IN NODAL ORDER  ***            ! 8
    NODE,  Radius r,   DOF_1,                         ! 9
       1  1.0000E+00  1.0000E+02                      !10
       2  1.2500E+00  7.1046E+01                      !11
       3  1.5000E+00  4.7356E+01                      !12
       4  1.7500E+00  2.7344E+01                      !13
       5  2.0000E+00  1.0000E+01                      !14
```

Figure 8.3 *Results from two quadratic axisymmetric elements*

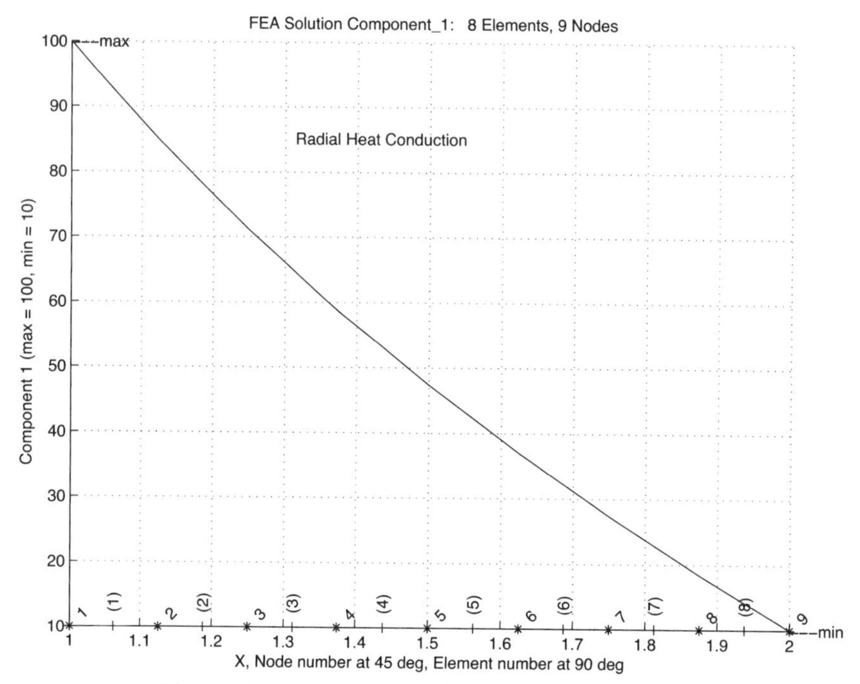

Figure 8.4 *Temperatures for eight linear elements*

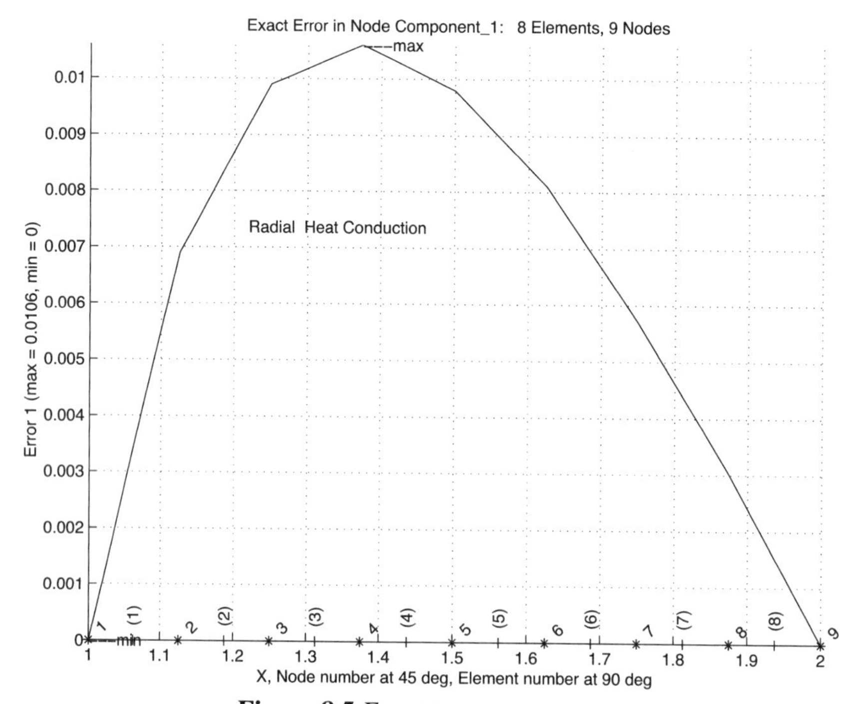

Figure 8.5 *Exact temperature error*

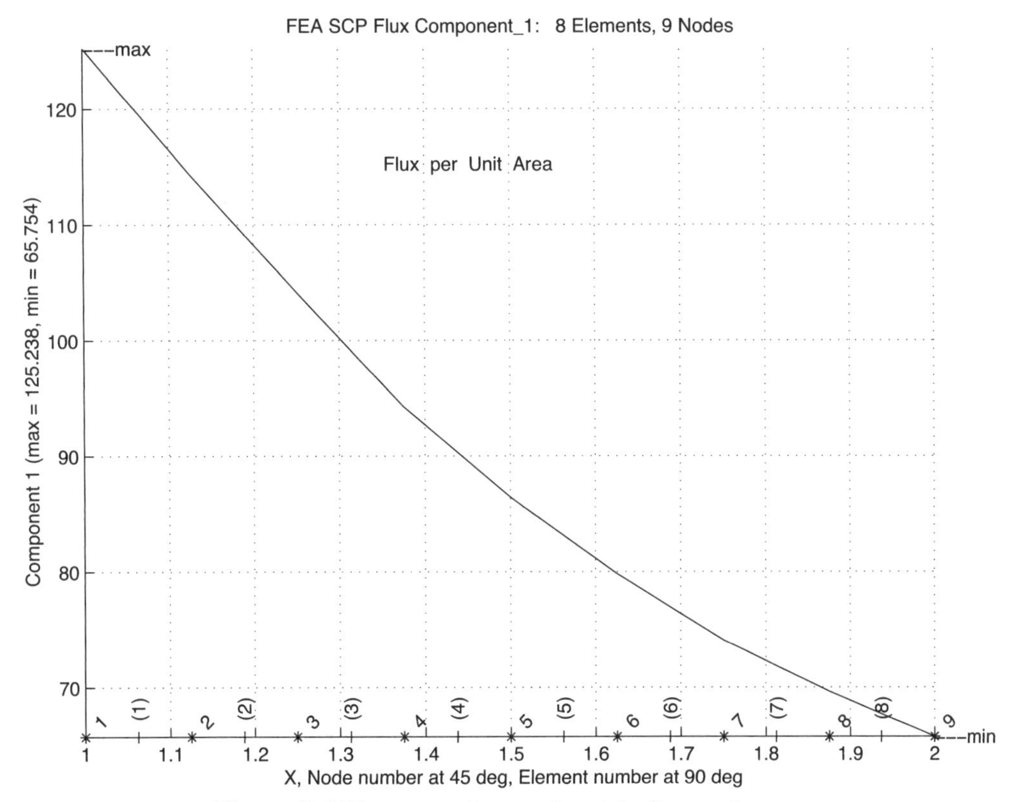

Figure 8.6 *Flux per unit area for eight linear elements*

reactions), q_1 and q_5, necessary to maintain the two prescribed surface temperatures. Substituting back into the first equation to recover the thermal reaction we obtain

$$\pi [\, 9(100) - 9(71.06) + 0 \,] = 818.3 = q_1 / k^e$$

entering at the inner radius. This compares quite well with the exact value of 8.15.8. The fifth equation gives q_5 an equal amount exiting at the outer radius. Therefore, in this problem the heat flux is always in the positive radial direction.

It should be noted that if we had used a higher order element then the integrals would have been much more complicated than the one-dimensional case. This is typical of most axisymmetric problems. Of course, in practice we use numerical integration to automate the evaluation of the element matrices, as described above. A typical implementation is shown in Fig. 8.1. A new consideration is that during the integration we must include the radius. This is done in line 31 using the radial position interpolated in line 26. The data for using two quadratic (3 noded) line elements to solve the above example is shown in Fig. 8.2, and the results are summarized in Fig. 8.3. There we see that the center temperature, line 12, is accurate to four significant figures and that the temperatures and reaction fluxes compare closely to the above four linear element model. The keyword *axisymmetric* (line 2 of Fig. 8.2) is not really used in formulating the element matrices as we have hard coded that knowledge in Fig. 8.1 but it is used in the

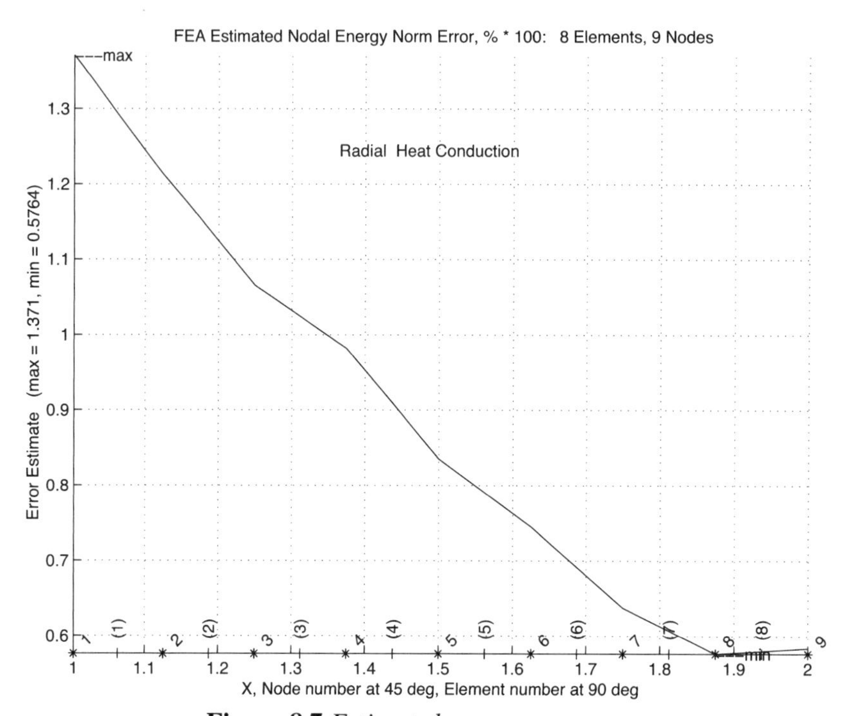

Figure 8.7 *Estimated energy norm error*

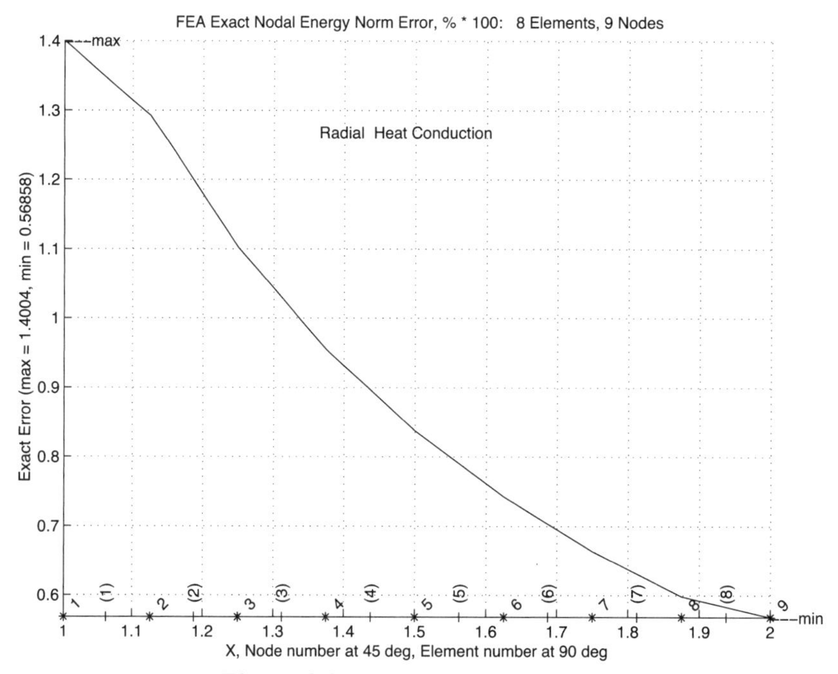

Figure 8.8 *Exact energy norm error*

```
! T = [T_in*ln(r_out/r) + T_out*ln(r/r_in)]/ln(r_out/r_in)        ! 1
                                                                   ! 2
*** REACTION RECOVERY ***                                          ! 3
   NODE, PARAMETER,      REACTION, EQUATION                        ! 4
      1,  DOF_1,        8.1640E+02     1                           ! 5
      9,  DOF_1,       -8.1640E+02     9                           ! 6
                                                                   ! 7
***  OUTPUT OF RESULTS AND EXACT VALUES IN NODAL ORDER  ***        ! 8
    NODE,  Radius r,   DOF_1,       EXACT1,                        ! 9
       1  1.0000E+00  1.0000E+02  1.0000E+02                       !10
       2  1.1250E+00  8.4714E+01  8.4707E+01                       !11
       3  1.2500E+00  7.1036E+01  7.1026E+01                       !12
       4  1.3750E+00  5.8662E+01  5.8651E+01                       !13
       5  1.5000E+00  4.7363E+01  4.7353E+01                       !14
       6  1.6250E+00  3.6968E+01  3.6960E+01                       !15
       7  1.7500E+00  2.7344E+01  2.7338E+01                       !16
       8  1.8750E+00  1.8383E+01  1.8380E+01                       !17
       9  2.0000E+00  1.0000E+01  1.0000E+01                       !18
                                                                   !19
** SUPER_CONVERGENT AVERAGED NODAL FLUXES & EXACT FLUXES **        !20
NODE,    Radius r,    FLUX_1,      EXACT1    (per unit area)       !21
   1  1.000E+00  1.252E+02   1.298E+02                             !22
   2  1.125E+00  1.144E+02   1.154E+02                             !23
   3  1.250E+00  1.042E+02   1.039E+02                             !24
   4  1.375E+00  9.426E+01   9.443E+01                             !25
   5  1.500E+00  8.645E+01   8.656E+01                             !26
   6  1.625E+00  7.982E+01   7.990E+01                             !27
   7  1.750E+00  7.408E+01   7.420E+01                             !28
   8  1.875E+00  6.964E+01   6.925E+01                             !29
   9  2.000E+00  6.575E+01   6.492E+01                             !30
                                                                   !31
-----------------------------------------                          !32
              ERROR IN          % ERROR IN                         !33
ELEMENT,      ENERGY_NORM,      ENERGY_NORM,                       !34
-----------------------------------------                          !35
      1       3.7078E+00        1.3710E+00                         !36
      2       2.8507E+00        1.0541E+00                         !37
      3       2.9117E+00        1.0766E+00                         !38
      4       2.3987E+00        8.8695E-01                         !39
      5       2.1187E+00        7.8341E-01                         !40
      6       1.9101E+00        7.0629E-01                         !41
      7       1.5382E+00        5.6878E-01                         !42
      8       1.5794E+00        5.8401E-01                         !43
```

Figure 8.9 *Exact and computed results in conducting cylinder*

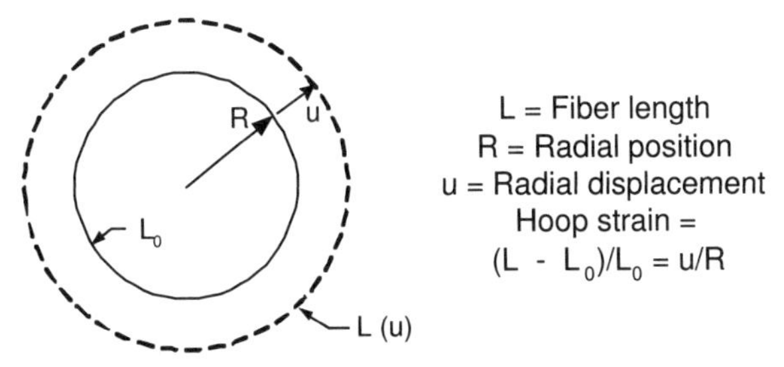

Figure 8.10 *Hoop strain due to radial displacement*

```
! B       = STRAIN-DISPLACEMENT MATRIX, (N_R_B, LT_FREE)            ! 1
! BODY    = BODY FORCE VECTOR, (N_SPACE)                            ! 2
! DGH     = GLOBAL DERIVS OF FUNCTIONS H, (N_SPACE, LT_N)           ! 3
! LT_FREE = NUMBER OF DEGREES OF FREEDOM PER ELEMENT                ! 4
! LT_N    = MAXIMUM NUMBER OF NODES FOR ELEMENT TYPE                ! 5
! N_R_B   = NUMBER OF ROWS IN B AND E MATRICES                      ! 6
! ...............................................................   ! 7
!  **  ELEM_SQ_MATRIX PROBLEM DEPENDENT STATEMENTS FOLLOW **        ! 8
! ...............................................................   ! 9
!       Cylindrical Stress Analysis, (see Section 8.3)              !10
!        (Stored as application source example 110)                 !11
 REAL(DP), PARAMETER :: TWO_PI = 6.2831853072d0                      !12
 REAL(DP)            :: CONST, DET                                   !13
 REAL(DP)            :: E_mod, P_ratio, Rho, Spin                    !14
 INTEGER             :: IP                                           !15
                                                                     !16
! Elem real prop: Young's modulus, Poisson's ratio, Density         !17
! Misc real prop: Spin about z-axis, radians per second             !18
  E_mod = GET_REAL_LP (1) ; P_ratio = GET_REAL_LP (2)               !19
  Rho   = GET_REAL_LP (3) ; Spin    = GET_REAL_MISC (1)             !20
                                                                     !21
!               CONSTITUTIVE LAW                                     !22
  E(1,1) = E_mod*(1 - P_ratio)/((1 + P_ratio)*(1 - 2*P_ratio))       !23
  E(2,1) = E_mod*P_ratio/((1 + P_ratio)*(1 - 2*P_ratio))             !24
  E(1,2) = E(2,1)  ;  E(2,2) = E(1,1)                                !25
                                                                     !26
!      STORE NUMBER OF POINTS FOR FLUX CALCULATIONS                  !27
  CALL STORE_FLUX_POINT_COUNT ! Save LT_QP                           !28
                                                                     !29
!-->   NUMERICAL INTEGRATION LOOP                                    !30
   DO IP = 1, LT_QP                                                  !31
     H   = GET_H_AT_QP (IP)    ! INTERPOLATION FUNCTIONS             !32
     XYZ = MATMUL (H, COORD)   ! FIND RADIUS (ISOPARAMETRIC)         !33
     DLH = GET_DLH_AT_QP (IP) ! FIND LOCAL DERIVATIVES               !34
     AJ = MATMUL (DLH, COORD) ! FIND JACOBIAN AT THE PT               !35
!         FORM INVERSE AND DETERMINATE OF JACOBIAN                   !36
     CALL INVERT_JACOBIAN (AJ, AJ_INV, DET, N_SPACE)                 !37
     CONST = TWO_PI * DET * WT(IP) * XYZ (1) ! 2 PI*|J|*w*R          !38
                                                                     !39
!      EVALUATE GLOBAL DERIVATIVES & STRAIN-DISPLACEMENT             !40
     DGH = MATMUL (AJ_INV, DLH)                                      !41
     B (1, :) = DGH (1, :)        ! DU/DR radial strain              !42
     B (2, :) = H (:) / XYZ (1)   ! U/R hoop strain                  !43
                                                                     !44
!      STIFFNESS MATRIX                                              !45
     S = S + CONST * MATMUL ((MATMUL (TRANSPOSE (B), E)), B)         !46
                                                                     !47
     BODY = - Rho * XYZ (1) * Spin **2 ! -Rho R Omega^2              !48
     C    = C + CONST * BODY (1) * H   ! CENTRIFUGAL RESULT          !49
                                                                     !50
!-->  SAVE COORDS, E AND B MATRIX, FOR POST PROCESSING               !51
     CALL STORE_FLUX_POINT_DATA (XYZ, E, B)                          !52
   END DO                                                            !53
! ** END ELEM_SQ_MATRIX PROBLEM DEPENDENT STATEMENTS **              !54
```

Figure 8.11 *Cylindrical stress with centrifugal loads*

geometric properties (volume and centroid) calculations that are provided for data validation and it would have triggered necessary calculations in the error estimation stage if keyword *no_error_est* had not been present. That overrides the default state of including error estimates, which is done in the next example.

As another comparison, now including error estimates, consider the linear element model run with eight equal elements. The nodal results are still accurate to about four significant figures but, of course, are less accurate than that inside the elements where the logarithmic exact value is being approximated by a straight line. The actual temperature values are shown in Fig.8.4 along with the exact temperature error in Fig. 8.5. Because of the boundary conditions used here remember that the temperatures are independent of the conductivity, but the flux is not. For a planar wall of the same thickness and same inner and outer wall essential boundary conditions the temperature would have been exactly linear with position and the total heat flux and the heat flux per unit area would have been constant. Here we see in Fig. 8.6 that the smoothed heat flux per unit area is not constant but drops off radially because it is passing through more material (a larger area) as the radius increases. The exact flux per unit area is given by $k(\theta_{in} - \theta_{out}) / [r \, ln(r_{out} / r_{in})]$.

The smoothed fluxes, per unit area, are compared to the element piecewise constant values in order to develop an error estimate. The estimated error, in the energy norm, is shown in Fig. 8.7. It was obtained by averaging the element values at the nodes. In this simple problem we have the exact solution so we can compare the estimate with the corresponding exact values in Fig. 8.8. Here we see that the SCP energy norm error estimate agrees with the exact energy norm error reasonably well. Referring back to Fig. 8.5 we observe that spatial distribution of the exact function error and the exact energy norm error can be quite different. The error in the solution is always exactly zero at nodes with essential boundary conditions. However, small distances away from such locations the function value error can increase very rapidly. In Fig. 8.9 we see (lines 5 and 6) that the heat flux reaction, over a total $2\pi r$ section, has changed very little. Remember that such reaction fluxes are obtained from the integral form and thus are more accurate than element fluxes which are extrapolated to the boundary. This can be seen again in the averaged nodal fluxes, per unit area, (lines 22-30) which when multiplied by the section areas give inner and outer values (which should be the same) of 787 and 826 for an average of 807 which represents an error of about one percent.

8.3 Cylindrical stress analysis

Another common problem is the analysis of an axisymmetric solid with axisymmetric loads and supports. This becomes a two-dimensional analysis that is very similar to plane strain analysis. The radial and axial displacement components will be denoted by u and v. These are the same unknowns used in the plane strain study. In addition to the previous strains there is another strain known as the *hoop strain*, ε_t, and a corresponding hoop stress, σ_t. The hoop strain results from the change in length of a fiber of material around the circumference of the solid. The definition of a normal strain is a change in length divided by the original length. The circumference at a typical radial position is $L = 2\pi r$. When such a point undergoes a radial displacement of u it occupies a new radial position of $(r + u)$, as shown in Fig. 8.10. It has a corresponding increase in circumference. The hoop strain becomes

$$\varepsilon_t = \frac{\Delta L}{L} = \frac{2\pi(r + u) - 2\pi r}{2\pi r} = \frac{u}{r}. \tag{8.8}$$

Then, our strains to be computed for the cylindrical analysis are denoted as

$$\varepsilon^T = [\varepsilon_r \quad \varepsilon_t] \tag{8.9}$$

and the corresponding stress components are

$$\sigma^T = [\sigma_r \quad \sigma_t]. \tag{8.10}$$

For an isotropic material the stress-strain law is like that for the plane strain case. In the axial, or z, direction of the assumed infinite cylinder one can either set the stress or strain to zero, but not both. They are related by:

$$\varepsilon_z = \frac{1}{E}(\sigma_z - \nu\sigma_r - \nu\sigma_t). \tag{8.11}$$

There is no shear stress or strain in the cylindrical case. In the case of an infinite cylinder the above formulation simplifies since $v = 0$ and $\partial/\partial z = 0$. Thus, we consider only the radial displacement, u, and the strains and stresses in the radial and hoop directions. This gives the two strain-displacement relations as

$$\boldsymbol{\varepsilon} = \mathbf{B}^e\,\mathbf{u}^e \tag{8.12}$$

where

$$\mathbf{B}^e = \begin{bmatrix} \partial\mathbf{H}/\partial r \\ \mathbf{H}/r \end{bmatrix}. \tag{8.13}$$

Note that this is the first time we have computed two secondary items (strains) from a single primary unknown. Now the stresses and the stress-strain law must have the same number of rows. The radial and hoop stresses are

$$\boldsymbol{\sigma} = \mathbf{E}\,\boldsymbol{\varepsilon}, \quad \mathbf{E} = \begin{bmatrix} E_{11} & E_{12} \\ E_{21} & E_{22} \end{bmatrix} \tag{8.14}$$

and for an isotropic material $E_{11} = E_{22} = E(1-\nu)/(1+\nu)(1-2\nu)$, while the off-diagonal terms are $E_{12} = E_{21} = E\nu/(1+\nu)(1-2\nu)$. Note that this constitutive law encounters problems for an 'incompressible' material where $\nu = 1/2$ and division by zero occurs. That forces one to employ an alternate theory. The stiffness matrix,

$$\mathbf{K}^e = 2\pi \int_{A^e} \mathbf{B}^{e^T}\mathbf{E}^e\mathbf{B}^e\, r\,dr\,dz\,,$$

can be expanded to the form

$$\mathbf{K}^e = 2\pi\,\Delta z \int_{L^e} \left[E_{11}\frac{\partial \mathbf{H}^T}{\partial r}\frac{\partial \mathbf{H}}{\partial r} + E_{12}\left(\frac{\partial \mathbf{H}^T}{\partial r}\mathbf{H} + \mathbf{H}^T\frac{\partial \mathbf{H}}{\partial r}\right)/r + E_{22}\mathbf{H}^T\mathbf{H}/r^2 \right] r\,dr. \tag{8.15}$$

The first integral we just evaluated and is given in Eqs. 8.4 and 8.6 if we let $\Delta z = 1$ and replace k with E_{11}. The second term we integrate by inspection since the r terms cancel. The result is

$$\mathbf{K}_{12}^{e} = 2\pi\,\Delta z\, E_{12}^{e}\begin{bmatrix} -1 & 0 \\ 0 & 1 \end{bmatrix}. \tag{8.16}$$

The remaining contribution is more difficult since it involves division by r. Assuming constant E_{22} we have

$$\mathbf{K}_{22} = 2\pi\,\Delta z\, E_{22}\int_{L^e} \frac{1}{r}\,\mathbf{H}^T\mathbf{H}\,dr \tag{8.17}$$

which requires analytic integration involving logarithms, or numerical integration. Using a one point (centroid) quadrature rule gives the approximation

$$\mathbf{K}_{22}^{e} = \frac{2\pi\,\Delta z\, E_{22}^{e} L^e}{2(r_1 + r_2)^e}\begin{bmatrix} 1 & 1 \\ 1 & 1 \end{bmatrix}. \tag{8.18}$$

For a cylinder the loading would usually be a pressure acting on an outer surface or an internal centrifugal load due to a rotation about the z-axis. For a pressure load the resultant force at a nodal ring is the pressure times the surface area. Thus, $F_{p_i} = 2\pi\,\Delta z\, r_i\, p_i$. As a numerical example consider a single element solution for a cylinder with an internal pressure of $p = 1$ ksi on the inner radius $r_1 = 10$ in. Assume $E = 10^4$ ksi and $\nu = 0.3$, and let the thickness of the cylinder be 1 in. Note that there is no essential boundary condition on the radial displacement. This is because the hoop effects prevent a rigid body radial motion. The numerical values of the above stiffness contributions are

$$\mathbf{K}_{11} = 2\pi\,\Delta z\,(1.413 \times 10^5)\begin{bmatrix} 1 & -1 \\ -1 & 1 \end{bmatrix}$$

$$\mathbf{K}_{12} = 2\pi\,\Delta z\,(5.769 \times 10^3)\begin{bmatrix} -1 & 0 \\ 0 & 1 \end{bmatrix}, \quad \mathbf{K}_{22} = 2\pi\,\Delta z\,(3.205 \times 10^2)\begin{bmatrix} 1 & 1 \\ 1 & 1 \end{bmatrix}$$

while the resultant force at the inner radius is $F_p = 2\pi\,\Delta z\; 10$. Assembling and canceling the common constant gives

$$10^5\begin{bmatrix} 1.35897 & -1.41026 \\ -1.41026 & 1.47436 \end{bmatrix}\begin{Bmatrix} u_1 \\ u_2 \end{Bmatrix} = \begin{Bmatrix} 10 \\ 0 \end{Bmatrix}.$$

Solving gives $\mathbf{u} = [9.9642 \quad 9.5309] \times 10^{-4}$ in. This represents a displacement error of about 8 percent and 9 percent, respectively, at the two nodes. The maximum radial stress equals the applied pressure. The stresses can be found from Eq. 8.14. The hoop strain at node 1 is $\varepsilon_t = u_1/r_1 = 9.964 \times 10^{-5}$ in/in. The finite element radial strain approximation is $\varepsilon_r = \partial H_1/\partial r\, u_1 + \partial H_2/\partial r\, u_2$. The constant radial strain approximation is

$$\varepsilon_r = \frac{-u_1 + u_2}{r_2 - r_1} = -4.333 \times 10^{-4}\ \text{in/in}.$$

Therefore, the hoop stress at the first node is

$$\sigma_t = E_{12}\varepsilon_r + E_{22}\varepsilon_t = -2.500 + 13.413 = 10.91\ \text{ksi}.$$

This compares well with the exact value of 10.52 ksi. Note that the inner hoop stress is

```
title 'Cylinder with pressure, no spin, Sec. 8.3'       ! 1
axisymmetric ! Problem is axisymmetric, x radius        ! 2
example  110 ! Source library example number            ! 3
data_set  01 ! Data set for example (this file)         ! 4
nodes      3 ! Number of nodes in the mesh              ! 5
elems      1 ! Number of elements in the system         ! 6
dof        1 ! Number of unknowns per node              ! 7
el_nodes   3 ! Maximum number of nodes per element      ! 8
space      1 ! Solution space dimension                 ! 9
b_rows     2 ! Number of rows in the B matrix           !10
shape      1 ! Element shape, 1=line, 2=tri, 3=quad     !11
remarks    5 ! Number of user remarks                   !12
gauss      3 ! Maximum number of quadrature point       !13
el_real    3 ! Number of real properties per element    !14
reals      1 ! Number of miscellaneous real properties !15
loads        ! An initial source vector is input        !16
el_no_col    ! All element column matrices are null     !17
no_error_est ! No SCP element error estimates           !18
quit ! keyword input, remarks follow                    !19
1 E = 10e4 ksi, Nu = 0.3, Internal pressure = 1 ksi     !20
2 Resultant = 62.832 kips, Unit axial length            !21
3 U_1_exact = 1.0824e-3   (1)                           !22
4 P = 1 ksi, r=10 in   *----*----*  r=11, P = 0         !23
5              Nodes    1    2    3                     !24
1    0    10.      ! begin nodes, bc flag, r            !25
2    0    10.5                                          !26
3    0    11.                                           !27
 1    1    2    3 ! element, connectivity               !28
 1 1.e5 0.3 0.    ! Elem, E, Nu, Rho                    !29
0.0               ! system spin rate                    !30
1    1   62.832   ! node, direction, force              !31
3    1   0.       ! terminate input with last force     !32
```

Figure 8.12 *Internal pressure load example*

```
***  OUTPUT OF RESULTS IN NODAL ORDER  ***                          ! 1
   NODE,  Radius r,   DOF_1,                                        ! 2
      1  1.0000E+01  9.9667E-04                                     ! 3
      2  1.0500E+01  9.7338E-04                                     ! 4
      3  1.1000E+01  9.5333E-04                                     ! 5
                                                                    ! 6
*** STRAIN COMPONENTS AT ELEMENT INTEGRATION POINTS ***             ! 7
ELEMENT, PT,  Radius r,  STRAIN_1,    STRAIN_2,                     ! 8
     1  1  1.0113E+01 -8.5488E-05  1.0404E-03                       ! 9
     1  2  1.0500E+01 -4.8508E-05  9.9792E-04                       !10
     1  3  1.0887E+01 -8.3185E-06  9.6295E-04                       !11
```

Figure 8.13 *Displacements and strains due to internal pressure* more than ten times the applied internal pressure. Since we have set ε_z to zero we should use Eq. 8.11 to determine the axial stress that results from the effect of Poisson's ratio. All three stresses would be used in evaluating a failure criterion like the Von Mises stress.

The implementation of this analysis via numerically integrated elements is shown in Fig. 8.11. There we have chosen to include another common loading case of a spinning cylinder (centrifugal load) to account for a body force vector. The first six lines note that the system is allowing for multiple rows in the arrays **B** and **E**. In solid mechanics the internal volumetric source terms are vectors, unlike the scalar heat source term in the previous example. Even though it has only one component we should always distinguish between scalars and vectors (and higher order tensors), thus line 2 reminds us the system has standard storage space for such optional vectors. The previous hand calculation is repeated here with a single quadratic element. The input data are shown in Fig. 8.12 which remarks that the exact inner displacement is 1.0824×10^{-3} inches. This gives less than 8 percent error in displacements, as noted in the output summary in Fig. 8.13, and compares closely to the simple hand estimate. The two strain components are noted to vary significantly over the element and that suggests the mesh is much too crude (as expected). The two average strain values of $[-4.744 \quad 1.000] \times 10^5$ in/in are close to the single linear element results above.

8.4 Exercises

1. In electrostatics the electrical potential, ϕ, is related to the charge density, ζ, and the permittivity of the material, ε. A coaxial cable can be represented in the radial direction by the equation

$$\frac{1}{r}\frac{d}{dr}(r\,\varepsilon\frac{d\phi}{dr}) + \zeta = 0$$

   Compare this to Eq. 8.1. A hollow coaxial cable is made from a hollow conducting core and an insulating outer layer with $\varepsilon_1 = 0.5$, $\varepsilon_2 = 2.0$ and charge densities of $\zeta_1 = 100$ and $\zeta_2 = 0$, respectively. The inner, interface, and outer radii are 5, 10, and 25 mm. The corresponding inner and outer potentials (boundary conditions) are $\phi = 500$ and $\phi = 0$, respectively. Compute the interface potential [analytical value is 918.29] using: a) the element formulation in Fig. 8.1 for two quadratic elements, b) the approximate closed form integration in Eqs. 8.6 and 7.

2. For the four linear element example in Section 8.2 determine and plot the element centroid fluxes. Obtain an 'eyeball' SCP fit of those values to get continuous nodal fluxes and plot them along with the exact flux.

8.5 Bibliography

[1] Buchanan, G.R., *Finite Element Analysis*, New York: McGraw-Hill (1995).
[2] Rockey, K.C., et al., *Finite Element Method – A Basic Introduction*, New York: Halsted Press (1975).
[3] Ross, C.T.F., *Finite Element Programs for Axisymmetric Problems in Engineering*, New York: Halsted Press (1984).
[4] Weaver, W.F., Jr. and Johnston, P.R., *Finite Elements for Structural Analysis*, Englewood Cliffs: Prentice Hall (1984).
[5] Zienkiewicz, O.C., *The Finite Element Method in Structural and Continuum Mechanics*, New York: McGraw-Hill (1967).
[6] Zienkiewicz, O.C. and Taylor, R.L., *The Finite Element Method,* 5th Edition, London: Butterworth-Heinemann (2000).

Chapter 9

General interpolation

9.1 Introduction

The previous sections have illustrated the heavy dependence of finite element methods on both spatial interpolation and efficient integrations. In a one-dimensional problem it does not make a great deal of difference if one selects a local or global coordinate system for the interpolation equations, because the inter-element continuity requirements are relatively easy to satisfy. That is not true in higher dimensions. To obtain practical formulations it is almost essential to utilize local coordinate interpolations. Doing this does require a small amount of additional work in relating the derivatives in the two coordinate systems.

9.2 Unit coordinate interpolation

The use of unit coordinates has been previously mentioned in Chapter 4. Here some of the procedures for deriving the interpolation functions in unit coordinates will be presented. Consider the three-node triangular element shown in Fig. 9.1. The local coordinates of its three nodes are (0, 0), (1, 0), and (0, 1), respectively. Once again we wish to utilize polynomial functions for our interpolations. In two dimensions the simplest complete polynomial has three constants. Thus, this linear function can be related to the three nodal quantities of the element. Assume the polynomial for some quantity, u, is defined as:

$$u^e(r, s) = d_1^e + d_2^e r + d_3^e s = \mathbf{P}(r, s)\, \mathbf{d}^e. \tag{9.1}$$

If it is valid everywhere in the element then it is valid at its nodes. Substituting the local coordinates of a node into Eq. 9.1 gives an identity between the $\mathbf{d}^e$ and a nodal value of $\mathbf{u}$. Establishing these identities at all three nodes gives

$$\begin{Bmatrix} u_1^e \\ u_2^e \\ u_3^e \end{Bmatrix} = \begin{bmatrix} 1 & 0 & 0 \\ 1 & 1 & 0 \\ 1 & 0 & 1 \end{bmatrix} \begin{Bmatrix} d_1^e \\ d_2^e \\ d_3^e \end{Bmatrix}$$

or

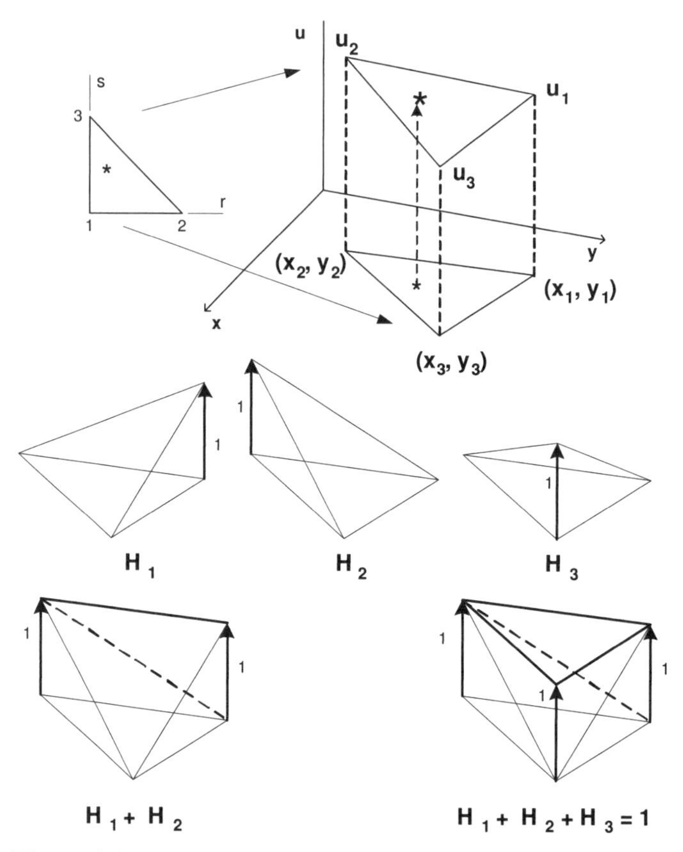

Figure 9.1 *Isoparametric interpolation on a simplex triangle*

$$\mathbf{u}^e = \mathbf{g}\,\mathbf{d}^e. \tag{9.2}$$

If the inverse exists, and it does here, this equation can be solved to yield

$$\mathbf{d}^e = \mathbf{g}^{-1}\mathbf{u}^e \tag{9.3}$$

and

$$u^e(r, s) = \mathbf{P}(r, s)\,\mathbf{g}^{-1}\,\mathbf{u}^e = \mathbf{H}(r, s)\mathbf{u}^e. \tag{9.4}$$

Here

$$\mathbf{g}^{-1} = \begin{bmatrix} 1 & 0 & 0 \\ -1 & 1 & 0 \\ -1 & 0 & 1 \end{bmatrix} \tag{9.5}$$

and

$$H_1(r, s) = 1 - r - s, \qquad H_2(r, s) = r, \qquad H_3(r, s) = s. \tag{9.6}$$

By inspection, one can see that the sum of these functions at all points in the local domain is unity. This is illustrated graphically at the bottom of Fig. 9.1. Typical coding for these relations and their local derivatives are shown as subroutines *SHAPE_3_T* and *DERIV_3_T* in Fig. 9.2. Similarly, for the unit coordinate bilinear quadrilateral mapping from $0 < (r, s) < 1$ one could assume that

$$u^e(r, s) = d_1^e + d_2^e r + d_3^e s + d_4^e rs \tag{9.7}$$

so that

$$\mathbf{g} = \begin{bmatrix} 1 & 0 & 0 & 0 \\ 1 & 1 & 0 & 0 \\ 1 & 1 & 1 & 1 \\ 1 & 0 & 1 & 0 \end{bmatrix} \tag{9.8}$$

and

```
SUBROUTINE SHAPE_3_T (S, T, H)                                 ! 1
! *-* *-* *-* *-* *-* *-* *-* *-* *-* *-* *-* *-* *-*         ! 2
! SHAPE FUNCTIONS FOR A THREE NODE UNIT TRIANGLE               ! 3
! *-* *-* *-* *-* *-* *-* *-* *-* *-* *-* *-* *-* *-*         ! 4
Use Precision_Module                                           ! 5
 IMPLICIT NONE                                                 ! 6
 REAL(DP), INTENT(IN)  :: S, T                                 ! 7
 REAL(DP), INTENT(OUT) :: H (3)                                ! 8
                                                               ! 9
! S,T = LOCAL COORDINATES OF THE POINT    3     T              !10
! H   = SHAPE FUNCTIONS                   . .   .              !11
! NODAL COORDS 1-(0,0)  2-(1,0)  3-(0,1)  1..2  0..S           !12
                                                               !13
  H (1) = 1.d0 - S - T                                         !14
  H (2) = S                                                    !15
  H (3) = T                                                    !16
END SUBROUTINE SHAPE_3_T                                       !17
                                                               !18
SUBROUTINE DERIV_3_T (S, T, DH)                                !19
! *-* *-* *-* *-* *-* *-* *-* *-* *-* *-* *-* *-* *-*         !20
! LOCAL DERIVATIVES OF A THREE NODE UNIT TRIANGLE              !21
!            SEE SUBROUTINE SHAPE_3_T                          !22
! *-* *-* *-* *-* *-* *-* *-* *-* *-* *-* *-* *-* *-*         !23
Use Precision_Module                                           !24
 IMPLICIT NONE                                                 !25
 REAL(DP), INTENT(IN)  :: S, T                                 !26
 REAL(DP), INTENT(OUT) :: DH (2, 3)                            !27
                                                               !28
! S,T     = LOCAL COORDINATES OF THE POINT                     !29
! DH(1,K) = DH(K)/DS                                           !30
! DH(2,K) = DH(K)/DT                                           !31
! NODAL COORDS ARE :  1-(0,0) 2-(1,0) 3-(0,1)                  !32
                                                               !33
  DH (1, 1) = - 1.d0                                           !34
  DH (1, 2) = 1.d0                                             !35
  DH (1, 3) = 0.d0                                             !36
  DH (2, 1) = - 1.d0                                           !37
  DH (2, 2) = 0.d0                                             !38
  DH (2, 3) = 1.d0                                             !39
END SUBROUTINE DERIV_3_T                                       !40
```

Figure 9.2 *Coding a linear unit coordinate triangle*

$$
\begin{aligned}
H_1(r,s) &= 1 - r - s + rs \\
H_2 &= r - rs \\
H_3 &= rs \\
H_4 &= s - rs.
\end{aligned}
\tag{9.9}
$$

However, for the quadrilateral it is more common to utilize the natural coordinates, as shown in Fig. 9.3. In that coordinate system $-1 \le a, b \le +1$ so that

$$
\mathbf{g} = \begin{bmatrix} 1 & -1 & -1 & 1 \\ 1 & 1 & -1 & -1 \\ 1 & 1 & 1 & 1 \\ 1 & -1 & 1 & -1 \end{bmatrix}
$$

and the alternate interpolation functions are

$$
H_i(a, b) = (1 + aa_i)(1 + bb_i)/4, \quad 1 \le i \le 4 \tag{9.10}
$$

where (a_i, b_i) are the local coordinates of node i. These four functions and their local derivatives can be coded as shown in Fig. 9.3.

Note that up to this point we have utilized the local element coordinates for interpolation. Doing so makes the geometry matrix, $\mathbf{g}$, depend only on element type instead of element number. If we use global coordinates then the geometric matrix, $\mathbf{g}^e$ is always dependent on the element number, e. For example, if Eq. 9.1 is written in physical coordinates then

$$
u^e(x, y) = d_1^e + d_2^e x + d_3^e y \tag{9.11}
$$

so when the identities are evaluated at each node the result is

$$
\mathbf{g}^e = \begin{bmatrix} 1 & x_1^e & y_1^e \\ 1 & x_2^e & y_2^e \\ 1 & x_3^e & y_3^e \end{bmatrix}. \tag{9.12}
$$

Inverting and simplifying the algebra gives the global coordinate equivalent of Eq. 9.6 for a specific element:

$$
H_i^e(x, y) = (a_i^e + b_i^e x + c_i^e y)/2A^e, \quad 1 \le i \le 3 \tag{9.13}
$$

where the geometric constants are

$$
\begin{aligned}
a_1^e &= x_2^e y_3^e - x_3^e y_2^e & b_1^e &= y_2^e - y_3^e & c_1^e &= x_3^e - x_2^e \\
a_2^e &= x_3^e y_1^e - x_1^e y_3^e & b_2^e &= y_3^e - y_1^e & c_2^e &= x_1^e - x_3^e \\
a_3^e &= x_1^e y_2^e - x_2^e y_1^e & b_3^e &= y_1^e - y_2^e & c_3^e &= x_2^e - x_1^e
\end{aligned}
\tag{9.14}
$$

and A^e is the area of the element, that is, $A^e = (a_1^e + a_2^e + a_3^e)/2$, or

$$
A^e = \left[x_1^e(y_2^e - y_3^e) + x_2^e(y_3^e - y_1^e) + x_3^e(y_1^e - y_2^e) \right] /2.
$$

These algebraic forms assume that the three local nodes are numbered counter-clockwise from an arbitrarily selected corner. If the topology is defined in a clockwise order then the area, A^e, becomes negative.

```
SUBROUTINE SHAPE_4_Q (R, S, H)                                    ! 1
! *-* *-* *-* *-* *-* *-* *-* *-* *-* *-* *-* *-* *-*             ! 2
! SHAPE FUNCTIONS OF A 4 NODE PARAMETRIC QUAD                     ! 3
!           IN NATURAL COORDINATES                                ! 4
! *-* *-* *-* *-* *-* *-* *-* *-* *-* *-* *-* *-* *-*             ! 5
Use Precision_Module                                              ! 6
 IMPLICIT NONE                                                    ! 7
 REAL(DP), INTENT(IN)  :: R, S                                    ! 8
 REAL(DP), INTENT(OUT) :: H (4)                                   ! 9
 REAL(DP)              :: R_P, R_M, S_P, S_M                      !10
                                                                  !11
! (R,S) = A POINT IN THE NATURAL COORDS        4---3              !12
! H     = LOCAL INTERPOLATION FUNCTIONS        |   |              !13
! H(I)  = 0.25d0*(1+R*R(I))*(1+S*S(I))         |   |              !14
! R(I)  = LOCAL R-COORDINATE OF NODE I         1---2              !15
! LOCAL COORDS, 1=(-1,-1)   3=(+1,+1)                             !16
                                                                  !17
  R_P = 1.d0 + R ; R_M = 1.d0 - R                                 !18
  S_P = 1.d0 + S ; S_M = 1.d0 - S                                 !19
  H (1) = 0.25d0*R_M*S_M                                          !20
  H (2) = 0.25d0*R_P*S_M                                          !21
  H (3) = 0.25d0*R_P*S_P                                          !22
  H (4) = 0.25d0*R_M*S_P                                          !23
END SUBROUTINE SHAPE_4_Q                                          !24
                                                                  !25
SUBROUTINE DERIV_4_Q (R, S, DELTA)                                !26
! *-* *-* *-* *-* *-* *-* *-* *-* *-* *-* *-* *-* *-*             !27
! LOCAL DERIVATIVES OF THE SHAPE FUNCTIONS FOR AN                 !28
! PARAMETRIC QUADRILATERAL WITH FOUR NODES                        !29
!              SEE SHAPE_4_Q                                      !30
! *-* *-* *-* *-* *-* *-* *-* *-* *-* *-* *-* *-* *-*             !31
Use Precision_Module                                              !32
 IMPLICIT NONE                                                    !33
 REAL(DP), INTENT(IN)  :: R, S                                    !34
 REAL(DP), INTENT(OUT) :: DELTA (2, 4)                            !35
 REAL(DP)              :: R_P, R_M, S_P, S_M                      !36
                                                                  !37
! DELTA(1,I) = DH/DR                                              !38
! DELTA(2,I) = DH/DS                                              !39
! H          = LOCAL INTERPOLATION FUNCTIONS                      !40
! (R,S)      = A POINT IN THE LOCAL COORDINATES                   !41
! HERE D(H(I))/DR = 0.25d0*R(I)*(1+S*S(I)), ETC.                  !42
                                                                  !43
  R_P = 1.d0 + R ; R_M = 1.d0 - R                                 !44
  S_P = 1.d0 + S ; S_M = 1.d0 - S                                 !45
  DELTA (1, 1) = -0.25d0 * S_M                                    !46
  DELTA (1, 2) =  0.25d0 * S_M                                    !47
  DELTA (1, 3) =  0.25d0 * S_P                                    !48
  DELTA (1, 4) = -0.25d0 * S_P                                    !49
  DELTA (2, 1) = -0.25d0 * R_M                                    !50
  DELTA (2, 2) = -0.25d0 * R_P                                    !51
  DELTA (2, 3) =  0.25d0 * R_P                                    !52
  DELTA (2, 4) =  0.25d0 * R_M                                    !53
END SUBROUTINE DERIV_4_Q                                          !54
```

Figure 9.3 *Coding a bilinear quadrilateral*

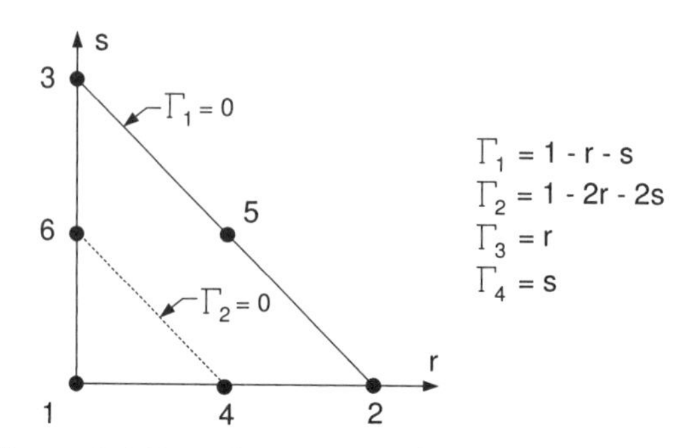

Figure 9.4 *Boundary curves through element nodes*

It would be natural at this point to attempt to utilize a similar procedure to define the four node quadrilateral in the same manner. For example, if Eq. 9.7 is written as

$$u^e(x, y) \;=\; d_1^e + d_2^e\, x + d_3^e\, y + d_4^e\, xy. \tag{9.15}$$

However, we now find that for a general quadrilateral the inverse of matrix $\mathbf{g}^e$ may not exist. This means that the global coordinate interpolation is in general very sensitive to the orientation of the element in global space. That is very undesirable. This important disadvantage vanishes only when the element is a rectangle. This global form of interpolation also yields an element that fails to satisfy the required interelement continuity requirements. These difficulties are typical of those that are encountered in two- and three-dimensions when global coordinate interpolation is utilized. Therefore, it is most common to employ the local coordinate mode of interpolation. Doing so also easily allows for the treatment of curvilinear elements. That is done with *isoparametric elements* that will be mentioned later.

It is useful to illustrate the lack of continuity that develops in the global coordinate form of the quadrilateral. First, consider the three-node triangular element and examine the interface or boundary where two elements connect. Along the interface between the two elements one has the geometric restriction that the edge is a straight line given by $y = m^b x + n^b$. The general form of the global coordinate interpolation functions for the triangle is $u(x, y) \;=\; d_1^e + d_2^e\, x + d_3^e\, y$ where the g_i are element constants. Along the typical interface this reduces to $u \;=\; d_1^e + d_2^e\, x \;+ d_3^e\,(m^b x + n^b)$, or simply $u = f_1 + f_2 x$. Clearly, this shows that the boundary displacement is a linear function of x. The two constants, f_i, could be uniquely determined by noting that $u(x_1) = u_1$ and $u(x_2) = u_2$. Since those two quantities are common to both elements the displacement, $u(x)$, will be continuous between the two elements. By way of comparison when the same substitution is made in Eq. 9.15 the resulting edge value for the quadrilateral element is $u \;=\; d_1^e + d_2^e\, x + d_3^e\,(m^b x + n^b) + d_4^e\, x\,(m^b x + n^b)$, or simply $u \;=\; f_1 + f_2\, x + f_3\, x^2$. This quadratic function cannot be uniquely defined by the two constants u_1 and u_2. Therefore, it is not possible to prove that the displacements will be continuous between elements. This is an undesirable feature of quadrilateral elements when formulated in global coordinates. If the quadrilateral interpolation is given in local coordinates such as Eq. 9.9 or Eq. 9.10, this problem does not occur. On the edge $s = 0$, Eq. 9.9 reduces to

$u = f_1 + f_2 r$. A similar result occurs on the edge $s = 1$. Likewise, for the other two edges $u = f_1 + f_2 s$. Thus, in local coordinates the element degenerates to a linear function on any edge, and therefore will be uniquely defined by the two shared nodal displacements. In other words, the local coordinate four node quadrilateral will be compatible with elements of the same type and with the three-node triangle. The above observations suggest that global coordinates could be utilized for the four-node element only so long as it is a rectangle parallel to the global axes.

The extension of the unit coordinates to the three-dimensional tetrahedra illustrated in Fig. 3.2 is straightforward. In the result given below

$$\begin{aligned} H_1(r, s, t) &= 1 - r - s - t \qquad & H_2(r, s, t) &= r \\ H_3(r, s, t) &= s & H_4(r, s, t) &= t, \end{aligned} \tag{9.16}$$

and comparing this to Eqs. 9.6, we note that the 2-D and 1-D forms are contained in the three-dimensional form. This concept was suggested by the topology relations shown in Fig. 3.2. The unit coordinate interpolation is easily extended to quadratic, cubic, or higher interpolation. The procedure employed to generate Eq. 9.6 can be employed. An alternate geometric approach can be utilized. We want to generate an interpolation function, H_i, that vanishes at the j-th node when $i \neq j$. Such a function can be obtained by taking the products of the equations of selected curves through the nodes on the element. For example, let $H_1(r, s) = C_1 \Gamma_1 \Gamma_2$ where the Γ_i are the equations of the lines are shown in Fig. 9.4, and where C_1 is a constant chosen so that $H_1(r_1, s_1) = 1$. This yields

$$H_1 = (1 - 3r - 3s + 2r^2 + 4rs + 2s^2).$$

Similarly, letting $H_4 = C_4 \Gamma_1 \Gamma_3$ gives $C_4 = 4$ and $H_4 = 4r(1 - r - s)$. This type of procedure is usually quite straightforward. However, there are times when there is not a unique choice of products, and then care must be employed to select the proper products. The resulting two-dimensional interpolation functions for the quadratic triangle are

$$\begin{aligned} H_1(r, s) &= 1 - 3r + 2r^2 - 3s + 4rs + 2s^2 \\ H_2(r, s) &= -r + 2r^2 \\ H_3(r, s) &= -s + 2s^2 \\ H_4(r, s) &= 4r - 4r^2 - 4rs \\ H_5(r, s) &= 4rs \\ H_6(r, s) &= 4s - 4rs - 4s^2. \end{aligned} \tag{9.17}$$

Once again, it is possible to obtain the one-dimensional quadratic interpolation on a typical edge by setting $s = 0$. Figure 9.5 shows the shape of the typical interpolation functions for a linear and quadratic triangular element.

Figure 9.6 illustrates the concept of Pascal's triangle for representing the complete polynomial terms in three dimensions. Beginning with the constant vertex (1), it can also be thought of as showing the polynomials that occur in the tetrahedron of linear, quadratic, cubic, and quartic degree, respectively, and the relative location of the nodes on the edges, faces, and interior of the tetrahedron. If one sets $z = 1$ then it can also show the relative nodes and polynomials for the triangular elements of linear, quadratic, cubic,

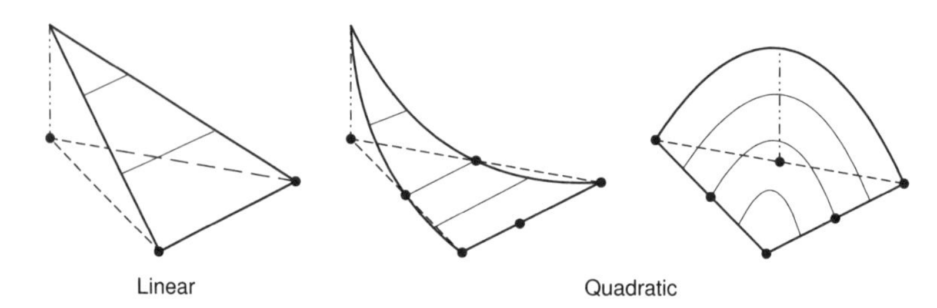

Figure 9.5 *Linear and quadratic triangle interpolation*

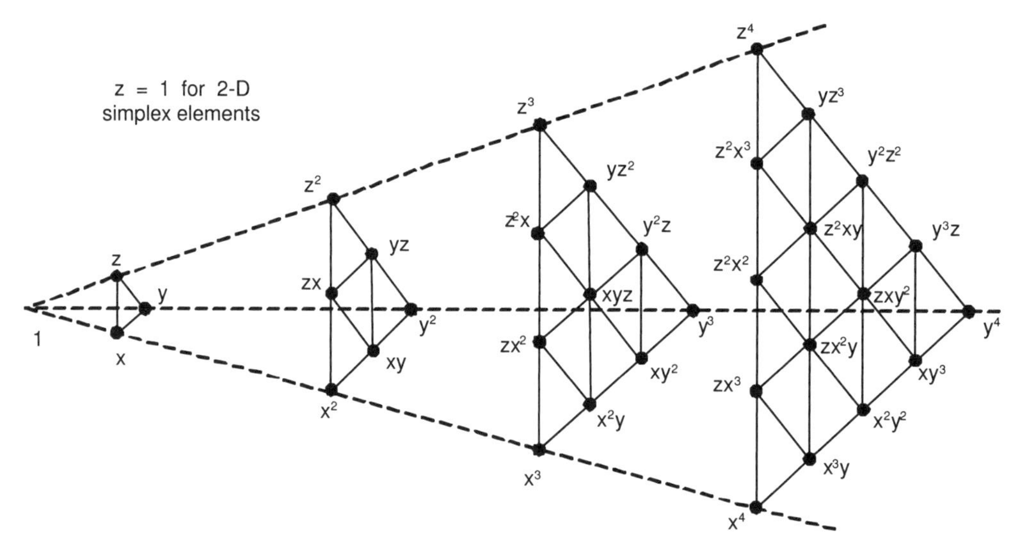

Figure 9.6 *The 2-D Pascal triangles and the 3-D simplex family*

and quartic degree from the left-most to the right-most triangles, respectively.

9.3 Natural coordinates

The natural coordinate formulations for the interpolation functions can be generated in a similar manner to that illustrated in Eq. 9.10. However, the inverse geometric matrix, $\mathbf{G}^{-1}$, may not exist. However, the most common functions have been known for several years and will be presented here in two groups. They are generally denoted as Lagrangian elements and as the Serendipity elements (see Tables 9.1 and 9.2). For the four-node quadrilateral element both forms yield Eq. 9.10. This is known as the bilinear quadrilateral since it has linear interpolation on its edges and a bilinear (incomplete quadratic) interpolation on its interior. This element is easily extended to the trilinear hexahedra of Table 9.2. Its resulting interpolation functions are

$$H_i\,(a, b, c) \;=\; (1 + aa_i)\,(1 + bb_i)\,(1 + cc_i)/8, \tag{9.18}$$

for $1 \le i \le 8$ where $(a_i,\ b_i,\ c_i)$ are the local coordinates of node i. On a given face, e.g.,

$c = \pm 1$, these degenerate to the four functions in Eq. 9.10 and four zero terms. For quadratic (or higher) edge interpolation, the Lagrangian and Serendipity elements are different. The Serendipity interpolation functions for the corner quadratic nodes are

$$H_i(a, b) = (1 + aa_i)(1 + bb_i)(aa_i + bb_i - 1)/4, \tag{9.19}$$

where $1 \le i \le 4$ and for the mid-side nodes

$$H_i(a, b) = a_i^2(1 - b^2)(1 + a_i a)/2 + b_i^2(1 - a^2)(1 + b_i b)/2, \ 5 \le i \le 8. \tag{9.20}$$

Other members of this family are listed in Tables 9.1 and 9.2. The two-dimensional Lagrangian functions are obtained from the products of the one-dimensional equations. The resulting quadratic functions are

$$H_1(a, b) = (a^2 - a)(b^2 - b)/4 \qquad H_6(a, b) = (a^2 + a)(1 - b^2)/2$$
$$H_2(a, b) = (a^2 + a)(b^2 - b)/4 \qquad H_7(a, b) = (1 - a^2)(b^2 + b)/2$$
$$H_3(a, b) = (a^2 + a)(b^2 + b)/4 \qquad H_8(a, b) = (a^2 - a)(1 - b^2)/2$$
$$H_4(a, b) = (a^2 - a)(b^2 + b)/4 \qquad H_9(a, b) = (1 - a^2)(1 - b^2)$$
$$H_5(a, b) = (1 - a^2)(b^2 - b)/2.$$

The typical shapes of these functions are shown in Fig. 9.7. The function $H_9(a, b)$ is referred to as a *bubble function* because it is zero on the boundary of the element and looks like a soap bubble blown up over the element. Similar functions are commonly used in hierarchical elements to be considered later. It is possible to mix the order of interpolation on the edges of an element. Figure 9.8 illustrates the Serendipity interpolation functions for quadrilateral elements that can be either linear, quadratic, or cubic on any of its four sides. Such an element is often referred to as a *transition element*. They can also be employed as *p-adaptive elements*. Those types of elements are sketched in Fig. 9.9. From the previous figures one will note that the supplied routines in the interpolation library generally start with the names SHAPE_ and DERIV_ and have the number of nodes and shape codes (L-line, T-triangle, Q-quadrilateral, H-hexahedron, P-pyramid or tetrahedron, and W-wedge) appended to those names. The class of elements shown in Fig. 9.9 is appended with the name L_Q_H because they can be determined for any of the three shapes. For elements of degree four or higher one needs to also include interior nodes for elements in Fig. 9.9 to form complete polynomials, or the rate of convergence will be decreased.

9.4 Isoparametric and subparametric elements

By introducing local coordinates to formulate the element interpolation functions we were able to satisfy certain continuity requirements that could not be satisfied by global coordinate interpolation. We will soon see that a useful by-product of this approach is the ability to treat elements with curved edges. At this point there may be some concern about how one relates the local coordinates to the global coordinates. This must be done since the governing integral is presented in global (physical) coordinates and it involves derivatives with respect to the global coordinates. This can be accomplished with the popular *isoparametric elements*, and *subparametric* elements.

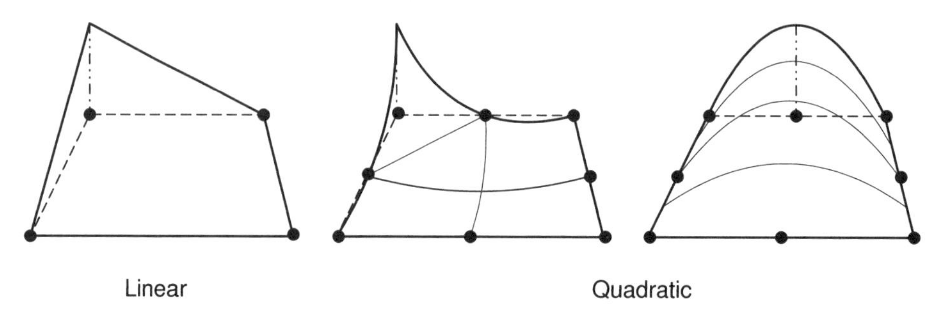

Figure 9.7 *Quadratic Serendipity quadrilateral interpolation*

```
Topology:        4 - 11 - 7 - 3
                 |            |
                 8      S     10
                 |     *R     |
                 12           6
                 |            |
                 1 -  5 - 9 - 2
```

If Cubic Side : $i = 5, 9$, or $6, 10$ or $7, 11$ or $8, 12$

$$H_i(r, s) = (1 - s^2)(1 + 9ss_i)(1 + rr_i)\,9/32$$

$$H_i(r, s) = (1 - r^2)(1 + 9rr_i)(1 + ss_i)\,9/32$$

If Quadratic Side : $i = 5,\ 6,\ 7$, or 8

$$H_i(r, s) = (1 + rr_i)(1 - s^2)/2$$

$$H_i(r, s) = (1 + ss_i)(1 - r^2)/2$$

$$H_j = 0, \quad j = i + 4$$

If Linear Side :

$$H_j = H_k = 0, \quad j = i + 4, \quad k = i + 8, \quad i = 1, 2, 3, \text{ or } 4$$

If Corners : $i = 1, 2, 3, 4 \qquad H_i(r, s) = (P_r + P_s)(1 + ss_i)/4$

See subroutine SHAPE_4_12_Q

Order of Side	$P_r,\ s_i = \pm 1$	$P_s,\ r_i = \pm 1$
Linear	1/2	1/2
Quadratic	$rr_i - 1/2$	$ss_i - 1/2$
Cubic	$(9r^2 - 5)/8$	$(9s^2 - 5)/8$

Figure 9.8 *Linear to cubic transition quadrilateral*

Table 9.1 *Serendipity quadrilaterals in natural coordinates*

Node Location a_i	b_i	Interpolation Functions $H_i(a, b)$	Name
± 1	± 1	$(1 + aa_i)(1 + bb_i)/4$	Q4
± 1	± 1	$(1 + aa_i)(1 + bb_i)(aa_i + bb_i - 1)/4$	Q8
± 1	0	$(1 + aa_i)(1 - b^2)/2$	
0	± 1	$(1 + bb_i)(1 - a^2)/2$	
± 1	± 1	$(1 + aa_i)(1 + bb_i)[9(a^2 + b^2) - 10]/32$	Q12
± 1	± 1/3	$9(1 + aa_i)(1 - b^2)(1 + 9bb_i)/32$	
± 1/3	± 1	$9(1 + bb_i)(1 - a^2)(1 + 9aa_i)/32$	
± 1	± 1	$(1 + aa_i)(1 + bb_i)[4(a^2 - 1)aa_i + 4(b^2 - 1)bb_i + 3aba_i b_i]/12$	Q16
± 1	0	$2(1 + aa_i)(b^2 - 1)(b^2 - aa_i)/4$	
0	± 1	$2(1 + bb_i)(a^2 - 1)(a^2 - bb_i)/4$	
± 1	± 1/2	$4(1 + aa_i)(1 - b^2)(b^2 + bb_i)/3$	
± 1/2	± 1	$4(1 + bb_i)(1 - a^2)(a^2 + aa_i)/3$	
0	0	$(a^2 - 1)(b^2 - 1)$	

Table 9.2 *Serendipity hexahedra in natural coordinates*

Node Location a_i	b_i	c_i	Interpolation Functions $H_i(a, b, c)$	Name
± 1	± 1	± 1	$(1 + aa_i)(1 + bb_i)(1 + cc_i)/8$	H8
± 1	± 1	± 1	$(1 + aa_i)(1 + bb_i)(1 + cc_i)(aa_i + bb_i + cc_i - 2)/8$	H20
0	± 1	± 1	$(1 - a^2)(1 + bb_i)(1 + cc_i)/4$	
± 1	0	± 1	$(1 - b^2)(1 + aa_i)(1 + cc_i)/4$	
± 1	± 1	0	$(1 - c^2)(1 + aa_i)(1 + bb_i)/4$	
± 1	± 1	± 1	$(1 + aa_i)(1 + bb_i)(1 + cc_i)[9(a^2 + b^2 + c^2) - 19]/64$	H32
± 1/3	± 1	± 1	$9(1 - a^2)(1 + 9aa_i)(1 + bb_i)(1 + cc_i)/64$	
± 1	± 1/3	± 1	$9(1 - b^2)(1 + 9bb_i)(1 + aa_i)(1 + cc_i)/64$	
± 1	± 1	± 1/3	$9(1 - c^2)(1 + 9cc_i)(1 + bb_i)(1 + aa_i)/64$	

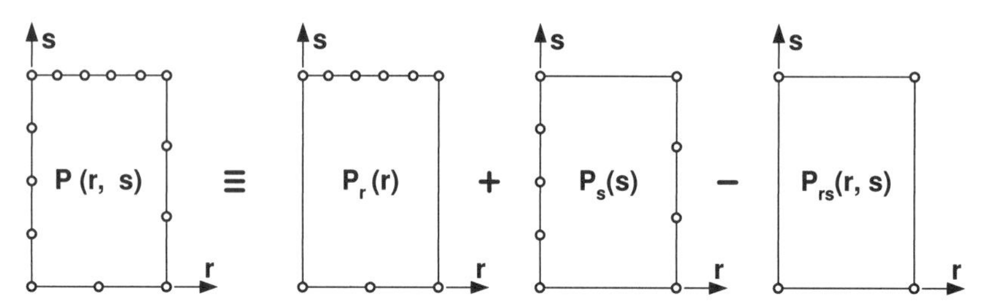

Figure 9.9 *Blended quadrilaterals of different edge degrees*

Isoparametric elements utilize a local coordinate system to formulate the element matrices. The local coordinates, say r, s, and t, are usually dimensionless and range from 0 to 1, or from -1 to 1. The latter range is usually preferred since it is directly compatible with the usual definition of abscissa utilized in numerical integration by Gaussian quadratures. The elements are called isoparametric since the same (iso) local coordinate parametric equations (interpolation functions) used to define any quantity of interest within the elements are also utilized to define the global coordinates of any point within the element in terms of the global spatial coordinates of the nodal points. If a lower order polynomial is used to describe the geometry then it is called a *subparametric element.* These are quite common when used with the newer hierarchical elements. Let the global spatial coordinates again be denoted by x, y, and z. Let the number of nodes per element be n_n. For simplicity, consider a single scalar quantity of interest, say $V(r, s, t)$. The value of this variable at any local point (r, s, t) within the element is assumed to be defined by the values at the n_n nodal points of the element (V_i^e, $1 \le i \le n_n$), and a set of interpolation functions ($H_i(r, s, t)$, $1 \le i \le n_n$). That is,

$$V(r, s, t) = \sum_{i=1}^{n_n} H_i(r, s, t)\, V_i^e = \mathbf{H}(\mathbf{r})\, \mathbf{V}^e, \tag{9.21}$$

where $\mathbf{H}$ is a row vector. Generalizing this concept, the global coordinates are defined with a geometric interpolation, or blending, function, $\mathbf{G}$. If it interpolates between n_x geometric data points then it is subparametric if $n_x < n_n$, isoparametric if $n_x = n_n$ so $\mathbf{G} = \mathbf{H}$, and superparametric if $n_x > n_n$. Blending functions typically use geometric data everywhere on the edge of the geometric element. The geometric interpolation, or blending, is denoted as: $x(r, s, t) = \mathbf{G}\,\mathbf{x}^e$, $y = \mathbf{G}\,\mathbf{y}^e$, and $z = \mathbf{G}\,\mathbf{z}^e$. Programming considerations make it desirable to write the last three relations as a position row matrix, $\mathbf{R}$, written in a partitioned form

$$\mathbf{R}(r, s, t) = \mathbf{G}(r, s, t)\, \mathbf{R}^e = \mathbf{G}\,[\,\mathbf{x}^e\ \mathbf{y}^e\ \mathbf{z}^e\,] \tag{9.22}$$

where the last matrix simply contains the spatial coordinates of the n_n nodal points incident with the element. If $\mathbf{G} = \mathbf{H}$, it is an isoparametric element. To illustrate a typical two-dimensional isoparametric element, consider a quadrilateral element with nodes at the four corners, as shown in Fig. 9.3. The global coordinates and local coordinates of a typical corner, i, are (x_i, y_i), and (r_i, s_i), respectively. The following local coordinate interpolation functions have been developed earlier for this element:

$$H_i(r, s) = \frac{1}{4}(1 + rr_i)(1 + ss_i), \qquad 1 \le i \le 4.$$

We interpolate any variable, V, as

$$V(r, s) = \mathbf{H}(r, s)\,\mathbf{V}^e = \lfloor H_1 \quad H_2 \quad H_3 \quad H_4 \rfloor \begin{Bmatrix} V_1 \\ V_2 \\ V_3 \\ V_4 \end{Bmatrix}^e .$$

Note that along an edge of the element ($r = \pm 1$ or $s = \pm 1$), these interpolation functions become linear and thus any of these three quantities can be uniquely defined by the two corresponding nodal values on that edge. If the adjacent element is of the same type (linear on the boundary), then these quantities will be continuous between elements since their values are uniquely defined by the shared nodal values on that edge. Since the variable of interest, V, varies linearly on the edge of the element, it is called the linear isoparametric quadrilateral although the interpolation functions are bilinear inside the element. If the (x, y) coordinates are also varying linearly with r or s on a side it means this element has straight sides.

For future reference, note that if one can define the interpolation functions in terms of the local coordinates then one can also define their partial derivatives with respect to the local coordinate system. For example, the local derivatives of the interpolation functions of the above element are

$$\partial H_i(r, s)/\partial r = r_i(1 + ss_i)/4, \qquad \partial H_i(r, s)/\partial s = s_i(1 + rr_i)/4.$$

In three dimensions ($n_s = 3$), let the array containing the local derivatives of the interpolation functions be denoted by **DL_H**, a $3 \times n_n$ matrix, where

$$\mathbf{DL\_H}(r, s, t) = \begin{bmatrix} \dfrac{\partial}{\partial r}\mathbf{H} \\[2ex] \dfrac{\partial}{\partial s}\mathbf{H} \\[2ex] \dfrac{\partial}{\partial t}\mathbf{H} \end{bmatrix} = \boldsymbol{\partial}_L \mathbf{H}. \tag{9.23}$$

Although x, y, and z can be defined in an isoparametric element in terms of the local coordinates, r, s, and t, a unique inverse transformation is not needed. Thus, one usually does not define r, s, and t in terms of x, y, and z. What one must have, however, are the relations between derivatives in the two coordinate systems. From calculus, it is known that the derivatives are related by the *Jacobian*. From the chain rule of calculus one can write, in general,

$$\frac{\partial}{\partial r} = \frac{\partial}{\partial x}\frac{\partial x}{\partial r} + \frac{\partial}{\partial y}\frac{\partial y}{\partial r} + \frac{\partial}{\partial z}\frac{\partial z}{\partial r}$$

with similar expressions for $\partial/\partial s$ and $\partial/\partial t$. In matrix form these identities become

$$\begin{Bmatrix} \frac{\partial}{\partial r} \\ \frac{\partial}{\partial s} \\ \frac{\partial}{\partial t} \end{Bmatrix} = \begin{bmatrix} \frac{\partial x}{\partial r} & \frac{\partial y}{\partial r} & \frac{\partial z}{\partial r} \\ \frac{\partial x}{\partial s} & \frac{\partial y}{\partial s} & \frac{\partial z}{\partial s} \\ \frac{\partial x}{\partial t} & \frac{\partial y}{\partial t} & \frac{\partial z}{\partial t} \end{bmatrix} \begin{Bmatrix} \frac{\partial}{\partial x} \\ \frac{\partial}{\partial y} \\ \frac{\partial}{\partial z} \end{Bmatrix} \tag{9.24}$$

where the square matrix is called the *Jacobian*. Symbolically, one can write the derivatives of a quantity, such as $V(r, s, t)$, which for convenience is written as $V(x, y, z)$ in the global coordinate system, in the following manner: $\boldsymbol{\partial}_L V = \mathbf{J}(r, s, t)\, \boldsymbol{\partial}_g V$, where $\mathbf{J}$ is the Jacobian matrix, and where the subscripts L and g have been introduced to denote local and global derivatives, respectively. Similarly, the inverse relation is

$$\boldsymbol{\partial}_g V = \mathbf{J}^{-1}\, \boldsymbol{\partial}_L V. \tag{9.25}$$

Thus, to evaluate global and local derivatives, one must be able to establish the Jacobian, $\mathbf{J}$, of the geometric mapping and its inverse, $\mathbf{J}^{-1}$. In practical application, these two quantities usually are evaluated numerically. Consider the first term in $\mathbf{J}$ that relates the geometric mapping: $\partial x / \partial r = \partial\,(\mathbf{G}\,\mathbf{x}^e)/\partial r = \partial \mathbf{G}/\partial r\ \mathbf{x}^e$. Similarly, for any component in Eq. 9.22 $\partial \mathbf{R}/\partial r = \partial(\mathbf{G}\,\mathbf{R}^e)/\partial r$. Repeating for all local directions, and noting that the $\mathbf{R}^e$ values are constant input coordinate data for the element, we find the identity that

$$\begin{bmatrix} \frac{\partial x}{\partial r} & \frac{\partial y}{\partial r} & \frac{\partial z}{\partial r} \\ \frac{\partial x}{\partial s} & \frac{\partial y}{\partial s} & \frac{\partial z}{\partial s} \\ \frac{\partial x}{\partial t} & \frac{\partial y}{\partial t} & \frac{\partial z}{\partial t} \end{bmatrix} = \begin{bmatrix} \frac{\partial}{\partial r}\mathbf{G} \\ \frac{\partial}{\partial s}\mathbf{G} \\ \frac{\partial}{\partial t}\mathbf{G} \end{bmatrix} \mathbf{R}^e$$

or, in symbolic form, the evaluation of the definition of the Jacobian within a specific element takes the form

$$\mathbf{J}^e(r, s, t) = \mathbf{DL\_G}(r, s, t)\, \mathbf{R}^e. \tag{9.26}$$

This numerically defines the Jacobian matrix, $\mathbf{J}$, at a local point inside a typical element in terms of the spatial coordinates of the element's nodes, $\mathbf{R}^e$, which is referenced by the name *COORD* in the subroutines, and the local derivatives, **DL_G**, of the geometric interpolation functions, $\mathbf{G}$. Thus, at any point (r, s, t) of interest, such as a numerical integration point, it is possible to define the values of $\mathbf{J}$, $\mathbf{J}^{-1}$, and the determinant of the Jacobian, $|\mathbf{J}|$. In practice, evaluation of the Jacobian is simply a matrix product, such as $AJ = MATMUL(DL\_G, COORD)$. We usually will consider two-dimensional problems. Then the Jacobian matrix is

$$\mathbf{J} = \begin{bmatrix} \frac{\partial x}{\partial r} & \frac{\partial y}{\partial r} \\ \frac{\partial x}{\partial s} & \frac{\partial y}{\partial s} \end{bmatrix}.$$

In general, the inverse Jacobian in two dimensions is

$$\mathbf{J}^{-1} = \frac{1}{|J|} \begin{bmatrix} \frac{\partial y}{\partial s} & -\frac{\partial y}{\partial r} \\ -\frac{\partial x}{\partial s} & \frac{\partial x}{\partial r} \end{bmatrix}, \quad \text{where} \quad |J| = x_{,r}\, y_{,s} - y_{,r}\, x_{,s}.$$

For future reference, note that by denoting $(\ )_{,r} = \partial(\ )/\partial r$, etc. the determinant and inverse of the three-dimensional Jacobian are

$$|\mathbf{J}| = x_{,r}(y_{,s}\, z_{,t} - y_{,t}\, z_{,s}) + x_{,s}(y_{,t}\, z_{,r} - y_{,r}\, z_{,t}) + x_{,t}(y_{,r}\, z_{,s} - y_{,s}\, z_{,r})$$

and

$$\mathbf{J}^{-1} = \begin{bmatrix} (y_{,s}\, z_{,t} - y_{,t}\, z_{,s}) & (y_{,t}\, z_{,r} - y_{,r}\, z_{,t}) & (y_{,r}\, z_{,s} - y_{,s}\, z_{,r}) \\ (x_{,t}\, z_{,s} - x_{,s}\, z_{,t}) & (x_{,r}\, z_{,t} - x_{,t}\, z_{,r}) & (x_{,s}\, z_{,r} - x_{,r}\, z_{,s}) \\ (x_{,s}\, y_{,t} - x_{,t}\, y_{,s}) & (x_{,t}\, y_{,r} - x_{,r}\, y_{,t}) & (x_{,r}\, y_{,s} - x_{,s}\, y_{,r}) \end{bmatrix} / |\mathbf{J}|.$$

Of course, one can in theory also establish the algebraic form of **J**. For simplicity consider the three-node isoparametric triangle in two dimensions. From Eq. 9.6 we note that the local derivatives of **G** are

$$\mathbf{DL\_G} = \begin{bmatrix} \partial \mathbf{G}/\partial r \\ \partial \mathbf{G}/\partial s \end{bmatrix} = \begin{bmatrix} -1 & 1 & 0 \\ -1 & 0 & 1 \end{bmatrix}. \tag{9.27}$$

Thus, the element has constant local derivatives since no functions of the local coordinates remain. Usually the local derivatives are also polynomial functions of the local coordinates. Employing Eq. 9.26 for a specific T3 element:

$$\mathbf{J}^e = \mathbf{DL\_G\,R}^e = \begin{bmatrix} -1 & 1 & 0 \\ -1 & 0 & 1 \end{bmatrix} \begin{bmatrix} x_1 & y_1 \\ x_2 & y_2 \\ x_3 & y_3 \end{bmatrix}^e$$

or simply

$$\mathbf{J}^e = \begin{bmatrix} (x_2 - x_1) & (y_2 - y_1) \\ (x_3 - x_1) & (y_3 - y_1) \end{bmatrix}^e \tag{9.28}$$

which is also constant. The determinant of this 2 × 2 Jacobian matrix is

$$|\mathbf{J}^e| = (x_2 - x_1)^e (y_3 - y_1)^e - (x_3 - x_1)^e (y_2 - y_1)^e = 2A^e,$$

which is twice the physical area of the element physical domain, Ω^e. For the above three-node triangle, the inverse relation is simply

$$\mathbf{J}^{e^{-1}} = \frac{1}{2A^e} \begin{bmatrix} (y_3 - y_1) & -(y_2 - y_1) \\ -(x_3 - x_1) & (x_2 - x_1) \end{bmatrix}^e = \frac{1}{2A^e} \begin{bmatrix} b_2 & b_3 \\ c_2 & c_3 \end{bmatrix}^e. \tag{9.29}$$

For most other elements it is common to form these quantities numerically by utilizing the numerical values of $\mathbf{R}^e$ given in the data. The use of the local coordinates in effect represents a change of variables. In this sense the Jacobian has another important function. The determinant of the Jacobian, $|\mathbf{J}|$, relates differential changes in the two coordinate systems, that is,

$$dL = dx = |\mathbf{J}|\, dr$$

$$da = dx\, dy = |\mathbf{J}|\, dr\, ds$$

$$dv = dx\, dy\, dz = |\mathbf{J}|\, dr\, ds\, dt$$

in one-, two-, and three-dimensional problems. When the local and physical spaces have the same number of dimensions we can write this symbolically as $d\,\Omega^e = |\mathbf{J}|\, d\,\square^e$.

The integral definitions of the element matrices usually involve the global derivatives of the quantity of interest. From Eq. 9.21 it is seen that the local derivatives of V are related to the nodal parameters by

$$\begin{Bmatrix} \dfrac{\partial V}{\partial r} \\ \dfrac{\partial V}{\partial s} \\ \dfrac{\partial V}{\partial t} \end{Bmatrix} = \begin{bmatrix} \dfrac{\partial}{\partial r}\mathbf{H} \\ \dfrac{\partial}{\partial s}\mathbf{H} \\ \dfrac{\partial}{\partial t}\mathbf{H} \end{bmatrix} \mathbf{V}^e,$$

or symbolically,

$$\boldsymbol{\partial}_L V(r, s, t) = \mathbf{DL\_H}(r, s, t)\, \mathbf{V}^e. \tag{9.30}$$

To relate the global derivatives of V to the nodal parameters, $\mathbf{V}^e$, one substitutes the above expression, and the geometry mapping Jacobian into Eq. 9.25 to obtain

$$\boldsymbol{\partial}_g V = \mathbf{J}^{-1}\, \mathbf{DL\_H}\, \mathbf{V}^e \equiv \mathbf{d}(r, s, t)\, \mathbf{V}^e,$$

where

$$\mathbf{d}(r, s, t) = \mathbf{J}(r, s, t)^{-1}\mathbf{DL\_H}(r, s, t). \tag{9.31}$$

The matrix $\mathbf{d}$ is very important since it relates the global derivatives of the quantity of interest to the quantity's nodal values. Note that it depends on both the Jacobian of the geometric mapping and the local derivatives of the solution interpolation functions. For the sake of completeness, note that $\mathbf{d}$ can be partitioned as

$$\mathbf{d}(r, s, t) = \begin{bmatrix} \mathbf{d}_x \\ \hline \mathbf{d}_y \\ \hline \mathbf{d}_z \end{bmatrix} = \begin{bmatrix} \dfrac{\partial}{\partial x}\mathbf{H} \\ \hline \dfrac{\partial}{\partial y}\mathbf{H} \\ \hline \dfrac{\partial}{\partial z}\mathbf{H} \end{bmatrix} = \partial_g \mathbf{H} \tag{9.32}$$

so that each row represents a derivative of the solution interpolation functions with respect to a global coordinate direction. Sometimes it is desirable to compute and store the rows of $\mathbf{d}$ independently. In practice the $\mathbf{d}$ matrix usually exists only in numerical form at selected points. Once again, it is simply a matrix product such as GLOBAL = MATMUL (AJ_INV, DL_H), where GLOBAL represents the physical derivatives of the parametric functions **H**. For the linear triangle **J**, **DL_G**, and **d** are all constant. Substituting the results from Eqs. 9.27 and 9.29 into 9.31 yields

$$\mathbf{d}^e = \frac{1}{2A^e}\begin{bmatrix} (y_2 - y_3) & (y_3 - y_1) & (y_1 - y_2) \\ (x_3 - x_2) & (x_1 - x_3) & (x_2 - x_1) \end{bmatrix}^e = \frac{1}{2A^e}\begin{bmatrix} b_1 & b_2 & b_3 \\ c_1 & c_2 & c_3 \end{bmatrix}^e . \quad (9.33)$$

As expected for a linear triangle, all the terms are constant. This element is usually referred to as the Constant Strain Triangle (CST). For Poisson problems $\mathbf{B}^e = \mathbf{d}^e$.

Any finite element analysis ultimately leads to the evaluation of the integrals that define the element and/or boundary segment matrices. The element matrices, $\mathbf{S}^e$ or $\mathbf{C}^e$, are usually defined by integrals of the symbolic form

$$\mathbf{I}^e = \int\int\int_{\Omega^e} \mathbf{F}^e(x, y, z)\, dx\, dy\, dz = \int_{-1}^{1}\int_{-1}^{1}\int_{-1}^{1} \tilde{\mathbf{F}}^e(r, s, t) \mid \mathbf{J}^e(r, s, t) \mid dr\, ds\, dt, \quad (9.34)$$

where $\mathbf{F}^e$ is usually the sum of products of other matrices involving the element interpolation functions, **H**, their derivatives, **d**, and problem properties. In practice, one would usually use numerical integration to obtain

$$\mathbf{I}^e = \sum_{i=1}^{n_q} W_i\, \tilde{\mathbf{F}}^e(r_i, s_i, t_i) \mid \mathbf{J}^e(r_i, s_i, t_i) \mid \quad (9.35)$$

where $\tilde{\mathbf{F}}^e$ and $\mid \mathbf{J} \mid$ are evaluated at each of the n_q integration points, and where (r_i, s_i, t_i) and W_i denote the tabulated abscissae and weights, respectively. It should be noted that this type of coding makes repeated calls to the interpolation functions to evaluate them at the quadrature points. If the element type is constant, then the quadrature locations would not change. Thus, these computations are repetitious. Since machines have larger memories today, it would be more efficient to evaluate the interpolation functions and their local derivatives once at each quadrature point and store those data for later use. This is done by adding an additional subscript to those arrays that correspond to the quadrature point number.

9.5 Hierarchical interpolation

In Sec. 4.6 we introduced the typical hierarchical functions on line elements and let the mid-point tangential derivatives from order m to order n be denoted by $m \rightarrow n$. The exact same functions can be utilized on each edge of a two-dimensional or three-dimensional hierarchical element. We will begin by considering quadrilateral elements, or the quadrilateral faces of a solid element. To apply the previous one-dimensional element to each edge of the element requires an arbitrary choice of which way we consider to the positive tangential direction. Our choice is to use the 'right hand rule' so that the tangential derivatives are taken counterclockwise around the element. In other words, if we circle the fingers of our right hand in the direction of the tangential circuit, our thumb points in the direction of the outward normal vector perpendicular to that face.

Usually a (sub-parametric) four node element will be used to describe the geometry of the element. The element starts with the standard isoparametric form of four nodal values to begin the hierarchical approximation of the function. As needed, tangential derivatives of the unknown solution are added as additional degrees of freedom. It is well known that it is desirable to have complete polynomials included in the interpolation polynomials. Thus, at some point it becomes necessary to add internal (bubble) functions at the centroid of the element. There is more than one way to go about doing this. The main question is does one want to use the function value at the centroid as a dof or just its higher derivatives? The latter is simpler to automate if we use the Q4 element.

Since the hierarchical derivative interpolation functions are all zero at both ends of their edge they will also be zero on their two adjoining edges of the quadrilateral. Thus, to use these functions on the interior of the Q4 element we must multiply them by a function that is unity on the edge where the hierarchical functions are defined and zero on the opposite parallel edge. From the discussion of isoparametric elements it should be clear that on each of the four sides the functions (in natural coordinates a, b) are

$$\begin{aligned} N^{(1)}(b) &= (1 - b)/2, \quad N^{(3)}(b) = (1 + b)/2 \\ N^{(2)}(a) &= (1 + a)/2, \quad N^{(4)}(a) = (1 - a)/2 \end{aligned} \tag{9.36}$$

respectively, where $N^{(i)}$ denotes the interpolation normal to side i. If T_{ij} denotes the hierarchical tangential interpolations on side i and node j, then their net interior contributions are $H_{ij}(a, b) = N^{(i)} T_{ij}$. That is, the p-th degree edge interpolation enrichments of the Q4 element are

$$\begin{array}{lll} \text{Side 1} & (b = -1) & H_p^{(1)}(a, b) = \frac{1}{2}(1 - b)\,\Psi_p(a) \\ \text{Side 2} & (a = 1) & H_p^{(2)}(a, b) = \frac{1}{2}(1 + a)\,\Psi_p(b) \\ \text{Side 3} & (b = 1) & H_p^{(3)}(a, b) = \frac{1}{2}(1 + b)\,\Psi_p(-a) \\ \text{Side 4} & (a = -1) & H_p^{(4)}(a, b) = \frac{1}{2}(1 - a)\,\Psi_p(-b) \end{array} \tag{9.37}$$

where the $\Psi_p(a) = [\,P_p(a) - P_{p-2}(a)\,]\,2p - 1, \quad p \geq 2$. They are normalized such that their p-th tangential derivative is unity. Note that there are $4(p - 1)$ such enrichments. Likewise, there are $(p - 2)(p - 3)/2$ internal enrichments for $p \geq 4$. They occur at the center (0, 0) of the element. Their degrees of freedom are the cross-partial derivatives $\partial^{\,p-2} / \partial a^j\, \partial b^k$, for $j + k = p - 2$, and $1 \leq j, k \leq p - 3$. The general form of the internal (centroid) enrichments are a product of 'bubble functions' and other functions

$$H_p^{(0)}(a, b) = (1 - a^2)(1 - b^2)\,P_{p-4-j}(a)\,P_j(b), \quad j = 0, 1, \ldots, p - 4, \tag{9.38}$$

where $P_j(a)$ is the Legendre polynomial of degree j given in Eq. 3.25. The number of internal degrees of freedom, n, are

p	4	5	6	7	8	9	10
n	1	2	3	4	5	6	7
Total	1	3	6	10	15	21	28

so that we see the number of internal terms corresponds to the number of coefficients in a complete polynomial of degree $(p-3)$. The n terms for degree 4 to 10 are given in Table 9.3. It can be shown that the above combinations are equivalent to a complete polynomial of degree p, plus the two monomial terms $a^p\, b, a\, b^p$ for $p \geq 2$. This boundary and interior enrichment of the Q4 element is shown in Fig. 9.10. There p denotes the order of the edge polynomial, n is the total number of degrees of freedom (interpolation functions), and c is the number of dof needed for a complete polynomial form. For a quadrilateral we note that the total number of shape functions on any side is $n = p + 1$ for $p \geq 1$, and the number of interior nodes is $n_i = (p-2)(p-3)/2$ for $p \geq 4$, and the total for the element is $n_t = (p-2)(p-3)/2 + 4p$, or simply $n_t = (p^2 + 3p + 6)/2$ for $p \geq 4$. Note that the number of dof grows rapidly and by the time $p = 9$ is reached the element has almost 15 times as many dof as it did originally.

At this point the reader should see that there is a very large number of alternate forms of this same element. Consider the case where an error estimator has predicted the need for a different polynomial order on each edge. This is called *anisotropic hierarchical p-enrichment*. For maximum value of $p = 8$ there are a total of 32 possible interpolation combinations, including the six uniform ones shown in Fig. 9.10. It is likely that future codes will take advantage of anisotropic enrichment, although very few do so today. If one is going to use a nine node quadrilateral (Q9) to describe the geometry then the same types of enrichments can be added to it. However, the Q4 form would have better orthogonality behavior, that is, it would produce square matrices that are more diagonally dominant. For triangular and tetrahedral elements one could generate different interpolation orders on each edge, and in the interior, by utilizing the enhancement procedures for Lagrangian elements to be described later. This is probably easier to do in baracentric coordinates.

Since these elements have so much potential power they tend to be relatively large in size, and/or distorted in shape, and small in number. That trend might begin to conflict with the major appeal of finite elements: the ability to match complicated shapes. Thus, the choice of describing the geometry (and its Jacobian) by isoparametric, or sub-parametric methods might be dropped in favor of other geometric modeling methods. That is, the user may want to exactly match an ellipse or circle rather than approximate it with a parametric curve. One way to do that is to employ *blending functions* such as Coon's functions to describe the geometry. To do this we use local analytical functions to describe each physical coordinate on the edge of the element rather than 2, 3, or 4 discrete point values as we did with isoparametric elements in the previous sections. Let (a, b) denote the quadrilateral's natural coordinates, $-1 \leq (a, b) \leq 1$. Consider only the x physical coordinate of any point in the element. Let the four corner values of x be denoted by X_i. Number the sides in a CCW manner also starting from the first (LLH) corner node. Let x_j be a function of the tangential coordinate describing x on side j. Then the *Coon's blending function* for the x-component of the geometry is:

$$\begin{aligned} x(a, b) = & \left[x_1(a)(1-b) + x_2(b)(1+a) + x_3(a)(1+b) + x_4(a)(1-a) \right]/2 \\ & - \sum_{i=1}^{4} x_i (1 + a a_i)(1 + b b_i)/4 \end{aligned} \tag{9.39}$$

Table 9.3 *Quadrilateral hierarchical internal functions*
$\Psi_p(a,b) = (1 - a^2)(1 - b^2)\, P_m(a)\, P_n(b), \quad p \geq 3$

p	m	n	j	k
4	0	0	1	1
5	1	0	1	2
	0	1	2	1
6	2	0	1	3
	1	1	2	2
	0	2	3	1
7	3	0	1	4
	2	1	2	3
	1	2	3	2
	0	3	4	1
8	4	0	1	5
	3	1	2	4
	2	2	3	3
	1	3	4	2
	0	4	5	1
9	5	0	1	6
	4	1	2	5
	3	2	3	4
	2	3	4	3
	1	4	5	2
	0	5	6	1
10	6	0	1	7
	5	1	2	6
	4	2	3	5
	3	3	4	4
	2	4	5	3
	1	5	6	2
	0	6	7	1

P_i = Legendre polynomial of degree i; *dof* $= \partial^{j+k} / \partial a^j \ \partial b^k$

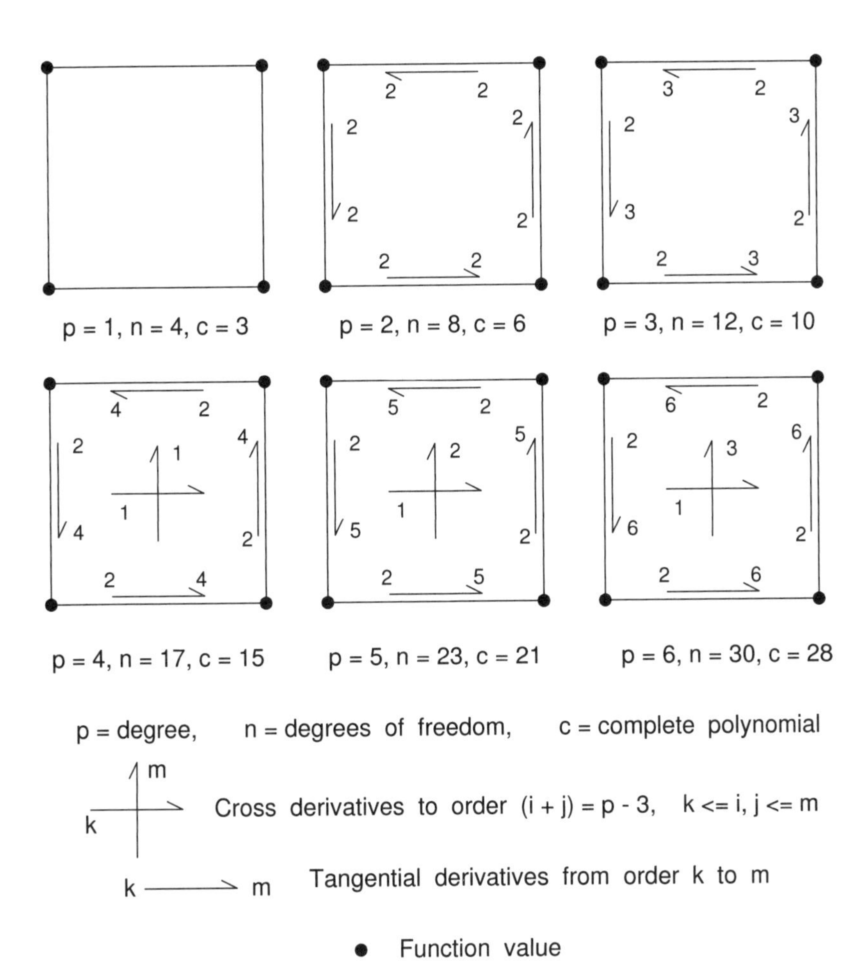

Figure 9.10 *Hierarchical enrichments of the Q4 element*

where (a_i, b_i) denote the local coordinates of the i-th corner. Since the term in brackets includes each corner twice (e.g., $x_1(1) = x_2(-1) = X_2$), the last summation simply subtracts off one full set of corner contributions.

The computational aspects of implementing the use of the tangential derivatives are not trivial. That is due to the fact that when multiple elements share an edge one must decide which one is moving in 'the' positive direction for that edge. One must establish some heuristic rule on how to handle the sign conflicts that can develop among different elements, or faces, on a common edge. The above suggested right hand rule means that edges share degrees of freedom, but view them as having opposite signs. These sign conflicts must be accounted for during the element assembly process, or by invoking a different rule when assigning equation numbers so that shared dof are always viewed as having the same sign when viewed from any face or element on that edge. One could, for example, take the tangential derivative to be acting from the end with the lowest node number toward the end with the higher node number. One must plan for these difficulties before developing a hierarchical program. However, the returns on such an investment of effort is clearly worth it many times over.

9.6 Differential geometry *

When the physical space is a higher dimension than the parametric space defining the geometry then the geometric mapping is no longer one-to-one and it is necessary to utilize the subject of *differential geometry.* This is covered in texts on vector analysis or calculus. It is also an introductory topic in most books on the mechanics of thin shell structures. Here we cover most of the basic topics except for the detailed calculation of surface curvatures. Every surface in a three-dimensional Cartesian coordinate system (x, y, z) may also be expressed by a pair of independent parametric coordinates (r, s) that lie on the surface. In our geometric parametric form, we have defined the x-coordinate as

$$x(r, s) = \mathbf{G}(r, s)\, \mathbf{x}^e. \tag{9.40}$$

The y- and z-coordinates are defined similarly. The components of the *position vector* to a point on the surface

$$\vec{R}(r, s) = x(r, s)\, \hat{\mathbf{i}} + y(r, s)\, \hat{\mathbf{j}} + z(r, s)\, \hat{\mathbf{k}}, \tag{9.41}$$

where $\hat{\mathbf{i}},\ \hat{\mathbf{j}},\ \hat{\mathbf{k}}$ are the constant unit base vectors, could be written in array form as

$$\mathbf{R}^T = [\, x \quad y \quad z \,] = \mathbf{G}(r, s)\, [\, \mathbf{x}^e\ \mathbf{y}^e\ \mathbf{z}^e \,]. \tag{9.42}$$

The local parameters (r, s) constitute a system of curvilinear coordinates for points on the physical surface. Equation 9.41 is called the *parametric equation* of a surface. If we eliminate the parameters (r, s) from Eq. 9.41, we obtain the familiar implicit form of the equation of a surface, $f(x, y, z) = 0$. Likewise, any relation between r and s, say $g(r, s) = 0$, represents a curve on the physical surface. In particular, if only one parameter varies while the other is constant, then the curve on the surface is called a *parametric curve*. Thus, the surface can be completely defined by a doubly infinite set of parametric curves, as shown in Fig. 9.11. We will often need the differential lengths, differential areas, tangent vectors, etc. We begin with differential changes in position on the surface. Since $\vec{R} = \vec{R}(r, s)$, we have

$$d\vec{R} = \frac{\partial \vec{R}}{\partial r}\, dr + \frac{\partial \vec{R}}{\partial s}\, ds \tag{9.43}$$

where $\partial \vec{R} / \partial r$ and $\partial \vec{R} / \partial s$ are the *tangent vectors* along the parametric curves. The physical distance, dl, associated with such a change in position on the surface is found from

$$(dl)^2 = dx^2 + dy^2 + dz^2 = d\vec{R} \cdot d\vec{R}. \tag{9.44}$$

This gives three contributions:

$$(dl)^2 = \left(\frac{\partial \vec{R}}{\partial r} \cdot \frac{\partial \vec{R}}{\partial r} \right) dr^2 + 2 \left(\frac{\partial \vec{R}}{\partial r} \cdot \frac{\partial \vec{R}}{\partial s} \right) dr\, ds + \left(\frac{\partial \vec{R}}{\partial s} \cdot \frac{\partial \vec{R}}{\partial s} \right) ds^2.$$

In the common notation of differential geometry this is called the *first fundamental form* of a surface, and is usually written as

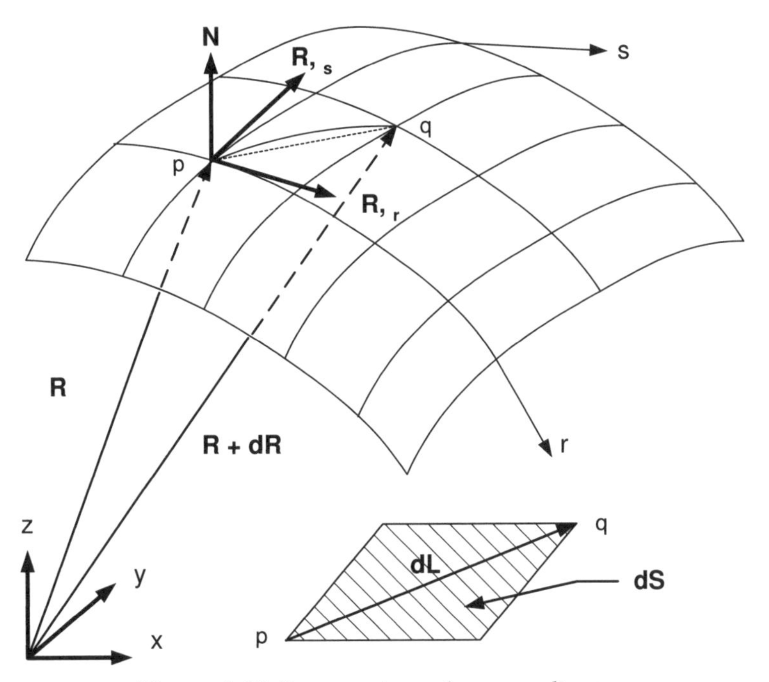

Figure 9.11 *Parametric surface coordinates*

$$(dl)^2 = E\,dr^2 + 2F\,dr\,ds + G\,ds^2 \tag{9.45}$$

where

$$E = \frac{\partial \vec{R}}{\partial r} \cdot \frac{\partial \vec{R}}{\partial r}, \qquad F = \frac{\partial \vec{R}}{\partial r} \cdot \frac{\partial \vec{R}}{\partial s}, \qquad G = \frac{\partial \vec{R}}{\partial s} \cdot \frac{\partial \vec{R}}{\partial s} \tag{9.46}$$

are called the first *fundamental magnitudes* (or metric tensor) of the surface. For future reference we will use this notation to note that the magnitudes of the surface tangent vectors are

$$\left| \frac{\partial \vec{R}}{\partial r} \right| = \sqrt{E}, \qquad \left| \frac{\partial \vec{R}}{\partial s} \right| = \sqrt{G}.$$

Of course, these magnitudes can be expressed in terms of the parametric derivatives of the surface coordinates, (x, y, z). For example, from Eq. 9.46,

$$F = \frac{\partial x}{\partial r}\frac{\partial x}{\partial s} + \frac{\partial y}{\partial r}\frac{\partial y}{\partial s} + \frac{\partial z}{\partial r}\frac{\partial z}{\partial s} \tag{9.47}$$

can be evaluated for an isoparametric surface by utilizing Eq. 9.42. Define a parametric surface gradient array given by

$$\mathbf{g} = \begin{bmatrix} \dfrac{\partial x}{\partial r} & \dfrac{\partial y}{\partial r} & \dfrac{\partial z}{\partial r} \\ \dfrac{\partial x}{\partial s} & \dfrac{\partial y}{\partial s} & \dfrac{\partial z}{\partial s} \end{bmatrix}. \tag{9.48}$$

The rows contain the components of the tangent vectors along the parametric r and s curves, respectively. In the notation of Eq. 9.26, this becomes

$$\mathbf{g}(r, s) = [\partial_l \, \mathbf{R}] = \mathbf{DL\_G\,R}^e = \Big[\ \partial_l \, \mathbf{G}(r, s)\ \Big]\Big[\ \mathbf{x}^e \ \mathbf{y}^e \ \mathbf{z}^e\ \Big]. \tag{9.49}$$

In other words, the surface gradient array at any point is the product of the parametric function derivatives evaluated at that point and the array of nodal data for the element of interest. The *metric array*, $\mathbf{m}$, is the product of the surface gradient and its transpose

$$\mathbf{m} \equiv \mathbf{g}\,\mathbf{g}^T = \begin{bmatrix} (x_{,r}^2 + y_{,r}^2 + z_{,r}^2) & (x_{,r}x_{,s} + y_{,r}y_{,s} + z_{,r}z_{,s}) \\ (x_{,r}x_{,s} + y_{,r}y_{,s} + z_{,r}z_{,s}) & (x_{,s}^2 + y_{,s}^2 + z_{,s}^2) \end{bmatrix} \tag{9.50}$$

where the subscripts denote partial derivatives with respect to the parametric coordinates. Comparing this relation with Eq. 9.46 we note that

$$\mathbf{m} = \begin{bmatrix} E & F \\ F & G \end{bmatrix} \tag{9.51}$$

contains the fundamental magnitudes of the surface. This surface metric has a determinant that is always positive. It is denoted in differential geometry as

$$|\mathbf{m}| \equiv H^2 = EG - F^2 > 0. \tag{9.52}$$

We can degenerate the differential length measure in Eq. 9.44 to the common special case where we are moving along a parametric curve, that is, $dr = 0$ or $ds = 0$. In the first case of $r = constant$, we have $(dl)^2 = G\,ds^2$ where dl is a physical differential length on the surface and ds is a differential change in the parametric surface. Then $dl = \sqrt{G}\,ds$ and likewise, for the parametric curve $s = constant$, $dl = \sqrt{E}\,dr$. The quantities $\sqrt{G}$ and $\sqrt{E}$ are known as the *Lame parameters*. They convert differential changes in the parametric coordinates to differential lengths on the surface when moving on a parametric curve. From Fig. 9.11 we note that the vector tangent to the parametric curves r and s are $\partial\vec{R}/\partial r$ and $\partial\vec{R}/\partial s$, respectively. While the isoparametric coordinates may be orthogonal, they generally will be non-orthogonal when displayed as parametric curves on the physical surface. The angle θ between the parametric curves on the surface can be found by using these tangent vectors and the definition of the dot product. Thus, $F \equiv \partial\vec{R}/\partial r \cdot \partial\vec{R}/\partial s = \sqrt{E}\sqrt{G}\,\mathrm{Cos}\,\theta$ and the angle at any point comes from

$$\mathrm{Cos}\,\theta = \frac{F}{\sqrt{E}\sqrt{G}}. \tag{9.53}$$

Therefore, we see that the parametric curves form an orthogonal curvilinear coordinate system on the physical surface only when $F = 0$. Only in that case does Eq. 9.44 reduce to the orthogonal form $(dl)^2 = E\,dr^2 + G\,ds^2$. The calculations of the most general relations between local parametric derivatives and global derivatives are shown in

Fig. 9.11. Later we will utilize the function PARM_GEOM_METRIC when computing fluxes or pressures on curved surfaces or edges.

Denote the parametric curve tangent vectors as $\vec{t}_r = \partial\vec{R}/\partial r$ and $\vec{t}_s = \partial\vec{R}/\partial s$. We have seen that the differential lengths in these two directions on the surface are $\sqrt{E}\, dr$ and $\sqrt{G}\, ds$. In a vector form, those lengths are $\vec{t}_r\, dr$ and $\vec{t}_s\, ds$, and they are separated by the angle θ. The corresponding differential surface area of the surface parallelogram is

```
FUNCTION PARM_GEOM_METRIC (DL_G, GEOMETRY) RESULT (FFM_ROOT)         ! 1
! * * * * * * * * * * * * * * * * * * * * * * * * * * * * * * *      ! 2
!  FUNDAMENTAL MAGNITUDE FROM PARAMETRIC TO GEOMETRIC SPACE           ! 3
! * * * * * * * * * * * * * * * * * * * * * * * * * * * * * * *      ! 4
 USE Elem_Type_Data    ! for LT_GEOM, LT_PARM                         ! 5
 USE System_Constants  ! for DP, N_SPACE                              ! 6
 IMPLICIT NONE                                                        ! 7
 REAL(DP), INTENT(IN) :: DL_G     (LT_PARM, LT_GEOM)                  ! 8
 REAL(DP), INTENT(IN) :: GEOMETRY (LT_GEOM, N_SPACE)                  ! 9
 REAL(DP) :: FFM, FFM_ROOT   ! first fundamental form data            !10
                                                                      !11
!             Automatic arrays                                        !12
 REAL(DP) :: METRIC (LT_PARM, LT_PARM)                                !13
 REAL(DP) :: P_GRAD (LT_PARM, N_SPACE)   ! Tangent vectors            !14
                                                                      !15
! GEOMETRY = COORDINATES OF THE ELEMENT'S GEOMETRIC NODES             !16
! DL_G     = LOCAL DERIVATIVES OF THE GEOMETRIC SHAPE FUNCTIONS       !17
! FFM      = DET(A), D_PHYSICAL = FFM * D_PARAMETRIC                  !18
! LT_GEOM  = NUMBER OF NODES DEFINING THE GEOMETRY                    !19
! LT_PARM  = DIMENSION OF PARAMETRIC SPACE FOR ELEMENT TYPE           !20
! METRIC   = 1-ST FUNDAMENTAL MAGNITUDE (METRIC MATRIX)               !21
! P_GRAD   = PARAMETRIC DERIVATIVES OF PHYSICAL SPACE                 !22
                                                                      !23
!  ESTABLISH PARAMETRIC GRADIENTS                                     !24
  P_GRAD = MATMUL (DL_G, GEOMETRY)   ! Tangent vectors                !25
                                                                      !26
!  FORM METRIC MATRIX                                                 !27
  METRIC = MATMUL (P_GRAD, TRANSPOSE (P_GRAD))                        !28
                                                                      !29
!  COMPUTE DETERMINANT OF METRIC MATRIX                               !30
  SELECT CASE (LT_PARM) ! size of parametric space                    !31
    CASE (1) ; FFM = METRIC (1, 1)                                    !32
    CASE (2) ; FFM = METRIC (1, 1) * METRIC (2, 2)          &         !33
                   - METRIC (1, 2) * METRIC (2, 1)                    !34
    CASE (3) ; FFM = METRIC(1,1)*( METRIC(2,2)*METRIC(3,3)  &         !35
                                 - METRIC(3,2)*METRIC(2,3)) &         !36
                   + METRIC(1,2)*(-METRIC(2,1)*METRIC(3,3)  &         !37
                                 + METRIC(3,1)*METRIC(2,3)) &         !38
                   + METRIC(1,3)*( METRIC(2,1)*METRIC(3,2)  &         !39
                                 - METRIC(3,1)*METRIC(2,2))           !40
    CASE DEFAULT ; STOP 'INVALID LT_PARM, P_GRAD_METRIC'              !41
  END SELECT ! LT_PARM                                                !42
  FFM_ROOT = SQRT (FFM)      ! CONVERT TO METRIC MEASURE              !43
END FUNCTION PARM_GEOM_METRIC                                         !44
```

Figure 9.12 *Computing the general metric tensor*

$$dS = (\sqrt{E}\, dr)\,(\sqrt{G}\, ds \,\text{Sin}\, \theta) = \sqrt{E}\sqrt{G}\, \text{Sin}\, \theta\, dr\, ds.$$

By substituting the relation between Cos θ and the surface metric, this simplifies to

$$dS^2 = EG\, \text{Sin}^2 \theta\, dr^2\, ds^2 = EG\,(1 - \text{Cos}^2 \theta)\, dr^2\, ds^2$$

$$dS^2 = (EG - F^2)\, dr^2\, ds^2,$$

or simply

$$dS = \sqrt{H}\, dr\, ds. \tag{9.54}$$

We also note that this calculation can be expressed as a vector cross product of the tangent vectors:

$$dS\,\vec{N} = \vec{t}_r \times \vec{t}_s\, dr\, ds$$

where $\vec{N}$ is a vector normal to the surface. We also note that the *normal vector* has a magnitude of

$$|\vec{N}| = |\vec{t}_r \times \vec{t}_s| = H.$$

Sometimes it is useful to note that the components of $\vec{N}$ are

$$\vec{N} = (y_{,r}\, z_{,s} - y_{,s}\, z_{,r})\,\hat{i} + (x_{,r}\, z_{,s} - x_{,s}\, z_{,r})\,\hat{j} + (x_{,r}\, y_{,s} - x_{,s}\, y_{,r})\,\hat{k}.$$

We often want the unit vector, $\vec{n}$, normal to the surface. It is

$$\vec{n} = \frac{\vec{N}}{H} = \frac{\vec{t}_r \times \vec{t}_s}{|\vec{t}_r \times \vec{t}_s|}.$$

9.7 Mass properties*

Mass properties and geometric properties are often needed in a design process. These computations provide a useful check on the model, and may also lead to reducing more complicated calculations by identifying geometrically equivalent elements. To illustrate the concept consider the following area, centroid, and inertia terms for a two-dimensional general curvilinear isoparametric element:

$$A = \int_A 1^2\, da, \quad A\bar{x} = \int_A x1\, da, \quad A\bar{y} = \int_A y1\, da \tag{9.55}$$

$$I_{xx} = \int_A y^2\, da, \quad -I_{xy} = \int_A xy\, da, \quad I_{yy} = \int_A x^2\, da, \quad I_{zz} = I_{xx} + Iyy.$$

From the parallel axis theorem we know that

$$\bar{I}_{xx} = I_{xx} - \bar{y}^2 A, \quad \bar{I}_{xy} = I_{xy} + \bar{x}\,\bar{y}\, A, \quad \bar{I}_{yy} = I_{yy} - \bar{x}^2 A, \quad \bar{I}_{zz} = \bar{I}_{xx} + \bar{I}_{yy}.$$

The corresponding two general inertia tensor definitions are

$$I_{ij} = \int_V (x_k\, x_k\, \delta_{ij} - x_i\, x_j)\, dV, \quad \bar{I}_{ij} = I_{ij} - (\bar{x}_k\, \bar{x}_k\, \delta_{ij} - \bar{x}_i\, \bar{x}_j)\, V \tag{9.56}$$

where x_i are the components of the position vector of a point in volume, V and δ_{ij} is the Kronecker delta. Typically, elements that have the same area, and inertia tensor, relative to the element centroid will have the same square matrix integral if the properties do not depend on physical coordinates (x, y).

We want to illustrate these calculations in a finite element context for a two-dimensional geometry. For the parametric form in local coordinates (r, s)

$$x(r, s) = \mathbf{G}(r, s)\, \mathbf{x}^e, \qquad y(r, s) = \mathbf{G}(r, s)\, \mathbf{y}^e$$

$$1 = \mathbf{G}(r, s)\, \mathbf{1} = \sum_i H_i(r, s)$$

where $\mathbf{1}$ is a vector of unity terms. Then the above measures become

$$A^e = \mathbf{1}^T \int_A^e \mathbf{G}^T\, \mathbf{G}\, d\,A\, \mathbf{1} = \mathbf{1}^T\, \mathbf{M}^e\, \mathbf{1}$$

where $\mathbf{M}^e$ is thought of as the element measure (or mass) matrix

$$A^e\, \bar{x}^{\,e} = \mathbf{1}^T\, \mathbf{M}^e\, \mathbf{x}^e, \qquad A^e\, \bar{y}^{\,e} = \mathbf{1}^T\, \mathbf{M}^e\, \mathbf{y}^e \tag{9.57}$$
$$I_{xx}^e = \mathbf{x}^{e^T}\, \mathbf{M}^e\, \mathbf{x}^e, \quad -I_{xy}^e = \mathbf{x}^{e^T}\, \mathbf{M}^e\, \mathbf{y}^e, \quad I_{yy}^e = \mathbf{y}^{e^T}\, \mathbf{M}^e\, \mathbf{y}^e.$$

The measure matrix is defined as:

$$\mathbf{M}^e = \int_A^e \mathbf{G}^T\, \mathbf{G}\, da = \int_\square \mathbf{G}^T\, \mathbf{G}\, |\mathbf{J}^e|\, d\,\square \tag{9.58}$$

where $\square$ denotes any non-dimensional parent domain (triangular or square) and $|\mathbf{J}^e|$ is the Jacobian of the transformation from $\square$ to A^e. For any straight sided triangular element it has a constant value of $|\mathbf{J}^e| = 2A^e$. Likewise, for a straight rectangular element or parallelogram element $|\mathbf{J}^e|$ is again constant. For a one-to-one geometric mapping, we always have the relation that

$$\mathbf{A}^e = \int_{A^e} d\,a = \int_\square |\mathbf{J}^e|\, d\,\square \tag{9.59}$$

so that when $\mathbf{J}^e$ is constant $A^e = |\mathbf{J}^e|\, \square_m$, and where here $\square_m$ is the measure (volume) of the non-dimensional parent domain. For example, for the unit coordinate triangle we have $\square_m = \frac{1}{2}$ so that we get $A^e = (2A^e)\,(\frac{1}{2})$, as expected. The calculation of the mass properties of each element and the total analysis domain is a data checking feature.

9.8 Interpolation error *

Here we will briefly outline some elementary error concepts in two-dimensions. From the Taylor expansion of a function, u, at a point (x, y) in two-dimensions:

$$u(x + h,\, y + k) = u(x, y) + \left[h\, \frac{\partial u}{\partial x}\,(x, y) + k\, \frac{\partial u}{\partial y}\,(x, y) \right]$$
$$+ \frac{1}{2!} \left[h^2\, \frac{\partial^2 u}{\partial x^2} + 2hk\, \frac{\partial^2 u}{\partial x\, \partial y} + k^2\, \frac{\partial^2 u}{\partial y^2} \right] + \cdots \tag{9.60}$$

The objective here is to show that if the third term is neglected, then the relations for a linear interpolation triangle are obtained. That is, we will find that the third term is proportional to the error between the true solution and the interpolated solutions. Consider a linear triangle whose maximum length in the $x-$ and $y-$directions are h and k, respectively. Let the three node numbers, given in CCW order, be i, j, and m. Employ Eq. 9.60 to estimate the nodal values u_j and u_m in terms of u_i:

$$u_j = u_i + \left[x_j \frac{\partial u}{\partial x}(x_i, y_i) + y_m \frac{\partial u}{\partial y}(x_i, y_i) \right].$$

The value of $\partial u(x_i, y_i)/\partial x$ can be found by multiplying the first relation by y_m, and subtracting the product of y_i and the second relation. The result is

$$\frac{\partial u}{\partial x}(x_i, y_i) = \frac{1}{2A}\left[u_i(y_j - y_m) + u_m(y_i - y_j) + u_j(y_m - y_i) \right]$$

where A is the area of the triangle. In a similar manner, if we compute this derivative at the other two nodes, we obtain

$$\frac{\partial u}{\partial x}(x_j, y_j) = \frac{\partial u}{\partial x}(x_m, y_m) = \frac{\partial u}{\partial x}(x_i, y_i).$$

That is, $\partial u/\partial x$ is a constant in the triangle. Likewise, $\partial u/\partial y$ is a constant. We will see later that a linear interpolation triangle has constant derivatives. Thus, these common elements will represent the first two terms in Eq. 9.60. Thus, the element error is proportional to the third term:

$$E \propto \left(h^2 \frac{\partial^2 u}{\partial x^2} + 2hk \frac{\partial^2 u}{\partial x \, \partial y} + k^2 \frac{\partial^2 u}{\partial y^2} \right) \tag{9.61}$$

where u is the exact solution, and h and k are the element size in x and y.

Once again, we would find that these second derivatives are related to the strain and stress gradients. If the strains (e.g., $\varepsilon_x = \partial u/\partial x$) are constant, then the error is small or zero. Before leaving these error comments, note that Eq. 9.61 could also be expressed in terms of the ratio (k/h). This is a measure of the relative shape of the element, and it is often called the *aspect ratio*. For an equilateral element, this ratio would be near unity. However, for a long narrow triangle, it could be quite large. Generally, it is best to keep the aspect ratio near unity (say < 5).

9.9 Element distortion*

The effects of distorting various types of elements can be serious, and most codes do not adequately validate data in this respect. As an example, consider a quadratic isoparametric line element. As shown in Fig. 9.13, let the three nodes be located in physical (x) space at points 0, ah, and h, where h is the element length, and $0 \le a \le 1$ is a location constant. The element is defined in a local unit space where $0 \le s \le 1$. The relation between x and s is easily shown to be

$$x(s) = h(4a - 1)\, s + h(2 - 4a)\, s^2$$

and the two coordinates have derivatives related by

$$\partial x/\partial s = h(4a - 1) + 4h(1 - 2a)\, s.$$

The Jacobian of the transformation, J, is the inverse relation; that is, $J = \partial s/\partial x$. The integrals required to evaluate the element matrices utilize this Jacobian. The mathematical principles require that J be positive definite. Distortion of the elements can cause J to go to zero or become negative. This possibility is easily seen in the present 1-D example. If one locates the interior ($s = 1/2$) node at the standard midpoint position,

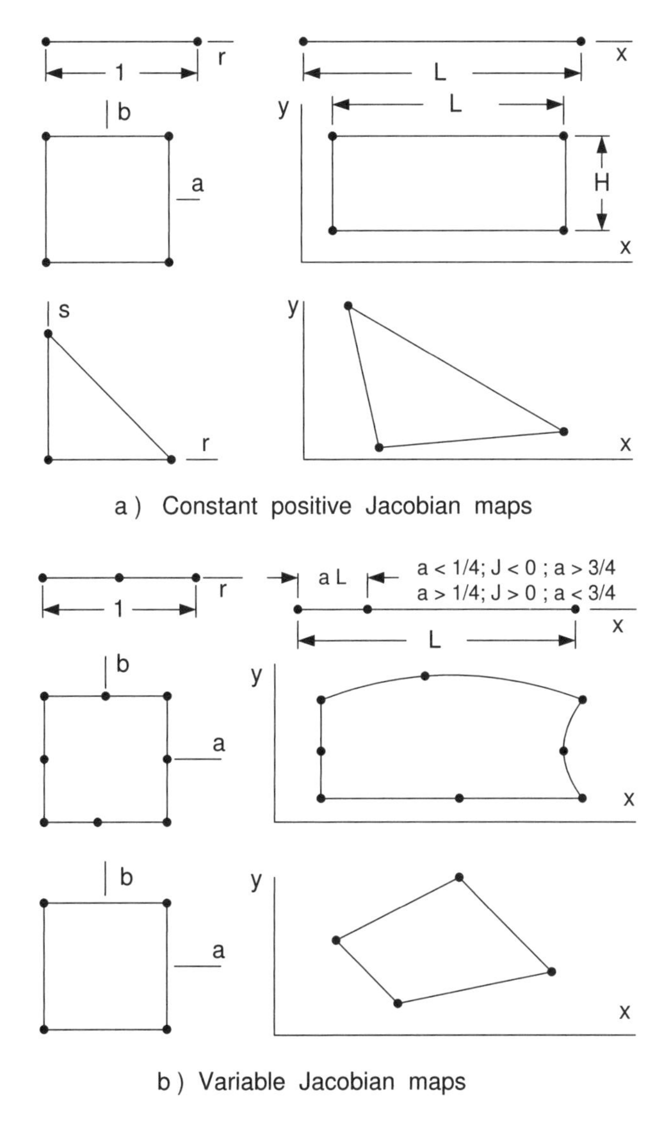

Figure 9.13 *Constant and variable Jacobian elements*

then $a = 1/2$ so that $\partial x / \partial s = h$ and J is constant throughout the element. Such an element is generally well formulated. However, if the interior node is distorted to any other position, the Jacobian will not be constant and the accuracy of the element may suffer. Generally, there will be points where $\partial x / \partial s$ goes to zero, so that the stiffness becomes singular due to division by zero. For slightly distorted elements, say $0.4 < a < 0.6$, the singular points lie outside the element domain. As the distortion increases, the singularities move to the element boundary, e.g., $a = 1/4$ or $a = 3/4$. Eventually, the distortions cause singularities of J inside the element. Such situations can cause poor stiffness matrices and very bad stress estimates, unless the true solution has the same singularity, as they do in linear fracture mechanics. In that special case these distorted elements are known as the quarter point element.

The effects of distortions of two- or three-dimensional elements are similar. For example, the edge of a quadratic element may have the non-corner node displaced in a similar way, or it may be moved normal to the line between the corners. Similar analytic singularities can be developed for such elements. However, the presence of singularities due to element distortions can easily be checked by numerical experiments. Several such analytic and numerical studies have led to useful criteria for checking the element geometry for undesirable effects. For example, envision a typical two- or three-dimensional quadratic element with a curved edge. Let L be the cord length of that edge, D the normal displacement of the mid-side node on that edge, and α the angle between the corner tangent and the cord line. Suggested ranges for linear elliptical problems are:

$$\text{warning range:} \quad 1/7 < D/L < 1/3, \qquad \alpha \le 30^\circ$$
$$\text{error range:} \quad 1/3 < D/L, \qquad \alpha \ge 53^\circ.$$

These values are obtained when only one edge is considered. If more than one edge of a single element causes a warning state, then the warnings should be considered more serious. Other parameters influence the seriousness of element distortion. Let R be a measure of the aspect ratio, that is, R is the ratio of the longest side to the shortest side. Let the minimum and maximum angles between corner cord lines be denoted by θ and γ, respectively. Define H, the lack of flatness, to be the perpendicular distance of a fourth node from the plane of the first three divided by the maximum side length. Then the following guidelines in Table 9.4 should be considered when validating geometric data for membrane or solid elements.

9.10 Space-time interpolation*

In Section 3.7 we addressed some of the aspects of space-time interpolation methods. In solving time dependent problems in three-dimensional space the main difficult is in visualization of the mesh and results. This is illustrated in Fig. 9.14 where

Table 9.4 *Geometric criteria for two- and three-dimensional elements*

Shape	*Warning state*	*Error state*
Triangle	$5 < R < 15$	$R > 15$
	$15^\circ < \theta < 30^\circ$	$\theta \le 15^\circ$
	$150^\circ < \gamma < 165^\circ$	$\gamma \ge 165^\circ$
Quadrilateral	$5 < R < 15$	$R \ge 15$
	$25^\circ < \theta < 45^\circ$	$\theta \le 25^\circ$
	$135^\circ < \gamma < 155^\circ$	$\gamma \ge 155^\circ$
	$10^{-5} < H < 10^{-2}$	$H \ge 10^{-2}$

Solids: The above limits on R, θ and γ are checked on each face.

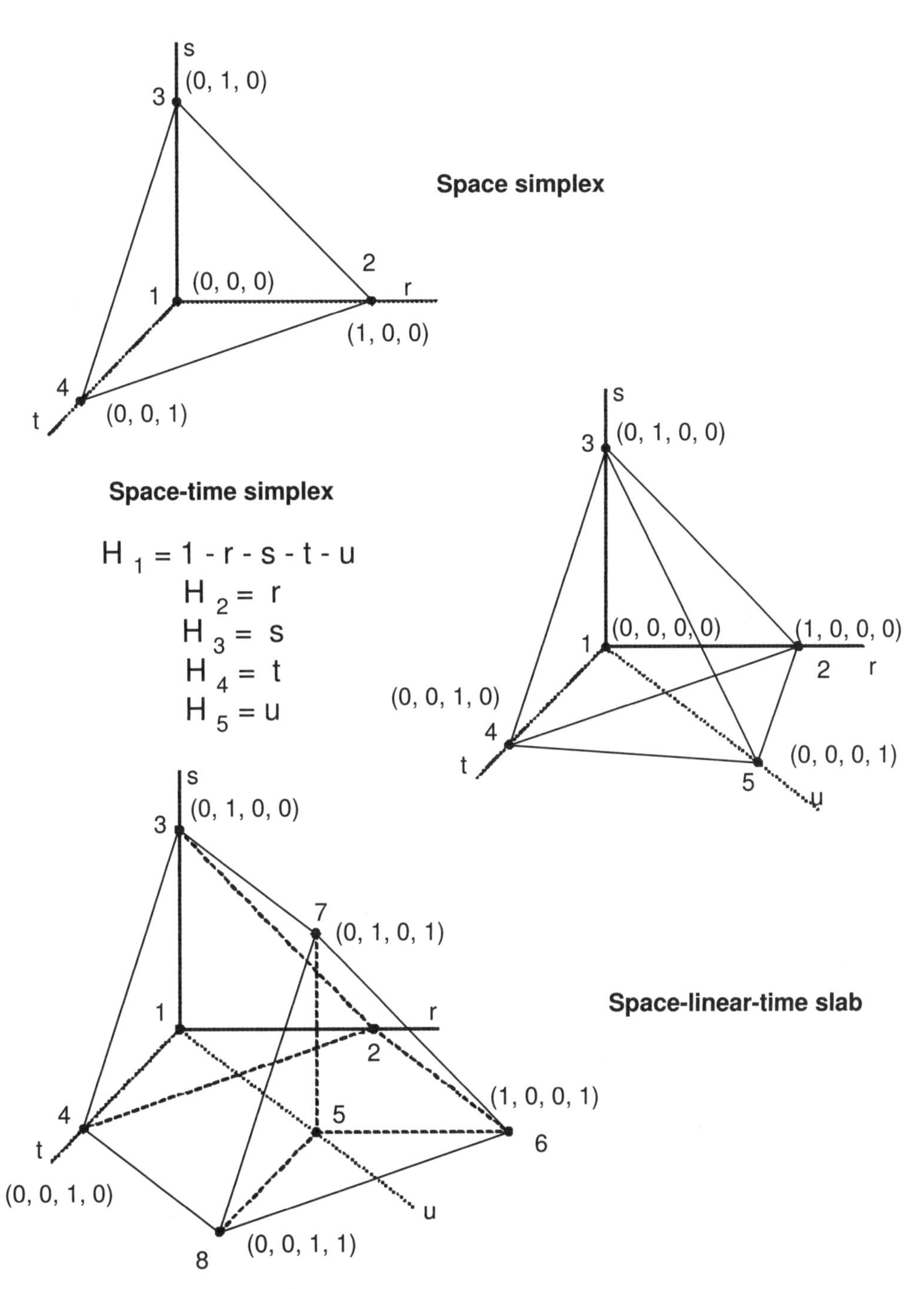

Figure 9.14 *Space-time forms for the solid simplex element*

the parametric solid tetrahedral element has been expanded into 4-D by adding a fourth parametric coordinate of u. If one wants a fully unstructured formulation in space-time then the element becomes a 5 noded simplex. However, if we want to view it as being structured so as to simply translate through a time slab we double its number of nodes from 4 in 3-D to 8 in a 4-D time slab. Of course, we would only have to generate the 3-D mesh and define the connectivity of the first four nodes.

One important advantage of space-time interpolation is that it automatically allows for elements that must significantly change shape or spatial position with time (that is, moving meshes). There are many published results where 2-D and 3-D space elements have been extended to space-time formulations. See for example the applications by Aziz and Monk [1], Behr [4], Bonnerot and Jamet [5], Dettmer and Peric [10], Gardner [12], Hansbo [13], Idesman [16], and Tezduyar [20-22] to cite a few.

One thing different about the space-time elements is in the calculation of their Jacobian matrix, which is now one dimension larger than in the pure space formulation. That is, it is a square matrix of size $(n_s + 1)$. Unlike Eq. 9.24 where we would allow the last column to compute how the physical space coordinate z varies with respect to all the element parametric coordinates, we know time will not depend on a spatial parametric coordinate. It will only depend on the non-dimensional time parametric coordinate (denoted by u in Fig. 9.14) and all but the last row of the right-most column of the Jacobian must be zero. Here, let τ denote time corresponding to the fourth parametric coordinate u. The generalization of the 3-D spatial Jacobian to 4-D space-time is

$$\begin{Bmatrix} \frac{\partial}{\partial r} \\ \frac{\partial}{\partial s} \\ \frac{\partial}{\partial t} \\ \frac{\partial}{\partial u} \end{Bmatrix} = \begin{bmatrix} \frac{\partial x}{\partial r} & \frac{\partial y}{\partial r} & \frac{\partial z}{\partial r} & 0 \\ \frac{\partial x}{\partial s} & \frac{\partial y}{\partial s} & \frac{\partial z}{\partial s} & 0 \\ \frac{\partial x}{\partial t} & \frac{\partial y}{\partial t} & \frac{\partial z}{\partial t} & 0 \\ \frac{\partial x}{\partial u} & \frac{\partial y}{\partial u} & \frac{\partial z}{\partial u} & \frac{\partial \tau}{\partial u} \end{bmatrix} \begin{Bmatrix} \frac{\partial}{\partial x} \\ \frac{\partial}{\partial y} \\ \frac{\partial}{\partial z} \\ \frac{\partial}{\partial \tau} \end{Bmatrix} \tag{9.62}$$

If the spatial nodes of the domain do not change with respect to time then the non-diagonal terms on the last row of the space-time Jacobian will also be zero. Otherwise it automatically includes a moving domain formulation. For linear interpolation in time the term $\partial \tau / \partial u$ in the space-time Jacobian will be $\Delta t/1$ for the unit coordinates of Fig. 9.14, or $\Delta t/2$ of the natural parametric coordinate from -1 to $+1$ is used for u.

9.11 Exercises

1. Use the subroutines in Fig. 3.7 to form similar functions for a C^1 rectangular element by taking a tensor product of the one-dimensional Hermite interpolation relations. This will be a 16 degree of freedom element since each node will have $u, \partial u / \partial x, \partial u / \partial y,$ and $\partial^2 u / \partial x \partial y$ as nodal unknowns. This element will not be C^1 if mapped to a quadrilateral shape. (Why not?)

2. Verify that for the H8 brick element in Table 9.2 that limiting its local coordinates to any one face, say $c = 1$, results in the interpolation functions not on that face becoming zero, and that the four non-zero interpolation functions on that face degenerate to those given for the Q4 quadrilateral in Table 9.1.

3. Create the local parametric derivatives ($\partial / \partial a$, etc.) of the interpolation functions for the: a) Q4 quadrilateral element of Table 9.1, b) the H8 hexahedra element of Table 9.2, c) the T6 triangular element of Eq. 9.17.

4. For a one-to-one geometric map the Jacobian matrix (of Eq. 9.26) is $\mathbf{J}^e = [\partial_L \mathbf{H}]\,[\mathbf{x}^e\ \mathbf{y}^e\ \mathbf{z}^e]$. For a 2-D quadrilateral (Q4) verify that in natural coordinates this simplifies to

$$\mathbf{J}^e(a,b) = \begin{bmatrix} H_{1,a} & H_{2,a} & H_{3,a} & H_{4,a} \\ H_{1,b} & H_{2,b} & H_{3,b} & H_{4,b} \end{bmatrix} \begin{bmatrix} x_1 & y_1 \\ x_2 & y_2 \\ x_3 & y_3 \\ x_4 & y_4 \end{bmatrix}^e$$

so that the Jacobian usually varies over the element with

$$[\partial_L \mathbf{H}] = \frac{1}{4} \begin{bmatrix} (b-1) & (1-b) & (1+b) & (-1-b) \\ (a-1) & (-1-a) & (1+a) & (1-a) \end{bmatrix}.$$

5. Verify that if the above Q4 element maps onto a rectangle, with its sides parallel to the global axes, of length L_x and height L_y then the Jacobian is constant at all points in the element.

6. If a Q4 element is mapped to a trapezoid having the four nodal coordinates of $\mathbf{x}^{e^T} = [0\ 2\ 2\ 0]$, and $\mathbf{y}^{e^T} = [0\ 0\ 2\ 1]$ verify that its Jacobian matrix is

$$\mathbf{J}^e(a,b) = \frac{1}{4} \begin{bmatrix} 4 & (1+b) \\ 0 & (3+a) \end{bmatrix}.$$

7. Sketch how you think an 8 noded parametric cube in 3-D parametric space would appear when extended to a time slab (with 16 nodes).

9.12 Bibliography

[1] Aziz, A.K. and Monk, P., "Continuous Finite Elements in Space and Time for the Heat Equation," *Math. Comp.*, **52**, pp. 255–274 (1989).

[2] Babuska, I., Griebel, M., and Pitkaranta, J., "The Problem of Selecting Shape Functions for a *p*-Type Finite Element," *Int. J. Num. Meth. Eng.*, **28**, pp. 1891–1908 (1989).

[3] Becker, E.B., Carey, G.F., and Oden, J.T., *Finite Elements – An Introduction*, Englewood Cliffs: Prentice Hall (1981).

[4] Behr, M., "Stablized Space-Time Finite Element Formulations for Free-Surface Flows," *Comm. for Num. Meth. in Engr.*, **11**, pp. 813–819 (2001).

[5] Bonnerot, R. and Jamet, P., "Numerical Computation of the Free Boundary for the Two-Dimensional Stefan Problem by Space-Time Finite Elements," *J. Computational Physics*, **25**, pp. 163–181 (1977).

[6] Bruch, J.C. Jr. and Zyvoloski, G., "Transient Two-Dimensional Heat Conduction Problems Solved by the Finite Element Method," *Int. J. Num. Meth. Eng.*, **8**, pp. 481–494 (1974).

[7] Ciarlet, P.G., *The Finite Element Method for Elliptical Problems*, Philadelphia, PA: SIAM (2002).

[8] Connor, J.C. and Brebbia, C.A., *Finite Element Techniques for Fluid Flow*, London: Butterworth (1976).

[9] Cook, R.D., Malkus, D.S., Plesha, N.E., and Witt, R.J., *Concepts and Applications of Finite Element Analysis*, New York: John Wiley (2002).

[10] Dettmer, W. and Peric, D., "An Analysis of the Time Integration Algorithms for the Finite Element Solutions of Incompressible Navier-Stokes Equations Based on a Stabilised Formulation," *Comp. Meth. Appl. Mech. Eng.*, **192**, pp. 1177–1226 (2003).

[11] El-Zafrany, A. and Cookson, R.A., "Derivation of Lagrangian and Hermite Shape Functions for Quadrilateral Elements," *Int. J. Num. Meth. Eng.*, **23**, pp. 1939–1958 (1986).

[12] Gardner, G.A., Gardner, L.R.T., and Cunningham, J., "Simulations of a Fox-Rabies Epidemic on an Island Using Space-Time Finite Elements," *Z. Naturforsch*, **45c**, pp. 1230–1240 (1989).

[13] Hansbo, P., "A Crank-Nicolson Type Space-Time Finite Element Method for Computing on Moving Meshes," *J. Comp. Physics*, **159**, pp. 274–289 (2000).

[14] Hu, K-K., Swartz, S.E., and Kirmser, P.G., "One Formula Generates N−th Order Shape Functions," *J. Eng. Mech.*, **110**(4), pp. 640–647 (1984).

[15] Hughes, T.J.R., *The Finite Element Method*, Englewood Cliffs: Prentice Hall (1987).

[16] Idesman, A., Niekamp, R., and Stein, E., "Finite Elements in Space and Time for Generalized Viscoelastic Maxwell Model," *Comp. Mech.*, (2001).

[17] Segerlind, L.J., *Applied Finite Element Analysis*, New York: John Wiley (1984).

[18] Silvester, P.P. and Ferrari, R.L., *Finite Elements for Electrical Engineers*, Cambridge: Cambridge University Press (1983).

[19] Szabo, B. and Babuska, I., *Finite Element Analysis*, New York: John Wiley (1991).

[20] Tezduyar, T.E. and Ganjoo, D.K., "Petrov-Galerkin Formulations with Weighting Functions Dependent Upon Spatial and Temporal Discretization," *Comp. Meth. Appl. Mech. Eng.*, **59**, pp. 47–71 (1986).

[21] Tezduyar, T.E., "Stabilized Finite Element Formulations for Incompressible Flow Computations," *Advances in Applied Mechanics*, **28**, pp. 1–44 (1991).

[22] Tezduyar, T.E., "Finite Element Methods for Flow Problems with Moving Boundaries and Interfaces," *Archives of Computational Methods in Engineering*, **8**, pp. 83–130 (2001).

[23] Zienkiewicz, O.C. and Morgan, K., *Finite Elements and Approximation*, Chichester: John Wiley (1983).

Chapter 10

Integration methods

10.1 Introduction

The finite element analysis techniques are always based on an integral formulation. At the very minimum it will always be necessary to integrate at least an element square matrix. This means that every coefficient function in the matrix must be integrated. In the following sections various methods will be considered for evaluating the typical integrals that arise. Most simple finite element matrices for two-dimensional problems are based on the use of linear triangular or quadrilateral elements. Since a quadrilateral can be divided into two or more triangles, only exact integrals over arbitrary triangles will be considered here. Integrals over triangular elements commonly involve integrands of the form

$$I = \int_A x^m y^n \, dx \, dy \tag{10.1}$$

where A is the area of a typical triangle. When $0 \le (m + n) \le 2$, the above integral can easily be expressed in closed form in terms of the spatial coordinates of the three corner points. For a right-handed coordinate system, the corners must be numbered in counter-clockwise order. In this case, the above integrals are given in Table 10.1. These integrals should be recognized as the area, and first and second moments of the area. If one had a volume of revolution that had a triangular cross-section in the $\rho - z$ plane, then one has

$$I = \int_V \rho f(\rho, z) \, d\rho \, dz \, d\phi = 2\pi \int_A \rho f(\rho, z) \, d\rho \, dz$$

so that similar expressions could be used to evaluate the volume integrals.

10.2 Unit coordinate integration

The utilization of global coordinate interpolation is becoming increasingly rare. However, as we have seen, the use of non-dimensional local coordinates is common. Thus we often see local coordinate polynomials integrated over the physical domain of an element. Sec. 4.3 presented some typical unit coordinate integrals in 1-D, written in exact closed form. These concepts can be extended to two- and three-dimensional

Table 10.1 *Exact integrals for a triangle*

m	n	$I = \int_A x^m y^n \, dx\, dy$
0	0	$\int dA = A = [\, x_1(y_2 - y_3) + x_2(y_3 - y_1) + x_3(y_1 - y_2)\,]/2$
0	1	$\int y\, dA = A\bar{y} = A(y_1 + y_2 + y_3)/3$
1	0	$\int x\, dA = A\bar{x} = A(x_1 + x_2 + x_3)/3$
0	2	$\int y^2\, dA = A(y_1^2 + y_2^2 + y_3^2 + 9\bar{y}^2)/12$
1	1	$\int xy\, dA = A(x_1\, y_1 + x_2\, y_2 + x_3\, y_3 + 9\overline{x}\,\overline{y})/12$
2	0	$\int x^2\, dA = A(x_1^2 + x_2^2 + x_3^2 + 9\bar{x}^2)/12$

elements. For example, consider an integration over a triangular element. It is known that for an element with a constant Jacobian

$$I = \int_A r^m\, s^n\, da = \frac{2A\,\Gamma(m+1)\,\Gamma(n+1)}{\Gamma(3+m+n)} \tag{10.2}$$

where Γ denote the Gamma function. Restricting consideration to positive integer values of the exponents, m and n, yields

$$I = 2\,A^e \frac{m!\,n!}{(2+m+n)\,!} = \frac{A^e}{K_{mn}}, \tag{10.3}$$

where ! denotes the factorial and K_{mn} is an integer constant given in Table 10.2 for common values of m and n. Similarly for the tetrahedral element

$$I^e = \int_{V^e} r^m\, s^n\, t^p\, dv = 6V^e \frac{m!\,n!\,p!}{(3+m+n+p)\,!}\,. \tag{10.4}$$

Thus, one notes that common integrals of this type can be evaluated by simply multiplying the element characteristic (i.e., global length, area, or volume) by known constants which could be stored in a data statement. To illustrate the application of these equations in evaluating element matrices, we consider the following example for the three node triangle in unit coordinates:

$$I = \int_{A^e} \mathbf{H}^T\, da = \int_{A^e} \begin{Bmatrix} (1 - r - s) \\ r \\ s \end{Bmatrix} da = \begin{Bmatrix} A^e - A^e/3 - A^e/3 \\ A^e/3 \\ A^e/3 \end{Bmatrix} = \frac{A^e}{3} \begin{Bmatrix} 1 \\ 1 \\ 1 \end{Bmatrix}.$$

$$\mathbf{I}_V = 2\pi \int_{A^e} \mathbf{H}^T \rho \, da = 2\pi \left(\int_{A^e} \mathbf{H}^T \mathbf{H} \, da \right) \rho^e = \frac{2\pi A^e}{12} \begin{bmatrix} 2 & 1 & 1 \\ 1 & 2 & 1 \\ 1 & 1 & 2 \end{bmatrix} \rho^e.$$

10.3 Simplex coordinate integration

A simplex region is one where the minimum number of vertices is one more than the dimension of the space. These were illustrated in Fig. 3.2. Some analysts like to define a set of *simplex coordinates* or *barycentric coordinates*. If there are N vertices

Table 10.2 *Denominator, K, for unit triangle* $I = \int_A r^m s^n \, da = A/K$

M	N: 0	1	2	3	4	5	6	7	8
0	1	3	6	10	15	21	28	36	45
1	3	12	30	60	105	168	252	360	495
2	6	30	90	210	420	756	1260	1980	2970
3	10	60	210	560	1260	2520	4620	7920	12870
4	15	105	420	1260	3150	6930	13860	25740	45045
5	21	168	756	2520	6930	16632	36036	72072	135135
6	28	252	1260	4620	13860	36036	84084	180180	360360
7	36	360	1980	7920	25740	72072	180180	411840	875160
8	45	495	2970	12870	45045	135135	360360	875160	1969110

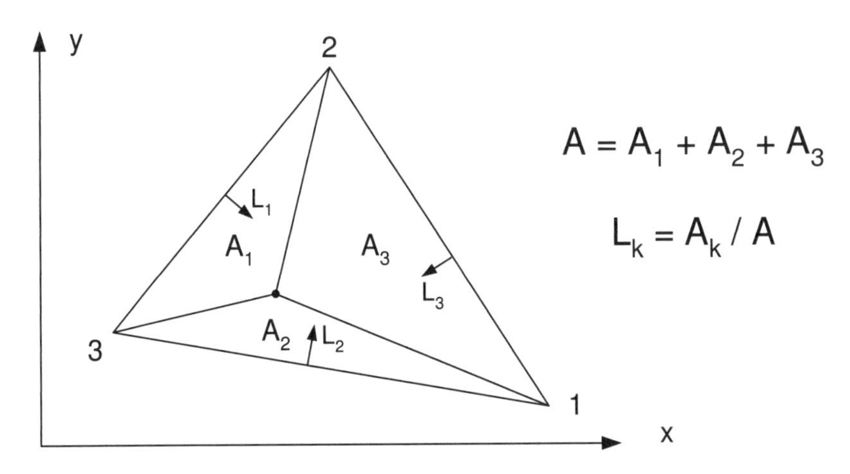

Figure 10.1 *Area coordinates*

then N non-dimensional coordinates, L_i, $1 \le i \le N$, are defined and constrained so that

$$1 = \sum_{i=1}^{N} L_i$$

at any point in space. Thus, they are not independent. However, they can be used to simplify certain recursion relations. In physical spaces these coordinates are sometimes called *line coordinates*, *area coordinates*, and *volume coordinates*. At a given point in the region we can define the simplex coordinate for node j, L_j, in a generalized manner. It is the ratio of the generalized volume from the point to all other vertices (other than j) and the total generalized volume of the simplex. This is illustrated in Fig. 10.1. If the simplex has a constant Jacobian (e.g., straight sides and flat faces), then the exact form of the integrals of the simplex coordinates are simple. They are

$$\begin{aligned} \int_L L_1^a \, L_2^b \, dL &= \frac{a!b!}{(a+b+1)!} \, (L) \\ \int_A L_1^a \, L_2^b \, L_3^c \, da &= \frac{a!b!c!}{(a+b+c+2)!} \, (2A) \\ \int_V L_1^a \, L_2^b \, L_3^c \, L_4^d \, dv &= \frac{a!b!c!d!}{(a+b+c+d+3)!} \, (6V) \, . \end{aligned} \tag{10.5}$$

The independent coordinates are those we have generally referred to as the unit coordinates of an element. Since a lot of references make use of barycentric coordinates it is useful to learn how to manipulate them correctly. The barycentric coordinates, say L_j, essentially measure the percent of total volume contained in the region from the face (lower dimensional simplex) opposite to node j to any point in the simplex. Therefore, $L_j \equiv 0$ when the point lies on the opposite face and $L_j \equiv 1$ when the point is located at node j. Clearly, the sum of all these volumes is the total volume of the simplex.

We have referred to the independent coordinates in the set as the unit coordinates. For simplex elements, the use of barycentric coordinates simplifies the algebra needed to define the interpolation functions; however, it complicates the calculation of their derivatives. Barycentric coordinates are often used to tabulate numerical integration rules for simplex domains.

For example, consider the three-dimensional case where $L_1 = r$, $L_2 = s$, $L_3 = t$, and $L_1 + L_2 + L_3 + L_4 = 1$. The interpolation functions for the linear tetrahedral (P4) are simply $G_j = L_j$. The expressions for the Lagrangian quadratic tetrahedral (P10) vertices are

$$\begin{aligned} G_1 &= L_1(2L_1 - 1) & G_2 &= L_2(2L_2 - 1) \\ G_3 &= L_3(2L_3 - 1) & G_4 &= L_4(2L_4 - 1) \end{aligned}$$

and the six mid-edge values are

$$\begin{aligned} G_5 &= 4\,L_1\,L_2 & G_6 &= 4\,L_1\,L_3 \\ G_7 &= 4\,L_1\,L_4 & G_8 &= 4\,L_2\,L_3 \\ G_9 &= 4\,L_3\,L_4 & G_{10} &= 4\,L_2\,L_4 \, . \end{aligned}$$

All the tetrahedra have the condition that

$$L_4 = 1 - L_1 - L_2 - L_3 = 1 - r - s - t$$

so that we can write the unit coordinate partial derivatives as

$$\frac{\partial L_j}{\partial r} = 1, 0, 0, -1, \qquad \frac{\partial L_j}{\partial s} = 0, 1, 0, -1, \qquad \frac{\partial L_j}{\partial t} = 0, 0, 1, -1$$

for $j = 1, 2, 3, 4$, respectively. The Jacobian calculation requires the derivatives of the geometric interpolation functions, **G**. Here we have

$$\frac{\partial \mathbf{G}}{\partial r} = \frac{\partial \mathbf{G}}{\partial L_1}\frac{\partial L_1}{\partial r} + \frac{\partial \mathbf{G}}{\partial L_2}\frac{\partial L_2}{\partial r} + \frac{\partial \mathbf{G}}{\partial L_3}\frac{\partial L_3}{\partial r} + \frac{\partial \mathbf{G}}{\partial L_4}\frac{\partial L_4}{\partial r}$$
$$= \frac{\partial \mathbf{G}}{\partial L_1} - \frac{\partial \mathbf{G}}{\partial L_4}.$$

Likewise,

$$\frac{\partial \mathbf{G}}{\partial s} = \frac{\partial \mathbf{G}}{\partial L_2} - \frac{\partial \mathbf{G}}{\partial L_4}, \qquad \frac{\partial \mathbf{G}}{\partial t} = \frac{\partial \mathbf{G}}{\partial L_3} - \frac{\partial \mathbf{G}}{\partial L_4}.$$

For a general simplex, we have

$$\partial_l \mathbf{G} = \partial_L \mathbf{G} - \mathbf{I}\frac{\partial \mathbf{G}}{\partial L}.$$

To illustrate these rules for derivatives, consider the linear triangle (T3) in barycentric coordinates ($n_n = 3$). The geometric interpolation array is

$$\mathbf{G} = \lfloor L_3 \quad L_1 \quad L_2 \rfloor$$

and the two independent local space derivatives are

$$\Delta = \partial_l \mathbf{G} = \begin{Bmatrix} \frac{\partial}{\partial r} \\ \frac{\partial}{\partial s} \end{Bmatrix} \mathbf{G} = \begin{Bmatrix} \frac{\partial}{\partial L_1} - \frac{\partial}{\partial L_3} \\ \frac{\partial}{\partial L_2} - \frac{\partial}{\partial L_3} \end{Bmatrix} \mathbf{G}$$

$$\Delta = \begin{bmatrix} (0-1) & (1-0) & (0-0) \\ (0-1) & (0-0) & (1-0) \end{bmatrix} = \begin{bmatrix} -1 & 1 & 0 \\ -1 & 0 & 1 \end{bmatrix},$$

which is the same as the previous result in Sec. 10.2.

If one is willing to restrict the elements to having a constant Jacobian (straight edges and flat faces), then the inverse global to barycentric mapping is simple to develop. Then the global derivatives that we desire are easy to write

$$\frac{\partial}{\partial x} = \sum_{j=1}^{n+1} \frac{\partial}{\partial L_j}\frac{\partial L_j}{\partial x},$$

where $\partial L_j / \partial x$ is a known value, say V_j. For example, in 1-D we have

$$\begin{Bmatrix} L_1 \\ L_2 \end{Bmatrix} = \frac{1}{L^e}\begin{bmatrix} x_2 & -1 \\ -x_1 & 1 \end{bmatrix}^e \begin{Bmatrix} 1 \\ x \end{Bmatrix},$$

and in 2-D

$$\begin{Bmatrix} L_1 \\ L_2 \\ L_3 \end{Bmatrix} = \frac{1}{2A^e} \begin{bmatrix} 2A_{23} & (y_2 - y_3) & (x_3 - x_2) \\ 2A_{13} & (y_3 - y_1) & (x_1 - x_3) \\ 2A_{12} & (y_1 - y_2) & (x_2 - x_1) \end{bmatrix} \begin{Bmatrix} 1 \\ x \\ y \end{Bmatrix}$$

where A_{ij} is the triangular area enclosed by the origin $(0, 0)$ and nodes i and j.

10.4 Numerical integration

In many cases it is impossible or impractical to integrate the expression in closed form and numerical integration must therefore be utilized. If one is using sophisticated elements, it is almost always necessary to use numerical integration. Similarly, if the application is complicated, e.g., the solution of a nonlinear ordinary differential equation, then even simple one-dimensional elements can require numerical integration. Many analysts have found that the use of numerical integration simplifies the programming of the element matrices. This results from the fact that lengthy algebraic expressions are avoided and thus the chance of algebraic and/or programming errors is reduced. There are many numerical integration methods available. Only those methods commonly used in finite element applications will be considered here.

10.4.1 Unit coordinate quadrature

Numerical quadrature in one-dimension was introduced in Sec. 5.4. There we saw that an integral is replaced with a summation of functions evaluated at tabulated points and then multiplied by tabulated weights. The same procedure applies to all numerical integration rules. The main difficulty is to obtain the tabulated data. For triangular unit coordinate regions the weights, W_i, and abscissae (r_i, s_i) are less well known. Typical points for rules on the unit triangle are shown in Fig. 10.2. It presents rules that yield points that are symmetric with respect to all corners of the triangle. These low order data are placed in subroutine *SYMRUL*. As before, one approximates an integral of $f(x, y) = F(r, s)$ over a triangle by

$$I = \int f(x, y)\, dx\, dy = \sum_{i=1}^{n} W_i\, F(r_i, s_i)\, |J_i|.$$

As a simple example of integration over a triangle, let $f = y$ and consider the integral over a triangle with its three vertices at $(0, 0)$, $(3, 0)$, and $(0, 6)$, respectively, in (x, y) coordinates. Then the area $A = 9$ and the Jacobian is a constant $|J| = 18$. For a three point quadrature rule the integral is thus given by

$$I = \sum_{i=1}^{3} W_i\, y_i\, |J_i|.$$

Since our interpolation defines $y(r, s) = y_1 + (y_2 - y_1)r + (y_3 - y_1)s = 0 + 0 + 6s$, the transformed integrand is $F(r, s) = 6s$. Thus, at integration point, i, $F(r_i, s_i) = 6s_i$. Substituting a three-point quadrature rule and factoring out the constant Jacobian gives $I = 18\,[\,(6(1/6))\,(1/6) + (6(1/6)(1/6) + (6(2/3))\,(1/6)\,] = 18$ which is the exact solution.

Table 10.3 *Symmetric quadrature for the unit triangle:*

$$\int_0^1 \int_0^{1-r} f(r, s)\, dr\, ds = \sum_{i=1}^{n} f(r_i, s_i) W_i$$

n	$p^\dagger$	i	r_i	s_i	W_i
1	1	1	1/3	1/3	1/2
3	2	1	1/6	1/6	1/6
		2	2/3	1/6	1/6
		3	1/6	2/3	1/6
4	3	1	1/3	1/3	–9/32
		2	3/5	1/5	25/96
		3	1/5	3/5	25/96
		4	1/5	1/5	25/96
7	4	1	0	0	1/40
		2	1/2	0	1/15
		3	1	0	1/40
		4	1/2	1/2	1/15
		5	0	1	1/40
		6	0	1/2	1/15
		7	1/3	1/3	9/40

P = Degree of Polynomial for exact integration.
See subroutine *DUNAVANT_UNIT_TRIANGLE_RULE*

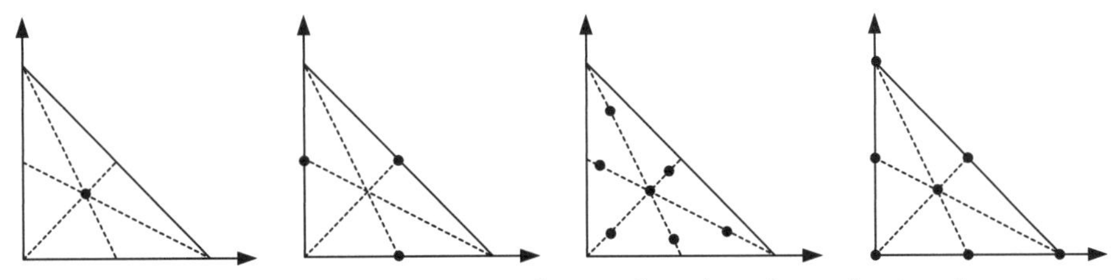

Figure 10.2 *Symmetric quadrature locations for unit triangle*

Table 10.3 gives a tabulation of symmetric quadrature rules over the unit triangle. Decimal versions are given in subroutine *SYMRUL* of values of n_q up to 13. A similar set of rules for extension to the three-dimensional tetrahedra in unit coordinates are given in Table 10.4 for polynomials up to degree four [7]. Quadrature rules for high degree polynomials on triangles have been published by Dunavant [5]. They are suitable for use

Table 10.4 *Quadrature for unit tetrahedra*

Number of points N	Degree of precision	Unit coordinates r_i	s_i	t_i	Weights W_i
1	1	1/4	1/4	1/4	1/6
4	2	a	b	b	1/24
		b	a	b	1/24
		b	b	a	1/24
		b	b	b	1/24
$a = (5+3\sqrt{5})/20 = 0.5854101966249685$					
$b = (5-\sqrt{5})/20 = 0.1381966011250105$					
5	3	1/4	1/4	1/4	– 4/30
		1/2	1/6	1/6	9/120
		1/6	1/2	1/6	9/120
		1/6	1/6	1/2	9/120
		1/6	1/6	1/6	9/120
11	4	1/4	1/4	1/4	– 74/5625
		11/14	1/14	1/14	343/45000
		1/14	11/14	1/14	343/45000
		1/14	1/14	11/14	343/45000
		1/14	1/14	1/14	343/45000
		a	a	b	56/2250
		a	b	a	56/2250
		a	b	b	56/2250
		b	a	a	56/2250
		b	a	b	56/2250
		b	b	a	56/2250
$a = (1+\sqrt{(5/14)})/4 = 0.3994035761667992$					
$b = (1-\sqrt{(5/14)})/4 = 0.1005964238332008$					
See subroutine KEAST_UNIT_TET_RULE					

with hierarchical elements. Those rules are given in Table 10.5 in area coordinates, since that form requires the smallest table size. Most of the lines are used multiple times by cycling through the area coordinates. The number N in the table indicates if the line is for the centroid, three symmetric points, or six symmetric locations. These data are expanded to their full form (up to 61 points for a polynomial of degree 17) in subroutine *DUNAVANT_UNIT_TRIANGLE_RULE*. The corresponding unit triangle coordinate data are also given in subroutine *D_Q_RULE*.

10.4.2 Natural coordinate quadrature

Here we assume that the coordinates are in the range of -1 to $+1$. In this space it is common to employ Gaussian quadratures. The one-dimensional rules were discussed in Sec. 5.4. For a higher number of space dimensions one obtains a multiple summation (tensor product) for evaluating the integral. For example, a typical integration in two dimensions

$$I = \int_{-1}^{1}\int_{-1}^{1} f(r, s)\, dr\, ds \approx \sum_{j=1}^{n} \sum_{k=1}^{n} f(r_j, s_k)\, W_j W_k$$

for n integration points in each dimension. This can be written as a single summation as

$$I \approx \sum_{i=1}^{m} f(r_i, s_i)\, W_i$$

where $m = n^2$, $i = j + (k-1)n$, and where $r_i = \alpha_j$, $s_i = \alpha_k$, and $W_i = W_j W_k$. Here α_j and W_j denote the tabulated one-dimensional abscissae and weights given in Sec. 5.4. A similar rule can be given for a three-dimensional region. The result of the above summation is given in Table 10.6. The extension of the 1-D data to the quadrilateral and hexahedra are done by subroutines *GAUSS_2D* and *GAUSS_3D* (see Fig. 10.3).

10.5 Typical source distribution integrals *

Previously we introduced the contributions of distributed source terms. For the C° continuity line elements we had

$$\mathbf{C}_Q^e = \int_{L^e} \mathbf{H}^{e^T} Q^e\, dx\, .$$

Similar forms occur in two-dimensional problems. Then typically one has

$$\mathbf{C}_Q^e = \int_{A^e} \mathbf{H}^{e^T} Q^e\, da\, .$$

If the typical source or forcing term, Q^e, varies with position we usually use the interpolation functions to define it in terms of the nodal values, Q^e, as

$$Q^e = \mathbf{H}^{e^T} \mathbf{Q}^e\, . \tag{10.6}$$

Thus, a common element integral for the consistent nodal sources is

$$\mathbf{C}_Q^e = \int_{\Omega^e} \mathbf{H}^{e^T}\, \mathbf{H}^e\, d\Omega\, \mathbf{Q}^e\, . \tag{10.7}$$

The previous sections present analytic and numerical methods for evaluating these integrals. Figure 10.4 shows the typical analytic results for the two and three node line integrals. For linear or constant source distributions the normalized nodal resultants are summarized in Fig. 10.5. Once one goes beyond the linear (two-node) element the consistent results usually differ from physical intuition estimates. Thus, you must rely on the mathematics or the summaries in the above figures. Many programs will numerically integrate the source distributions for any element shape. If the source acts on an area shaped like the parent element (constant Jacobian) then we can again easily evaluate the integrals analytically. For a uniform source over an area the consistent nodal contributions for triangles and quadrilaterals are shown in Figs. 10.6 and 10.7, respectively. Note that the Serendipity families can actually develop negative contributions. Triangular and Lagrangian elements do not have that behavior for uniform sources. Of course, a general loading can be treated by numerical integration.

```
SUBROUTINE GAUSS_3D (M_QP, N_IP, PT, WT)                              ! 1
! * * * * * * * * * * * * * * * * * * * * * * * * *                   ! 2
!      USE 1-D GAUSSIAN DATA TO GENERATE                              ! 3
!        QUADRATURE DATA FOR A CUBE                                   ! 4
! * * * * * * * * * * * * * * * * * * * * * * * * *                   ! 5
Use Precision_Module                                                  ! 6
 IMPLICIT  NONE                                                       ! 7
 INTEGER,  INTENT(IN)   :: M_QP, N_IP                                 ! 8
 REAL(DP), INTENT(OUT)  :: PT (3, N_IP), WT (N_IP)                    ! 9
 REAL(DP) :: GPT (M_QP), GWT (M_QP) ! Automatic Arrays                !10
 INTEGER  :: I, J, K, L, N_GP       ! Loops                           !11
                                                                      !12
! M_QP = NUMBER OF TABULATED 1-D POINTS                               !13
! N_IP = M_QP**3 = NUMBER OF 3-D POINTS                               !14
! GPT  = TABULATED 1-D QUADRATURE POINTS                              !15
! GWT  = TABULATED 1-D QUADRATURE WEIGHTS                             !16
! PT   = CALCULATED COORDS  IN A CUBE                                 !17
! WT   = CALCULATED WEIGHTS IN A CUBE                                 !18
                                                                      !19
  N_GP = M_QP                                                         !20
  CALL GAUSS_COEFF (N_GP, GPT, GWT) !  GET 1-D DATA                   !21
                                                                      !22
!  LOOP OVER GENERATED POINTS                                         !23
  K = 0                                                               !24
  DO L = 1, N_GP                                                      !25
    DO I = 1, N_GP                                                    !26
      DO J = 1, N_GP                                                  !27
        K = K + 1                                                     !28
        WT (K) = GWT (I) * GWT (J) * GWT (L)                          !29
        PT (1, K) = GPT (J)                                           !30
        PT (2, K) = GPT (I)                                           !31
        PT (3, K) = GPT (L)                                           !32
      END DO                                                          !33
    END DO                                                            !34
  END DO                                                              !35
END SUBROUTINE GAUSS_3D                                               !36
```

Figure 10.3 *Gaussian rules for a cube*

Table 10.5 *Dunavant quadrature for area coordinate triangle*

P	N	Wt	L_1	L_2	L_3
1	1	1.000000000000000	0.333333333333333	0.333333333333333	0.333333333333333
2	3	0.333333333333333	0.666666666666667	0.166666666666667	0.166666666666667
3	1	–0.562500000000000	0.333333333333333	0.333333333333333	0.333333333333333
	3	0.520833333333333	0.600000000000000	0.200000000000000	0.200000000000000
4	3	0.223381589678011	0.108103018168070	0.445948490915965	0.445948490915965
	3	0.109951743655322	0.816847572980459	0.091576213509771	0.091576213509771
5	1	0.225000000000000	0.333333333333333	0.333333333333333	0.333333333333333
	3	0.132394152788506	0.059715871789770	0.470142064105115	0.470142064105115
	3	0.125939180544827	0.797426985353087	0.101286507323456	0.101286507323456
6	3	0.116786275726379	0.501426509658179	0.249286745170910	0.249286745170910
	3	0.050844906370207	0.873821971016996	0.063089014491502	0.063089014491502
	6	0.082851075618374	0.053145049844817	0.310352451033784	0.636502499121399
7	1	–0.149570044467682	0.333333333333333	0.333333333333333	0.333333333333333
	3	0.175615257433208	0.479308067841920	0.260345966079040	0.260345966079040
	3	0.053347235608838	0.869739794195568	0.065130102902216	0.065130102902216
	6	0.077113760890257	0.048690315425316	0.312865496004874	0.638444188569810
8	1	0.144315607677787	0.333333333333333	0.333333333333333	0.333333333333333
	3	0.095091634267285	0.081414823414554	0.459292588292723	0.459292588292723
	3	0.103217370534718	0.658861384496480	0.170569307751760	0.170569307751760
	3	0.032458497623198	0.898905543365938	0.050547228317031	0.050547228317031
	6	0.027230314174435	0.008394777409958	0.263112829634638	0.728492392955404
9	1	0.097135796282799	0.333333333333333	0.333333333333333	0.333333333333333
	3	0.031334700227139	0.020634961602525	0.489682519198738	0.489682519198738
	3	0.077827541004774	0.125820817014127	0.437089591492937	0.437089591492937
	3	0.079647738927210	0.623592928761935	0.188203535619033	0.188203535619033
	3	0.025577675658698	0.910540973211095	0.044729513394453	0.044729513394453
	6	0.043283539377289	0.036838412054736	0.221962989160766	0.741198598784498
10	1	0.090817990382754	0.333333333333333	0.333333333333333	0.333333333333333
	3	0.036725957756467	0.028844733232685	0.485577633383657	0.485577633383657
	3	0.045321059435528	0.781036849029926	0.109481575485037	0.109481575485037
	6	0.072757916845420	0.141707219414880	0.307939838764121	0.550352941820999
	6	0.028327242531057	0.025003534762686	0.246672560639903	0.728323904597411
	6	0.009421666963733	0.009540815400299	0.066803251012200	0.923655933587500

P = Degree of complete polynomial exactly integrated, N = Number of cyclic uses
Wt = Weight at point, L_j = Area coordinates at the point
(See subroutine *D_Q_RULE* for $P \le 17$)

Table 10.6 *Gaussian quadrature on a quadrilateral*

$$\int_{-1}^{1}\int_{-1}^{1} f(r,s)\,dr\,ds = \sum_{i=1}^{n} f(r_i, s_i)\,W_i$$

n	i	r_i	s_i	W_i
1	1	0	0	4
4	1	$-1/\sqrt{3}$	$-1/\sqrt{3}$	1
	2	$+1/\sqrt{3}$	$-1/\sqrt{3}$	1
	3	$-1/\sqrt{3}$	$+1/\sqrt{3}$	1
	4	$+1/\sqrt{3}$	$+1/\sqrt{3}$	1
9	1	$-\sqrt{3/5}$	$-\sqrt{3/5}$	25/81
	2	0	$-\sqrt{3/5}$	40/81
	3	$+\sqrt{3/5}$	$-\sqrt{3/5}$	25/81
	4	$-\sqrt{3/5}$	0	40/81
	5	0	0	64/81
	6	$+\sqrt{3/5}$	0	40/81
	7	$-\sqrt{3/5}$	$+\sqrt{3/5}$	25/81
	8	0	$+\sqrt{3/5}$	40/81
	9	$+\sqrt{3/5}$	$+\sqrt{3/5}$	25/81

10.6 Minimal, optimal, reduced and selected integration *

Since the numerical integration of the element square matrix can represent a large part of the total cost it is desirable to use low order integration rules. Care must be taken when selecting the *minimal order* of integration. Usually the integrand will contain global derivatives so that in the limit, as the element size h approaches zero, the integrand can be assumed to be constant, and then only the integral $I = \int dv = \int |J| dr\, ds\, dt$ remains to be integrated exactly. Such a rule could be considered the minimal order. However, the order is often too low to be practical since it may lead to a rank deficient element (and system) square matrix, if the rule does not exactly integrate the equations. Typical integrands involve terms such as the strain energy density per unit volume: $\mathbf{B}^T\mathbf{D}\mathbf{B}/2$.

Let n_q denote the number of element integration points while m_I represents the number of independent relations at each integration point; then the rank of the element is $n_q \times m_I$. Generally, m_I corresponds to the number of rows in $\mathbf{B}$ in the usual symbolic integrand $\mathbf{B}^T\mathbf{D}\mathbf{B}$. For a typical element, we want $n_q \times (m_i - m_c) \geq n_i$, where m_c represents the number of element constraints, if any. For a non-singular system matrix a similar expression is $n_e \times (n_q \times m_i - m_c) \geq n_d - m_r$, where $m_r \geq 1$ denotes the number of nodal parameter restraints. These relations can be used as guides in selecting a minimal value of n_q. Consider a problem involving a governing integral statement with m-th order derivatives. If the interpolation (trial) functions are complete polynomials of

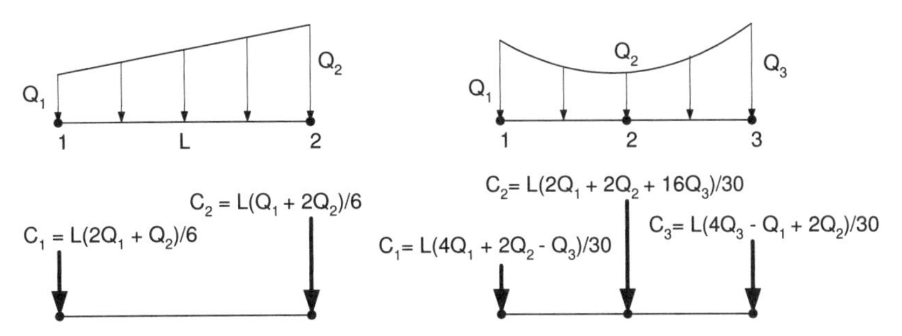

Figure 10.4 *General consistent line sources*

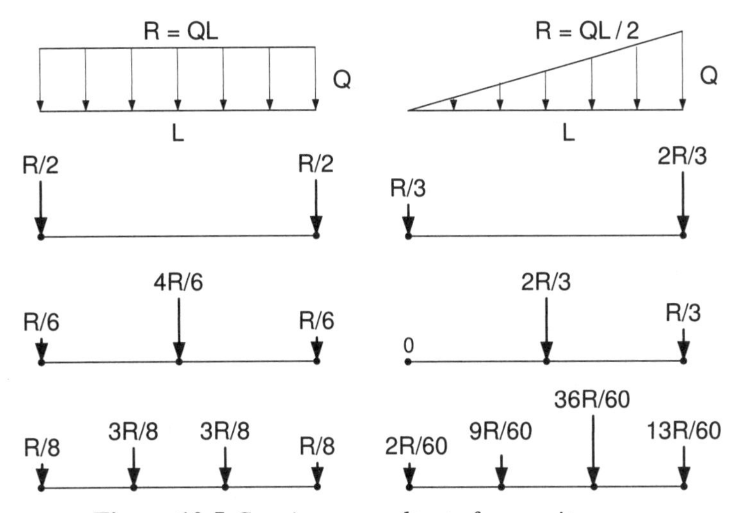

Figure 10.5 *Consistent resultants for a unit source*

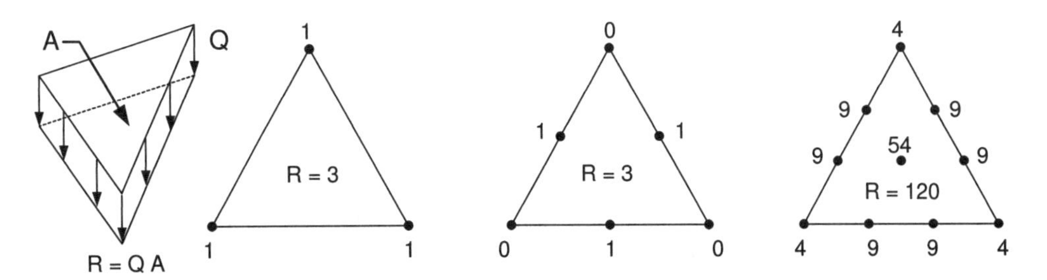

Figure 10.6 *Resultants for a constant source on a triangle*

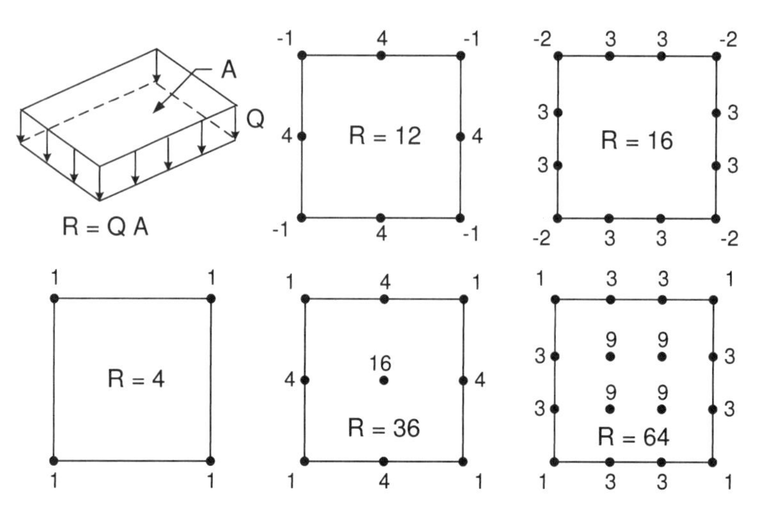

Figure 10.7 *Resultants for a constant source rectangle*

order p then to maintain the theoretical convergence rate n_q should be selected [14] to give accuracy of order $0(h^{2(p-m)+1})$. That is, to integrate polynomial terms of order $(2p - m)$ exactly.

It has long been known that a finite element model gives a stiffness which is too high. Using reduced integration so as to underestimate the element stiffness has been accepted as one way to improve the results. These procedures have been investigated by several authors including Zienkiewicz [14], Zienkiewicz and Hinton [14], Hughes, Cohen and Haroun [8] and Malkus and Hughes [12]. Reduced integration has been especially useful in problems with constraints, such as material incompressibility. A danger of low order integration rules is that *zero energy modes* may arise in an element. That is, the element energy is $\mathbf{D}^{eT}\,\mathbf{S}^e\,\mathbf{D}^e = 0$ for $\mathbf{D}^e \neq 0$. Usually these zero energy modes, $\mathbf{D}^e$, are incompatible with the same modes in an adjacent element. Thus, the assembly of elements may have no zero energy modes (except for the standard *rigid body* modes). Cook [3] illustrates that an eigen-analysis of the element square matrix can be used as a check since zero eigenvalues correspond to zero energy modes.

The integrand usually involves derivatives of the function of interest. Many solutions require the post-solution calculation of these derivatives for auxiliary calculations. Thus a related question is which points give the most accurate estimates for those derivatives. These points are often called *optimal points* or Barlow points. Their locations have been derived by Barlow [1, 2] and Moan [13]. The optimal points usually are the common quadrature points. For low order elements the optimal points usually correspond to the minimal integration points. This is indeed fortunate. As discussed in Chapter 1, it is possible in some cases to obtain exact derivative estimates from the optimal points. Barlow considered line elements, quadrilaterals and hexahedra while Moan considered the triangular elements. The points were found by assuming that the p-th order polynomial solution, in a small element, is approximately equal to the $(p + 1)$

order exact polynomial solution. The derivatives of the two forms were equated and the coordinates of points where the identity is satisfied were determined. For triangles the optimal rules are the symmetric rules involving 1, 4, 7, and 13 points. For machines with small word lengths the 4 and 13 point rules may require higher precision due to the negative centroid weights. Generally, all interior point quadrature rules can be used to give more accurate derivative estimates. The derivatives of the interpolation functions are least accurate at the nodes. Later we will show how patch methods can be used to generate much more accurate derivatives at the nodes.

For element formulations involving element constraints, or penalties, it is now considered best to employ selective integration rules [14]. For penalty formulations it is common to have equations of the form $(\mathbf{S}_1 + \alpha \mathbf{S}_2)\,\mathbf{D} = \mathbf{C}$ where the constant $\alpha \rightarrow \infty$ in the case where the penalty constraint is exactly satisfied. In the limit as $\alpha \rightarrow \infty$ the system degenerates to $\mathbf{S}_2\,\mathbf{D} = 0$, where the solution approaches the trivial result, $\mathbf{D} = \mathbf{0}$. To obtain a non-trivial solution in this limit it is necessary for $\mathbf{S}_2$ to be singular. Therefore, the two contributing element parts $\mathbf{S}_1^e$ and $\mathbf{S}_2^e$ are *selectively* integrated. That is, $\mathbf{S}_2^e$ is under integrated so as to be rank deficient (singular) while $\mathbf{S}_1^e$ is integrated with a rule which renders $\mathbf{S}_1$ non-singular. Typical applications of selective integration were cited above and include problems such as plate bending where the bending contributions are in $\mathbf{S}_1^e$ while the shear contributions are in $\mathbf{S}_2^e$.

10.7 Exercises

1. Explain why in Tables 10.3, 10.4, 10.5, and 10.6 and in Fig. 10.3 the sum of the weights are exactly 1/2, 1/6, 1, 4, and 8, respectively.

2. Assume a constant Jacobian and numerically evaluate the matrices:

$$a)\ \mathbf{C}^e = \int_{\Omega^e} \mathbf{H}^T\, dx, \quad b)\ \mathbf{M}^e = \int_{\Omega^e} \mathbf{H}^T\, \mathbf{H}\, dx,$$

$$c)\ \mathbf{S}^e = \int_{\Omega^e} \frac{d\mathbf{H}^T}{dx}\, \frac{d\mathbf{H}}{dx}\, dx, \quad d)\ \mathbf{U}^e = \int_{\Omega^e} \mathbf{H}^T\, \frac{d\mathbf{H}}{dx}\, dx.$$

   for: a) a unit right angle triangle, b) a unit square, based on linear interpolation.

3. Confirm the nodal resultants of Fig. 10.4.

4. Confirm the nodal resultants of Fig. 10.5.

5. Confirm the nodal resultants of Fig. 10.6.

6. Confirm the nodal resultants of Fig. 10.7.

10.8 Bibliography

[1] Barlow, J., "Optimal Stress Locations in Finite Element Models," *Int. J. Num. Meth. Eng.*, **10**, pp. 243–251 (1976).

[2] Barlow, J., "More on Optimal Stress Points — Reduced Integration, Element Distortions and Error Estimation," *Int. J. Num. Meth. Eng.*, **28**, pp. 1487–1504 (1989).

[3] Cook, R.D., *Concepts and Applications of Finite Element Analysis*, New York: John Wiley (1974).

[4] Cools, R., "An Encyclopedia of Cubature Formulas," *J. Complexity*, **19**, pp. 445–453 (2003).

[5] Dunavant, D.A., "High Degree Efficient Symmetrical Gaussian Quadrature Rules for the Triangle," *Int. J. Num. Meth. Eng.*, **21**, pp. 1129–1148 (1985).

[6] Felippa, C.A., "A Compendium of FEM Integration Formulas for Symbolic Work," *Engineering Computations*, **21**(8), pp. 867–890 (2004).

[7] Gellert, M. and Harbord, R., "Moderate Degree Cubature Formulas for 3–D Tetrahedral Finite-Element Approximations," *Comm. Appl. Num. Meth.*, **7**, pp. 487–495 (1991).

[8] Hughes, T.J.R., Cohen, M., and Haroun, M., "Reduced and Selective Integration Techniques in the Finite Element Analysis of Plates," *Nuclear Eng. Design*, **46**(1), pp. 203–222 (1978).

[9] Hughes, T.J.R., *The Finite Element Method*, Englewood Cliffs: Prentice Hall (1987).

[10] Irons, B.M. and Ahmad, S., *Techniques of Finite Elements*, New York: John Wiley (1980).

[11] Keast, P., "Moderate Degree Tetrahedral Quadrature Formulas," *Comm. Appl. Num. Meth.*, **55**, pp. 339–348 (1986).

[12] Malkus, D.S. and Hughes, T.J.R., "Mixed Finite Element Methods – Reduced and Selective Integration Techniques," *Comp. Meth. Appl. Mech. Eng.*, **15**(1), pp. 63–81 (1978).

[13] Moan, T., "Orthogonal Polynomials and Best Numerical Integration Formulas on a Triangle," *Zamm*, **54**, pp. 501–508 (1974).

[14] Zienkiewicz, O.C. and Hinton, E., "Reduced Integration Smoothing and Non-Conformity," *J. Franklin Inst.*, **302**(6), pp. 443–461 (1976).

Chapter 11

Scalar fields

11.1 Introduction

The physical behavior governing a variety of problems in engineering can be described as scalar field problems. That is, where a scalar quantity varies over a continuum. We usually need to compute the value of the scalar quantity, its gradient, and sometimes its integral over the solution domain. Typical applications of scalar fields include: electrical conduction, heat transfer, irrotational fluid flow, magnetostatics, seepage in porous media, torsion stress analysis, etc. Often these problems are governed by the well known Laplace and Poisson differential equations. The analytic solution of these equations in two- and three-dimensional field problems can present a formidable task, especially in the case where there are complex boundary conditions and irregularly shaped regions. The finite element formulation of this class of problems by using Galerkin or variational methods has proven to be a very effective and versatile approach to the solution. Previous difficulties associated with irregular geometry and complex boundary conditions are virtually eliminated. The following development will be introduced through the details of formulating the solution to the steady-state heat conduction problem. The approach is general, however, and by redefining the physical quantities involved the formulation is equally applicable to other problems involving the Poisson equation.

11.2 Variational formulation

We can obtain from any book on heat transfer the governing differential equation for steady and un-steady (transient) state heat conduction. The most general form of the heat conduction equation, in the material principal coordinate directions is the transient three-dimensional equation:

$$\frac{\partial}{\partial x}(k_x \frac{\partial \theta}{\partial x}) + \frac{\partial}{\partial y}(k_y \frac{\partial \theta}{\partial y}) + \frac{\partial}{\partial z}(k_z \frac{\partial \theta}{\partial z}) + Q = \frac{\partial}{\partial t}(\rho c_p \theta) \qquad (11.1)$$

where, k_x, k_y, k_z = thermal conductivity coefficients, θ = temperature, Q = heat generation per unit volume, ρ = density, and c_p = specific heat at constant pressure. If we focus our attention to the two-dimensional ($\partial / \partial z = 0$) steady-state ($\partial / \partial t = 0$) problem, such as Fig. 11.1, the governing equation becomes

$$\frac{\partial}{\partial x}(k_x \frac{\partial \theta}{\partial x}) + \frac{\partial}{\partial y}(k_y \frac{\partial \theta}{\partial y}) + Q = 0 \tag{11.2}$$

in which k_x, k_y, and Q are known. Equations 11.1 or 2, along with the boundary (and initial) conditions specify the problem completely. The most commonly encountered boundary conditions are those in which the temperature, θ, is specified on the boundary,

$$\theta = \theta(s) \text{ on } \Gamma_D,$$

or the normal heat flux into the boundary, q_s, is specified:

$$k_x \frac{\partial \theta}{\partial x} n_x + k_y \frac{\partial \theta}{\partial y} n_y + q_s = k_n \frac{\partial \theta}{\partial n} + q_s = 0 \text{ on } \Gamma_q$$

or a normal heat flux due to convection:

$$k_n \frac{\partial \theta}{\partial n} + h(\theta - \theta_\infty) = 0, \quad \text{on } \Gamma_h \tag{11.3}$$

where n_x and n_y are the direction cosines of the outward normal to the boundary surface, q_s represents the known heat flux per unit of surface, and $h(\theta - \theta_\infty)$ is the convection heat loss per unit area due to a convection coefficient h and a surrounding fluid at a temperature of θ_∞. Only one of these two last two items is non-zero on a particular surface. Note that the last two surface integrals could be written in a more general form if we combine them into a *mixed* or *Robin* condition written as:

$$k_n \frac{\partial \theta}{\partial n} + h\,\theta + g = 0, \tag{11.4}$$

where g is either a known influx (when $h = 0$), or $h\,\theta_\infty$ on a convection surface.

In Section 2.13.2 we illustrated how to apply the Galerkin method to this equation. As stated previously, an alternative formulation to the above heat conduction problem is possible using the calculus of variations. It has been shown that if a variational form exists for a differential equation then both the Galerkin form and the Euler variational form will yield exactly the same element matrix definitions. Euler's theorem of the calculus of variations states that if the integral

$$I(u) = \int_\Omega f(x, y, z, u, \frac{\partial u}{\partial x}, \frac{\partial u}{\partial y}, \frac{\partial u}{\partial z})\, d\Omega + \int_\Gamma (gu + hu^2/2)\, d\Gamma \tag{11.5}$$

is to be minimized, the necessary and sufficient condition for this minimum to be reached is that the unknown function $u(x, y, z)$ satisfy the following differential equation

$$\frac{\partial}{\partial x}\frac{\partial f}{\partial(\partial u/\partial x)} + \frac{\partial}{\partial y}\frac{\partial f}{\partial(\partial u/\partial y)} + \frac{\partial}{\partial z}\frac{\partial f}{\partial(\partial u/\partial z)} - \frac{\partial f}{\partial u} = 0 \tag{11.6}$$

within the region Ω, provided u satisfies the essential boundary conditions on Γ_D and

$$n_x k_x \frac{\partial \theta}{\partial x} + n_y k_y \frac{\partial \theta}{\partial y} + n_z k_z \frac{\partial \theta}{\partial z} + g + h\theta = 0 = k_n \frac{\partial \theta}{\partial n} + g + h\theta$$

on the remainder of Γ. We can verify that the minimization of the volume integral

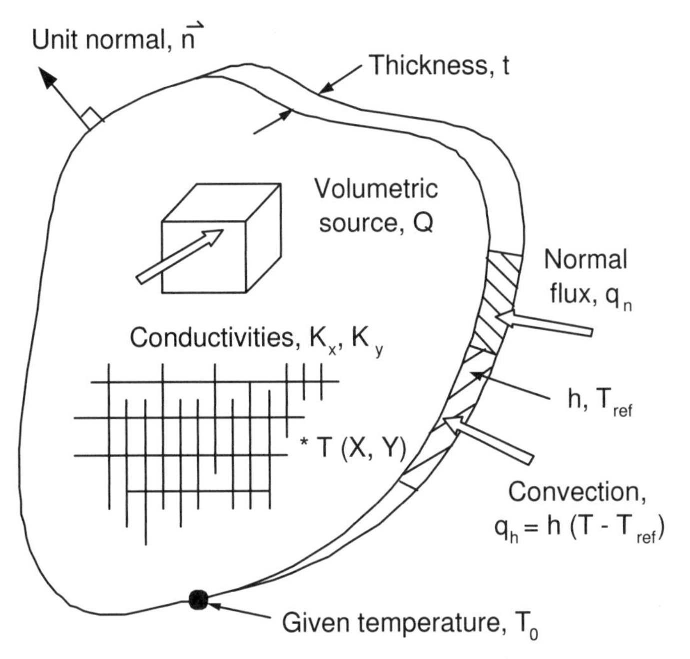

Figure 11.1 *An anisotropic heat transfer region*

$$I = \int_{\Omega} \left[\frac{1}{2} \left\{ k_x (\frac{\partial \theta}{\partial x})^2 + k_y (\frac{\partial \theta}{\partial y})^2 + k_z (\frac{\partial \theta}{\partial z})^2 \right\} - Q\theta \right] d\Omega \tag{11.7}$$

$$+ \int_{\Gamma} \left[g\theta + h\theta^2/2 \right] d\Gamma$$

leads directly to the formulation equivalent to Eq. 11.2 for the steady-state case. It should also be noted that the surface Γ will be split into different regions for each distinct set of surface input. One of those segments will usually be a Dirichlet region, Γ_D, and that surface integral represents the unknown resultant reaction fluxes at the nodes that get lumped into the RHS of the algebraic system. The functional volume contribution is

$$f = \tfrac{1}{2} \left\{ k_x (\frac{\partial \theta}{\partial x})^2 + k_y (\frac{\partial \theta}{\partial y})^2 + k_z (\frac{\partial \theta}{\partial z})^2 \right\} - Q\theta.$$

Thus, if f is to be minimized it must satisfy Eq. 11.6. Here

$$\frac{\partial f}{\partial(\partial\theta/\partial x)} = k_x \frac{\partial \theta}{\partial x}, \quad \frac{\partial f}{\partial(\partial\theta/\partial y)} = k_y \frac{\partial \theta}{\partial y}, \quad \frac{\partial f}{\partial(\partial\theta/\partial z)} = k_z \frac{\partial \theta}{\partial z}, \quad \frac{\partial f}{\partial \theta} = -Q$$

so Eq. 11.6 results in

$$\frac{\partial}{\partial x}(k_x \frac{\partial \theta}{\partial x}) + \frac{\partial}{\partial y}(k_y \frac{\partial \theta}{\partial y}) + \frac{\partial}{\partial z}(k_z \frac{\partial \theta}{\partial z}) + Q = 0$$

verifying that the function f does lead to correct steady state formulation, if the boundary conditions are also satisfied. Euler also stated that the natural boundary condition associated with Eq. 11.5 on a surface with a unit normal vector $\vec{n}$ is

$$n_x \left\{ \frac{\partial f}{\partial(\partial u / \partial x)} \right\} + n_y \left\{ \frac{\partial f}{\partial(\partial u / \partial y)} \right\} + n_z \left\{ \frac{\partial f}{\partial(\partial u / \partial z)} \right\} + g + hu = 0$$

on the boundary where the value of u is not prescribed. If both g and h are non-zero this type of boundary condition is a Robin, or mixed, condition since it imposes a linear combination on the solution and the normal gradient on part of the boundary.

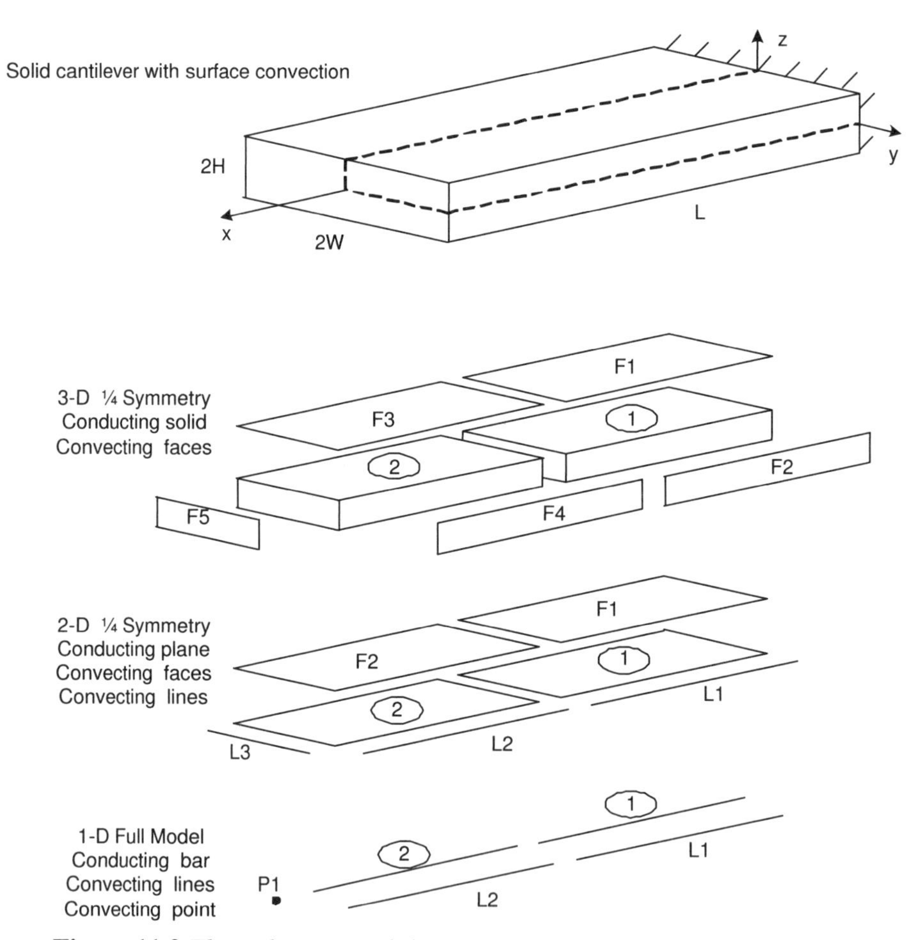

Figure 11.2 *Three-dimensional thermal models and their approximations*

The element and boundary matrices arising from Eq. 11.7 and the Galerkin method will be identical. Therefore we have the tools to build non-homogeneous, anisotropic thermal models of solids of complex geometry. As illustrated in Fig. 11.2 this will often require combining the effects of conduction elements and convection elements at shared nodes. Today commercial codes can quickly generate and solve fine meshes for 3-D thermal studies. However, it is still common to approximate some 3-D solids by 2-D models as seen in that figure. If we allow the specified thickness to vary from element to element we sometimes call this a $2\,\frac{1}{2}$-D model. The convection elements in a 2-D model can occur over the face of a 2-D conduction element and/or along an edge having a specified thickness. Likewise, 1-D approximations can have perimeter convection effects, as line elements, combined with the 1-D conduction elements. They can also have convection over an end area which is represented as a point convection element. Later we will see that a single computer implementation can handle all the combinations shown in the above figure. It is still recommended that various types of finite element models be compared to each other as a means for validating the results to problems for which the answer is not known. Sometimes a model that can be solved by hand gives a useful validation of results from a commercial code. For simplicity, in the next section we look at 2-D models that can yield matrices that can be manipulated in closed form, and then return later to numerically integrated elements for general 3-D use.

11.3 Element and boundary matrices

From Eqs. 11.1 and 11.7 it is clearly seen that the two-dimensional functional required for the steady-state analysis is

$$I = \int_A \left[\frac{1}{2} \left\{ k_x \left(\frac{\partial \theta}{\partial x}\right)^2 + k_y \left(\frac{\partial \theta}{\partial y}\right)^2 \right\} - Q\theta \right] t\, dA + \int_\Gamma (g\theta + h\theta^2/2) t\, ds \qquad (11.8)$$

where t is the thickness of the domain. We will proceed in exactly the same manner as we did for the previous variational formulations. That is, we will assume that the area integral is the sum of the integrals over the element areas. Likewise, the boundary integral where the temperature is not specified is assumed to be the sum of the boundary segment integrals. Thus, $I = \sum_e I^e + \sum_b I^b$ where the element contributions are

$$I^e = \int_{A^e} \left[\frac{1}{2} \left\{ k_x \left(\frac{\partial \theta}{\partial x}\right)^2 + k_y \left(\frac{\partial \theta}{\partial y}\right)^2 \right\} - Q\theta \right] t\, dA$$

and the boundary segment contributions are

$$I^b = \int_{\Gamma^b} (g\theta + h\theta^2/2) t\, ds.$$

If we make the usual interpolation assumptions in the element and on its typical edge then we can express these quantities as

$$I^e = \frac{1}{2} \mathbf{D}^{e^T} \mathbf{S}^e \mathbf{D}^e - \mathbf{D}^{e^T} \mathbf{C}^e, \qquad I^b = \frac{1}{2} \mathbf{D}^{b^T} \mathbf{S}^b \mathbf{D}^b - \mathbf{D}^{b^T} \mathbf{C}^b.$$

Here the element matrices are the orthotropic conduction square matrix

$$\mathbf{S}^e = \int_{A^e} (k_x^e \, \mathbf{H}_x^{e^T} \mathbf{H}_x^e + k_y^e \, \mathbf{H}_y^{e^T} \mathbf{H}_y^e) t^e \, da = \int_{A^e} \mathbf{B}^{e^T} \mathbf{E}^e \mathbf{B}^e t^e \, da \tag{11.9}$$

the internal source vector

$$\mathbf{C}_Q^e = \int_{A^e} \mathbf{H}^{e^T} Q^e t^e \, da \tag{11.10}$$

the surface square matrix and source vector from convection

$$\mathbf{S}_h^b = \int_{\Gamma^b} h^b \mathbf{H}^{b^T} \mathbf{H}^b t^b \, ds, \quad \mathbf{C}_h^b = \int_{\Gamma^b} \theta_\infty^b h^b \mathbf{H}^{b^T} t^b \, ds \tag{11.11}$$

and/or the source vector due to a specified inward heat flow

$$\mathbf{C}_q^b = \int_{\Gamma^b} q^b \mathbf{H}^{b^T} t^b \, ds \tag{11.12}$$

where $\mathbf{H}$ denotes the shape functions and $\mathbf{H}_x = \partial \mathbf{H} / \partial x$, etc., are rows of the solution interpolation gradient, $\mathbf{B}^e$, and where the general constitutive matrix is $\mathbf{E}$. The symmetric array $\mathbf{E}^e$ reduces to its diagonal orthotropic form when $k_{xy} = 0 = k_{yx}$, and to the common isotropic form, $\mathbf{E}^e = k\mathbf{I}$, when $k_{xx} = k_{yy} = k$. For this class of problem there is only one unknown temperature per node. Once again, if $\mathbf{D}$ denotes all of these unknowns then $\mathbf{D}^e \subset \mathbf{D}$ and $\mathbf{D}^b \subset \mathbf{D}$. Figure 11.3 illustrates where these typical conduction, convection, and source terms are inserted in the algebraic equations of thermal equilibrium.

Likewise, Figure 11.4 reminds us that a Dirichlet (essential) boundary condition assigns a value to one of the degrees of freedom and introduces unknown reaction source terms; that a Neumann (flux) condition only contributes known terms to the source vector; while a Robin (mixed) boundary condition couples the flux linearly to the boundary value and introduces known terms into both the system square matrix and source vector.

Many analysis problems have features that allow the analyst to reduce greatly the cost of FEA through the use of *symmetry, anti-symmetry,* or *cyclic symmetry* and *coupled nodes*. The system equations are sparse and can often be described by the bandwidth, say B, and the total number of equations, say M. Solving the equations is very expensive and has an operations count that is proportional to the product $B^2 M$. The storage required by the system is proportional to BM. Whenever possible we try to reduce these two parameters. Partial models allow us to reduce them very easily and still generate all the information we require. For example, a half-symmetry model would usually reduce both B and M by a factor of two and thereby reduce the solution costs by a factor of eight and reduce the storage requirement by a factor of four.

When a model has a plane of geometric, material, and support symmetry, it is not usually necessary to analyze the whole model. Conditions of symmetry or anti-symmetry in the source terms can be applied to the planes of symmetry in order to produce a partial model that includes the effects of the other removed parts of the complete model. In order to employ a partial analysis involving a symmetry plane, it is necessary for the following quantities to be symmetric with respect to that plane: the geometry, the material regions, the material properties, and the essential boundary conditions (on the

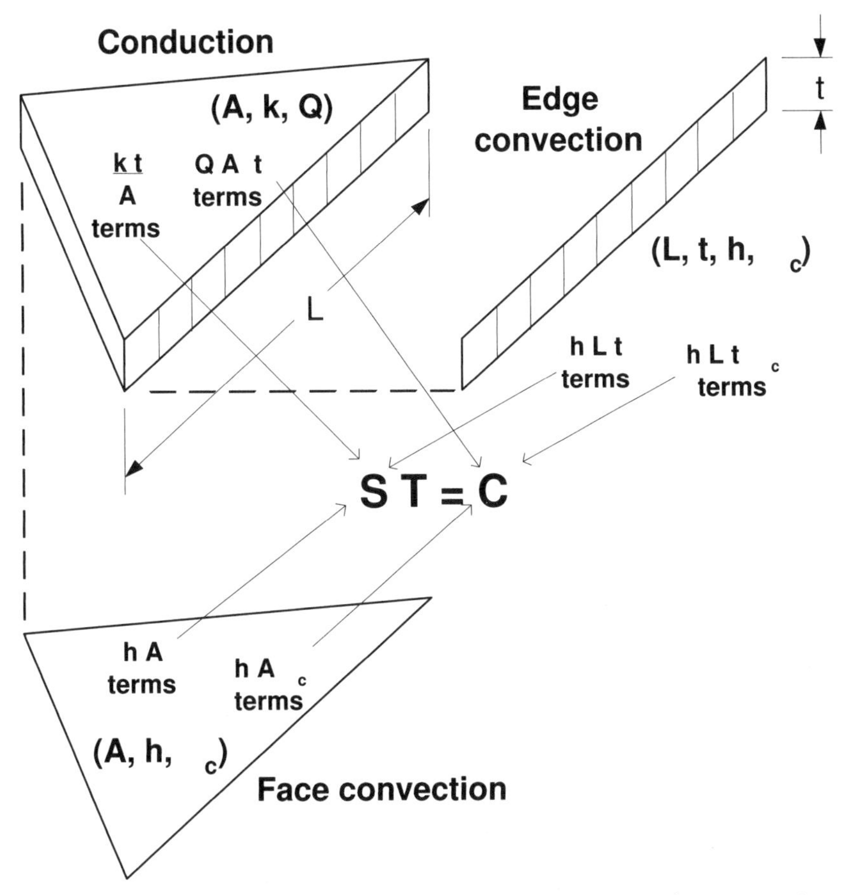

Figure 11.3 *Contributions from conducting, convecting and source regions*

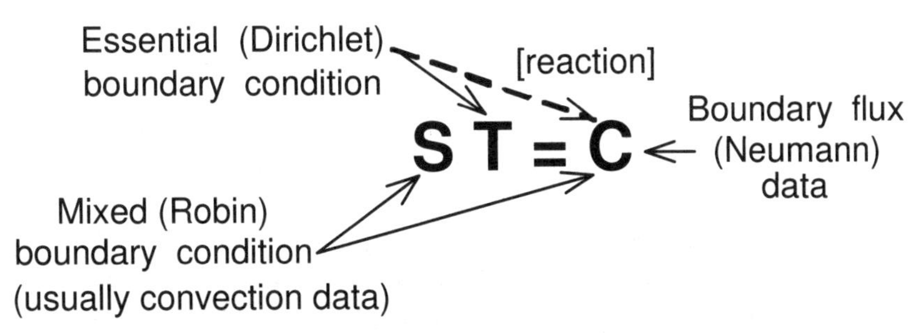

Figure 11.4 *Contributions of boundary conditions to algebraic system*

temperature). If these conditions are not quite fulfilled, then the analyst will have to exercise judgement before selecting a partial model. Even if a full model is selected for eventual use, an approximate partial model can give useful insight and aid in planning the details of the full model to be run later.

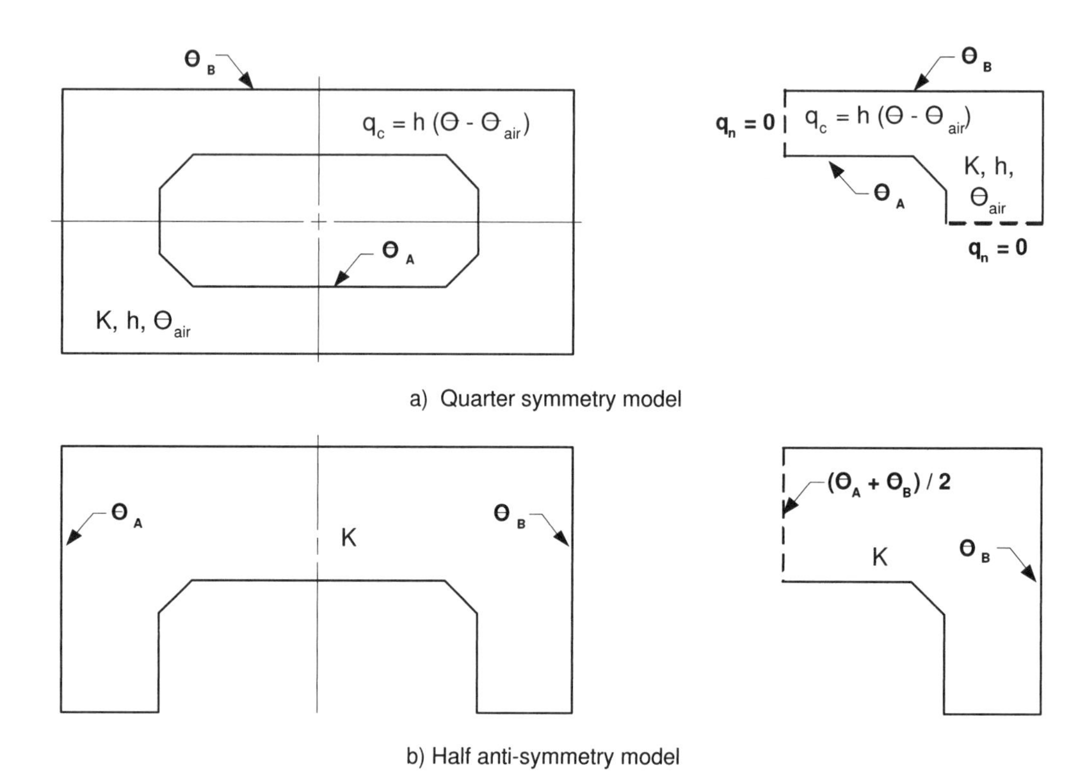

Figure 11.5 *Typical symmetry and anti-symmetry thermal models*

In thermal models any symmetry plane has a zero heat flux and temperature gradient normal to that plane. That is because the temperatures are mirror images of each other and have the same sign when moving normal to the plane. The state of zero heat flux normal to a boundary is a natural boundary condition in a finite element analysis (but not in finite differences). We obtain that condition by default. If you desire a different type of condition to apply on a boundary, then we must prescribe either the temperature (the essential condition) or a different nonzero normal flux. We cannot give both. When you prescribe one of them, the other becomes a 'reaction' to be determined from the final result. Thermal anti-symmetry means the temperatures approach the plane with temperature increments of opposite signs but equal magnitudes. Thus, the temperature must be a known mean temperature on a plane of anti-symmetry. When that essential boundary condition is imposed, the necessary normal heat flow through that plane can be found as a 'reaction' from the final solution.

To give some specific examples of these concepts, some typical thermal and stress problems will be represented. Figure 11.5 shows a planar rectangular region with homogeneous and isotropic conduction properties, k, and internal and external specified edge temperatures, θ_A and θ_B. In addition, it has free convection, q_h, over its face to a fluid with homogeneous convection properties, h and θ_∞. Such a system has double symmetry, which allows us to employ a one-quarter model with consistent boundary conditions. Doing this cuts M by a factor of 4 and reduces B by at least a factor of 2 and possibly much more. Thus, the cost drops by at least a factor of 16. The alternate partial model form still requires the essential conditions on θ_A and θ_B and the face convection data for q_h. The change is that the heat flow is zero on the new boundary lines formed by the symmetry planes. This is a natural condition in FEA and requires no input data (other than the geometry of the new line). That figure also shows a simple anti-symmetric domain where it is relatively clear to determine the value of the middle temperature to be assigned as an essential boundary condition. The temperatures on either side of the centerline are both either positive or negative increments of the relative temperature change. Commercial visualization codes often have the ability to plot a full region from a partial analysis region and the identification of the symmetry and anti-symmetry planes.

11.4 Linear triangular element

If we select the three node (linear) triangle then the element interpolation functions, $\mathbf{H}^e$, are given in unit coordinates by Eq. 9.6 and in global coordinates by Eqs. 9.13 and 9.14. From either set of equations we note that

$$\begin{aligned} \mathbf{H}_x^e &= \partial \mathbf{H}^e / \partial x = \lfloor b_1 \;\; b_2 \;\; b_3 \rfloor^e / 2A^e = \mathbf{d}_x \\ \mathbf{H}_y^e &= \partial \mathbf{H}^e / \partial y = \lfloor c_1 \;\; c_2 \;\; c_3 \rfloor^e / 2A^e = \mathbf{d}_y . \end{aligned} \tag{11.13}$$

Since these are constant we can evaluate the integral by inspection if the conductivities are also constant:

$$\mathbf{S}^e = \frac{k_x^e t^e}{4A^e} \begin{bmatrix} b_1 b_1 & b_1 b_2 & b_1 b_3 \\ b_2 b_1 & b_2 b_2 & b_2 b_3 \\ b_3 b_1 & b_3 b_2 & b_3 b_3 \end{bmatrix}^e + \frac{k_y^e t^e}{4A^e} \begin{bmatrix} c_1 c_1 & c_1 c_2 & c_1 c_3 \\ c_2 c_1 & c_2 c_2 & c_2 c_3 \\ c_3 c_1 & c_3 c_2 & c_3 c_3 \end{bmatrix}^e . \tag{11.14}$$

This is known as the *element conductivity matrix.* Note that this allows for different conductivities in the x- and y- directions. Equations 11.14 show that the conduction in the x-direction depends on the size of the element in the y-direction, and vice versa. If the internal heat generation, Q, is also constant then Eq. 11.10 yields:

$$\mathbf{C}_Q^{e\,T} = \frac{Q^e A^e t^e}{3} \lfloor 1 \;\; 1 \;\; 1 \rfloor . \tag{11.15}$$

This internal source vector shows that a third of the internal heat generated, $Q^e A^e t^e$, is equally lumped to each of the three nodes. On a typical boundary segment the edge interpolation can also be given by a linear form. The exact integrals can be evaluated for a constant Jacobian. For example, if the coefficient, h, and the surrounding temperature, T_∞^b, are constant then the boundary segment square and column matrices are:

$$\mathbf{S}_h^b = \frac{h^b L^b t^b}{6} \begin{bmatrix} 2 & 1 \\ 1 & 2 \end{bmatrix}, \quad \mathbf{C}_h^b = \frac{T_\infty^b h^b L^b t^b}{2} \begin{Bmatrix} 1 \\ 1 \end{Bmatrix}, \tag{11.16}$$

where $L^b t^b$ represents the surface area over which the convection occurs. Similarly if a constant normal flux, q, is given over a similar surface area then the resultant boundary flux vector is

$$\mathbf{C}_q^b = \frac{q^b L^b t^b}{2}\begin{Bmatrix} 1 \\ 1 \end{Bmatrix}. \tag{11.17}$$

In this case half the total normal flux is lumped at each of the two nodes on the segment.

```
! ............................................................   ! 1
!    ***  ELEM_SQ_MATRIX PROBLEM DEPENDENT STATEMENTS FOLLOW ***   ! 2
! ............................................................   ! 3
!           (Stored as application source example 204.)           ! 4
                                                                   ! 5
! Linear Triangle, K_xx U,xx + 2 K_xy U,xy + K_yy U,yy + Q = 0     ! 6
  REAL(DP) :: X_I, X_J, X_K, Y_I, Y_J, Y_K      ! Global coordinates ! 7
  REAL(DP) :: A_I, A_J, A_K, B_I, B_J, B_K      ! Standard geometry  ! 8
  REAL(DP) :: C_I, C_J, C_K, X_CG, Y_CG, TWO_A ! Standard geometry  ! 9
  REAL(DP) :: THICK                              ! Element thickness  !10
                                                                   !11
!   DEFINE NODAL COORDINATES, CCW: I, J, K                         !12
   X_I = COORD (1,1) ;    X_J = COORD (2,1) ;    X_K = COORD (3,1)  !13
   Y_I = COORD (1,2) ;    Y_J = COORD (2,2) ;    Y_K = COORD (3,2)  !14
                                                                   !15
!   DEFINE CENTROID COORDINATES (QUADRATURE POINT)                 !16
   X_CG = (X_I + X_J + X_K)/3.d0 ;  Y_CG = (Y_I + Y_J + Y_K)/3.d0  !17
                                                                   !18
!   GEOMETRIC PARAMETERS: H_I (X,Y) = (A_I + B_I*X + C_I*Y)/TWO_A  !19
   A_I = X_J * Y_K - X_K * Y_J ; B_I = Y_J - Y_K ; C_I = X_K - X_J  !20
   A_J = X_K * Y_I - X_I * Y_K ; B_J = Y_K - Y_I ; C_J = X_I - X_K  !21
   A_K = X_I * Y_J - X_J * Y_I ; B_K = Y_I - Y_J ; C_K = X_J - X_I  !22
                                                                   !23
!   CALCULATE TWICE ELEMENT AREA                                   !24
   TWO_A = A_I + A_J + A_K               ! = B_J*C_K - B_K*C_J also !25
                                                                   !26
!   DEFINE 2 BY 3 GRADIENT MATRIX, B (= DGH)                       !27
   B (1, 1:3) = (/ B_I, B_J, B_K /) / TWO_A  ! DH/DX, row 1        !28
   B (2, 1:3) = (/ C_I, C_J, C_K /) / TWO_A  ! DH/DY, row 2        !29
                                                                   !30
!   DEFINE PROPERTIES: 1-K_xx, 2-K_yy, 3-K_xy, 4-Source, 5-thick  !31
   E (1, 1) = GET_REAL_LP (1) ; E (1, 2) = GET_REAL_LP (3)        !32
   E (2, 2) = GET_REAL_LP (2) ; E (2, 1) = E (1, 2) ; THICK = 1   !33
   IF ( EL_REAL >= 5 ) THICK = GET_REAL_LP (5)                    !34
   E = E * THICK  ! for proper flux recovery                      !35
                                                                   !36
!   CONDUCTION MATRIX, WITH CONSTANT JACOBIAN (t in E)            !37
   S = MATMUL ( TRANSPOSE (B), MATMUL (E, B) ) * TWO_A * 0.5d0    !38
                                                                   !39
!   SOURCE VECTOR: C(1:3) = SOURCE_PER_UNIT_AREA * AREA / 3       !40
   C = GET_REAL_LP (4) * THICK * TWO_A / 6.d0                     !41
                                                                   !42
!  SAVE ONE POINT RULE TO AVERAGING, OR ERROR ESTIMATOR           !43
   LT_QP = 1 ; CALL STORE_FLUX_POINT_COUNT ! Save LT_QP           !44
   CALL STORE_FLUX_POINT_DATA ( (/ X_CG, Y_CG /), E, B )          !45
                                                                   !46
!  End of application dependent code 204.my_el_sq_inc            !47
!  ***  END ELEM_SQ_MATRIX PROBLEM DEPENDENT STATEMENTS ***        !48
```

Figure 11.6 *Anisotropic linear triangle conduction element*

If one wished to code this simple element in closed form it is very easy to do as shown in Fig. 11.6. There it is assumed that each element has four or five real, or floating point, properties of K_{xx}, K_{yy}, plus the anisotropic value K_{xy} not used above, and the source per unit area of Q. There the thickness is assumed to be unity for all elements unless it is provided as an optional fifth property. If one wanted to allow more general element families, then numerical integration would be required and the coding is a little longer, as we will see shortly. If we wish to allow for only a constant normal flux along any straight line edge segment then it is quite simple to implement Eq. 11.12 in closed form, as shown in Fig. 11.7. A common use of two-dimensional models is to predict the temperature in thin cooling fins. Then Eq. 11.11 would be applied over the face(s) of the element to define the convection matrices, which represent the most common kind of mixed, or Robin, boundary conditions. Again, for the linear triangle the closed form equations are easy to implement, if we assume constant data over each mixed boundary segment (face). The implementation is given in Fig. 11.8. We will shortly illustrate these matrix definitions with some simple applications. These codes are saved as example 204.

11.5 Linear triangle applications

Since the three node triangle is so widely used in finite element analysis we will examine its more common applications in detail.

11.5.1 Internal source

Consider a uniform square of material that has its exterior perimeter maintained at a constant temperature while its interior generates heat at a constant rate. We note that the solution will be symmetric about the square's centerlines as well as about its two diagonals. This means that we only need to utilize one-eighth of the region in the analysis. For simplicity we will assume that the material is homogeneous and isotropic so $k_x = k_y = k$. The planes of symmetry have zero normal heat flux, $q = 0$. That condition is a natural boundary condition in a finite element analysis. That is true since $\mathbf{C}_q$ in Eq. 11.12 is identically zero when the normal flux, q, is zero. The remaining essential condition is that of the known external boundary temperature as shown in Fig. 11.9. For this model we have first selected a crude mesh suitable for a hand solution, then we will give results for a finer mesh using the *MODEL* program. As shown in the figure we will use four elements and six nodes. The last three nodes have the known temperature and the first three are the unknown internal temperatures. For this homogeneous region the data are:

Element	k^e	Q^e	*Topology*	t^e
1	8	6	1, 2, 3	1
2	8	6	2, 4, 5	1
3	8	6	5, 3, 2	1
4	8	6	3, 5, 6	1

From the geometry in the figure we determine that the element geometric properties from Eq. 9.14 are:

```
! .............................................................  ! 1
! ***  SEG_COL_MATRIX PROBLEM DEPENDENT STATEMENTS FOLLOW ***    ! 2
! .............................................................  ! 3
!           (Stored as application source example 204.)         ! 4
! Given normal flux on a straight boundary segment (BS) edge:    ! 5
! Standard form: -K_n * U,n = Q_NORMAL_SEG, where Q_NORMAL_SEG   ! 6
! is from control keywords normal_flux, flux_thick, or via       ! 7
! optional flux segment real properties: 1-flux, 2-thickness     ! 8
 REAL(DP) :: EDGE_L, THICK ! Edge length, thickness              ! 9
                                                                 !10
   ! Get the edge length, and thickness of edge                  !11
   THICK = 1    ! Default in all cases                           !12
   IF ( FLUX_THICK /= 1.d0    ) THICK = FLUX_THICK ! line only   !13
   EDGE_L = SQRT ( (COORD(2,1) - COORD(1,1)) **2 &               !14
                 + (COORD(2,2) - COORD(1,2)) **2 )               !15
                                                                 !16
   ! Override keyword option via segment properties              !17
   IF ( SEG_REAL > 0 ) Q_NORMAL_SEG = GET_REAL_SP (1)            !18
   IF ( SEG_REAL > 1 ) THICK        = GET_REAL_SP (2)            !19
                                                                 !20
   C (1) = Q_NORMAL_SEG * THICK * EDGE_L / 2                     !21
   C (2) = Q_NORMAL_SEG * THICK * EDGE_L / 2                     !22
!  End of application dependent code 204.my_seg_col_inc          !23
```

Figure 11.7 *Straight linear edge flux source element*

```
! .............................................................  ! 1
!  ***  MIXED_SQ_MATRIX PROBLEM DEPENDENT STATEMENTS FOLLOW ***  ! 2
! .............................................................  ! 3
!           (Stored as application source example 204.)         ! 4
! Global CONVECT_COEF set by keyword convect_coef is available   ! 5
! Global CONVECT_TEMP set by keyword convect_temp is available   ! 6
! Standard form: -K_n * U,n = CONVECT_COEF ( U - CONVECT_TEMP)   ! 7
                                                                 ! 8
! Linear Triangle Boundary Face Convection Matrices              ! 9
  REAL(DP) :: X_I, X_J, X_K, Y_I, Y_J, Y_K  ! Global coordinates !10
  REAL(DP) :: A_I, A_J, A_K, TWO_A          ! Standard geometry  !11
                                                                 !12
!   DEFINE NODAL COORDINATES, CCW: I, J, K                       !13
   X_I = COORD (1,1) ;   X_J = COORD (2,1) ;   X_K = COORD (3,1) !14
   Y_I = COORD (1,2) ;   Y_J = COORD (2,2) ;   Y_K = COORD (3,2) !15
                                                                 !16
!   GEOMETRIC PARAMETERS, TWICE ELEMENT AREA                     !17
   A_I = X_J * Y_K - X_K * Y_J ; A_J = X_K * Y_I - X_I * Y_K     !18
   A_K = X_I * Y_J - X_J * Y_I ; TWO_A = A_I + A_J + A_K         !19
                                                                 !20
!   Get convection data from keyword or optional properties      !21
   IF ( MIXED_REAL > 0 ) THEN   ! override keyword               !22
     CONVECT_COEF = GET_REAL_MX (1)  ! convection coefficient    !23
     CONVECT_TEMP = GET_REAL_MX (2)  ! convection temperature    !24
   END IF                            ! properties supplied       !25
                                                                 !26
!   FACE CONVECTION SQUARE MATRIX, WITH CONSTANT JACOBIAN        !27
   S = CONVECT_COEF * TWO_A / 24                           &     !28
       * RESHAPE ( (/ 2, 1, 1, 1, 2, 1, 1, 1, 2 /), (/3, 3/))    !29
                                                                 !30
!   FACE CONVECTION SOURCE VECTOR                                !31
   C = CONVECT_TEMP * CONVECT_COEF * TWO_A / 6 * (/ 1, 1, 1 /)   !32
!  ***  END MIXED_SQ_MATRIX PROBLEM DEPENDENT STATEMENTS ***     !33
```

Figure 11.8 *Face convection for a linear triangle segment*

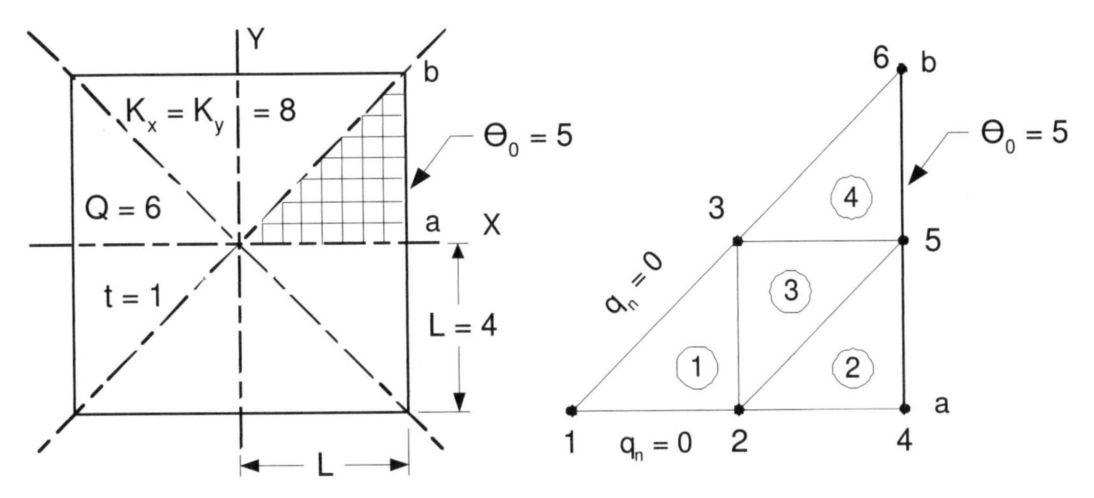

Figure 11.9 *A one-eighth symmetry model of a square*

```
title "CONDUCTION EXAMPLE, T3, INTERNAL SOURCE"                       ! 1
example  204 ! Application source code library numbe                   ! 2
b_rows     2 ! Number of rows in the B (operator) matrix               ! 3
dof        1 ! Number of unknowns per node                             ! 4
el_nodes   3 ! Maximum number of nodes per element                     ! 5
elems      4 ! Number of elements in the system                        ! 6
gauss      1 ! Maximum number of quadrature points                     ! 7
nodes      6 ! Number of nodes in the mesh                             ! 8
shape      2 ! Element shape, 1=line, 2=tri, 3=quad, 4=hex             ! 9
space      2 ! Solution space dimension                                !10
el_homo      ! Element properties are homogeneous                      !11
el_real    4 ! Number of real properties per element                   !12
remarks    9 ! Number of user remarks, e.g. property names             !13
end          ! Terminate the keyword control, remarks follow           !14
Conduction example of Section 11.4                 6                   !15
K_x = K_y = 8, Q = 6, K_xy = 0                   / |                   !16
L_1_4 = L_4_6 = 4, Thickness = 1                /  |                   !17
with 1/8 symmetry, so natural BC on            /(4)|                   !18
edges 1_6 and 1_4 ia q_n = 0                 3-----5                   !19
EBC on edge 4_6 is T = 5                    / |(3)/|                   !20
K_x T,xx + 2K_xy T,xy + K_y T,yy + Q = 0 /   | /  |                   !21
[gauss > 0 turns on flux averaging        /(1)|/(2)|                   !22
and possibly post-processing]            1----2----4                   !23
   1     0 0.   0. ! node, ebc flag, x, y                              !24
   2     0 2.   0. ! node, ebc flag, x, y                              !25
   3     0 2.   2. ! node, ebc flag, x, y                              !26
   4     1 4.   0. ! node, ebc flag, x, y                              !27
   5     1 4.   2. ! node, ebc flag, x, y                              !28
   6     1 4.   4. ! node, ebc flag, x, y                              !29
  1  1  2  3       ! elem, three nodes                                 !30
  2  2  4  5       !                                                   !31
  3  2  5  3                                                           !32
  4  3  5  6                                                           !33
   4     1 5.      ! node, dof, value of EBC                           !34
   5     1 5.                                                          !35
   6     1 5.                                                          !36
  1 8. 8. 0. 6.    ! elem, K_x, K_y, K_xy, Q (homogeneous)             !37
```

Figure 11.10 *Sample data for square bar thermal analysis*

	$e = 1, 2, 4$			$e = 3$		
i	1	2	3	1	2	3
b_i	−2	2	0	2	−2	0
c_i	0	−2	2	0	2	−2
	$A^e = 2$			$A^e = 2$		

From Eq. 11.14 the conduction square matrix for elements 1, 2, and 4 are

$$\mathbf{S}^e = \frac{8(1)}{4(2)}\begin{bmatrix} 4 & -4 & 0 \\ -4 & 4 & 0 \\ 0 & 0 & 0 \end{bmatrix} + \frac{8(1)}{4(2)}\begin{bmatrix} 0 & 0 & 0 \\ 0 & 4 & -4 \\ 0 & -4 & 4 \end{bmatrix} = \begin{bmatrix} 4 & -4 & 0 \\ -4 & 8 & -4 \\ 0 & -4 & 4 \end{bmatrix}. \tag{11.18}$$

Since element 3 results from a 180° rotation of element 1, it happens to have exactly the same S^e. Assembling the four element matrices gives the six system equations $\mathbf{S\,D} = \mathbf{C}$ where

$$\mathbf{S} = \begin{bmatrix} +4 & -4 & 0 & 0 & 0 & 0 \\ -4 & (+8+4+4) & (-4-4) & -4 & 0 & 0 \\ 0 & (-4-4) & (+4+8+4) & 0 & (-4-4) & 0 \\ 0 & -4 & 0 & +8 & -4 & 0 \\ 0 & 0 & (-4-4) & -4 & (+4+4+8) & -4 \\ 0 & 0 & 0 & 0 & -4 & +4 \end{bmatrix}$$

and

$$\mathbf{C} = \frac{QAt}{3}\begin{Bmatrix} 1 \\ 1+1+1 \\ 1+1+1 \\ 1 \\ 1+1+1 \\ 1 \end{Bmatrix} + \begin{Bmatrix} 0 \\ 0 \\ 0 \\ q_4 \\ q_5 \\ q_6 \end{Bmatrix} = \begin{Bmatrix} 4 \\ 12 \\ 12 \\ 4 \\ 12 \\ 4 \end{Bmatrix} + \begin{Bmatrix} 0 \\ 0 \\ 0 \\ q_4 \\ q_5 \\ q_6 \end{Bmatrix}.$$

In the above vector the qs are the nodal heat flux reactions required to maintain the specified external temperature. Since the last three equations have essential boundary conditions applied we can reduce the first three to

$$\begin{bmatrix} 4 & -4 & 0 \\ -4 & 16 & -8 \\ 0 & -8 & 16 \end{bmatrix}\begin{Bmatrix} \theta_1 \\ \theta_2 \\ \theta_3 \end{Bmatrix} = \begin{Bmatrix} 4 \\ 12 \\ 12 \end{Bmatrix} - \theta_4\begin{Bmatrix} 0 \\ -4 \\ 0 \end{Bmatrix} - \theta_5\begin{Bmatrix} 0 \\ 0 \\ -8 \end{Bmatrix} - \theta_6\begin{Bmatrix} 0 \\ 0 \\ 0 \end{Bmatrix}. \tag{11.19}$$

Substituting the data that the exterior surface temperatures are $\theta_4 = \theta_5 = \theta_6 = 5$ yields the reduced source term

$$\mathbf{C}^* = \begin{Bmatrix} 4 \\ 12 \\ 12 \end{Bmatrix} + \begin{Bmatrix} 0 \\ 20 \\ 0 \end{Bmatrix} + \begin{Bmatrix} 0 \\ 0 \\ 40 \end{Bmatrix} = \begin{Bmatrix} 4 \\ 32 \\ 52 \end{Bmatrix}.$$

Solving for the interior temperatures using the inverse

$$\mathbf{S}^{*-1} = \frac{1}{512}\begin{bmatrix} 192 & 64 & 32 \\ 64 & 64 & 32 \\ 32 & 32 & 48 \end{bmatrix}$$

and multiplying by $\mathbf{C}^*$ yields:

$$\mathbf{\Theta}^* = \begin{Bmatrix} 8.750 \\ 7.750 \\ 7.125 \end{Bmatrix} = \begin{Bmatrix} \theta_1 \\ \theta_2 \\ \theta_3 \end{Bmatrix}.$$

Substituting these values into the original system equations will give the exterior nodal heat flux values (*thermal reactions*) required by this problem. For example, the fourth row reaction equation yields: $-4\theta_2 + 8\theta_4 - 4\theta_5 = -4(7.75) + 8(5) - 4(5) = 4 + q_4$, or simply $-15 = q_4$. The other two nodal fluxes are $q_5 = -29$, $q_6 = -4$. These sum to -48. The internal heat generated was $\sum_e Q^e A^e t^e = +48$, which is equal and opposite.

```
*** INPUT SOURCE RESULTANTS ***                                  ! 1
ITEM        SUM            POSITIVE        NEGATIVE              ! 2
   1     4.8000E+01     4.8000E+01     0.0000E+00                ! 3
                                                                 ! 4
*** REACTION RECOVERY ***                                        ! 5
NODE, PARAMETER,       REACTION, EQUATION                        ! 6
   4,   DOF_1,       -1.5000E+01        4                        ! 7
   5,   DOF_1,       -2.9000E+01        5                        ! 8
   6,   DOF_1,       -4.0000E+00        6                        ! 9
                                                                 !10
REACTION RESULTANTS                                              !11
PARAMETER,     SUM            POSITIVE       NEGATIVE            !12
 DOF_1,      -4.8000E+01   0.0000E+00  -4.8000E+01               !13
                                                                 !14
***  OUTPUT OF RESULTS IN NODAL ORDER  ***                       !15
NODE,  X-Coord,      Y-Coord,      DOF_1,                        !16
   1  0.0000E+00  0.0000E+00  8.7500E+00                         !17
   2  2.0000E+00  0.0000E+00  7.7500E+00                         !18
   3  2.0000E+00  2.0000E+00  7.1250E+00                         !19
   4  4.0000E+00  0.0000E+00  5.0000E+00                         !20
   5  4.0000E+00  2.0000E+00  5.0000E+00                         !21
   6  4.0000E+00  4.0000E+00  5.0000E+00                         !22
                                                                 !23
*** SUPER_CONVERGENT AVERAGED NODAL FLUXES ***                   !24
NODE,    X-Coord,     Y-Coord,      FLUX_1,       FLUX_2,        !25
   1  0.0000E+00  0.0000E+00 -3.3333E-01 -4.5833E+00             !26
   2  2.0000E+00  0.0000E+00 -7.3333E+00 -2.0833E+00             !27
   3  2.0000E+00  2.0000E+00 -4.8333E+00 -2.0833E+00             !28
   4  4.0000E+00  0.0000E+00 -1.4333E+01  4.1667E-01             !29
   5  4.0000E+00  2.0000E+00 -1.1833E+01  4.1667E-01             !30
   6  4.0000E+00  4.0000E+00 -9.3333E+00  4.1667E-01             !31
```

Figure 11.11 *Results for square bar thermal analysis*

Thus, we have verified that the generated heat equals the heat outflow. Of course, this must be true for all steady state heat conduction problems. Note that while we started with six equations from the integral formulation only three were independent equations for the unknown temperatures. The other equations were independent equations for the thermal reactions necessary to maintain the essential boundary conditions on the temperature. One does not have to assemble and solve the reaction set but it is a recommended procedure. The input data for this example are given in Fig. 11.10 and selected outputs are given in Fig. 11.11.

Replacing the previous mesh with one of 64 elements and applying the *MODEL* code gives the temperature results shown as contours and a surface in Figs. 11.12 and 13, respectively. In the first we get a visual check that the contours appear perpendicular to the insulated boundaries and parallel to the Dirichlet boundary. From Fig. 11.13 we get the impression that the temperature gradient and flux would also be smooth. However, they are constant in each element and must be made continuous by the SCP fit as seen in Fig. 11.14, which is a 2-D generalization of Fig. 2.21 in that the element flux values are constant. From the symmetry boundary conditions we expect zero heat flux at the $(0, 0)$. The continuous flux, in Fig. 11.15, misses that so more refinement is needed. The discontinuous element flux vectors are in Fig. 11.16 while the continuous nodal flux vectors are shown in Fig. 11.17. The last figure should have the vectors parallel to the insulated boundaries and they seem to do that reasonably well. The energy norm error estimates in Fig. 11.18 exceed 2 percent and are about ten times larger than we usually want. The projected maximum element sizes for a new mesh to reduce those errors are shown in Fig. 11.19.

The same four element T3 mesh was used in the text by Kwon [14], a former student of mine, to analyze a half symmetry triangular region, of Fig. 11.20, with given normal flux on the right edge, a null essential boundary condition along the bottom, and the inclined edge insulated. The modified input data and selected output are shown in Figs. 11.21 and 22, respectively. In this case the heat flux in (+) normal to the right edge was 2 per unit length, acting over the side length of 4, for a resultant input of 8. Here there were only two such flux line segments (with a local nodal resultant at each end of 2). Therefore, the source effects were simply lumped at the three nodes by inspection. The keyword *loads*, at line 12, flagged the presence of such sources and their numeric values were read in lines 38-40. (Such values are terminated by reading the source at the last degree of freedom, which is usually zero.)

For this class of problem we expect that the sum of the external sources will be equal and opposite to the reactions at the essential boundary conditions. Figure 11.22 shows that the expected result is obtained, as are the expected temperatures. The lumping of the heat flux was easy only because the flux was constant and we could get the boundary length it acted on by inspection. Usually we would have to supply the programming and data input necessary to formulate the boundary flux resultant arrays, $\mathbf{C}_q^e$. For the edge of the T3 element we need only the two nodal contributions. Implementing such a 'boundary flux segment' takes relatively little coding as shown in Fig. 11.7, but it requires much more flexibility in the data input options and the ability to recover those data. In most of the previous examples we considered only a mesh with a

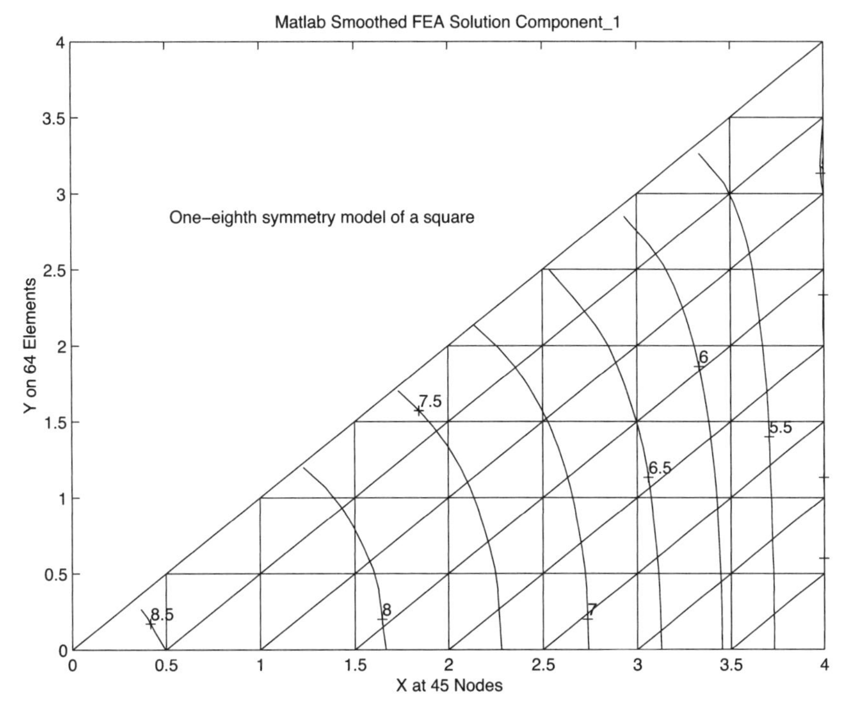

Figure 11.12 *Temperatures on the square segment*

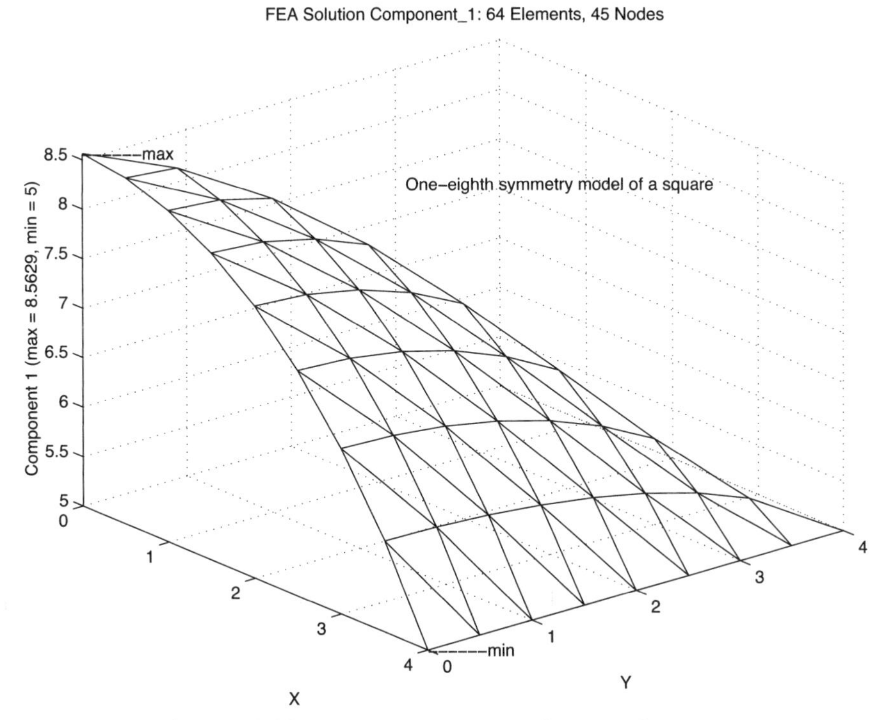

Figure 11.13 *Temperature carpet plot over the square*

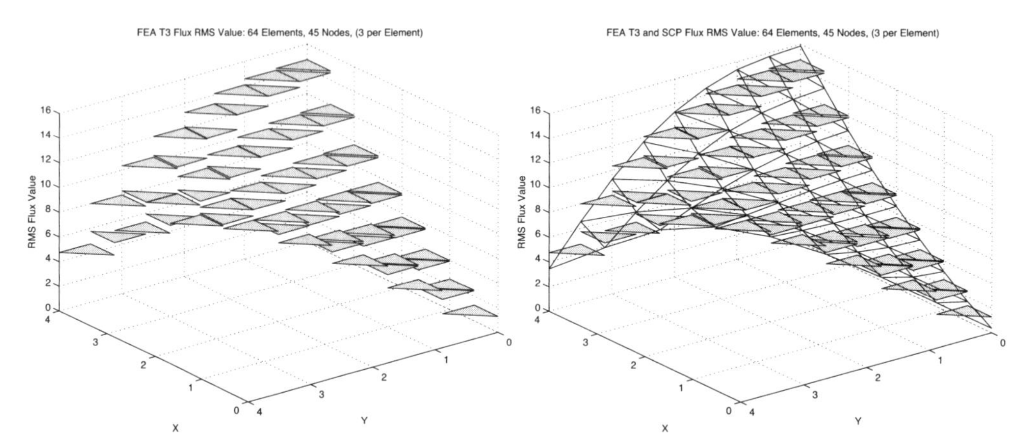

Figure 11.14 *Constant T3 fluxes and their SCP fit*

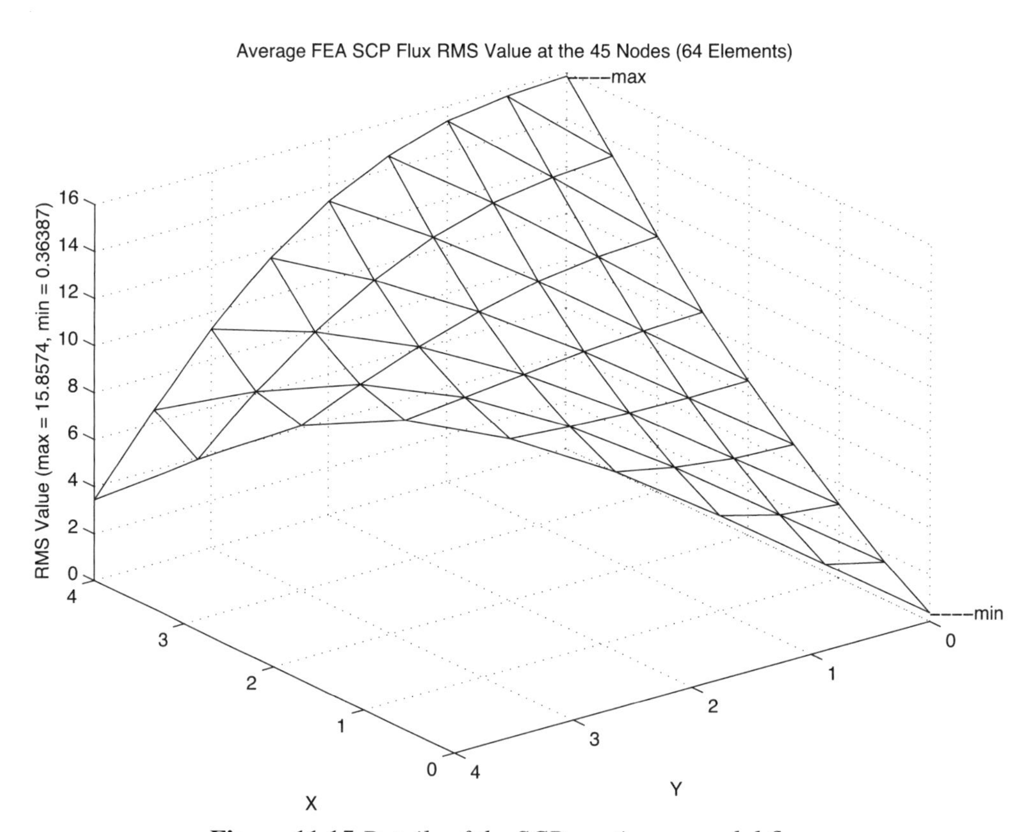

Figure 11.15 *Details of the SCP continuous nodal fluxes*

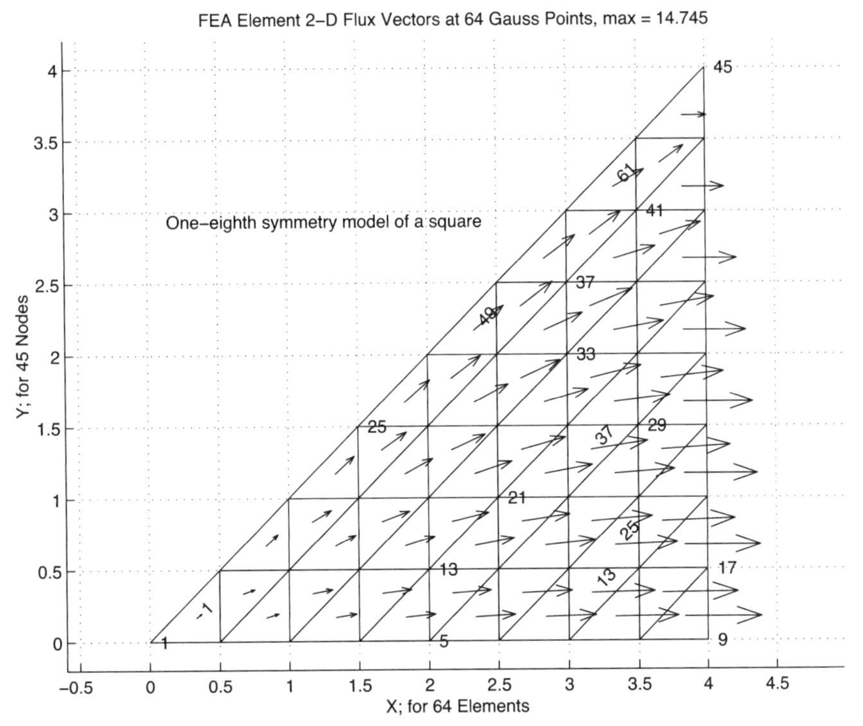

Figure 11.16 *Constant element flux vectors*

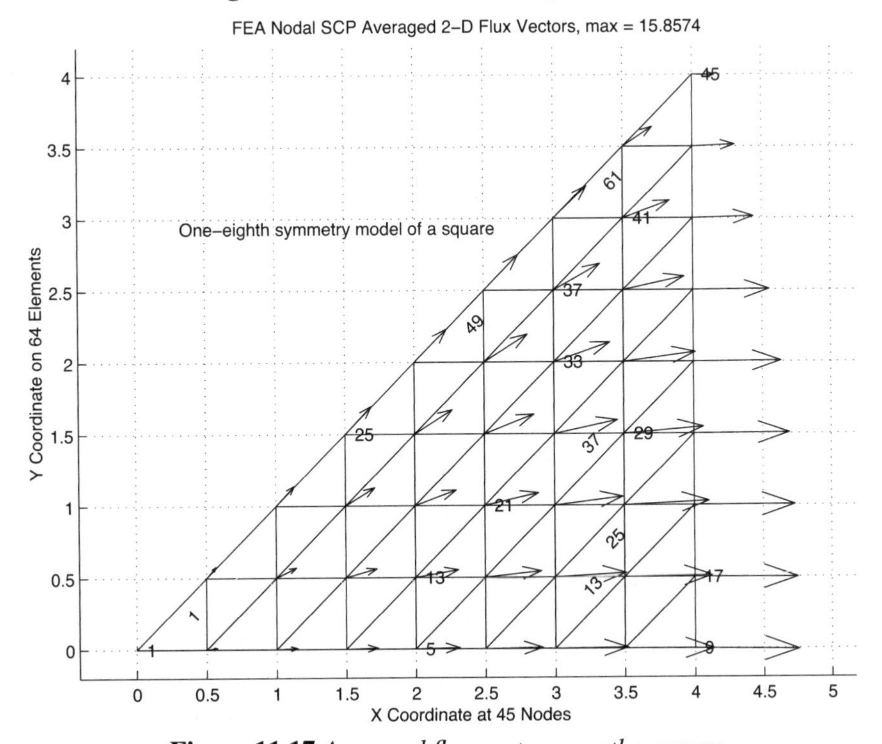

Figure 11.17 *Averaged flux vector over the square*

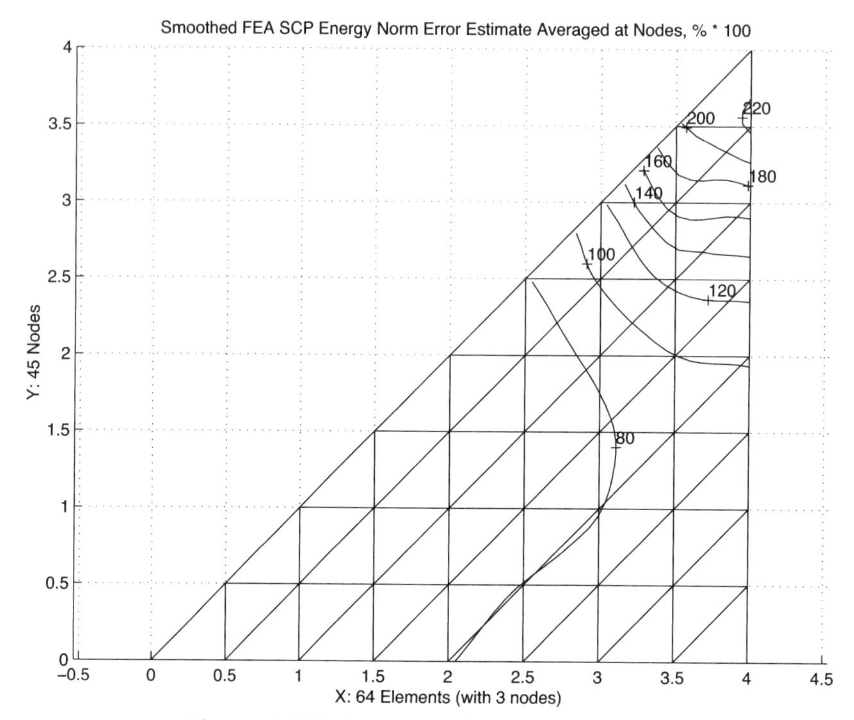

Figure 11.18 *Contours of energy norm error*

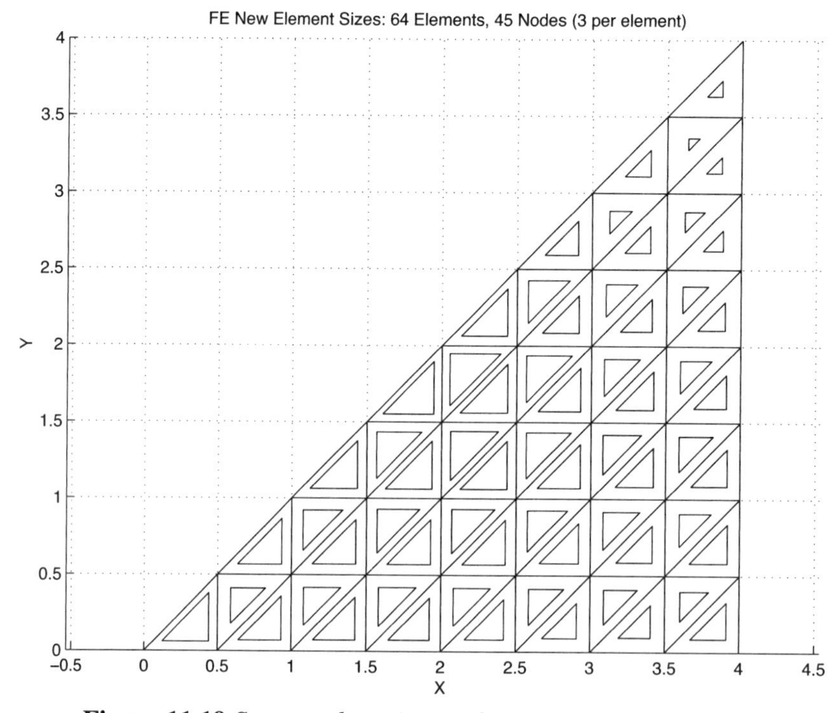

Figure 11.19 *Suggested maximum element sizes for new mesh*

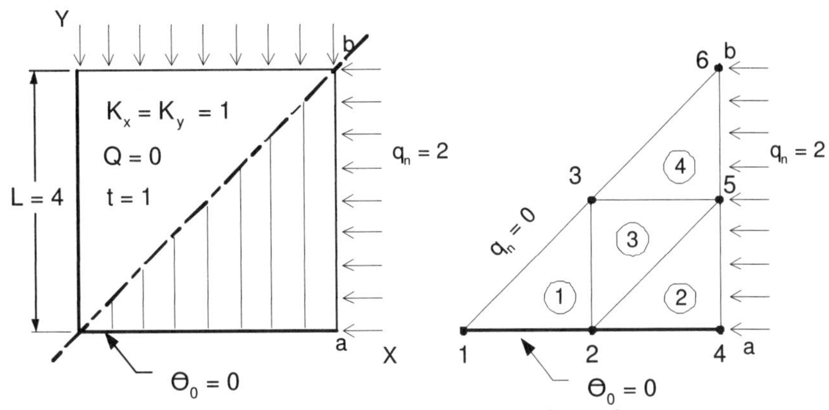

Figure 11.20 *Square with two flux sides*

```
title "T3 Conduction with given flux. Kwon example 5.4.1"      ! 1
example 204 ! Application source code library numbe            ! 2
b_rows    2 ! Number of rows in the B (operator) matrix        ! 3
dof       1 ! Number of unknowns per node                      ! 4
el_nodes  3 ! Maximum number of nodes per element              ! 5
elems     4 ! Number of elements in the system                 ! 6
gauss     1 ! Maximum number of quadrature points              ! 7
nodes     6 ! Number of nodes in the mesh                      ! 8
space     2 ! Solution space dimension                         ! 9
el_homo     ! Element properties are homogeneous               !10
el_real   4 ! Number of real properties per element            !11
loads       ! An initial source vector is input                !12
remarks   9 ! Number of user remarks, e.g. property names      !13
end         ! Terminate the keyword control, remarks follow    !14
Kwon example 5.4.1, conduction with given flux.  6 <- q_n      !15
K_x = K_y = 1, 0 = 6, K_xy = 0                  / | <- q_n     !16
L_1_4 = L_4_6 = 4, Thickness = 1               /  | <- q_n     !17
Edge 1_3_6 is insulated, so q_n = 0.          /(4)| <- q_n     !18
Edge 4_5_6 has q_n =2, per unit length.     3-----5 <- q_n     !19
EBC on edge 1_2_4 is T = 0                 / |(3)/| <- q_n     !20
K_x T,xx + 2K_xy T,xy + K_y T,yy + Q = 0 /   | /  | <- q_n     !21
Solution gives T_3 = 3, T_6 = 11.      /(1)|/(2)| <- q_n       !22
Edge 4_5_6 lumped flux = 2, 4, 2.     1----2----4 <- q_n       !23
   1    1 0.   0. ! node, ebc flag, x, y                       !24
   2    1 2.   0.                                              !25
   3    0 2.   2.                                              !26
   4    1 4.   0.                                              !27
   5    0 4.   2.                                              !28
   6    0 4.   4.                                              !29
  1  1  2  3       ! elem, three nodes                         !30
  2  2  4  5                                                   !31
  3  2  5  3                                                   !32
  4  3  5  6                                                   !33
   1    1 0.       ! node, dof, value of EBC                   !34
   2    1 0.                                                   !35
   4    1 0.                                                   !36
  1 1. 1. 0. 0.    ! elem, K_x, K_y, K_xy, Q (homogeneous)     !37
4 1 2.0  ! node, dof, lumped heat flux                         !38
5 1 4.0  ! node, dof, lumped heat flux                         !39
6 1 2.0  ! node, dof, lumped heat flux                         !40
```

Figure 11.21 *Triangular region with an edge flux*

single type of element. But it is common to mix elements like triangles, quadrilaterals, and line elements, so long as they have compatible edge interpolations. If we are going to allow multiple types of elements to be mixed then we must be able to define the number of nodes, shape, integration rule, properties, etc. for each. We will also split how we think about their purpose. Most will simply be thought of as 'standard elements', while others will exist to treat 'boundary flux segments' or to treat 'mixed (Robin) segment regions'. Their nodal connectivities and properties will be input in that order. The following examples will introduce some of the new free format keyword controls that *MODEL* employs to distinguish between these element types and the amount of data the user wishes to assign to each (if any). Table 11.1 lists most of the controls that can be selected to define combinations of element types.

To extend the previous lumped flux example to one that uses the source in Fig. 11.7 we must modify the prior data to allow the combination of some conduction elements with two flux boundary segment elements. The new controls and corresponding data are given in Fig. 11.23. Since a constant normal flux is quite common a control option *normal_flux* (line 15) is made available to the user for assigning a value to a global variable, Q_NORMAL_SEG, that can be used in application dependent arrays. Should the normal flux data not be constant everywhere one could define a different constant on each edge by using properties defined for each boundary segment (BS). The new data, in Fig. 11.23, yield exactly the same temperature results as before. The keyword *segments* 2 (line 14) caused the boundary integral calculations of Fig. 11.7 to be invoked, and told the system that two segments needed to be read (at lines 38-39). Expanding the mesh size to include 64 conduction triangles and 8 edge flux boundary segments yields the mesh (note right side) and the temperatures of Figs. 11.24 and 25, respectively.

11.5.2 Face convection

Two-dimensional models that combine the conduction through an area with convection from one or both of its faces often approximate cooling fins. We refer to such regions (points, lines, or surfaces) as 'mixed segments' or Robin segments. For the linear triangle it is again practical to write the matrices, defined in Eq. 11.11, in closed form and they were given in Fig. 11.8 for constant data as:

$$\mathbf{S}_h^b = \frac{h^b A^b}{12}\begin{bmatrix} 2 & 1 & 1 \\ 1 & 2 & 1 \\ 1 & 1 & 2 \end{bmatrix}, \quad \mathbf{C}_h^b = \frac{h_b A^b \Theta_\infty^b}{3}\begin{Bmatrix} 1 \\ 1 \\ 1 \end{Bmatrix}. \tag{11.20}$$

Again, it is not uncommon for the convection data, h^b, θ_∞, to be constant. To allow for that condition, two user keyword control options are provided: *convect_coef* and *convect_temp*. The second one (the surrounding fluid temperature) defaults to zero. They can be used to set the corresponding values of two global variables (with the same names) for possible use in application dependent matrices, as was done in Fig. 11.8. Should the mixed segment (MX) data not be constant everywhere one could define different constants on each face by using properties defined for each mixed segment.

```
*** INITIAL FORCING VECTOR DATA ***                                  ! 1
NODE   PARAMETER      VALUE     EQUATION                             ! 2
   4          1   2.00000E+00        4                               ! 3
   5          1   4.00000E+00        5                               ! 4
   6          1   2.00000E+00        6                               ! 5
*** RESULTANTS ***                                                   ! 6
COMPONENT          SUM         POSITIVE       NEGATIVE               ! 7
 IN_1,        8.0000E+00    8.0000E+00     0.0000E+00                ! 8
                                                                     ! 9
*** REACTION RECOVERY ***                                            !10
NODE, PARAMETER,        REACTION, EQUATION                           !11
   1,  DOF_1,          0.0000E+00      1                             !12
   2,  DOF_1,         -3.0000E+00      2                             !13
   4,  DOF_1,         -5.0000E+00      4                             !14
REACTION RESULTANTS                                                  !15
PARAMETER,      SUM            POSITIVE       NEGATIVE               !16
 DOF_1,      -8.0000E+00    0.0000E+00   -8.0000E+00                 !17
                                                                     !18
***  OUTPUT OF RESULTS IN NODAL ORDER  ***                           !19
NODE,  X-Coord,      Y-Coord,      DOF_1,                            !20
   1  0.0000E+00  0.0000E+00  0.0000E+00                             !21
   2  2.0000E+00  0.0000E+00  0.0000E+00                             !22
   3  2.0000E+00  2.0000E+00  3.0000E+00                             !23
   4  4.0000E+00  0.0000E+00  0.0000E+00                             !24
   5  4.0000E+00  2.0000E+00  6.0000E+00                             !25
   6  4.0000E+00  4.0000E+00  1.0000E+01                             !26
```

Figure 11.22 *Selected results for triangle with an edge flux*

Table 11.1 *Typical keywords for multiple element types*

```
WORD,       VALUES ! REMARKS                                     [DEFAULT]
area_thick    1.5  ! Global thickness of 2-D domain                    [1]
el_segment      3  ! Maximum nodes on element boundary segment         [0]
el_types        1  ! Number of different types of elements             [1]
type_parm     2 1  ! Parametric space for each element type            [d]
type_nodes    3 2  ! Number of analysis nodes for element types        [2]
type_gauss    1 2  ! Number of Gauss points in each element type       [0]

segments        1  ! Number of element segments with flux input        [0]
normal_flux    5.  ! Constant normal flux on all flux segments         [0]
flux_thick    1.2  ! Thickness of all flux load lines or points        [1]
seg_int         1  ! Number of integer properties per segment          [0]
seg_pt_flux     1  ! Segment flux components input at flux nodes       [1]
seg_real        3  ! Number of real properties per segment             [0]
seg_thick     1.2  ! Thickness of all flux and mixed segments          [1]

mixed_segs      1  ! Number of mixed boundary condition segments       [0]
mixed_int       0  ! Number of integer properties per mixed_bc         [0]
mixed_real      3  ! Number of real properties per mixed_bc            [0]
convect_coef   1.  ! Convection coefficient on all mixed segments      [0]
convect_temp   1.  ! Convection temperature on all mixed segments      [0]
convect_thick  2.  ! Thickness of all convection lines or points       [1]
convect_vary       ! Convection, different on all mixed segments       [F]

type_shape    2 2  ! Shape code of each element type                   [1]
type_geom     3 2  ! Number of geometric nodes for type       [type_nodes]
```

```
title "T3 Conduction with given flux. Via flux elements"        ! 1
example  204 ! Application source code library numbe            ! 2
remarks    9 ! Number of user remarks, e.g. property names      ! 3
b_rows     2 ! Number of rows in the B (operator) matrix        ! 4
dof        1 ! Number of unknowns per node                      ! 5
elems      4 ! Number of elements in the system                 ! 6
nodes      6 ! Number of nodes in the mesh                      ! 7
space      2 ! Solution space dimension                         ! 8
el_homo      ! Element properties are homogeneous               ! 9
el_real    4 ! Number of real properties per element            !10
el_types     2 ! Number of different types of elements          !11
type_nodes 3 2 ! Number of analysis nodes for element types     !12
type_shape 2 1 ! Shape code of each element type                !13
segments     2 ! Number of element segments with flux input     !14
normal_flux 2. ! Constant normal flux on all flux segments      !15
el_segment   2 ! Maximum nodes on element boundary segment      !16
no_error_est   ! Do NOT compute SCP element error estimates     !17
end ! Terminate keyword control input, remarks follow           !18
Kwon example 5.4.1, conduction with given flux.  6 <- q_n       !19
K_x = K_y = 1, 0 = 6, K_xy = 0                  / | <- q_n      !20
L_1_4 = L_4_6 = 4, Thickness = 1               /  | <- q_n      !21
Edge 1_3_6 is insulated, so q_n = 0.          /(4)| <- q_n      !22
Edge 4_5_6 has q_n =2, per unit length.     3-----5 <- q_n      !23
EBC on edge 1_2_4 is T = 0                 / |(3)/| <- q_n      !24
K_x T,xx + 2K_xy T,xy + K_y T,yy + Q = 0 /  | /  | <- q_n      !25
Solution gives T_3 = 3, T_6 = 11.       /(1)|/(2)| <- q_n      !26
Edges 4_5, 5_6 normal flux             1----2----4 <- q_n      !27
   1    1 0.   0. ! node, ebc flag, x, y                        !28
   2    1 2.   0.                                               !29
   3    0 2.   2.                                               !30
   4    1 4.   0.                                               !31
   5    0 4.   2.                                               !32
   6    0 4.   4. ! last node                                   !33
1  1 1  2  3  ! standard elem, el_type, three nodes             !34
2  1 2  4  5  ! standard elem, el_type, three nodes             !35
3  1 2  5  3  ! standard elem, el_type, three nodes             !36
4  1 3  5  6  ! standard elem, el_type, three nodes             !37
1 2 4 5 ! flux segment, el_type, two edge nodes                 !38
2 2 5 6 ! flux segment, el_type, two edge nodes                 !39
 1    1 0.  ! node, dof, value of EBC                           !40
 2    1 0.  ! node, dof, value of EBC                           !41
 4    1 0.  ! node, dof, value of EBC                           !42
1 1. 1. 0. 0. ! elem, K_x, K_y, K_xy, Q (homogeneous)           !43
```

Figure 11.23 *Combining edge flux segments with conduction*

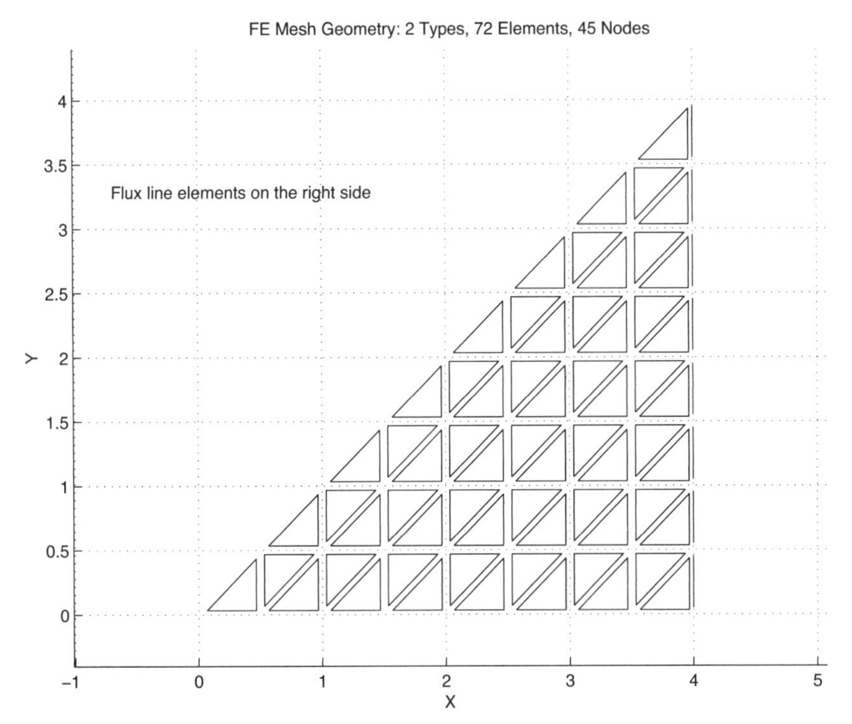

Figure 11.24 *Exploded conduction triangles and flux line segments*

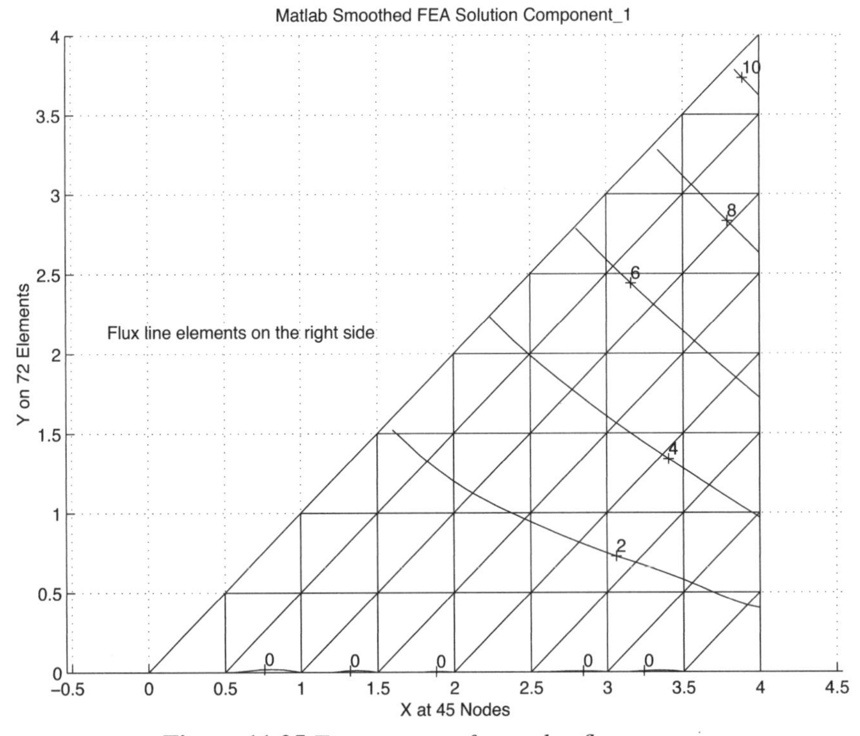

Figure 11.25 *Temperatures from edge flux sources*

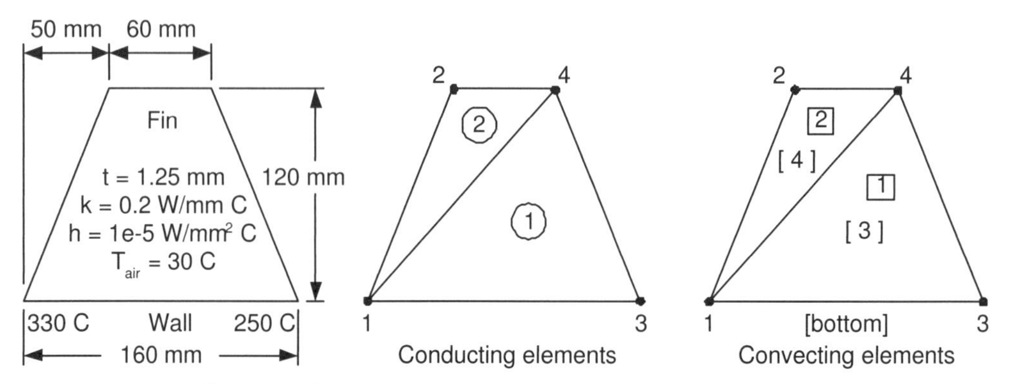

Figure 11.26 *A cooling fin with air convection on its faces*

As a simple example of a cooling fin we will consider the trapezoidal fin [2] given in detail by Allaire. He used a two linear triangle model that had convection on the top and bottom faces, and a linearly varying given temperature on the wall edge. The given problem is shown in Fig. 11.26. The corresponding data to invoke the face convection segments, as well as the standard conduction, are given in Fig. 11.27. The keyword *mixed_segs* 4 (line 14) is one way to cause the convection calculations of Fig. 11.8 to be invoked and the four sets of connectivity data to be read (lines 40-43). Two convecting face elements were on top of the fin and two were on the bottom. The model is so crude there is little precision in the temperatures in the selected output, which is given in Fig. 11.28. That figure also shows the numerical values of the matrices from each conducting and convecting element should the reader wish to verify their calculation. The system reactions, on lines 46-52, are significant. They show that to maintain the temperatures given in the two essential boundary conditions a total of 57.73 W of heat must flow into the fin. Shortly, when we consider the post-processing calculation for the convection heat loss we will find that exactly the same amount of heat flow is convected away. That is a reminder that a finite element is always flux conserving, when the fluxes are calculated properly (contrary to a common misconception).

When numerical integration is used *MODEL* lists the flux at each quadrature point. Here we have used a closed form expression for $\mathbf{S}^e$, but it corresponds to a one-point integration rule (as seen in Table 9.3). The necessary flux data was saved for the centroid of each element (in lines 17, 35, and 45 of Fig. 11.6). Then the nodal temperatures were gathered and multiplied by $-\mathbf{E}^e\mathbf{B}^e$ to yield the flux vector $\mathbf{q}^e = -\mathbf{E}^e\mathbf{B}^e\mathbf{T}^e$ at the point. If we try to use the element gradients (since the centroid is far from the boundary) to estimate the heat flow we get about 40.3 W for 42 percent flux error. For a more accurate heat loss calculation we would need to evaluate Eq. 7.37 by summing over each convection (mixed) segment. Such a post-processing implementation for the linear triangle convecting face is given in Fig. 11.29. The above flux recovery and the convection losses are given in Fig. 11.30. There (in line 13) we see the convection loss matches the thermal reactions of 57.73 W (line 52 of Fig. 11.28), as expected. The heat balances and element heat flux vectors for this crude mesh are shown in Fig. 11.31.

```
title "Fin face convection, Allaire, Basics FEM p. 343-353 2T3"     ! 1
example  204 ! Application source code library number               ! 2
elems      2 ! Number of elements in the system                     ! 3
nodes      4 ! Number of nodes in the mesh                          ! 4
space      2 ! Solution space dimension                             ! 5
b_rows     2 ! Number of rows in the B (operator) matrix            ! 6
dof        1 ! Number of unknowns per node                          ! 7
el_homo      ! Element properties are homogeneous                   ! 8
el_real    5 ! Number of real properties per element                ! 9
remarks   13 ! Number of user remarks, e.g. property names          !10
convect_coef 1.e-5 ! Convection coefficient on all mixed segments   !11
convect_temp    30 ! Convection temperature on all mixed segments   !12
el_segment       3 ! Maximum nodes on element boundary segment      !13
mixed_segs       4 ! Number of mixed boundary condition segments    !14
el_types         2 ! Number of different types of elements          !15
type_gauss     1 1 ! Number of Gauss points in each element type    !16
type_nodes     3 3 ! Number of nodes on each element type           !17
type_shape     2 2 ! Shape code of each element type                !18
post_mixed         ! Post-process mixed segments, create n_file2    !19
end ! Terminate the keyword control, remarks follow                 !20
Trapezoidal fin: 160 mm and 60 mm by 120 mm high (no edge conv.)    !21
Units: x,y-mm,  k-W/mm C, h-W/mm^2 C, t-C, Q-W/mm^3, q-W/mm^2       !22
Face convection uses constant convect_coef, convect_temp            !23
Conduction: K_xx U,xx + 2K_xy U,xy + K_yy U,yy + Q = 0              !24
PROP(1) = CONDUCTIVITY K_XX, PROP(2) = CONDUCTIVITY K_YY            !25
PROP(3) = CONDUCTIVITY K_XY, PROP(4) = SOURCE PER UNIT AREA         !26
PROP(5) = THICKNESS (DEFAULT 1.0), here 1.25mm                      !27
Convection: -K_n * U,n = CONVECT_COEF ( U - CONVECT_TEMP)           !28
Optional mixed properties: 1=CONVECT_COEF, 2=CONVECT_TEMP           !29
The conduction reaction total of 57.73 W is equal and opposite      !30
to the convection loss integral of 57.73 W, (actual is about        !31
57.66 W) BUT using element gradients gives 40.34 W for 42 %         !32
flux error.  (Reverse the signs on flux listings.)                  !33
 1   1    0.0    0.0 ! node, bc-flag,x, y                           !34
 3   1  160.0    0.0                                                !35
 2   0   50.0  120.0                                                !36
 4   0  110.0  120.0                                                !37
1 1    1 3 4 ! elem, el_type, 3 nodes. conducting                   !38
2 1    2 1 4 ! elem, el_type, 3 nodes. conducting                   !39
1 2    1 3 4 ! face, el_type, 3 nodes. convecting, top              !40
2 2    2 1 4 ! face, el_type, 3 nodes. convecting, top              !41
3 2    1 3 4 ! face, el_type, 3 nodes. convecting, bottom           !42
4 2    2 1 4 ! face, el_type, 3 nodes. convecting, bottom           !43
  1  1  330. ! node, dof, value                                     !44
  3  1  250. ! node, dof, value                                     !45
 1 0.2 0.2 0.0 0.0 1.25 ! el, k_x, k_y, K_xy, Q, thick              !46
```

Figure 11.27 *Cooling fin element types data*

```
BEGINNING STANDARD ELEMENT ASSEMBLY  (debug output)             ! 1
S matrix:                                 ! conduction          ! 2
ROW/COL       1          2          3                           ! 3
   1  1.10E-01 -5.79E-02 -5.21E-02                              ! 4
   2 -5.79E-02  1.73E-01 -1.15E-01                              ! 5
   3 -5.21E-02 -1.15E-01  1.67E-01                              ! 6
S matrix:                                 ! conduction          ! 7
ROW/COL       1          2          3                           ! 8
   1  4.60E-01 -1.15E-01 -3.45E-01                              ! 9
   2 -1.15E-01  6.25E-02  5.21E-02                              !10
   3 -3.45E-01  5.21E-02  2.93E-01                              !11
BEGINNING MIXED_BC SEGMENTS ASSEMBLY                            !12
S matrix:                              ! convect top            !13
ROW/COL       1          2          3                           !14
   1  1.60E-02  8.00E-03  8.00E-03                              !15
   2  8.00E-03  1.60E-02  8.00E-03                              !16
   3  8.00E-03  8.00E-03  1.60E-02                              !17
C matrix:                              ! convect top            !18
ROW/COL       1          2          3                           !19
   1  9.60E-01  9.60E-01  9.60E-01                              !20
S matrix:                              ! convect top            !21
ROW/COL       1          2          3                           !22
   1  6.00E-03  3.00E-03  3.00E-03                              !23
   2  3.00E-03  6.00E-03  3.00E-03                              !24
   3  3.00E-03  3.00E-03  6.00E-03                              !25
C matrix:                              ! convect top            !26
ROW/COL       1          2          3                           !27
   1  3.60E-01  3.60E-01  3.60E-01                              !28
S matrix:                               ! convect bottom        !29
ROW/COL       1          2          3                           !30
   1  1.60E-02  8.00E-03  8.00E-03                              !31
   2  8.00E-03  1.60E-02  8.00E-03                              !32
   3  8.00E-03  8.00E-03  1.60E-02                              !33
C matrix:                               ! convect bottom        !34
ROW/COL       1          2          3                           !35
   1  9.60E-01  9.60E-01  9.60E-01                              !36
S matrix:                               ! convect bottom        !37
ROW/COL       1          2          3                           !38
   1  6.00E-03  3.00E-03  3.00E-03                              !39
   2  3.00E-03  6.00E-03  3.00E-03                              !40
   3  3.00E-03  3.00E-03  6.00E-03                              !41
C matrix:                               ! convect bottom        !42
ROW/COL       1          2          3                           !43
   1  3.60E-01  3.60E-01  3.60E-01                              !44
                                                                !45
*** REACTION RECOVERY ***                                       !46
NODE, PARAMETER,      REACTION, EQUATION                        !47
   1,  DOF_1,       3.9927E+01     1                            !48
   3,  DOF_1,       1.7806E+01     3                            !49
REACTION RESULTANTS                                             !50
PARAMETER,     SUM           POSITIVE      NEGATIVE             !51
 DOF_1,      5.7733E+01  5.7733E+01   0.0000E+00                !52
                                                                !53
***  OUTPUT OF RESULTS IN NODAL ORDER  ***                      !54
NODE,  X-Coord,     Y-Coord,     DOF_1,                         !55
  1  0.0000E+00  0.0000E+00  3.3000E+02                         !56
  2  5.0000E+01  1.2000E+02  2.0556E+02                         !57
  3  1.6000E+02  0.0000E+00  2.5000E+02                         !58
  4  1.1000E+02  1.2000E+02  1.7817E+02                         !59
```

Figure 11.28 *Selected convecting fin results*

```
!  ......................................................          ! 1
!  *** POST_PROCESS_MIXED PROBLEM DEPENDENT STATEMENTS FOLLOW ***  ! 2
!  ......................................................          ! 3
!           (Stored as application source example 204.)            ! 4
! Global CONVECT_COEF set by keyword convect_coef is available     ! 5
! Global CONVECT_TEMP set by keyword convect_temp is available     ! 6
! H_INTG (LT_N) Integral of interpolation functions, H, available  ! 7
                                                                   ! 8
! Linear triangle face convection heat loss recover                ! 9
  REAL(DP) :: X_I, X_J, X_K, Y_I, Y_J, Y_K     ! Global coordinates !10
  REAL(DP) :: A_I, A_J, A_K, B_I, B_J, B_K     ! Standard geometry  !11
  REAL(DP) :: C_I, C_J, C_K, X_CG, Y_CG, TWO_A ! Standard geometry  !12
  REAL(DP), SAVE :: Q_LOSS, TOTAL       ! Face and total heat loss  !13
                                                                    !14
  LOGICAL, SAVE :: FIRST = .TRUE.  ! printing                       !15
                                                                    !16
   IF ( FIRST ) THEN                    ! first call                !17
     FIRST = .FALSE. ; WRITE (6, 5)     ! print headings            !18
     5 FORMAT ('*** CONVECTION HEAT LOSS ***', /, &                 !19
     &        'ELEMENT    HEAT_LOST')                               !20
     TOTAL = 0.d0                                                   !21
   END IF ! first call                                              !22
                                                                    !23
!   DEFINE NODAL COORDINATES, CCW: I, J, K                          !24
   X_I = COORD (1,1) ;    X_J = COORD (2,1) ;    X_K = COORD (3,1)  !25
   Y_I = COORD (1,2) ;    Y_J = COORD (2,2) ;    Y_K = COORD (3,2)  !26
                                                                    !27
!   GEOMETRIC PARAMETERS: H_I (X,Y) = (A_I + B_I*X + C_I*Y)/TWO_A   !28
   A_I = X_J * Y_K - X_K * Y_J ; B_I = Y_J - Y_K ; C_I = X_K - X_J  !29
   A_J = X_K * Y_I - X_I * Y_K ; B_J = Y_K - Y_I ; C_J = X_I - X_K  !30
   A_K = X_I * Y_J - X_J * Y_I ; B_K = Y_I - Y_J ; C_K = X_J - X_I  !31
                                                                    !32
!   CALCULATE TWICE ELEMENT AREA                                    !33
   TWO_A = A_I + A_J + A_K               ! = B_J*C_K - B_K*C_J also  !34
                                                                    !35
!   HEAT LOST FROM THIS FACE: Integral over face of h * (T - T_inf) !36
   H_INTG (1:3) = TWO_A / 6            ! Integral of H array        !37
   D (1:3) = D(1:3) - CONVECT_TEMP   ! Temp difference at nodes     !38
   Q_LOSS = CONVECT_COEF * DOT_PRODUCT (H_INTG, D)  ! Face loss     !39
   TOTAL  = TOTAL + Q_LOSS             ! Running total              !40
                                                                    !41
   PRINT '(I6, ES15.5)', IE, Q_LOSS                                 !42
   IF ( IE == N_MIXED ) PRINT *, 'TOTAL = ', TOTAL                  !43
!  *** END POST_PROCESS_MIXED PROBLEM DEPENDENT STATEMENTS ***      !44
```

Figure 11.29 *Convecting mixed segment heat loss recovery*

```
*** FLUX COMPONENTS AT ELEMENT INTEGRATION POINTS ***              ! 1
ELEMENT, PT,  X-Coord,     Y-Coord,     FLUX_1,     FLUX_2          ! 2
      1  1  9.0000E+01  4.0000E+01  1.2500E-01  2.0172E-01          ! 3
      2  1  5.3333E+01  8.0000E+01  1.1412E-01  2.1169E-01          ! 4
                                                                   ! 5
*** CONVECTION HEAT LOSS ***                                       ! 6
 ELEMENT      HEAT_LOST                                            ! 7
     1      2.13815E+01                                            ! 9
     2      7.48482E+00                                            !10
     3      2.13815E+01                                            !11
     4      7.48482E+00                                            !12
 TOTAL =   57.73267       <=== note <===                           !13
```

Figure 11.30 *Additional convecting fin results*

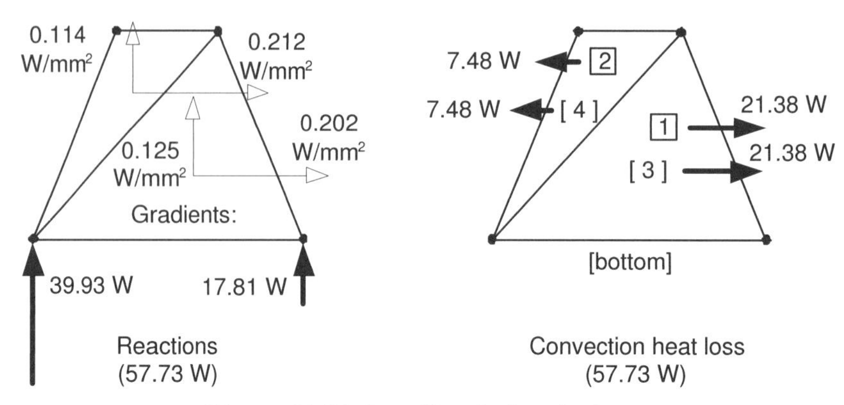

Figure 11.31 *A cooling fin heat balance*

With the numerically integrated formulations we have the ability to use any number of the linear, quadratic, and cubic elements in the *MODEL* library. We simply must generate more data to improve the accuracy. Mesh generators are used for that purpose. Here we will divide each edge with 5 nodes leading to a total of 25 nodes. Then we could use 32 T3, 16 Q4, 8 T6, 4 Q9, or 2 T15 elements in the mesh. Here we will graphically summarize the linear triangles and cubic quadrilateral results. The meshes, temperature contours, and the smoothed heat flux vectors, from the SCP, are shown in Fig. 11.32. The flux vectors should be parallel to the three outer edges since we assumed them insulated. If we revised the data to include edge convection on those three edges we would get more correct results, but they are so thin it probably is not worth the effort. The fin temperature distribution is given as a surface in Fig. 11.33. There the dashed vertical lines can be used with the left (z-axis) scale to obtain local nodal values.

The most important aspect of selecting various element types is assuring that the proper number of quadrature points are selected for each element or segment shape and polynomial degree. Here the conduction element integrand is a lower degree polynomial than for the convection segment. For the T3 element we specified a 1 point rule for conduction, and a 3 point rule for the convection matrix. The most important data changes for this Q9 element calculation are given in Fig. 11.34, while selected output results are in Fig. 11.35. In the latter we note that the conduction reactions and the convection heat loss results are again equal and opposite (lines 26 and 57), but slightly lower than in the very crude two element model.

Examining the energy norm error estimates for the linear triangle mesh in Fig. 11.32 (left) it exceeds 9 percent, as shown in Fig.11.36 so the projected mesh refinement is obtained as illustrated in Fig. 11.37. Employing those suggested element sizes to create a new mesh, with 215 T3 elements, one gets the temperature surface shown in Fig. 11.38. Note that the minimum temperature has changed from about 210 C to about 203 C. Since the surrounding air temperature is low (30 C) there is a corresponding reduction of total convection heat loss to 55.48 W. Of course, the essential boundary condition reactions resultant is equal and opposite to that value. One might find these relatively small changes in temperatures and heat flows to be enough to cease refining the solution.

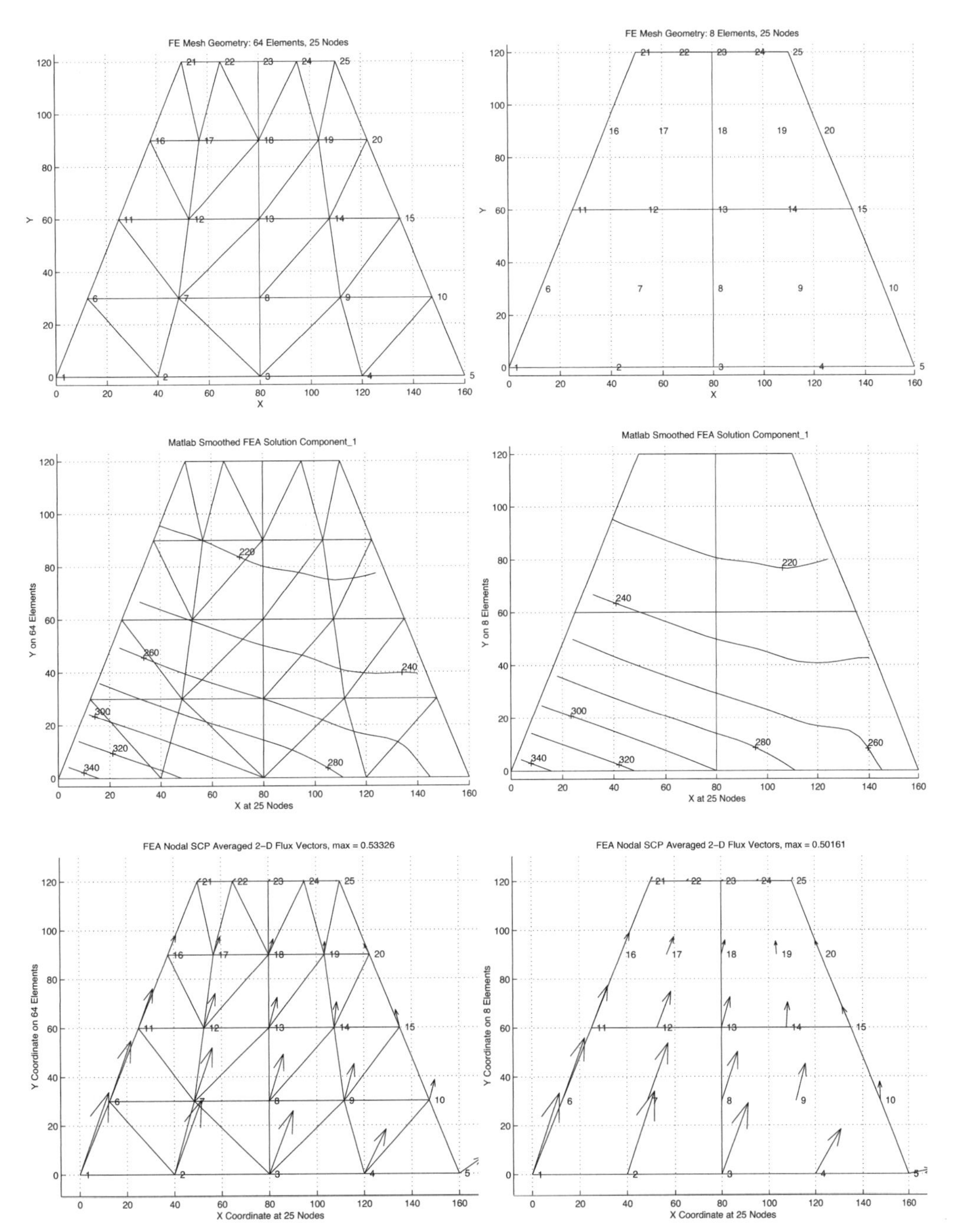

Figure 11.32 *Mesh, temperature, and smoothed flux vectors for the T3 and Q9 fin*

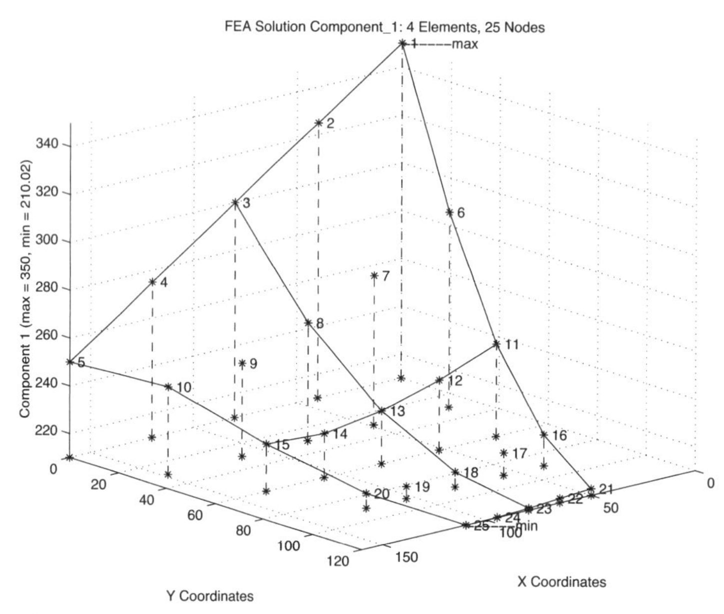

Figure 11.33 *Carpet-graph of fin temperatures, for Q9 elements*

```
title "Fin face convection, Allaire, Four Q9 elements"          ! 1
example        209 ! Application source code library number     ! 2
elems            4 ! Number of elements in the system           ! 3
nodes           25 ! Number of nodes in the mesh                ! 4
convect_coef 2.e-5 ! Convection coefficient on mixed segments   ! 5
el_segment       9 ! Maximum nodes on element boundary segment  ! 6
mixed_segs       4 ! Number of mixed boundary condition segments ! 7
type_gauss    9 16 ! Number of Gauss points in each element type ! 8
type_nodes    9  9 ! Number of nodes on each element type       ! 9
type_shape    3  3 ! Shape code of each element type            !10
. . .                                                           !11
Here h is doubled for two face convection on 4 segment.         !12
    1    1    0.0000    0.0000 ! node, bc-flag,x, y             !13
    2    1   40.0000    0.0000                                  !14
. . .                                                           !15
   24    0   95.0000  120.0000                                  !16
   25    0  110.0000  120.0000 ! last node                      !17
1 1  1  3 13 11  2  8 12  6  7 ! elem, type, 9 nodes. conducting !18
2 1  3  5 15 13  4 10 14  8  9 ! conducting                     !19
3 1 11 13 23 21 12 18 22 16 17 ! conducting                     !20
4 1 13 15 25 23 14 20 24 18 19 ! conducting                     !21
1 2  1  3 13 11  2  8 12  6  7 ! convecting                     !22
2 2  3  5 15 13  4 10 14  8  9 ! convecting                     !23
3 2 11 13 23 21 12 18 22 16 17 ! convecting                     !24
4 2 13 15 25 23 14 20 24 18 19 ! convecting                     !25
  1  1  3.50000E+02   ! bc along wall                           !26
  2  1  3.25000E+02                                             !27
  3  1  3.00000E+02                                             !28
  4  1  2.75000E+02                                             !29
  5  1  2.50000E+02                                             !30
1  0.2 0.2 0.0 0.0 1.25 ! el, kx, ky, kxy, Q, thick             !31
```

Figure 11.34 *Data changes for the Q9 model*

```
TITLE: "Fin face convection, Allaire, Four Q9 elements"                 ! 1
. . .                                                                   ! 2
*** MIXED BOUNDARY CONDITION SEGMENTS ***                               ! 3
 NOTE: CONSTANT CONVECTION ON ALL MIXED SEGMENTS USING                  ! 4
 CONVECTION COEFFICIENT =    2.0000000000000002E-05                     ! 5
 CONVECTION TEMPERATURE =   30.0000000000000000                         ! 6
SEGMENT, TYPE, 9 NODES ON THE SEGMENT                                   ! 7
      1     2   1    3   13   11    2    8   12    6    7               ! 8
      2     2   3    5   15   13    4   10   14    8    9               ! 9
      3     2  11   13   23   21   12   18   22   16   17               !10
      4     2  13   15   25   23   14   20   24   18   19               !11
. . .                                                                   !12
 *** SYSTEM GEOMETRIC PROPERTIES ***                                    !13
VOLUME   =  1.32000E+04                                                 !14
CENTROID =  8.00000E+01  5.09091E+01                                    !15
. . .                                                                   !16
*** REACTION RECOVERY ***                                               !17
NODE, PARAMETER,      REACTION, EQUATION                                !18
   1,  DOF_1,        6.4546E+00     1                                   !19
   2,  DOF_1,        2.5437E+01     2                                   !20
   3,  DOF_1,        1.1474E+01     3                                   !21
   4,  DOF_1,        1.4669E+01     4                                   !22
   5,  DOF_1,       -4.2698E-01     5                                   !23
REACTION RESULTANTS                                                     !24
PARAMETER,     SUM           POSITIVE      NEGATIVE                     !25
 DOF_1,      5.7608E+01   5.8035E+01   -4.2698E-01                      !26
                                                                        !27
*** EXTREME VALUES OF THE NODAL PARAMETERS ***                          !28
PARAMETER          MAXIMUM,    NODE       MINIMUM,    NODE              !29
 DOF_1,        3.5000E+02,      1   2.1002E+02,     25                  !30
. . .                                                                   !31
 *** SUPER_CONVERGENT AVERAGED NODAL FLUXES ***                         !32
NODE,    X-Coord,     Y-Coord,      FLUX_1,       FLUX_2,               !33
   1  0.0000E+00  0.0000E+00  1.5958E-01  4.6654E-01                    !34
   2  4.0000E+01  0.0000E+00  1.5970E-01  4.7551E-01                    !35
   3  8.0000E+01  0.0000E+00  1.5709E-01  4.0524E-01                    !36
   4  1.2000E+02  0.0000E+00  1.5177E-01  2.5573E-01                    !37
   5  1.6000E+02  0.0000E+00  1.4372E-01  2.6974E-02                    !38
   6  1.2500E+01  3.0000E+01  1.3393E-01  3.6177E-01                    !39
   7  4.8396E+01  3.0000E+01  1.2167E-01  3.3147E-01                    !40
   8  8.0000E+01  3.0000E+01  9.6232E-02  2.8267E-01                    !41
. . .                                                                   !42
  19  1.0340E+02  9.0000E+01 -4.2837E-03  7.5790E-02                    !43
  20  1.2250E+02  9.0000E+01 -3.3886E-02  7.8877E-02                    !44
  21  5.0000E+01  1.2000E+02  1.2224E-02  2.4631E-02                    !45
  22  6.5000E+01  1.2000E+02  1.3099E-02  1.1080E-02                    !46
  23  8.0000E+01  1.2000E+02  1.2498E-02  5.4736E-03                    !47
  24  9.5000E+01  1.2000E+02  1.0419E-02  7.8118E-03                    !48
  25  1.1000E+02  1.2000E+02  6.8624E-03  1.8095E-02                    !49
                                                                        !50
*** CONVECTION HEAT LOSS ***                                            !51
ELEMENT     HEAT_LOST                                                   !52
     1    2.02094E+01                                                   !53
     2    1.79895E+01                                                   !54
     3    9.83119E+00                                                   !55
     4    9.57818E+00                                                   !56
 TOTAL =    57.6082      <=== note <===                                 !57
```

Figure 11.35 *Selected cubic quadrilateral (Q9) fin results*

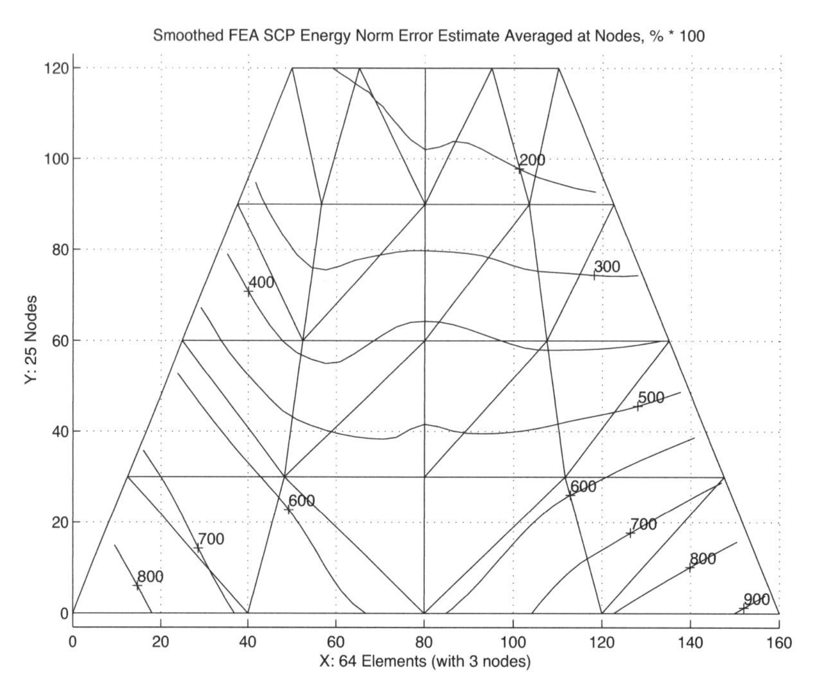

Figure 11.36 *Error norm levels for the 32 T3 model (max 9 percent)*

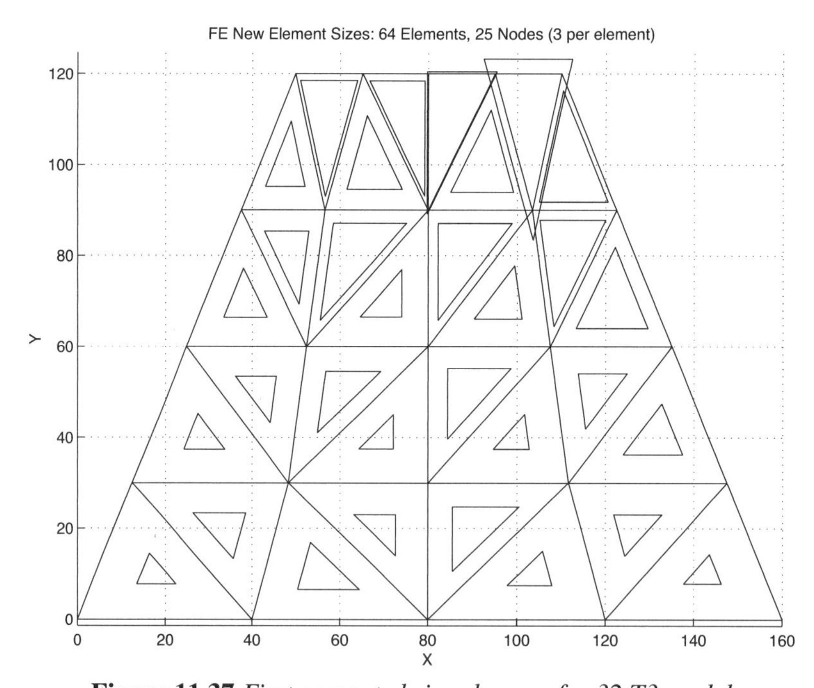

Figure 11.37 *First computed size changes for 32 T3 model*

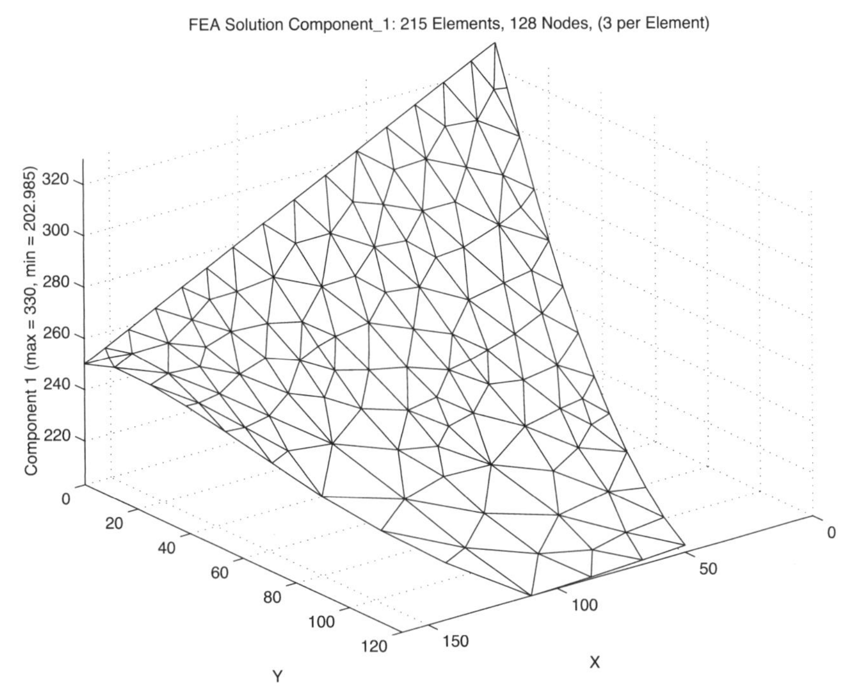

Figure 11.38 *Temperature surface for revised T3 model*

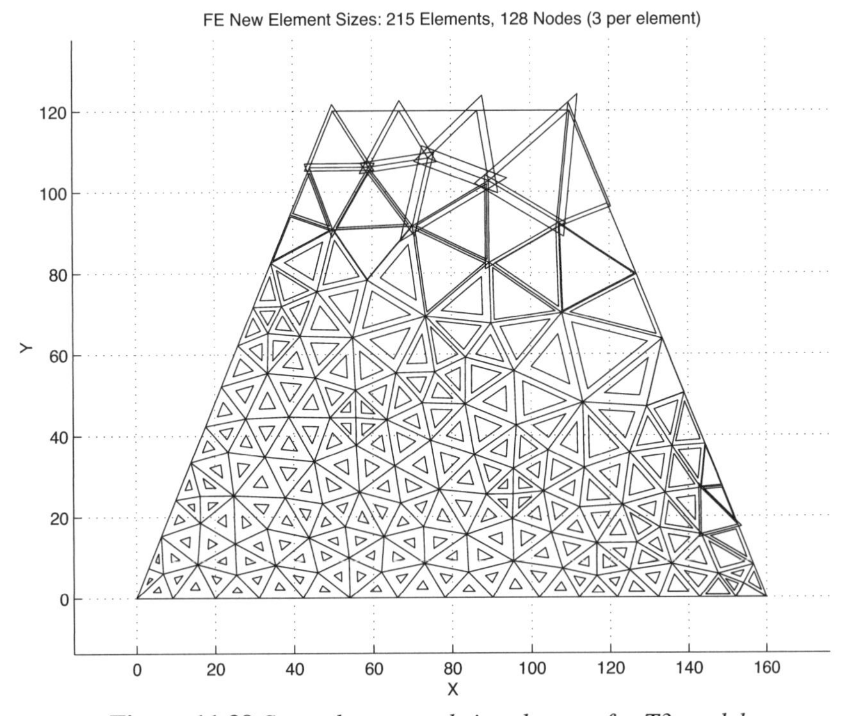

Figure 11.39 *Second computed size changes for T3 model*

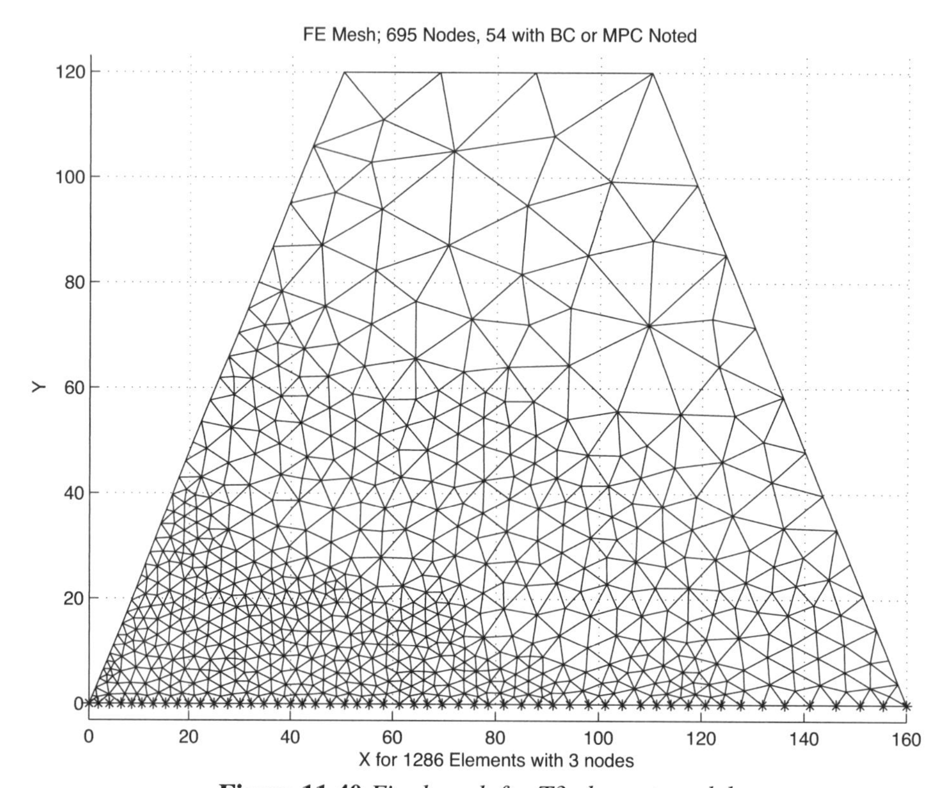

Figure 11.40 *Final mesh for T3 element model*

However, that mesh still has more than a 2.2 percent error level so another mesh size adjustment is computed. It is shown in Fig. 11.39 where again regions of high error project the need for elements much smaller than the current mesh and some elements near the free edge might be enlarged. The next stage of mesh generation creates a model with 695 nodes and 1286 T3 elements as shown in Fig. 11.40. One continues in this way until an acceptable error estimate is obtained.

11.6 Bilinear rectangles *

When quadrilateral elements have a parallelogram shape in physical space, their Jacobian is constant. If it takes the form of a rectangle with sides parallel to the global axes, then the Jacobian matrix is a constant diagonal matrix that allows the easy analytic evaluation of the element matrices. Consider a Q4 element mapped into a rectangular element that is parallel to the global axes so that the lengths parallel to the x and y axes are L_x and L_y, respectively. This mapping gives a constant Jacobian:

$$x = \bar{x} + a\,L_x/2\,, \qquad \frac{\partial x}{\partial a} = \frac{L_x}{2}\,, \qquad \frac{\partial x}{\partial b} \equiv 0\,,$$

$$y = \bar{y} + b\,L_y/2\,, \qquad \frac{\partial y}{\partial a} = 0\,, \qquad \frac{\partial y}{\partial b} = \frac{L_y}{2}\,,$$

$$\mathbf{J} = \tfrac{1}{2}\begin{bmatrix} L_x & 0 \\ 0 & L_y \end{bmatrix}, \qquad |J| = \frac{L_x\,L_y}{4} = \frac{A^e}{4}\,.$$

By inspection of the linear geometry mappings (or from the inverse Jacobian) we see $\partial/\partial x = \partial/\partial a\ \partial a/\partial x = 2/L_x\ \partial/\partial a,$ and likewise $\partial/\partial y = 2/L_y\ \partial/\partial b$ so that the typical term in the condition matrix due to k_x is

$$\mathbf{S}_{x_{i,j}} = \frac{4}{L_x^2}\int_{\Omega^e} k_x^e\ H_{i,a}\ H_{j,a}\ d\Omega, \tag{11.21}$$

but $d\,\Omega = |J|\ d\,\square$ so that if k_x is constant

$$\mathbf{S}_{x_{i,j}} = \frac{k_x^e\ L_y^e}{L_x^e}\int_{\square} H_{i,a}\ H_{j,a}\ d\square.$$

The interpolation functions are $H_j = (1 + a_j\,a)\,(1 + b_j\,b)/4$, so that a typical local derivative is $H_{j,a} = a_j\ (1 + b_j\ b)/4$ and the integrand becomes

$$H_{i,a}\ H_{j,a} = a_i\ a_j\Big[\,1 + (b_i + b_j)\ b + b_i\ b_j\ b^2\,\Big]/16. \tag{11.22}$$

Invoking numerical integration in $\square$ gives

$$\mathbf{S}_{x_{i,j}} = \frac{a_i\ a_j\ k_x^e\ L_y^e}{16\ L_x^e}\sum_{q=1}^{Q}\Big[\,1 + (b_i + b_j)\,b_q + b_i\,b_j\,b_q^2\,\Big]\,w_q.$$

Since this expression is quadratic in b, we need to pick only two points in the b direction. Likewise, for similar terms in the $\mathbf{S}_y$ matrix, we need two points in the a direction. Using the $Q = 4$ rule, we note that $w_q = 1$ and is constant, and that two $b_q = 1/\sqrt{3},$ while two are $-\,1/\sqrt{3}$. Thus, the linear terms cancel so that

$$\mathbf{S}_{x_{i,j}} = \frac{a_i\,a_j\ k_x^e\ L_y^e}{12\,L_x^e}\Big[\,3 + b_i\,b_j\,\Big]. \tag{11.23}$$

For the chosen local numbering, we have

j	a_j	b_j
1	−1	−1
2	1	−1
3	1	1
4	−1	1

so that the conduction matrix contributions are

$$\mathbf{S}_x = \frac{k_x^e L_y^e}{6 L_x^e} \begin{bmatrix} 2 & -2 & -1 & 1 \\ -2 & 2 & 1 & -1 \\ -1 & 1 & 2 & -2 \\ 1 & -1 & -2 & 2 \end{bmatrix} \tag{11.24}$$

which agrees with the exact integration. Likewise, for a constant k_y^e:

$$\mathbf{S}_y = \frac{k_y^e L_x^e}{6 L_y^e} \begin{bmatrix} 2 & 1 & -1 & -2 \\ 1 & 2 & -2 & -1 \\ -1 & -2 & 2 & 1 \\ -2 & -1 & 1 & 2 \end{bmatrix}. \tag{11.25}$$

Note that the sum of the terms in each row and each column is zero. This is typical for Lagrangian interpolation and serves as a useful visual check when doing hand calculations. The typical convention square (capacity) matrix term is

$$\begin{aligned} \mathbf{M}_{i,j} &= \int_{\Omega^e} \zeta^e H_i H_j \, d\Omega \\ &= \frac{L_x^e L_y^e}{64} \int_{\square^e} \zeta^e (1 + a_i a)(1 + a_j a)(1 + b_i b)(1 + b_j b) \, d\square \end{aligned} \tag{11.26}$$

so for constant ζ^e per unit area

$$\mathbf{M}_{i,j} = \frac{\zeta^e L_x^e L_y^e}{64} \int_{\square} [(1 + (a_i + a_j) a + a_i a_j a^2)(1 + (b_i + b_j) b + b_i b_j b^2)] \, d\square.$$

Again the four point rule is valid, and since $a_q = \pm 1/\sqrt{3}$, the linear terms cancel. Since $w_q \equiv 1$, we get

$$\mathbf{M}_{i,j} = \frac{\zeta^e L_x^e L_y^e}{64} [\, 4(1 + a_i a_j/3)(1 + b_j b_i/3) \,]$$

or

$$\mathbf{M} = \frac{\zeta^e L_x^e L_y^e}{36} \begin{bmatrix} 4 & 2 & 1 & 2 \\ 2 & 4 & 2 & 1 \\ 1 & 2 & 4 & 2 \\ 2 & 1 & 2 & 4 \end{bmatrix}. \tag{11.27}$$

In this last matrix the sum of all the terms in the square brackets is 36, which assures that the total 'mass', $\zeta^e L_x^e L_y^e$, is properly accounted for. This is true for Lagrangian interpolation but not for Hermite elements. The edge boundary matrices are the same as for the linear triangle, since both are linear along their edges.

11.7 General 2-d elements

If we select a higher order element such as the isoparametric quadrilateral then some of the above integrations are more difficult to evaluate. In the notation of Chapter 9, Eq. 9.32 with $\mathbf{d}_x = \partial \mathbf{H}/\partial x$, etc. the conduction contribution becomes

$$\mathbf{S}^e = \int_{A^e} \left(k_x^e \mathbf{d}_x^{e^T} \mathbf{d}_x^e + k_y^e \mathbf{d}_y^{e^T} \mathbf{d}_y^e \right) t^e \, dA = \int_{A^e} \mathbf{B}^{e^T} \mathbf{E}^e \mathbf{B}^e \, t^e \, dA. \tag{11.28}$$

If we allow for general quadrilaterals and/or curved sides then we will need numerical integration. Thus, we write

$$\mathbf{S}^e = \sum_{i=1}^{n_q} W_i \, |J^{e_i}| (k_{x_i}^e \, \mathbf{d}_{x_i}^{e^T} \, \mathbf{d}_{x_i}^e + k_{y_i}^e \, \mathbf{d}_{y_i}^{e^T} \, \mathbf{d}_{y_i}^e) \, t_i^e . \tag{11.29}$$

Typically, we would place the conductivities, k_x^e etc., in the constitutive matrix, $\mathbf{E}^e$, especially if they are fully anisotropic. In that case the integration is more clearly cast into a form involving matrix products:

$$\mathbf{S}^e = \sum_{i=1}^{n_q} W_i \, |J^{e_i}| (\mathbf{B}^{e^T} \, \mathbf{E}^e \, \mathbf{B}^e) \, t_i^e . \tag{11.30}$$

Similar expressions are available for the mass (or capacity) matrix and source vectors.

11.8 Numerically integrated arrays

The processes given above can be generalized to handle any curvilinear 1-, 2-, or 3-dimensional element by using general interpolation libraries, quadrature rule libraries, and numerical integration. First we will consider the conduction matrix and volumetric source vector as shown in Fig. 11.41 which accepts any of the elements in the *MODEL* library. Next, we will sometimes have a known flux acting across the normal to a given segment of the boundary. This is usually called a Neumann condition. It degenerates to a natural boundary condition when the flux is zero because the integral of zero is zero and there is no need to invoke the calculation. Since the flux can act normal to any segment of the mesh the general differential geometry considerations, of Sec. 8.6, may come into play. That requires a more general numerical integration process because the geometric mappings are not always one to one. That is, the normal flux can occur on a doubly curved 3-dimensional surface, a flat surface in the analysis plane, a curved edge, or simply a point. Thus, a full and more expensive use of differential geometry replaces a simple (one to one) Jacobian calculation shown before.

In this case only a column vector (that may degenerate to a scalar) must be computed as given in Fig. 11.42. There, in line 26, the general parametric measure is computed via function *PARM_GEOM_METRIC* (in Fig. 9.12) instead of the determinant of the usual Jacobian matrix shown in previous examples. Of course, they give the same results when the space mapping is one to one (straight line to straight line, or flat area to flat area). Finally, we will consider a pair of mixed or Robin type of matrices. Here they will represent heat convection data. This similar coding always also forms a square matrix and a column vector (which may degenerate to scalars) as listed in Fig. 11.43. It includes logic to identify either a globally constant set of convection data, or a different set of convection constants on each segment. One could also allow the convection data to be input as nodal properties if they vary over the segments but those details are omitted here to save space. Again *PARM_GEOM_METRIC*, line 42, allows for a general non-flat convection surface, or convection along a non-straight space curve.

```
! .............................................................  ! 1
!    ***  ELEM_SQ_MATRIX PROBLEM DEPENDENT STATEMENTS FOLLOW ***   ! 2
! .............................................................  ! 3
!    ANISOTROPIC POISSON EQUATION IN 1-, 2-, or 3-DIMENSIONS       ! 4
!            VIA NUMERICALLY INTEGRATED ELEMENTS                   ! 5
!       (K_ij * U,i),j + Q = 0;  1 <= (i,j) <= N_SPACE             ! 6
 REAL(DP) :: CONST, DET, THICK          ! integration              ! 7
 REAL(DP) :: SOURCE                     ! data                     ! 8
 INTEGER  :: IP                         ! loop counter             ! 9
!           (Stored as application source example 209.)           !10
! Convection added in MIXED_SQ_MATRIX, known flux in SEG_COL_MATRIX !11
                                                                   !12
! Conduction properties assumed, (order in GET_REAL_LP (n)):       !13
! 1-D problem, K_xx, Q, Thickness                                  !14
! 2-D problem, K_xx, K_yy, K_xy, Q, Thickness                      !15
! 3-D problem, K_xx, K_yy, K_zz, K_xy, K_xz, K_yz, Q               !16
  CALL POISSON_ANISOTROPIC_E_MATRIX (E) ! for 1-, 2-, or 3-D       !17
                                                                   !18
!   CHECK FOR KEYWORD GLOBAL CONSTANTS: scalar_source & area_thick !19
  SOURCE = 0 ; IF ( SCALAR_SOURCE /= 0.d0 ) SOURCE = SCALAR_SOURCE !20
  THICK  = 1 ; IF ( AREA_THICK    /= 1.d0 ) THICK  = AREA_THICK    !21
  IF ( EL_REAL > 0 ) THEN      ! Get local element constant values, !22
    SELECT CASE (N_SPACE)      ! for source or thickness           !23
      CASE (1) ; IF ( EL_REAL > 1 ) SOURCE = GET_REAL_LP (2)       !24
                 IF ( EL_REAL > 2 ) THICK  = GET_REAL_LP (3)       !25
      CASE (2) ; IF ( EL_REAL > 3 ) SOURCE = GET_REAL_LP (4)       !26
                 IF ( EL_REAL > 4 ) THICK  = GET_REAL_LP (5)       !27
      CASE (3) ; IF ( EL_REAL > 6 ) SOURCE = GET_REAL_LP (7)       !28
    END SELECT ! for properties options                            !29
  END IF ! element data provided                                   !30
                                                                   !31
  CALL STORE_FLUX_POINT_COUNT ! Save LT_QP, for post-processing    !32
                                                                   !33
  DO IP = 1, LT_QP            ! NUMERICAL INTEGRATION LOOP          !34
    H   = GET_H_AT_QP (IP)     ! EVALUATE INTERPOLATION FUNCTIONS   !35
    XYZ = MATMUL (H, COORD)    ! FIND GLOBAL PT, ISOPARAMETRIC      !36
    DLH = GET_DLH_AT_QP (IP)   ! FIND LOCAL DERIVATIVES, dH / dr    !37
    AJ  = MATMUL (DLH, COORD) ! FIND JACOBIAN AT THE PT            !38
    CALL INVERT_JACOBIAN (AJ, AJ_INV, DET, N_SPACE)      ! inverse !39
    IF ( AXISYMMETRIC ) THICK = TWO_PI * XYZ (1) ! via axisymmetric !40
    CONST  = DET * WT(IP) * THICK                                  !41
                                                                   !42
    DGH = MATMUL (AJ_INV, DLH)          ! Physical gradient, dH / dx !43
    B   = COPY_DGH_INTO_B_MATRIX (DGH) ! B = DGH                   !44
                                                                   !45
!   VARIABLE VOLUMETRIC SOURCE, via keyword use_exact_source       !46
!   Defaults to file my_exact_source_inc if no exact_case key      !47
    IF ( USE_EXACT_SOURCE )  CALL         & ! analytic Q           !48
         SELECT_EXACT_SOURCE (XYZ, SOURCE) ! via exact_case key    !49
    C = C + CONST * SOURCE * H          ! source resultant         !50
                                                                   !51
!     CONDUCTION SQUARE MATRIX (THICKNESS IN E)                    !52
    S = S + CONST * MATMUL ((MATMUL (TRANSPOSE (B), E)), B)        !53
                                                                   !54
!-->  SAVE COORDS, E AND DERIVATIVE MATRIX, FOR POST PROCESSING    !55
    CALL STORE_FLUX_POINT_DATA (XYZ, (E * THICK), B)               !56
  END DO                                                           !57
!   ***  END ELEM_SQ_MATRIX PROBLEM DEPENDENT STATEMENTS ***       !58
```

Figure 11.41 *The generalized volumetric contributions*

```
! .................................................              ! 1
!   ***  SEG_COL_MATRIX PROBLEM DEPENDENT STATEMENTS FOLLOW ***    ! 2
! .................................................              ! 3
! Given normal flux on an element face or edge:                  ! 4
! Standard form: -K_n * U,n = Q_NORMAL_SEG (in System_Constants)  ! 5
!          (Stored as application source example 209.)           ! 6
 INTEGER  :: IQ                      ! loops                     ! 7
 REAL(DP) :: CONST, DET, THICK       ! flux area                 ! 8
  ! Get normal flux from keyword, or segment property            ! 9
  IF ( SEG_REAL > 0 ) Q_NORMAL_SEG= GET_REAL_SP (1)              !10
                                                                 !11
  THICK = 1   ! Default flux line segment real property # 2      !12
  IF ( SEG_REAL > 1 ) THICK = GET_REAL_SP (2)                    !13
                                                                 !14
  IF ( LT_N > 1 ) THEN ! Not a point value                       !15
                                                                 !16
    DO IQ = 1, LT_QP ! NUMERICAL INTEGRATION LOOP                !17
                                                                 !18
      H = GET_H_AT_QP (IQ) ! BOUNDARY INTERPOLATION FUNCTIONS    !19
!       FIND GLOBAL COORD, XYZ = H*COORD (ISOPARAMETRIC)         !20
      XYZ = MATMUL (H, COORD)                                    !21
!        FIND LOCAL DERIVATIVES                                  !22
      DLH = GET_DLH_AT_QP (IQ)   ! dH / dr                       !23
                                                                 !24
!        FORM DETERMINATE OF GENERALIZED JACOBIAN                !25
      DET   = PARM_GEOM_METRIC (DLH, COORD) ! dX / dr            !26
      IF ( AXISYMMETRIC ) THICK = TWO_PI * XYZ (1) ! keyword     !27
      CONST = DET * WT(IQ) * THICK                               !28
                                                                 !29
!       GET NORMAL FLUX COMPONENT                                !30
      IF ( USE_EXACT_FLUX ) CALL SELECT_EXACT_NORMAL_FLUX &      !31
           (XYZ, Q_NORMAL_SEG) ! via keyword use_exact_flux      !32
      C = C + Q_NORMAL_SEG * CONST * H     ! Source vector       !33
    END DO                                                       !34
                                                                 !35
  ELSE ! This is a point value                                   !36
                                                                 !37
    IF ( USE_EXACT_FLUX ) CALL SELECT_EXACT_NORMAL_FLUX &        !38
         (COORD (1, :), Q_NORMAL_SEG) ! via use_exact_flux       !39
    C (1) = Q_NORMAL_SEG                                         !40
                                                                 !41
  END IF ! boundary segment type                                 !42
! End application dependent flux 202.my_seg_col_inc_2            !43
```

Figure 11.42 *Computing a Neumann flux contribution*

If one wants to later recover the convection heat loss as a physical check or to better understand the problem then the necessary integral of $\mathbf{H}^b$ is saved, in line 58, for later use with the constant segment convection properties. Keyword *post_mixed* is used in the data to activate a file unit (*N_FILE*2) for these purposes. After a solution has been obtained that keyword also causes the convection heat loss calculations, in Fig. 11.44, to be carried out, as discussed in Eq. 7.37 and Section 11.5.2.

These numerically integrated forms for evaluating the matrices were applied to the previous linear triangle examples and gave exactly the same results as the explicit coded forms in Figs. 11.6-8. A one-point rule was used in most cases but the face convection square matrix required a three-point rule since it involved the product of two linear functions, which resulted in the integrand being a quadratic polynomial. The quadrature based formulation allows for curved element boundaries or other reasonable varying Jacobians. They allow use of any point, 1-, 2-, or 3-dimensional element in the library.

11.9 Strong diagonal gradient SCP test case

This problem is defined in terms of the Poisson equation where the solution has been chosen in advance to give zero values on the boundary of a unit square and to exhibit a relatively sharp transition of gradients along a region near the diagonal of the square domain. This test case has been used by Oden 18 and Zienkiewicz and Zhu 23 for testing various error estimators. The exact solution is given by

$$u(x, y) = xy(1 - x)(1 - y)\,Tan^{-1}(\alpha(\xi - \xi_0))$$

where $\xi = (x + y)/\sqrt{2}$ with $\xi_0 = 0.8$ and $\alpha = 20$. The contours of the exact solution are plotted in Fig. 11.45, and the corresponding exact gradient contours are in Fig. 11.46. When it is substituted into Poisson's equation the algebraic definition of the source term per unit area, Q, is obtained and placed in routine to be used in the numerical integration of the element matrices. Since the source term is rather complicated this represents a case where one may want to have different subroutines for evaluating the element square and column matrices since the latter requires more quadrature points than the former. The initial mesh for this problem is a rather crude one consisting of 50 identical 6 noded (complete quadratic) elements. They are shown in Fig. 11.47 along with the flags on the boundary which indicate the nodes where the null essential boundary conditions were applied. A crude mesh is chosen so that the reader can see the differences in the exact and approximate solutions and verify that they are decreased as the model is later refined. Usually we do not know the exact result and must rely on the computed error measures when deciding to stop an analysis.

When the solution is computed with this mesh one gets the contour levels shown in Fig. 11.48 which are compared to corresponding exact levels in Fig. 11.49. The wiggles appearing in the contours provide an 'eyeball' check that indicates that the mesh is too crude in this region and will need refining if it is an important part of the solution domain. In this case the the region of wiggles is where the contours are the closest together which means the solution is rapidly changing and the mesh is indeed too coarse to be accepted. Of course, our error estimator will lead us to the same conclusion but it is still wise to employ a little common sense along the way.

Post-processing for the gradients at the quadrature points of the elements yields the distribution of flux vectors shown in Fig. 11.50. While we could compare contour values of the gradient components it makes more sense in this case to use the true shape two-dimensional flux vectors. Here the vectors represent the item obtained from Eq. 11.8 by using a standard post-processing technique once the $\mathbf{\Phi}^e$ of the element have been gathered from the system $\mathbf{\Phi}$ and multiplied by the matrices $\mathbf{E}^e\ \mathbf{B}^e$ at each quadrature point in the element. They are the data that are processed in the SCP method to obtain continuous nodal flux estimates.

In this case we have selected element based patches using all adjacent element neighbors, as shown in Fig. 11.2. The resulting SCP nodal flux vectors are shown in

```
!    ***  MIXED_SQ_MATRIX PROBLEM DEPENDENT STATEMENTS FOLLOW ***    ! 1
! ..............................................................    ! 2
! Mixed or Robin boundary condition, Standard form:                  ! 3
!   K_n * U,n + ROBIN_1_SEG * U + ROBIN_2_SEG = 0                    ! 4
!           (Stored as application source example 209.)              ! 5
 INTEGER  :: IQ                   ! Integration loop                 ! 6
 REAL(DP) :: CONST, DET, THICK ! face area                           ! 7
                                                                     ! 8
  IF ( N_FILE2 > 0 ) H_INTG = 0 ! via post_mixed to post-process     ! 9
 !  GET ROBIN TERMS, IF GLOBAL CONSTANTS                             !10
 !    Set in keywords robin_square and robin_column, or via          !11
 !    convect_coef, convect_temp for convection special case         !12
  THICK = 1.d0            ! if an edge                               !13
  IF ( CONVECTION ) THEN ! constant convection all segments          !14
    ROBIN_1_SEG =  CONVECT_COEF                                      !15
    ROBIN_2_SEG = -CONVECT_COEF * CONVECT_TEMP                       !16
    IF ( LT_PARM < 2 ) THICK =  CONVECT_THICK ! point or line        !17
  END IF ! Globally constant convection                              !18
                                                                     !19
 !  GET ROBIN TERMS, IF LOCAL SEGMENT CONSTANTS                      !20
  IF ( MIXED_REAL > 0 ) THEN ! are local data                        !21
    IF ( CONVECT_VARY ) THEN ! data are coeff, temperature, thick    !22
      ROBIN_1_SEG = GET_REAL_MX (1) ! convection coeff               !23
      ROBIN_2_SEG = 0.d0            ! default temperature effect     !24
      IF (MIXED_REAL > 1) ROBIN_2_SEG = -ROBIN_1_SEG   &             !25
                          * GET_REAL_MX (2) ! coeff*temp             !26
      IF (MIXED_REAL > 2 .AND. LT_PARM < 2) THICK = GET_REAL_MX (3)  !27
    ELSE ! a non-convection Robin condition                          !28
      ROBIN_1_SEG = GET_REAL_MX (1) ; ROBIN_2_SEG = 0.d0             !29
      IF (MIXED_REAL > 1) ROBIN_2_SEG = GET_REAL_MX (2)              !30
      IF (MIXED_REAL > 2 .AND. LT_PARM < 2) THICK = GET_REAL_MX (3)  !31
    END IF ! convection or general Robin data                        !32
  END IF ! local data                                                !33
                                                                     !34
  IF ( LT_N > 1 ) THEN ! Not a point, must integrate line or surf    !35
    DO IQ = 1, LT_QP            ! NUMERICAL INTEGRATION LOOP         !36
      H   = GET_H_AT_QP (IQ)    ! BOUNDARY INTERPOLATION FUNCTIONS   !37
      XYZ = MATMUL (H, COORD)   ! FIND GLOBAL COORD, (ISOPARAMETRIC) !38
      DLH = GET_DLH_AT_QP (IQ) ! FIND LOCAL DERIVATIVES, dH / dr     !39
                                                                     !40
!        FORM DETERMINATE OF GENERALIZED JACOBIAN, Fig 9.12          !41
      DET   = PARM_GEOM_METRIC (DLH, COORD)         ! dX / dr        !42
      IF ( AXISYMMETRIC ) THICK = TWO_PI * XYZ (1) ! axisymmetric    !43
      CONST = DET * WT(IQ) * THICK                                   !44
      IF ( N_FILE2 > 0 ) H_INTG = H_INTG + H * DET * WT(IQ)          !45
                                                                     !46
!       FORM MIXED ARRAYS                                            !47
      S = S + ROBIN_1_SEG * CONST * OUTER_PRODUCT (H, H) ! Sq        !48
      C = C - ROBIN_2_SEG * CONST * H                    ! Source    !49
    END DO                                                           !50
                                                                     !51
  ELSE ! This is a point value                                       !52
    IF ( AXISYMMETRIC ) THICK = TWO_PI * COORD (1, 1)                !53
    S (1, 1) =  ROBIN_1_SEG * THICK ; C (1) = -ROBIN_2_SEG * THICK   !54
    IF ( N_FILE2 > 0 ) H_INTG = THICK ! actually area at pt, input   !55
                                                                     !56
  END IF ! boundary segment type                                     !57
  IF ( N_FILE2 > 0 ) WRITE (N_FILE2) H_INTG ! via post_mixed         !58
! End mixed condition BC my_mixed_sq_inc                             !59
```

Figure 11.43 *General Robin or convection contributions*

```
! ..........................................................      ! 1
!  *** POST_PROCESS_MIXED PROBLEM DEPENDENT STATEMENTS FOLLOW ***   ! 2
! ..........................................................      ! 3
!          (Stored as application source example 209.)            ! 4
! Global CONVECT_COEF  set by keyword convect_coeff is available   ! 5
! Global CONVECT_TEMP  set by keyword convect_temp  is available   ! 6
! H_INTG (LT_N) Integral of interpolation functions, H, available ! 7
                                                                  ! 8
!              convection heat loss recovery                      ! 9
 REAL(DP) :: THICK                         ! line width           !10
 REAL(DP), SAVE :: Q_LOSS, TOTAL = 0.d0 ! Face and total heat loss !11
 LOGICAL,  SAVE :: FIRST = .TRUE.          ! printing             !12
                                                                  !13
  IF ( FIRST ) THEN                        ! first call           !14
    FIRST = .FALSE. ; WRITE (6, 5)         ! print headings       !15
    5 FORMAT ('*** CONVECTION HEAT LOSS ***', /, &                !16
    &         'ELEMENT    HEAT_LOST')                             !17
  END IF ! first call                                             !18
                                                                  !19
  IF ( N_FILE2 > 0 ) THEN ! H already integrated in MIXED SQ.     !20
    READ (N_FILE2) H_INTG ! via keyword post_mixed                !21
  ELSE                                                            !22
    PRINT *,'WARNING: NEED post_mixed IN CONTROL'                 !23
    N_WARN = N_WARN + 1                                           !24
    STOP 'POST_PROCESS_MIXED 208,209, or 302 DATA NOT SAVED'      !25
  END IF ! H_INTG                                                 !26
                                                                  !27
!  GET ROBIN TERMS, IF GLOBAL CONSTANTS                           !28
!    Set in keywords robin_square and robin_column, or via        !29
!    convect_coef, convect_temp for convection special case       !30
  THICK = 1.d0           ! if an edge                             !31
  IF ( CONVECTION ) THEN ! constant convection all segments       !32
    ! Then CONVECT_COEF, CONVECT_TEMP, CONVECT_THICK global       !33
    IF ( LT_PARM < 2 ) THICK = CONVECT_THICK ! point or line      !34
  END IF ! Globally constant convection                           !35
                                                                  !36
 !  GET ROBIN TERMS, IF LOCAL SEGMENT CONSTANTS                   !37
  IF ( MIXED_REAL > 0 ) THEN ! are local data                     !38
    IF ( CONVECT_VARY ) THEN ! data: coeff, temperature, thick    !39
      CONVECT_COEF = GET_REAL_MX (1) ! convection coeff           !40
      CONVECT_TEMP = 0.d0            ! default temperature        !41
      IF (MIXED_REAL > 1) CONVECT_TEMP = GET_REAL_MX (2) ! temp   !42
      IF (MIXED_REAL > 2 .AND. LT_PARM < 2) THICK = GET_REAL_MX (3) !43
    ELSE ! a non-convection Robin condition                       !44
      PRINT *,'WARNING: NEED KEY convect_coef OR convect_vary '   !45
      N_WARN = N_WARN + 1                                         !46
      STOP 'POST_PROCESS_MIXED 208,209, or 302 KEYWORD ERROR'     !47
    END IF                                                        !48
  END IF ! local data                                             !49
                                                                  !50
!   HEAT LOST FROM THIS FACE: Integral over face of h * (T - T_inf) !51
   D (1:LT_N) = D(1:LT_N) - CONVECT_TEMP  ! Temp difference at nodes !52
   Q_LOSS = CONVECT_COEF  * DOT_PRODUCT (H_INTG, D)  ! Face loss  !53
   IF ( LT_PARM < 2 ) Q_LOSS = Q_LOSS * THICK        ! line or point !54
   TOTAL  = TOTAL + Q_LOSS                           ! Running total !55
   PRINT '(I6, ES15.5)', IE, Q_LOSS                  ! Each segment  !56
   IF ( IE == N_MIXED ) PRINT *, 'TOTAL = ', TOTAL   ! Last segment  !57
!  *** END POST_PROCESS_MIXED PROBLEM DEPENDENT STATEMENTS ***    !58
```

Figure 11.44 *General convection heat loss post-processing*

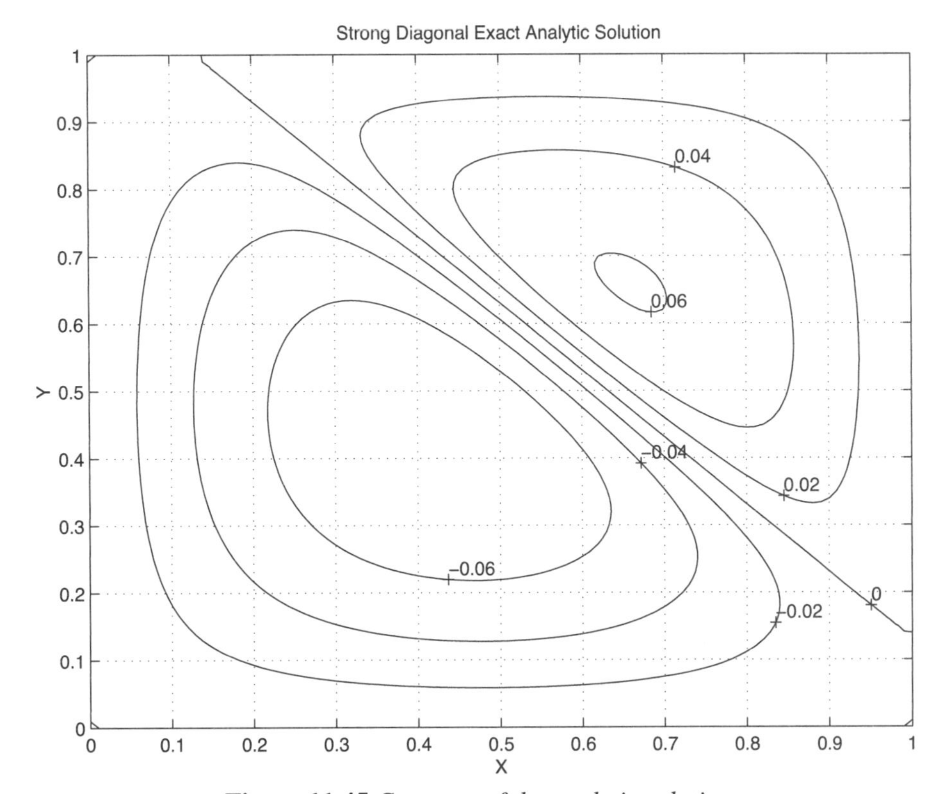

Figure 11.45 *Contours of the analytic solution*

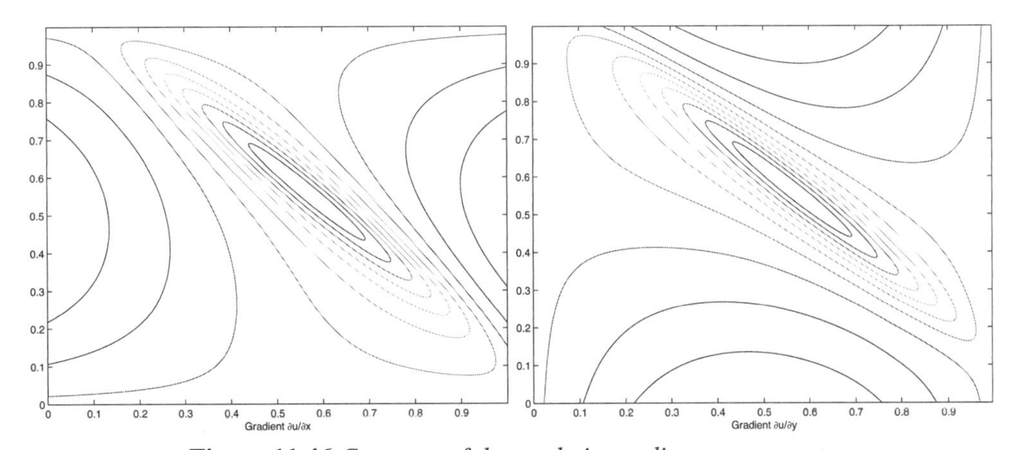

Figure 11.46 *Contours of the analytic gradient components*

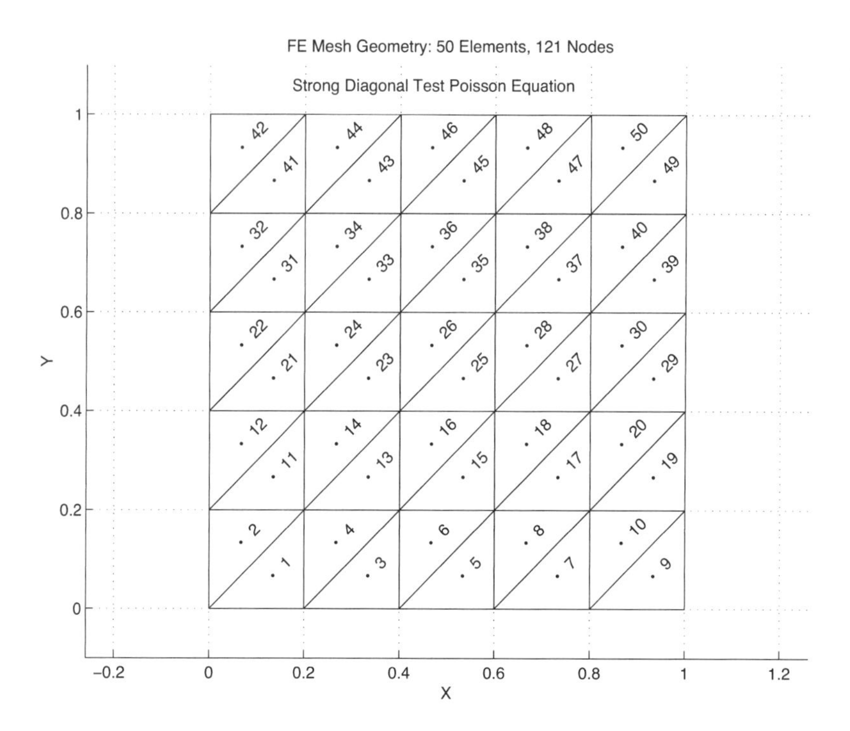

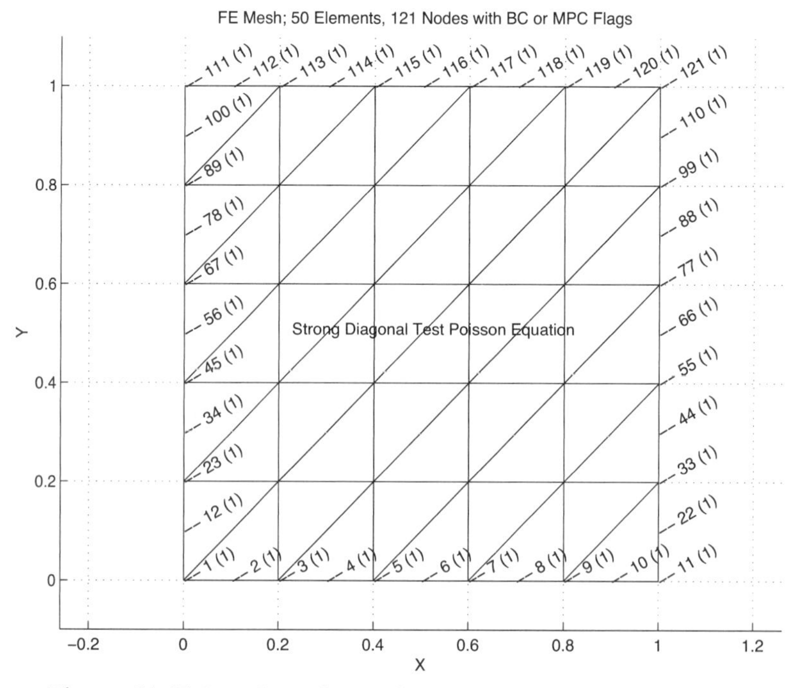

Figure 11.47 *Initial quadratic element mesh and Dirichlet nodes*

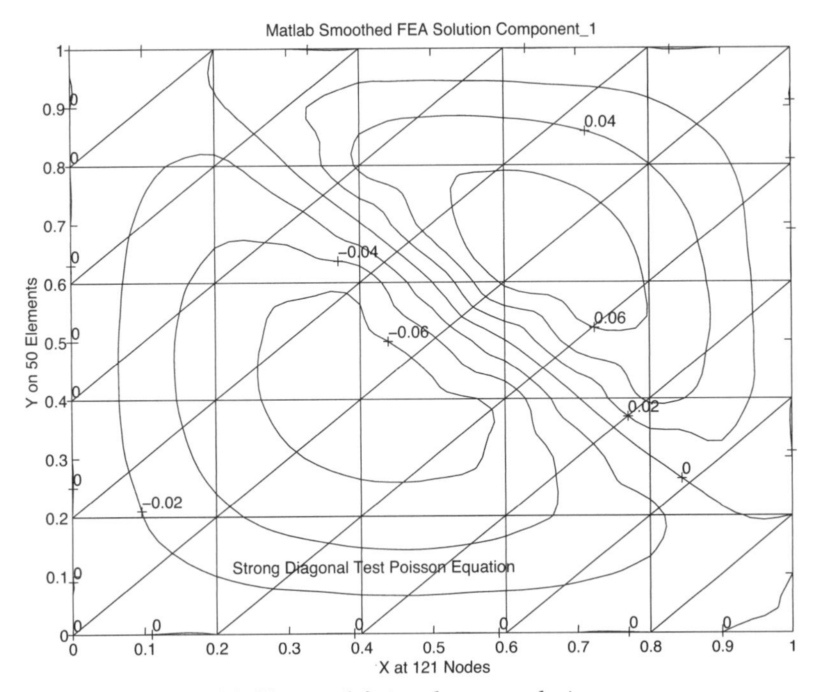

Figure 11.48 *Initial finite element solution contours*

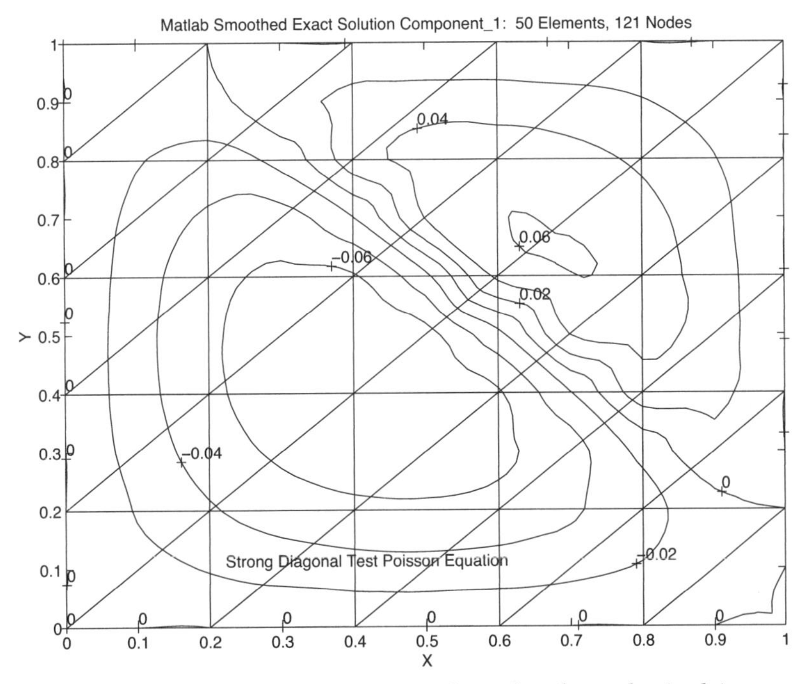

Figure 11.49 *Exact solution evaluated at the nodes (only)*

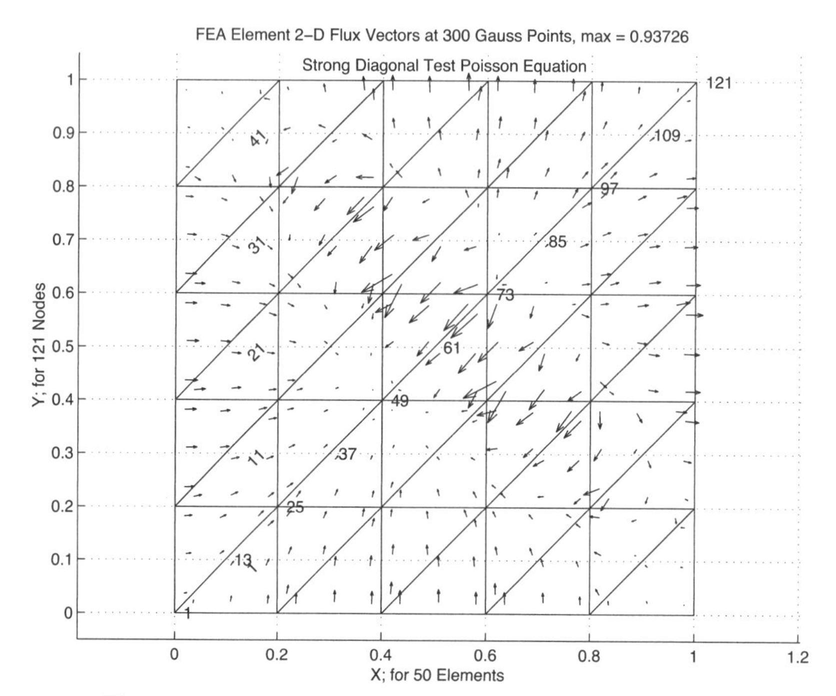

Figure 11.50 *Initial element quadrature point flux vectors*

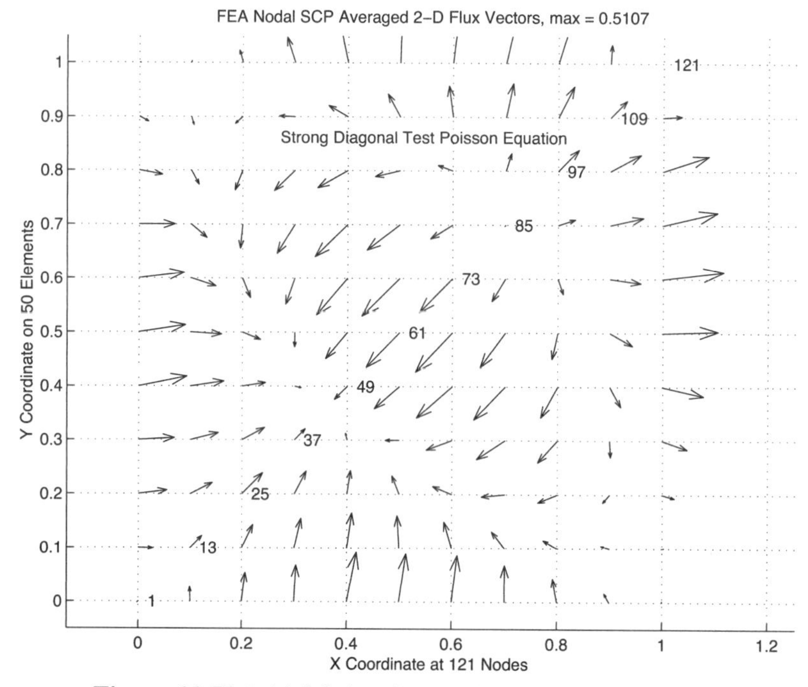

Figure 11.51 *Initial finite element SCP nodal flux vectors*

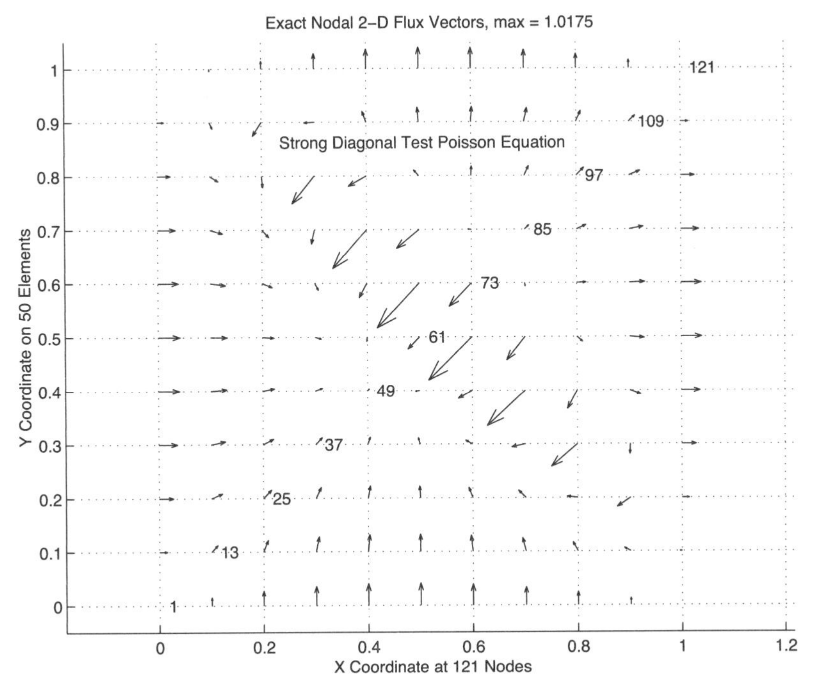

Figure 11.52 *Exact nodal flux vectors*

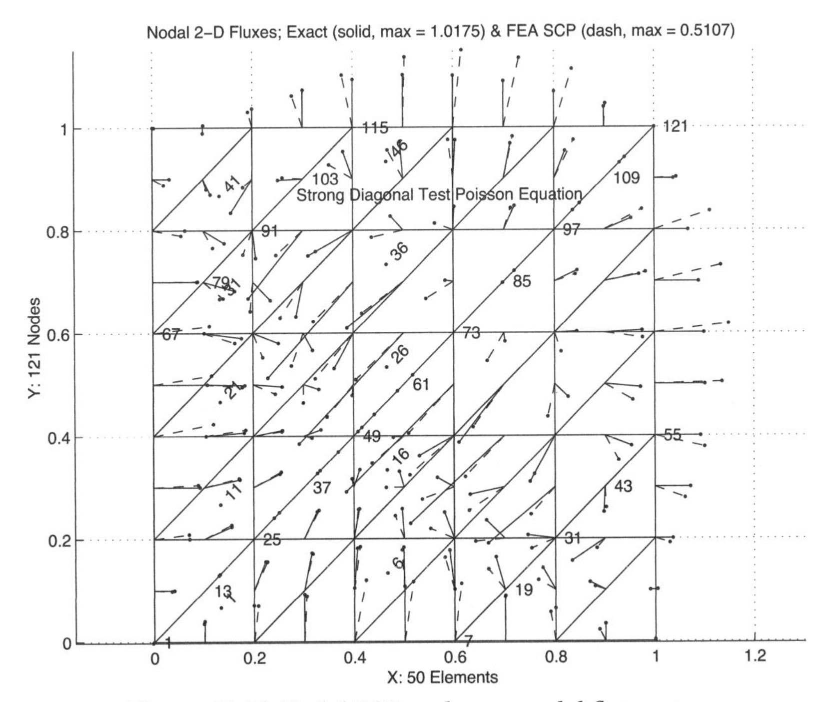

Figure 11.53 *Nodal SCP and exact nodal flux vectors*

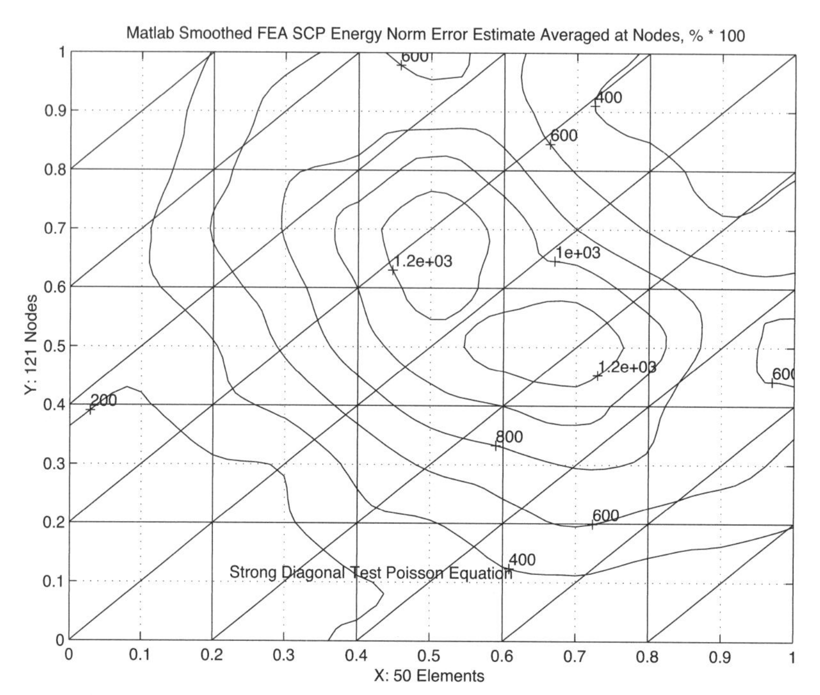

Figure 11.54 *Initial finite element energy norm error estimate*

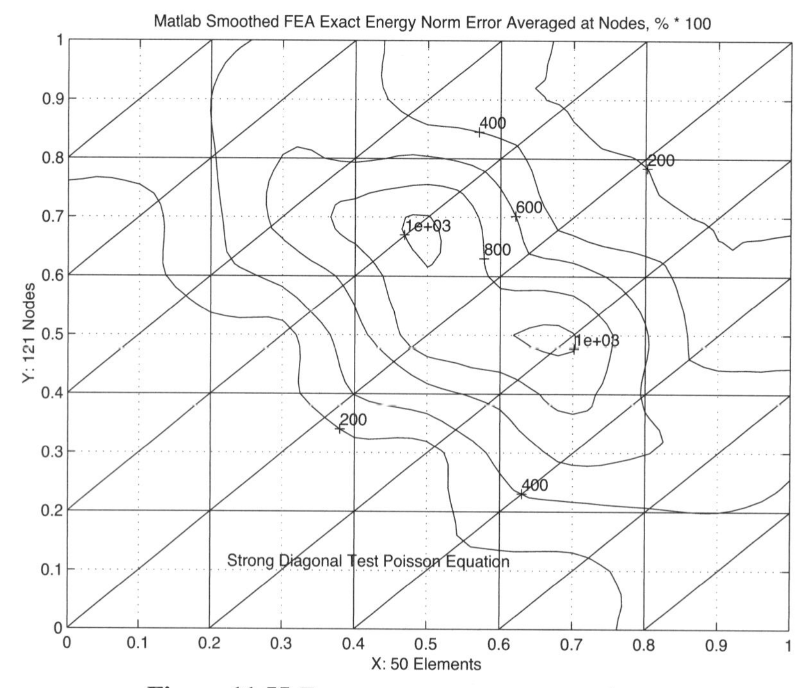

Figure 11.55 *Exact energy norm error estimate*

Fig. 11.51. That figure represents the continuous nodal fluxes, **a** computed in Eqs. 2.45 and 6.1, that will be interpolated to serve as a way to define the flux error estimate at any point in Eqs. 2.46 and 5.20. These vectors represent the results of the averaging obtained from the SCP. They can be used in a typical post-processing activity and/or further processed to yield an error indicator. If the error in the energy is to be computed then we use the last version of Eqs. 5.16 and 6.3, but with $\boldsymbol{\sigma}$ replaced with $\boldsymbol{\sigma}^*$. The integrals are carried out at the element level in a process similar to that used in formulating the element matrices.

An important difference at this stage is that more quadrature points are probably needed, because of the polynomial degree of **P** in Eq. 2.45 is of higher degree than the **B** matrix, and this will require the computation of the standard flux estimate, $\mathbf{E}^e\,\mathbf{B}_j^e\,\mathbf{u}^e$, rather than recovering the saved values of **B** that were used in the SCP averaging process. That is an additional cost that must be paid to extend the SCP averaging process to an error indicator.

The corresponding exact nodal flux vectors are given in Figs. 11.52. They are not shown to the same scale. The crude mesh is estimating the maximum flux value to be 0.5107 versus a maximum of 1.0175 from the exact values. This is because the crude mesh is spanning the diagonal region where the exact flux changes from very large to very small fluxes. As we refine the mesh we expect these flux vectors to approach each other. In Fig. 11.53 we see the error (magnitude and direction) in the continuous SCP nodal fluxes and the exact nodal fluxes. In this case they are seen at the same scale. The solid line gives the magnitude and direction of the exact vector, while the dashed line gives the approximate vector. Wherever they are in agreement one should only be able to see the solid line. They start from the same point where they were computed and each ends with a dot instead of an arrowhead so as to reduce the clutter as the mesh is refined. Here we see some vectors (along the horizontal, vertical and 45 degree diagonal) agree on direction but still have significant differences in magnitudes. Most other vectors also have large errors in directions. It is reassuring to see these differences vanish as the error indicator gets smaller.

The comparison of the SCP energy norm error estimate and the exact error in the energy norm is seen by examining Figs. 11.54 and 55. Even though they have slightly different shapes it is reassuring that even with this crude mesh the SCP error estimating process is giving very similar locations for the highest error. The SCP estimate is higher than the exact values but we expect them to approach the same values as the mesh is refined. The same two energy norm errors are represented as surfaces in Figs. 11.56 and 57. It is informative to see how the energy norm estimate we will compute compares to the true error in the value of the solution variable, $\boldsymbol{\phi}$. Those values are shown as contours and a surface plot in Figs. 11.58 and 59, respectively. While representing a different measure of error, with different units, both forms are showing the peak error occurring in the same general regions. Note that the exact function error is zero on the (entire) boundary where essential boundary conditions are applied while the exact and SCP energy norm estimates of the error are not zero in those regions. It is common for energy based error estimates to have their largest error in such regions. This suggests that while the function values are accurate there the gradient is not. That is another reason why one may want to use somewhat smaller element sizes near the boundaries.

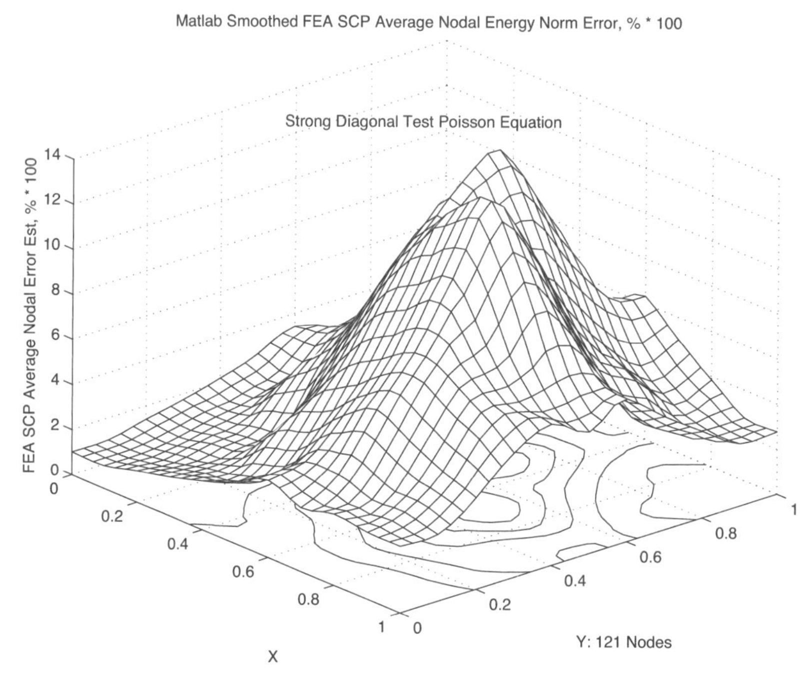

Figure 11.56 *Surface of SCP energy norm error estimate*

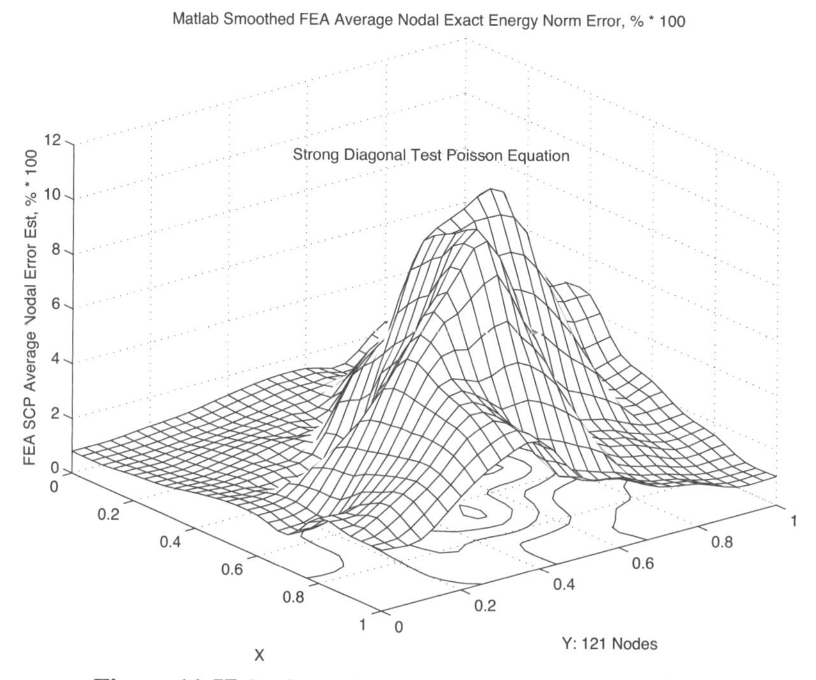

Figure 11.57 *Surface of exact energy norm error estimate*

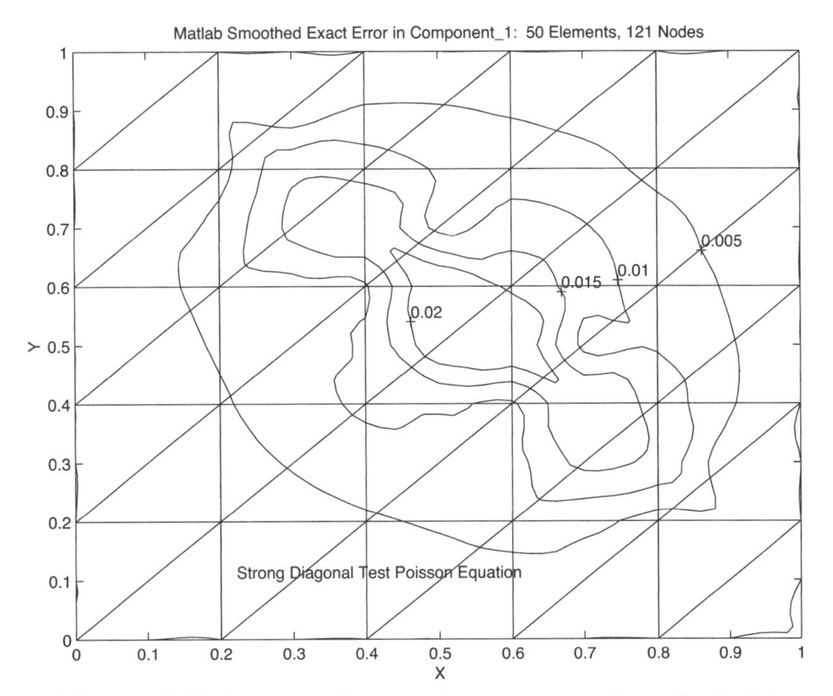

Figure 11.58 *Contours of exact solution error at the nodes (only)*

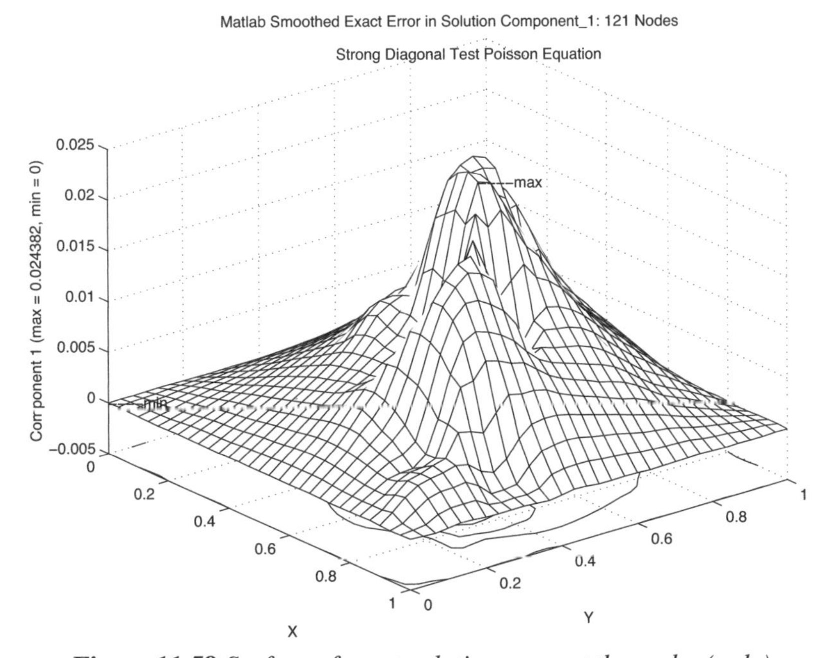

Figure 11.59 *Surface of exact solution error at the nodes (only)*

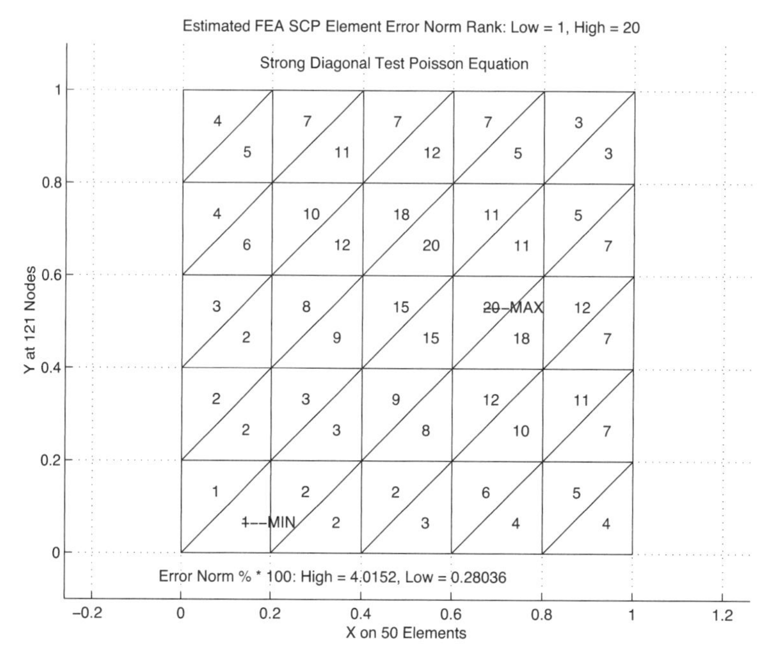

Figure 11.60 *Relative ranking of element error estimates*

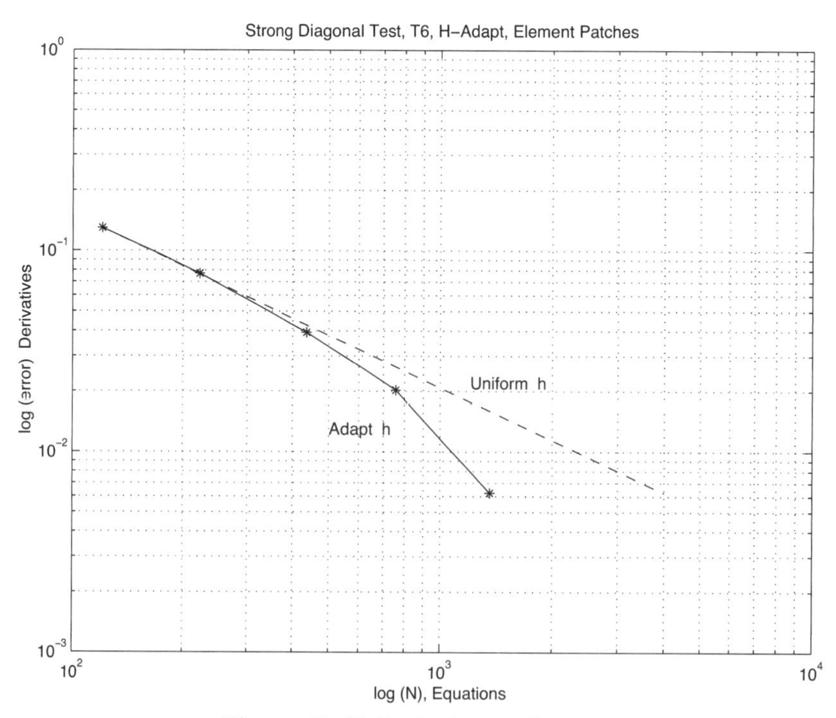

Figure 11.61 *Reduction in the error*

The relative ranking of the error levels in the uniform mesh is given in Fig. 11.60. The SCP error estimating process yields a refinement parameter that can be used to select new element sizes that can be passed as input data into an automatic mesh generation program such as that given by Huang and Usmani [10]. That process can be repeated to develop a series of solutions that approach the specified level of acceptable error in the energy norm. The number of equations involved in this series of meshes was 121, 223, 436, 757, and 1360, respectively. The reduction in the error norm is shown in Fig. 11.61 as a function of the number of equations solved. Note that to obtain the same error reduction with a uniform mesh refinement would have required about 4000 unknowns compared to the 1360 in the last mesh. For this problem all three choices for the type of patch definition gave the same error estimates and mesh refinements. The four such meshes that followed from the initial uniform mesh are shown in Fig. 11.62.

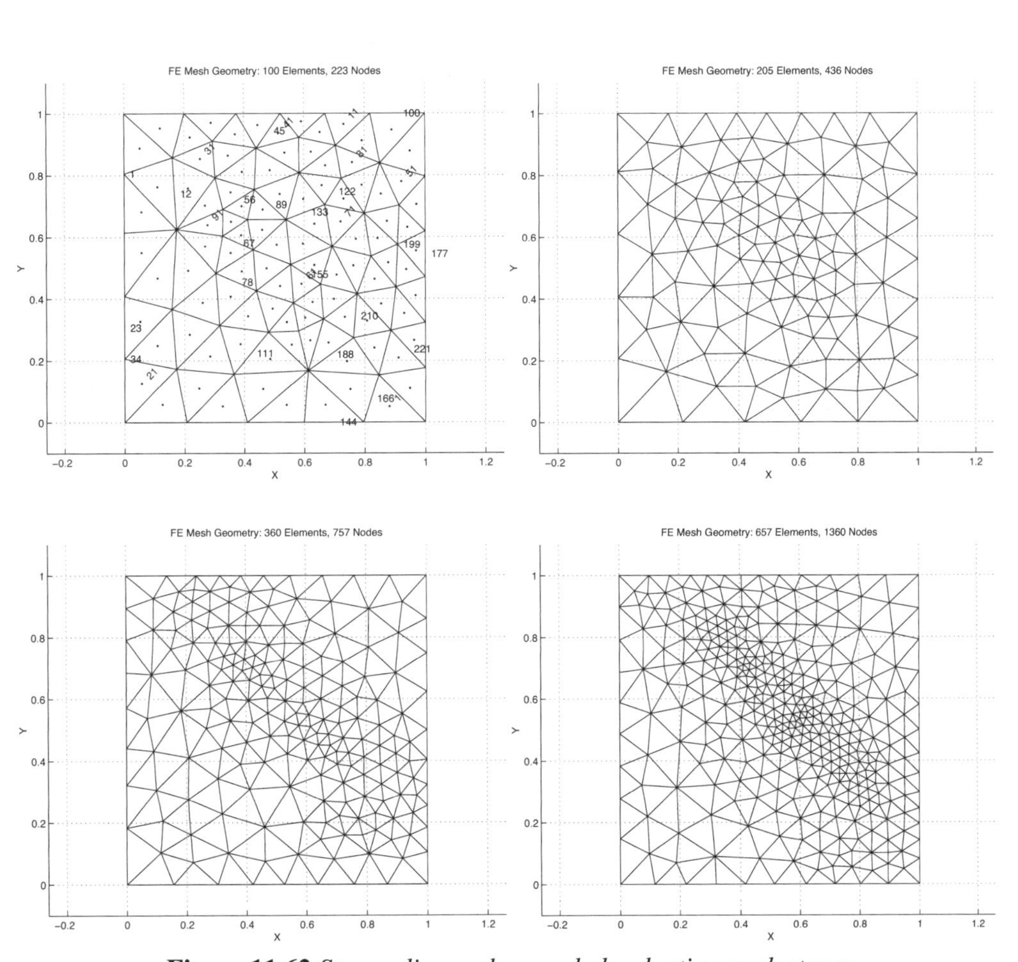

Figure 11.62 *Strong diagonal example h-adaptive mesh stages*

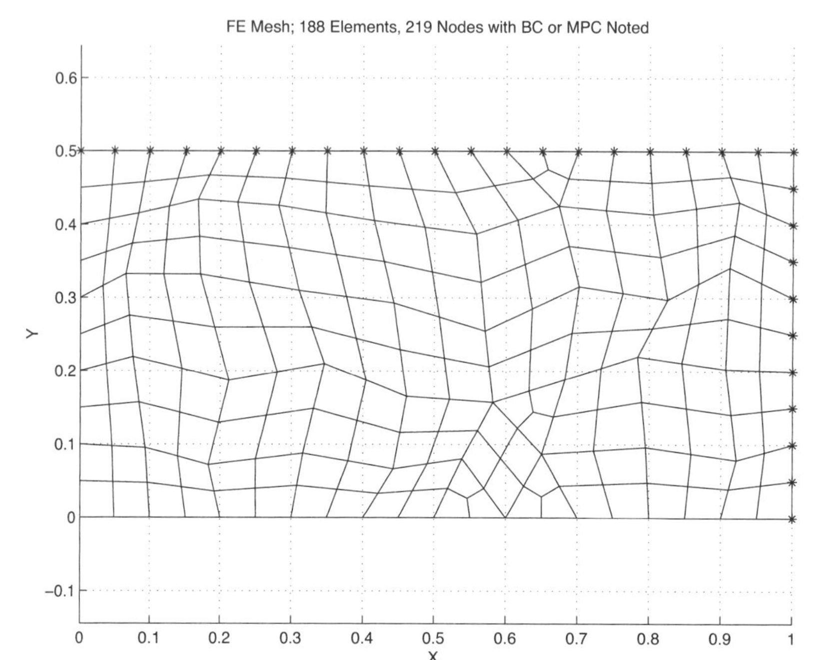

Figure 11.63 *Quarter symmetry model of orthotropic conduction*

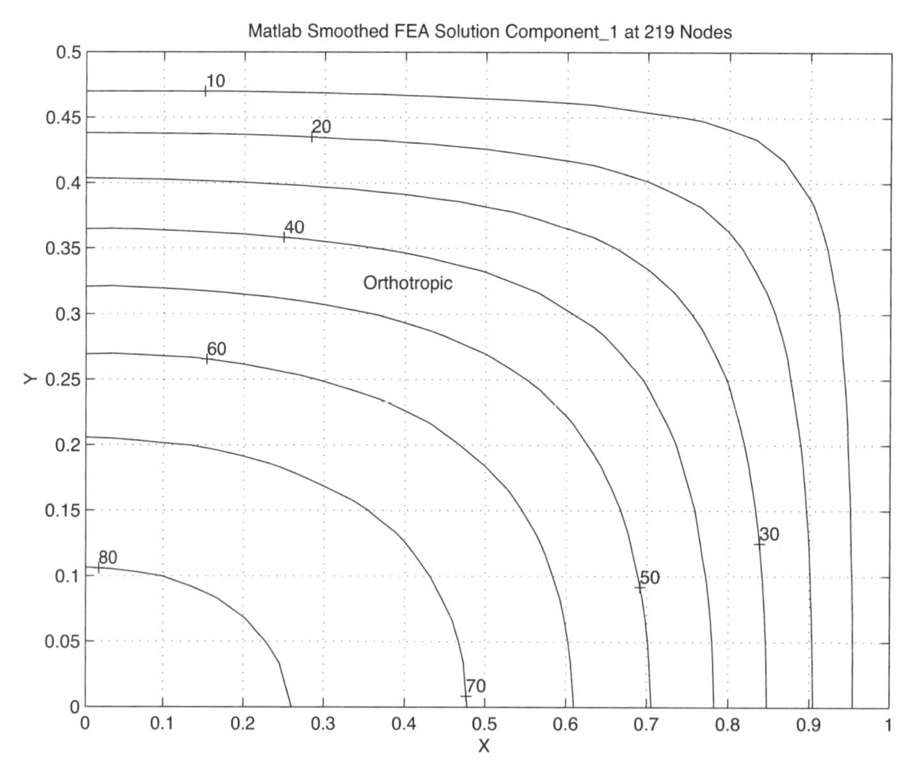

Figure 11.64 *Temperature,* $K_x = 2, K_y = 1.237, Q = 1000$

11.10 Orthotropic conduction

We have seen that the finite element method automatically includes the ability to have directionally dependent (anisotropic) material properties. However we have not illustrated how much such properties can influence a solution. Carslaw and Jaeger [5] have given an exact infinite series solution for an orthotropic rectangular region (*exact_case* 19) with a constant internal heat source, $Q = 1000$, and with a temperature of zero around its edge. Consider a material that has thermal conductivity values of $K_x = 2.0, K_y = 1.2337$, and $K_{xy} = 0$. The problem has symmetry of the geometry, boundary conditions, source term, and conductivities with respect to both the y-axis and the x-axis. That is, one can employ a quarter symmetry model to study this problem. The top right segment is shown with a mesh of Q4 quadrilateral elements in Fig. 11.63. It also has asterisks marking those nodes with the null essential boundary conditions. For a region that has full lengths of 2.0 and 1.0, in the x- and y-directions, respectively, the analytic solution at the center point, (0, 0), is 83.72 degrees. The above mesh yielded a corresponding value of 83.77 degrees. That differed significantly from the 70.31 degrees that is computed if one assumes an isotropic material with a $K = 1.617$ value which is the average of the above orthotropic conductivities. Had the orthotropic conductivities been reversed (largest in the y-direction) then the center temperature would be 60.13 degrees so we see about a 30 percent variation in the peak temperature as various principal material directions are considered in our study. The temperature contours for these three cases are shown in Figs. 11.64-66.

The exact flux vectors for the isotropic material are shown in Fig. 11.67. The FEA vectors obtained from the SCP post-averaging look almost identical so they are not plotted. To emphasize the difference between the FEA and exact fluxes the differences in the two sets of vectors are shown in Fig. 11.68, to a different scale. The biggest differences occur along the edges and symmetry planes and at the top right corner where both flux vector components are approaching zero. We can see the magnitude of the heat flow in the exact and SCP flux contour plots in Figs. 11.69 and 70, respectively. Even with this crude mesh the agreement in the isotropic heat flow is quite good. Considering the two orthotropic material choices we see that the heat flow changes relatively little, as Figs. 11.71 and 72 illustrate. The difference between the exact energy norm error and the SCP estimate is also quite close, as will be discussed in Chapter 12. The estimated new mesh sizes do not change much here with the small deviations from isotropic properties. The isotropic case suggests a relatively uniform mesh refinement as seen in the suggested new grid in Fig. 11.73.

An interesting question is how well does the energy norm error estimate compare to the exact energy norm error, and how does the exact energy norm error compare to the exact error in the computed primary variable. We will illustrate the answers for the orthotropic case where $K_x > K_y$. For this crude mesh there is little correlation between the peak error in the solution value (Fig. 11.74) and the peak error in the energy norm. Figures 11.75 and 76 show the exact and SCP estimated error as measured in the energy norm. The agreement in both the relative location of the peak error and the values of the local errors are unusually good.

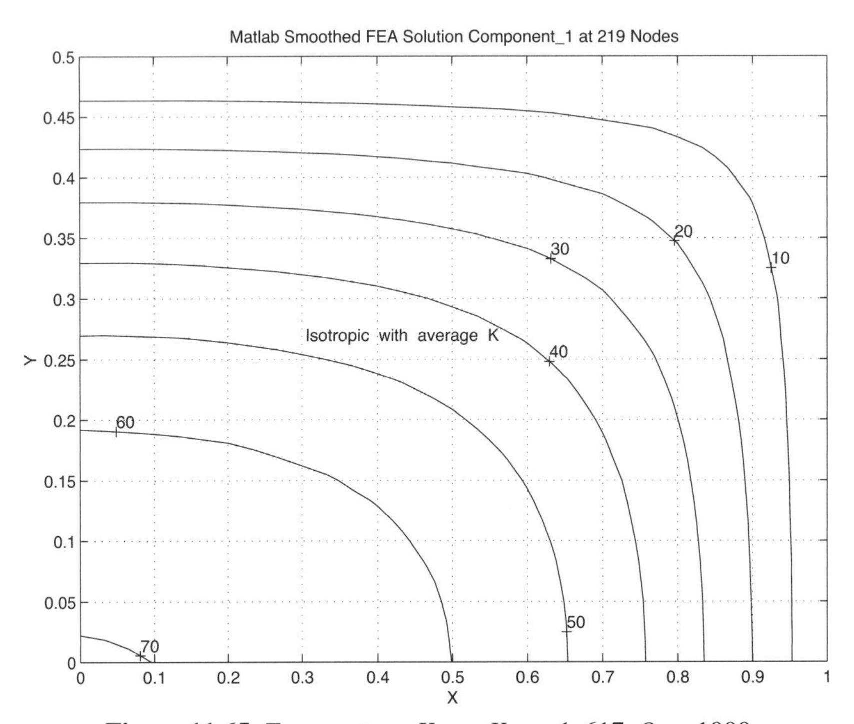

Figure 11.65 *Temperature,* $K_x = K_y = 1.617, Q = 1000$

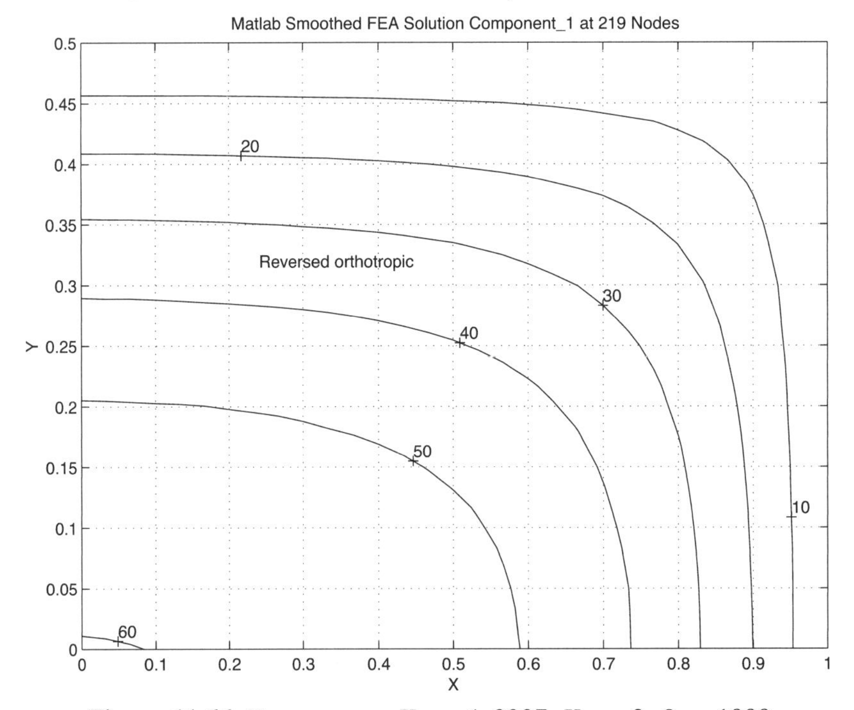

Figure 11.66 *Temperature,* $K_x = 1.2337, K_y = 2, Q = 1000$

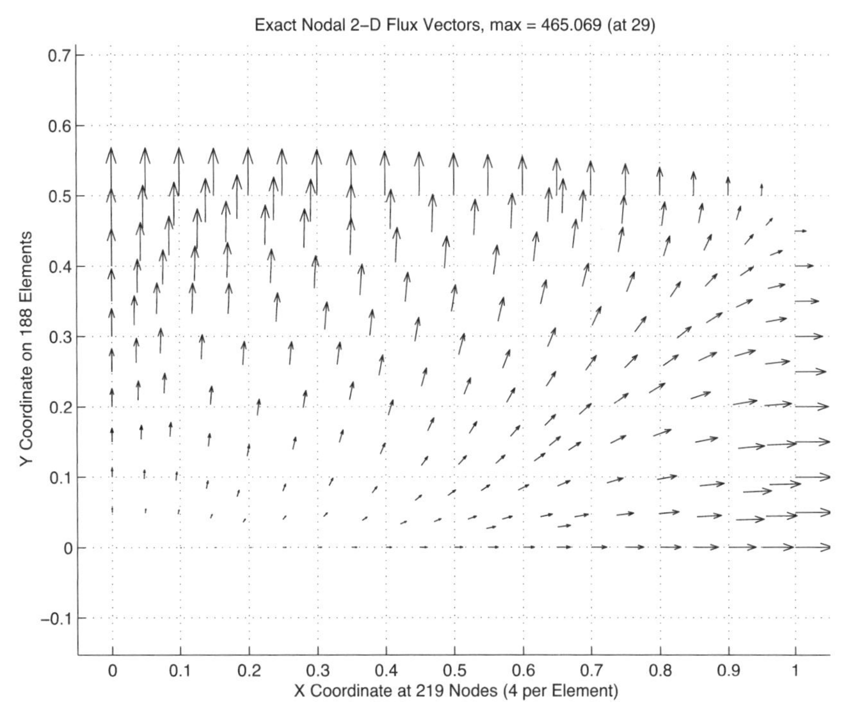

Figure 11.67 *Exact isotropic heat flux vectors*

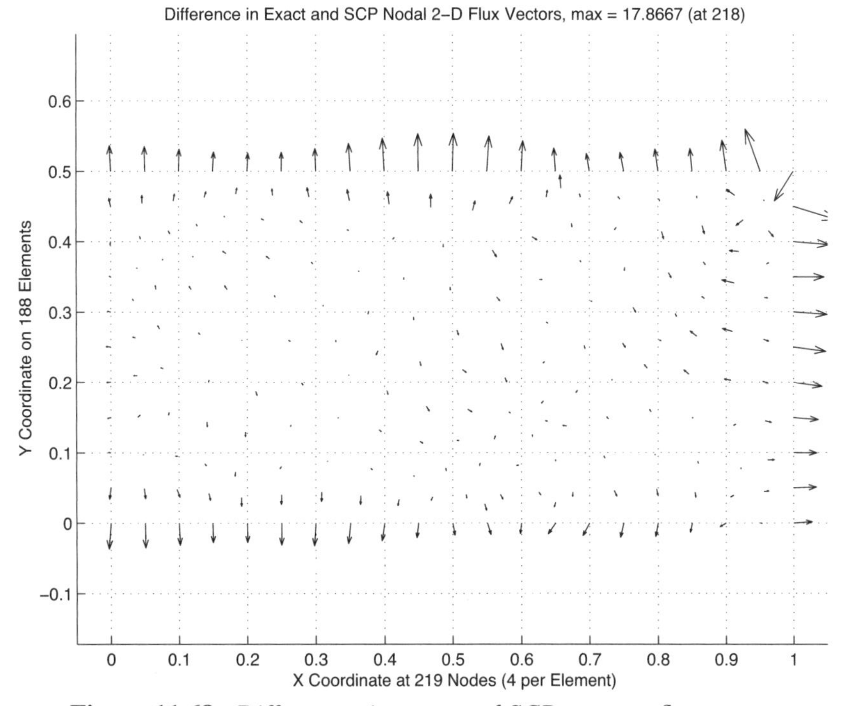

Figure 11.68 *Differences in exact and SCP average flux vectors*

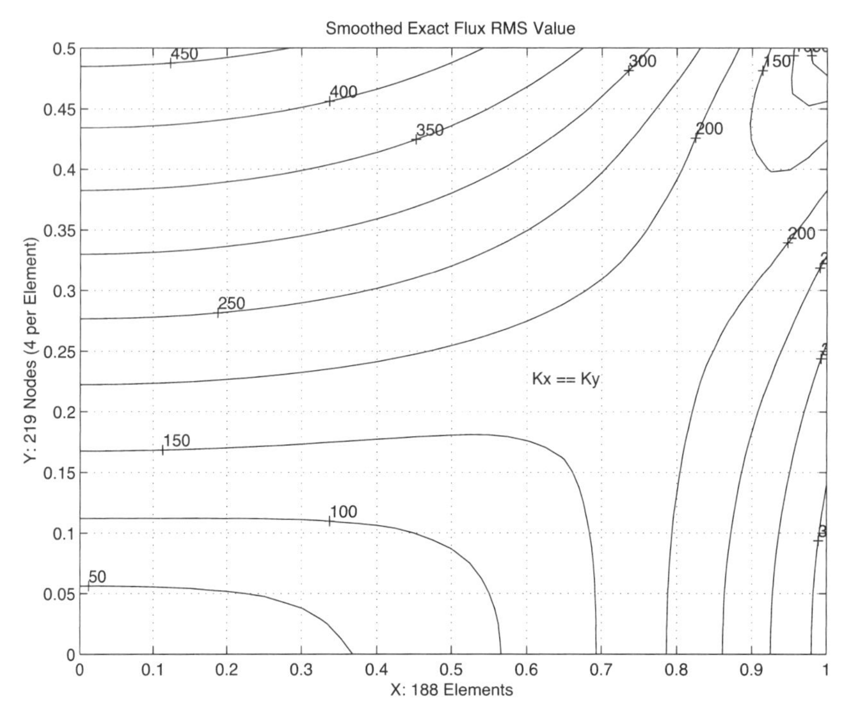

Figure 11.69 *Exact isotropic heat flux value contours*

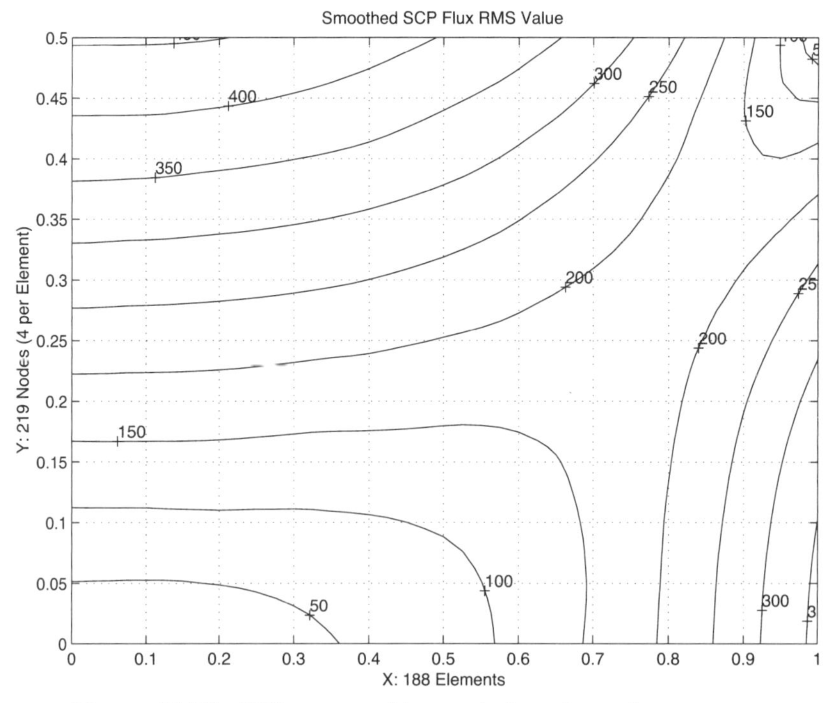

Figure 11.70 *SCP averaged isotropic heat flux value contours*

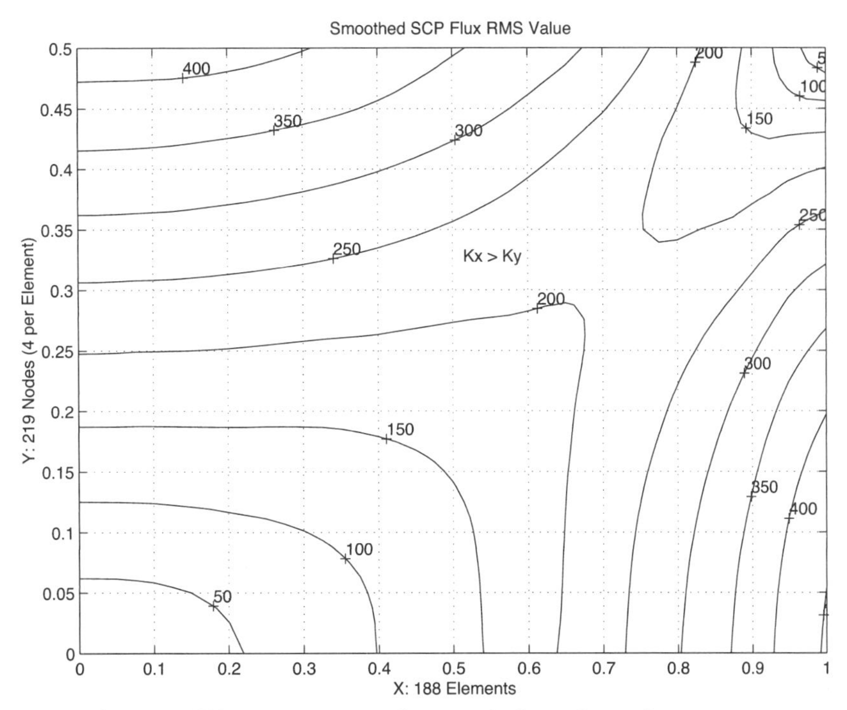

Figure 11.71 *SCP averaged Kx > Ky heat flux value contours*

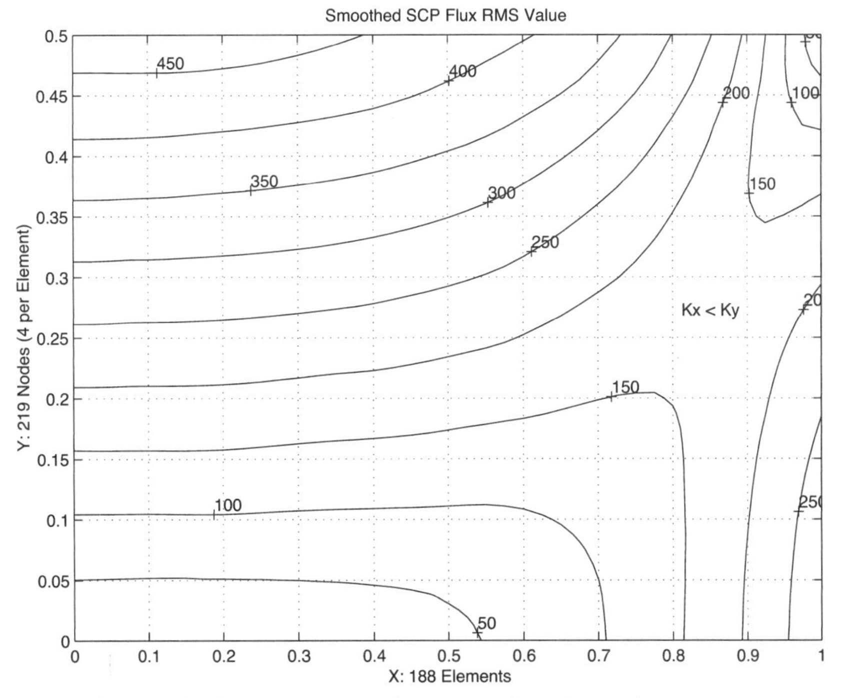

Figure 11.72 *SCP averaged Kx < Ky heat flux value contours*

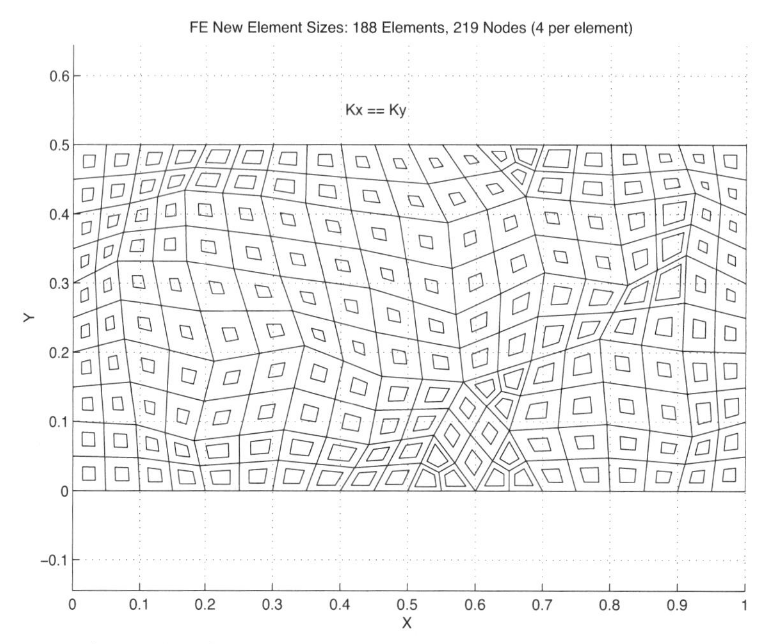

Figure 11.73 *Suggested new mesh for isotropic material*

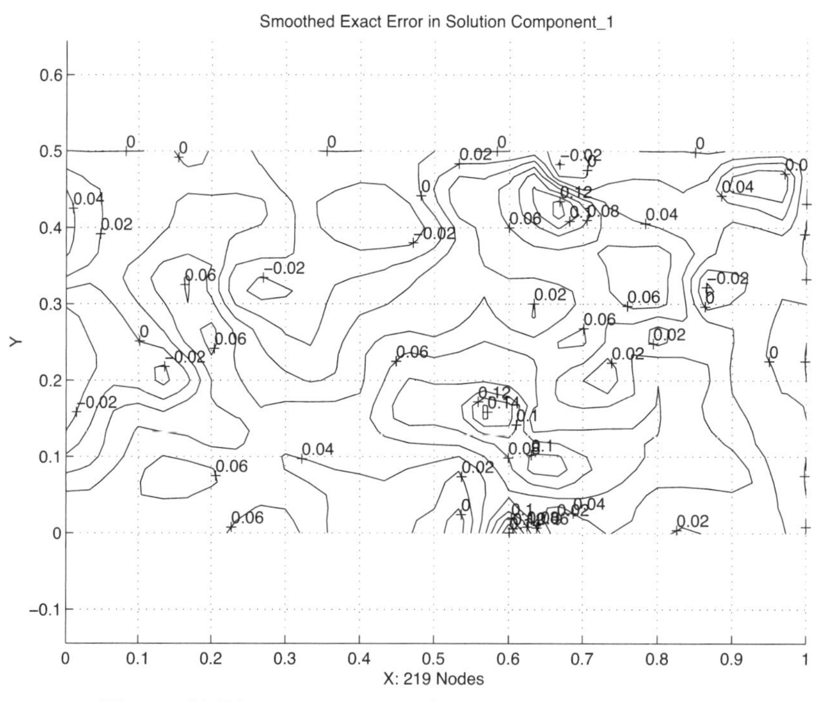

Figure 11.74 *Exact error in the solution value, Kx < Ky*

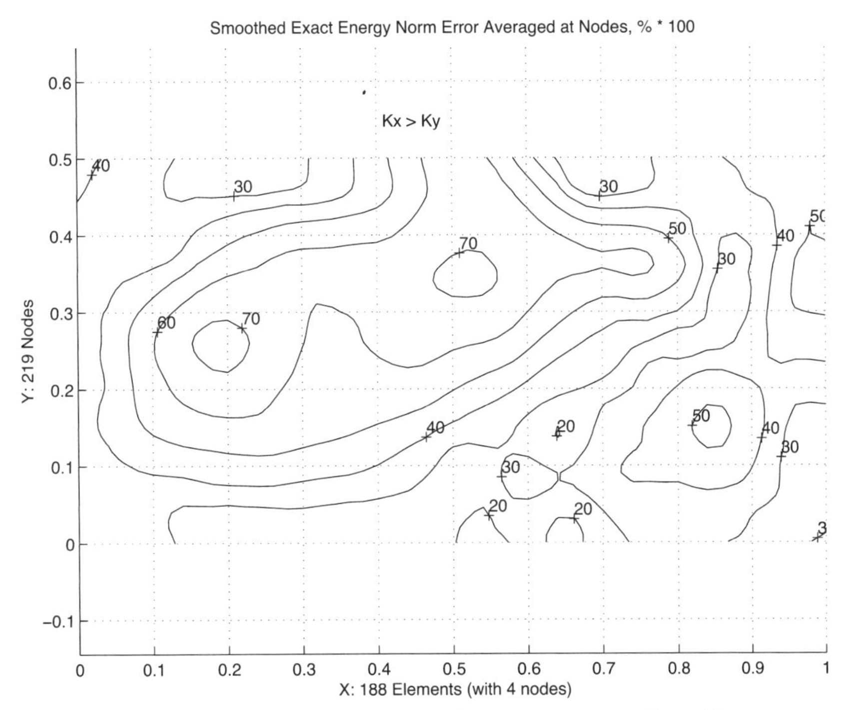

Figure 11.75 *Exact error in the energy norm, Kx > Ky*

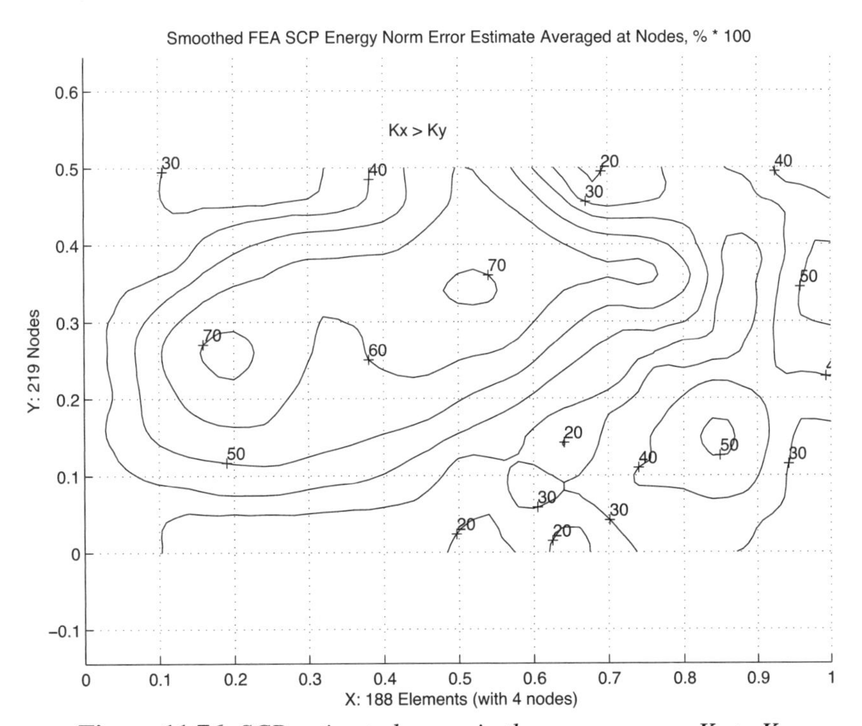

Figure 11.76 *SCP estimated error in the energy norm, Kx > Ky*

11.11 Axisymmetric conductions

For a general steady state orthotropic axisymmetric material we just recast the governing differential equation, Eq. 11.2, into cylindrical coordinates:

$$\frac{1}{r}\, k_r \frac{\partial \theta}{\partial r} + k_r \frac{\partial^2 \theta}{\partial r^2} + k_z \frac{\partial^2 \theta}{\partial z^2} + Q = 0$$

or re-arranging

$$\frac{\partial}{\partial r}(k_r r \frac{\partial \theta}{\partial r}) + \frac{\partial}{\partial z}(k_z r \frac{\partial \theta}{\partial z}) + Q\, r = 0 \tag{11.31}$$

Comparing this form to Eq. 11.2 we see that one difference is that one could consider substituting the values $(k_r r)$, $(k_z r)$, and $(Q\, r)$ as modified conductivities and source term in the previous formulation. That is, material constants would now become spatially varying, but that is no problem to include in a numerically integrated version like those given in the previous section. However, the change in coordinates also means that we must change the differential volume $d\Omega$, to $2\pi\, r\, dr\, dz$. Some analysts like to use a one radian segment rather than the full 2π body. When specifying resultant point (ring) flux data you need to know which basis is being employed. Carrying out the usual Galerkin process the integral definitions of the element and segment arrays are almost identical to those in Eq.s 11.10-11, except that x and y are replaced by r and z, respectively, while the differential volume and surface areas change from $t^e\, da$ and $t^b\, ds$ to the corresponding axisymmetric measures of $2\pi\, r\, da$ and $2\pi\, r\, ds$. Namely,

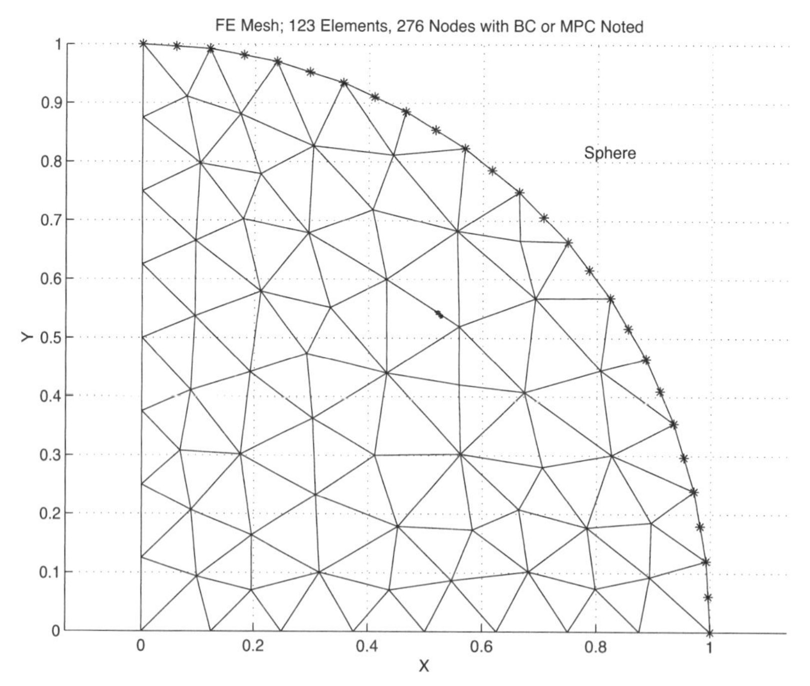

Figure 11.77 *Half sphere model and boundary condition nodes*

```
title "Kreyszig unit sphere with EBC'                          ! 1
exact_case 22 ! Exact analytic solution                        ! 2
axisymmetric  ! Problem is axisymmetric, x radius, y C/L       ! 3
b_rows      2 ! Number of rows in the B (operator) matrix      ! 4
dof         1 ! Number of unknowns per node                    ! 5
el_nodes    6 ! Maximum number of nodes per element            ! 6
elems     123 ! Number of elements in the system               ! 7
gauss       7 ! Maximum number of quadrature points            ! 8
nodes     276 ! Number of nodes in the mesh                    ! 9
shape       2 ! Element shape, 1=line, 2=tri, 3=quad, 4=hex    !10
space       2 ! Solution space dimension                       !11
el_homo       ! Element properties are homogeneous             !12
el_real     4 ! Number of real properties per element          !13
pt_list       ! List the answers at each node point            !14
list_exact    ! List given exact answers at nodes, etc         !15
save_new_mesh ! Save new element sizes for adaptive mesher     !16
example   202 ! Source library example number                  !17
remarks     3 ! Number of user remarks, e.g. property names    !18
end ! Terminate the keyword control inputs, remarks follow     !19
The surface temperature BC is T=cosine(angle_from_Z)^2         !10
so it varies from 1 to zero. R_max = 1 = Z_max.                !21
T_exact=(R^2 + Z^2)*( cos(ang)^2 - third) + third              !22
  1     1  6.02683E-02  9.96354E-01 ! node bc_flag, R, Z       !23
  2     1  0.00000E+00  1.00000E+00                            !24
  3     0  3.95504E-02  9.55542E-01                            !25
  4     0  0.00000E+00  9.37500E-01                            !26
   ...                                                         !27
271     0  7.98446E-01  7.32400E-02 ! node bc_flag, R, Z       !28
272     0  8.75000E-01  0.00000E+00                            !29
273     0  8.36723E-01  3.66200E-02                            !30
274     0  7.50000E-01  0.00000E+00                            !31
275     0  7.74223E-01  3.66200E-02                            !32
276     0  8.12500E-01  0.00000E+00 ! last node                !33
  1   272   259   262   261   260   264 ! elem, 6 nodes        !34
  2   274   272   271   276   273   275                        !35
  3   268   274   271   270   275   269                        !36
  4   259   246   262   248   247   260                        !37
  5   265   274   268   267   270   266                        !38
   ...                                                         !39
120    45    17    21    18    16    22 ! elem, 6 nodes        !40
121    37    45    21    36    22    23                        !41
122    33    21    10    20    11    12                        !42
123    21     6    10     7     5    11 ! last elem            !43
  2      1   1.00000E+00 ! node, dof, exact bc value           !44
  1      1   9.93937E-01                                       !45
  6      1   9.85471E-01                                       !46
  8      1   9.65196E-01                                       !47
 17      1   9.42728E-01                                       !48
 19      1   9.09384E-01                                       !49
 29      1   8.74255E-01                                       !50
   ...                                                         !51
201      1   8.94005E-02 ! node, dof, exact bc value           !52
222      1   5.72720E-02                                       !53
224      1   3.35888E-02                                       !54
246      1   1.45291E-02                                       !55
248      1   4.84775E-03                                       !56
259      1   1.00000E+00 ! last EBC                            !57
1 1. 1. 0. 0. ! el kr kz krz Q   (homogeneous elements)        !58
```

Figure 11.78 *Partial sphere input data*

```
 TITLE: "Kreyszig unit sphere with EBC"                                     ! 1
                                                                            ! 2
**** OPTIONS: (DEFAULT) VALUE ****                                          ! 3
AXISYMMETRIC DOMAIN:          0=FALSE, 1=TRUE ..(0)    1                    ! 4
                                                                            ! 5
*** SYSTEM GEOMETRIC PROPERTIES ***                                         ! 6
VOLUME   =  2.08676E+00                                                     ! 7
CENTROID =  5.88333E-01  3.74543E-01                                        ! 8
                                                                            ! 9
*** OUTPUT OF RESULTS AND EXACT VALUES IN NODAL ORDER ***                   !10
NODE,  Radius r,    Axial z,     DOF_1,       EXACT1,                       !11
   1  6.0268E-02  9.9635E-01  9.9394E-01  9.9394E-01                        !12
   2  0.0000E+00  1.0000E+00  1.0000E+00  1.0000E+00                        !13
   3  3.9550E-02  9.5554E-01  9.4152E-01  9.4152E-01                        !14
   4  0.0000E+00  9.3750E-01  9.1927E-01  9.1927E-01                        !15
   ...                                                                      !16
 274  7.5000E-01  0.0000E+00  1.4583E-01  1.4583E-01                        !17
 275  7.7422E-01  3.6620E-02  1.3442E-01  1.3442E-01                        !18
 276  8.1250E-01  0.0000E+00  1.1328E-01  1.1328E-01                        !19
                                                                            !20
*** FLUX COMPONENTS AT ELEMENT INTEGRATION POINTS ***                       !21
ELEM, PT,  Radius r,    Axial z,     FLUX_1,      FLUX_2,                   !22
   1  1  9.2269E-01  3.0937E-02 -3.5661E+00  2.3913E-01                     !23
   1  2  8.9095E-01  4.3634E-02 -3.3251E+00  3.2568E-01                     !24
   1  3  9.3485E-01  5.5422E-03 -3.6607E+00  4.3406E-02                     !25
   1  4  9.4226E-01  4.3634E-02 -3.7190E+00  3.4444E-01                     !26
   1  5  9.7651E-01  9.4004E-03 -3.9943E+00  7.6904E-02                     !27
   1  6  9.0206E-01  7.4009E-02 -3.4085E+00  5.5929E-01                     !28
   1  7  8.8949E-01  9.4004E-03 -3.3141E+00  7.0049E-02                     !29
   ...                                                                      !30
 123  1  1.2462E-01  9.2829E-01 -6.5057E-02  9.6919E-01                     !31
 123  2  1.2630E-01  9.0186E-01 -6.6820E-02  9.5426E-01                     !32
 123  3  1.4331E-01  9.3536E-01 -8.6027E-02  1.1230E+00                     !33
 123  4  1.0426E-01  9.4767E-01 -4.5535E-02  8.2776E-01                     !34
 123  5  1.2178E-01  9.7314E-01 -6.2120E-02  9.9280E-01                     !35
 123  6  9.2934E-02  9.1631E-01 -3.6177E-02  7.1340E-01                     !36
 123  7  1.5916E-01  8.9544E-01 -1.0611E-01  1.1940E+00                     !37
                                                                            !38
 *** SUPER_CONVERGENT AVERAGED NODAL FLUXES ***                             !39
NODE,    Radius r,    Axial z,     FLUX_1,      FLUX_2,                     !40
   1  6.0268E-02  9.9635E-01 -1.5215E-02  5.0306E-01                        !41
   2  0.0000E+00  1.0000E+00  7.9862E-07  7.1783E-08                        !42
   3  3.9550E-02  9.5554E-01 -6.5518E-03  3.1661E-01                        !43
   4  0.0000E+00  9.3750E-01  1.1258E-07  2.2950E-08                        !44
   5  9.9819E-02  9.5190E-01 -4.1737E-02  7.9601E-01                        !45
   ...                                                                      !46
 273  8.3672E-01  3.6620E-02 -2.9326E+00  2.5670E-01                        !47
 274  7.5000E-01  0.0000E+00 -2.3562E+00  2.4802E-08                        !48
 275  7.7422E-01  3.6620E-02 -2.5109E+00  2.3752E-01                        !49
 276  8.1250E-01  0.0000E+00 -2.7653E+00  4.2908E-09                        !50
                                                                            !51
MAXIMUM ELEMENT ENERGY ERROR OF 7.01E-01 OCCURS IN ELEM 8                   !52
MAXIMUM ENERGY ERROR DENSITY OF 3.69E+00 OCCURS IN ELEM 63                  !53
```

Figure 11.79 *Selected sphere output results*

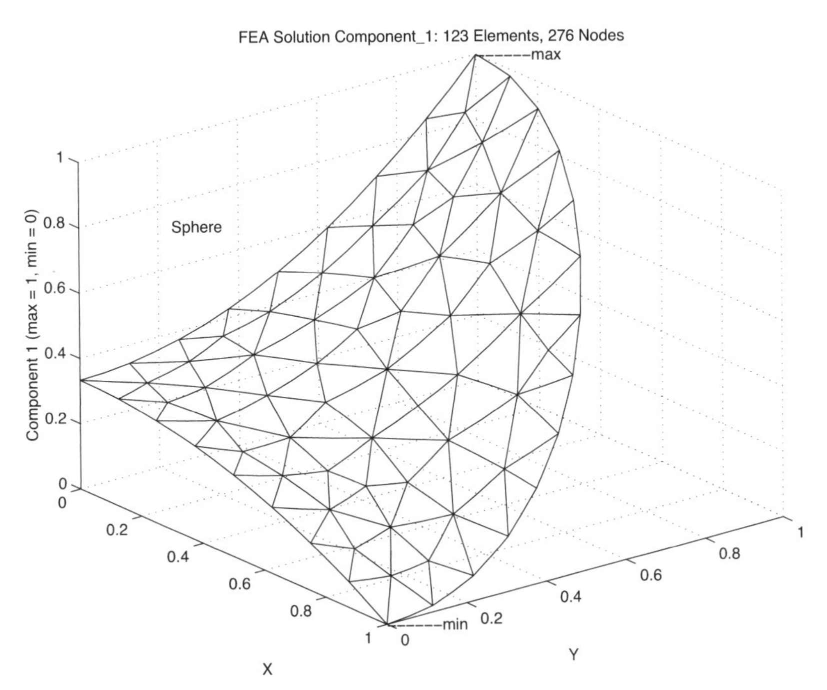

Figure 11.80 *Carpet plot of finite element temperature*

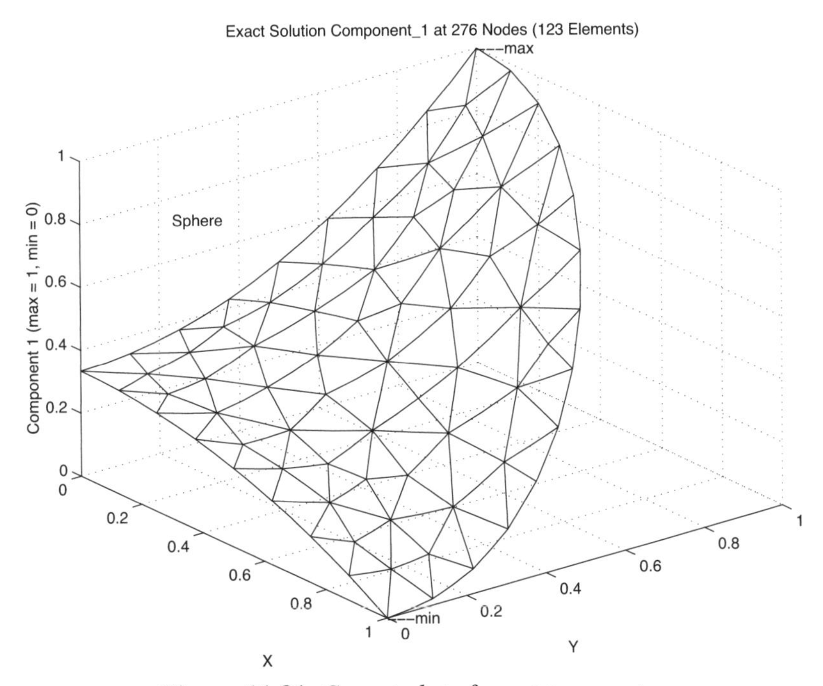

Figure 11.81 *Carpet plot of exact temperature*

$$\mathbf{S}^e = \int_{A^e} (k_r^e\, \mathbf{H}_r^{e^T} \mathbf{H}_r^e + k_z^e\, \mathbf{H}_z^{e^T} \mathbf{H}_z^e)\, 2\pi\, r\, da \tag{11.32}$$

$$\mathbf{C}_Q^e = \int_{A^e} \mathbf{H}^{e^T} Q^e 2\pi\, r\, da$$

$$\mathbf{S}_h^b = \int_{\Gamma^b} h^b \mathbf{H}^{b^T} \mathbf{H}^b 2\pi\, r\, ds, \quad \mathbf{C}_h^b = \int_{\Gamma^b} \theta_\infty^b h^b \mathbf{H}^{b^T} 2\pi\, r\, ds$$

$$\mathbf{C}_q^b = \int_{\Gamma^b} q^b \mathbf{H}^{b^T} 2\pi\, r\, ds$$

In other words we could use the previous formulations, but now require the thickness to be $t = 2\pi r$ at any point. Then we need to activate the option in the previous numerical integration coding in Figs. 11.41-43 in the numerical integration loop, after forming the Jacobian, such as:

```
IF ( AXISYMMETRIC ) THICK = TWO_PI * XYZ (1) ! via key axisymmetric  !27
```

where *TWO_PI* is a global program parameter, and *AXISYMMETRIC* is a global logical variable that is true only if the keyword *axisymmetric* appears in the control data. Then the r coordinate is the first component of the coordinates of the point, $XYZ(1)$.

As an axisymmetric heat transfer example we will consider the temperature of a solid homogeneous unit sphere where the temperature on the surface is specified to be unity at the north and south poles and decreases to zero at the equator. We employ a half-symmetry model as shown in Fig. 11.77. The mesh was created by an automatic mesh generator as the first step in an adaptive analysis. The mesh consists of quadratic triangles with six nodes (the T6 element). They have curved edges where they approximate the surface of the sphere. The conduction matrix involves the products of the gradients (linear polynomials) and the radius, which is varying at least in a linear fashion (for straight-sided elements) or quadratically in general. Thus the square matrix is at least a cubic polynomial. From Table 9.3, a cubic polynomial requires four points. Near the surface the Jacobian is not constant so a seven-point rule is used.

Portions of the input data are shown in Fig. 11.78 where the main new control item is the keyword *axisymmetric* that flags the need for an extension of the integration rules in forming the conduction matrix, and the domain geometry properties. The exact result is known, exact_case 22, and is used for comparison purposes in the output list and plots. The computed temperatures agree with the exact values to several significant figures. Selected output results are given in Fig. 11.79. Note that the volume, provided as a data checking aid, is in error by less than one percent. It is inexact because we have approximated a circular arc by eight parabolic (three-noded) segments. Carpet plots of the two solutions are given in Figs. 11.80-81. The error estimator, developed previously, suggests a new size for each element. They are plotted on top of the original mesh in Fig. 11.82.

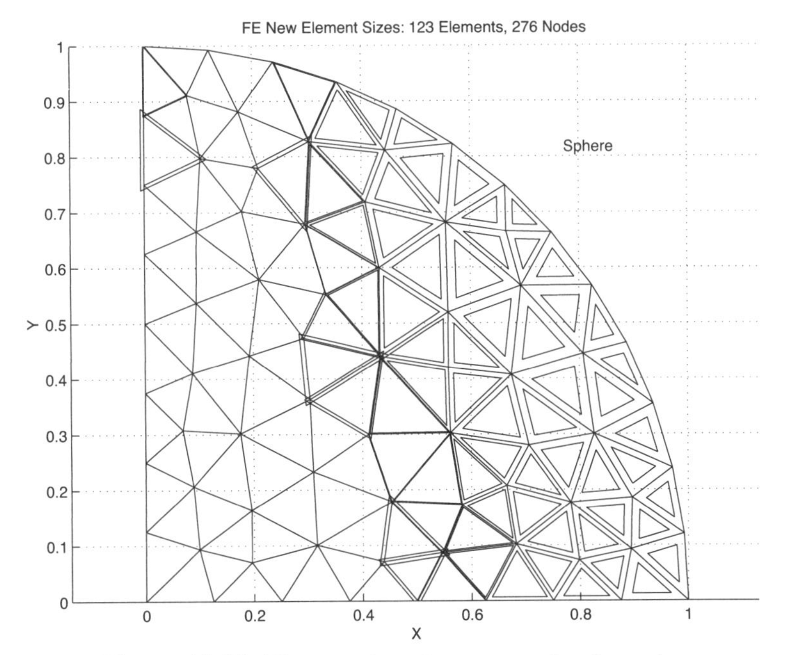

Figure 11.82 *Element sizes for next mesh adaptation*

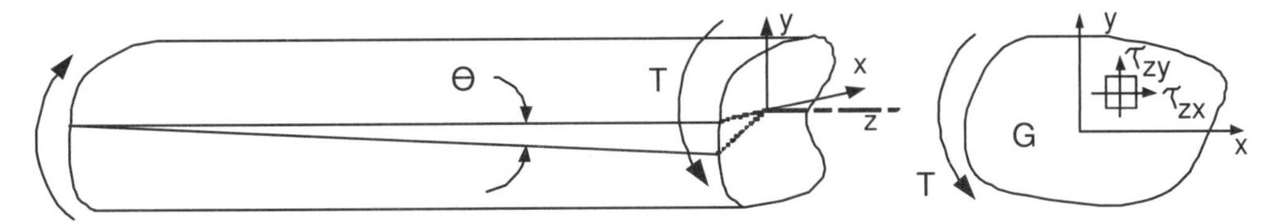

Figure 11.83 *Torsion of a constant cross-section bar*

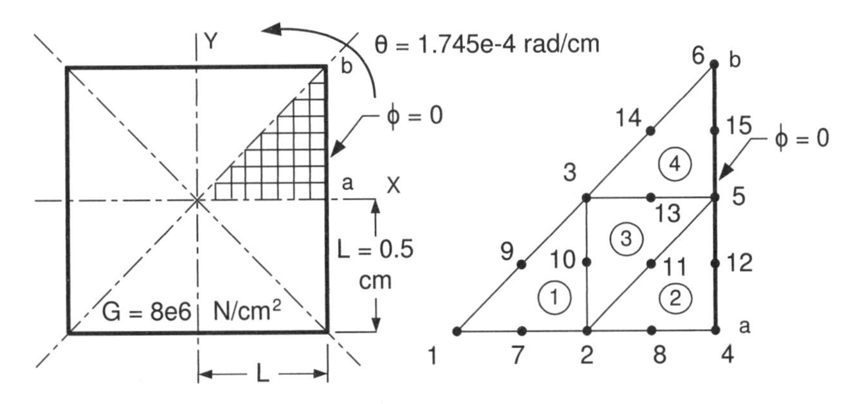

Figure 11.84 *Torsion of a square shaft*

11.12 Torsion

Most finite element formulations of stress analysis problems employ an energy formulation rather than beginning with the differential equations of equilibrium and applying the Galerkin method. It has been shown that both approaches yield identically the same element matrices. However, there are a few cases where a different approach is also useful. One such method is to employ a stress function, $\Phi(x, y)$, whose derivatives define the mechanical stresses that appear in the equation of equilibrium. This lets us recast those equations into another set of differential equations, and boundary conditions whose solution is known, or more easily computed. The approach illustrated here is only valid for singly connected domains (cross-sections without holes). If holes exist in the cross-section one must either use a different formulation or employ a numerical trick where the hole interiors are crudely meshed and assigned a material with a shear modulus, G, that is nearly zero. The case in point here is the torsion of a straight elastic bar of arbitrary cross-sectional area, shown in Fig. 11.83. The bar is subjected to a twisting moment, or torque, T, that causes an angle of twist per unit length of Θ. The twist per unit length is assumed to be small compared to its value squared. Assuming the stress function defined the shear stress components, in the cross-sectional plane, by $\tau_{zx} = \partial\Phi / \partial y$ and $\tau_{zy} = \partial\Phi / \partial x$. Then the governing differential equation is:

$$\frac{1}{G}\frac{\partial^2 \Phi}{\partial x^2} + \frac{1}{G}\frac{\partial^2 \Phi}{\partial y^2} + 2\Theta = 0 \tag{11.33}$$

in the domain of the cross-section, Ω with the essential boundary condition that $\Phi = 0$ on its boundary. In the above, G denotes the elastic shear modulus of the material. For a homogeneous material we could usually divide by G but we wish to allow for non-homogeneous bars so we must keep it with the differential operator. Based on the previous chapter we recognize this as the Poisson equation. We previously defined how to implement its solution by the finite element method. Here the terms just take on new meaning and the post-processing will change. Before, the gradient vector components were directly proportional to the heat flux vector. But here different signs are associated with each component of the recovered gradient. Also the x component of stress is related to the y component of the gradient, and vice versa. Another post-processing aspect is that once the solution Φ is known its integral over the cross-section is related to the applied torque causing the twist by

$$T = 2\int_{\Omega} \Phi d\Omega. \tag{11.34}$$

It turns out that these equations are related to another problem known as the 'soap film' analogy. There we visualized a thin soap film inflated over the cross-sectional shape by a constant pressure. Then the height of the soap film is proportional to the value of Φ. Also, the slope of the soap film is proportional to the shear stress at the same point, but the shear stress acts in a direction perpendicular to that slope. Finally, the volume under the membrane is proportional to the applied torque. This lets us visualize the expected results for the two common shapes of a circular and rectangular cross-section. For a circular bar the shear stress is zero at the center and maximum and constant along its circumference. The distribution of shear stress is more complicated for a rectangular shape. It is also zero at the center point, but the maximum shear stress occurs at the two

```
title "TORSION OF A SQUARE BAR, 1/8 SYMMETRY, (4 ELEMS)"          ! 1
nodes     15 ! Number of nodes in the mesh                        ! 2
elems      4 ! Number of elements in the system                   ! 3
dof        1 ! Number of unknowns per node                        ! 4
el_nodes   6 ! Maximum number of nodes per element                ! 5
space      2 ! Solution space dimension                           ! 6
b_rows     2 ! Number of rows in the B (operator) matrix          ! 7
shape      2 ! Element shape, 1=line, 2=tri, 3=quad, 4=hex        ! 8
remarks   12 ! Number of user remarks                             ! 9
gauss      7 ! Maximum number of quadrature point                 !10
el_types   1 ! Number of different types of elements              !11
el_real    1 ! Number of real properties per element              !12
reals      1 ! Number of miscellaneous real properties            !13
el_homo      ! Element properties are homogeneous                 !14
example 205 ! Application source code library number              !15
post_el      ! Require post-processing, create n_file1            !16
quit ! keyword input, remarks follow                              !17
1 SEGERLIND, 2ND ED. EXAMPLE P. 102, U1=217, U2=159, U4=125,      !18
2 AND T=21.9 VIA LINEAR ELEMENTS (T THEORY = 24.5). Keyword       !19
3 post_el turns on the torque recovery and stress listing.        !20
4                       / 6   Torque, T, twice the solution       !21
5 Mesh:               /   :   integral is reported after the      !22
6                   14    15  integral reported after the         !23
7 C/L             /   (3) :   Here T=24.41 N-cm, and              !24
8 |             4 -- 13 -- 5  U1=205.9, U2=160.3, U4=126.7 cm     !25
9 |           /  : (4)   /  :                                     !26
0 v     9    10     11     12  Max shear stress = 853.6 N/cm^2    !27
1   / (1)    : /   (2)    :   x=0.47 & y=0.025 cm (theory max     !28
2 1--   7 --- 2 --- 8 --- 3 <--- C/L value = 942 N/cm^2, @ 3)     !29
   1    0 0.       0.         ! node, bc_flag, x, y  (cm)         !30
   2    0 0.25     0.                                             !31
   3    1 0.5      0.                                             !32
   4    0 0.25     0.25                                           !33
   5    1 0.5      0.25                                           !34
   6    1 0.5      0.5                                            !35
   7    0 0.125    0.                                             !36
   8    0 0.375    0.                                             !37
   9    0 0.125    0.125                                          !38
  10    0 0.25     0.125                                          !39
  11    0 0.375    0.125                                          !40
  12    1 0.5      0.125                                          !41
  13    0 0.375    0.25                                           !42
  14    0 0.375    0.375                                          !43
  15    1 0.5      0.375                                          !44
     1  1    2    4    7   10    9  ! elem, six nodes             !45
     2  2    3    5    8   12   11                                !46
     3  4    5    6   13   15   14                                !47
     4  2    5    4   11   13   10                                !48
  3    1 0.     ! node, dof, essential bc value                   !49
  5    1 0.                                                       !50
  6    1 0.                                                       !51
 12    1 0.                                                       !52
 15    1 0.                                                       !53
  1 8.e6        ! elem, shear_modulus (homogeneous) N/cm^2        !54
 1.745e-4       ! angle of twist (global) radians/cm              !55
```

Figure 11.85 *Data for the torsion model*

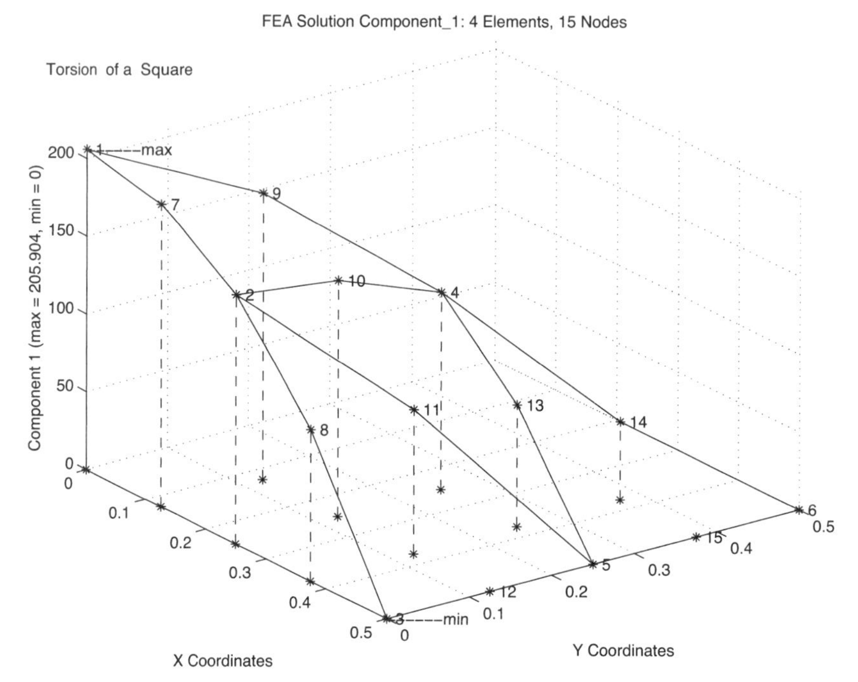

Figure 11.86 *Stress function amplitude*

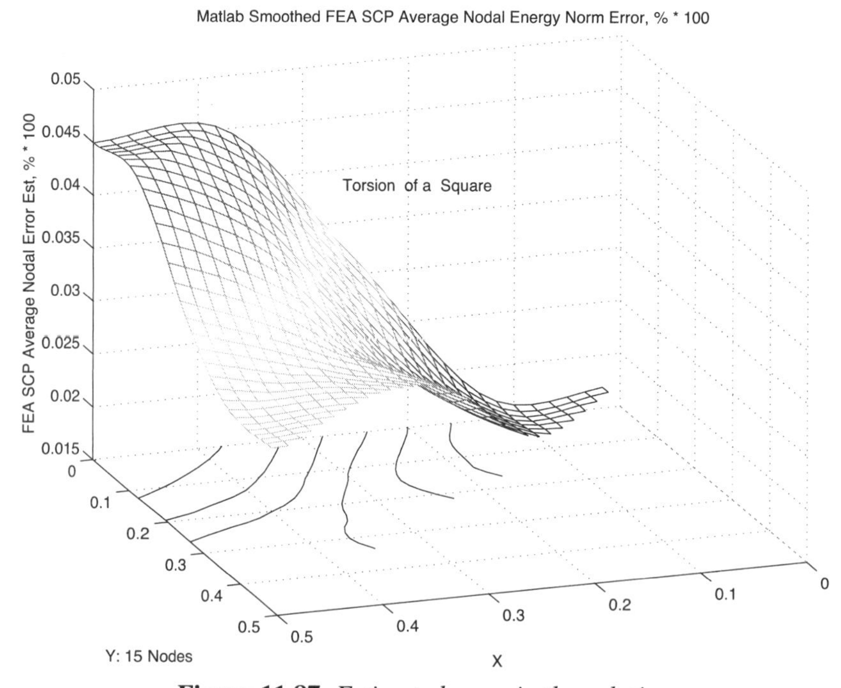

Figure 11.87 *Estimated error in the solution*

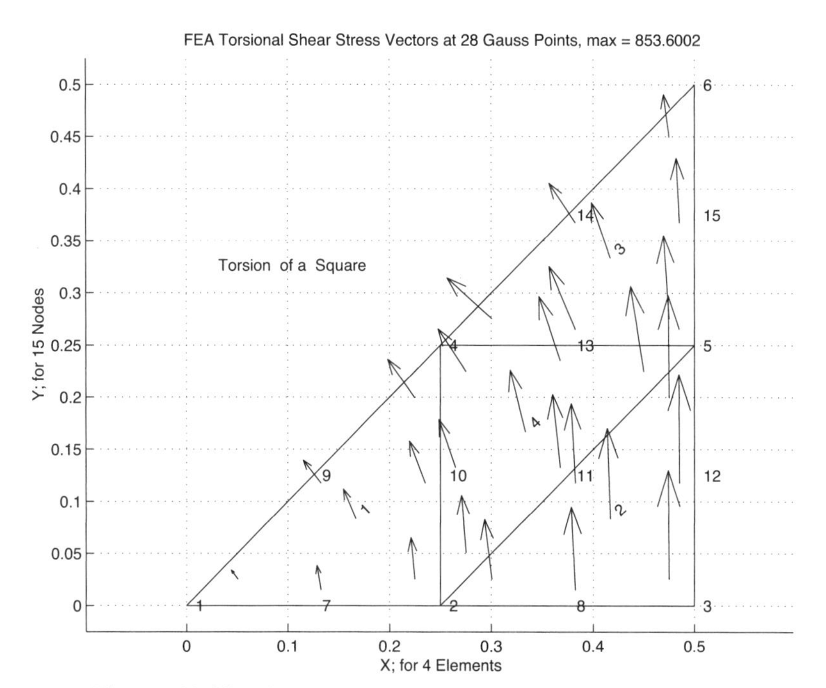

Figure 11.88 *Shear stress vectors at the quadrature points*

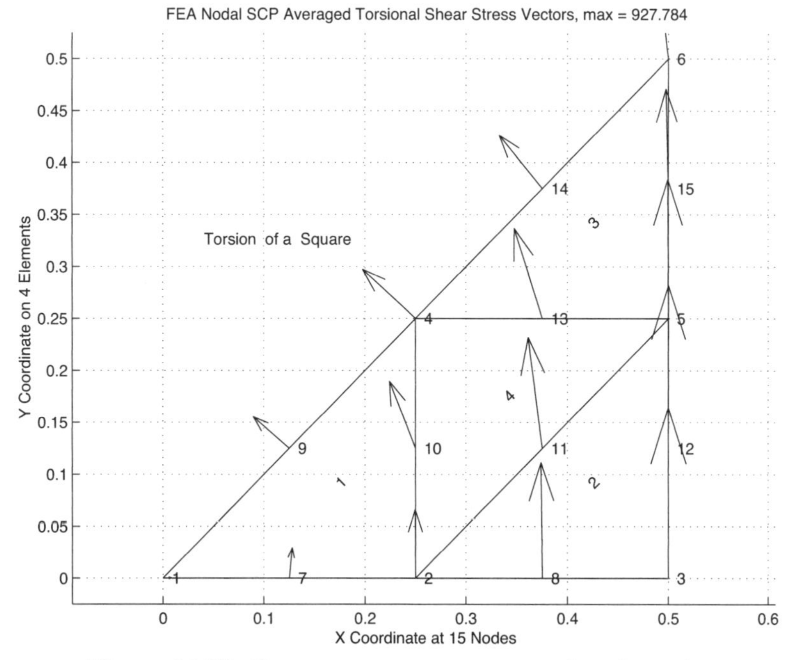

Figure 11.89 *Shear stress vectors averaged at the nodes*

```
! .......................................................        ! 1
!  **  ELEM_SQ_MATRIX PROBLEM DEPENDENT STATEMENTS FOLLOW **     ! 2
! .......................................................        ! 3
!    TORSION (POISSON EQUATION) OF TWO-DIMENSIONAL SHAPE         ! 4
!    (Stored as application source example number 205.)          ! 5
 REAL(DP), PARAMETER :: ZERO = 2 * TINY (1.d0)                   ! 6
 REAL(DP)            :: CONST, DET, SOURCE                       ! 7
 INTEGER             :: IP                                       ! 8
! 1/G *(U,xx + U,yy) + Q = 0, Example 205, Q = 2*Twist_Angle     ! 9
! Shear modulus  = el real property 1   = GET_REAL_LP (1)        !10
! Angle of twist = misc real property 1 = GET_REAL_MISC (1)      !11
                                                                 !12
!-->  DEFINE ELEMENT PROPERTIES                                  !13
  CALL APPLICATION_E_MATRIX (IE, XYZ, E)  ! diagonal 1/G         !14
  SOURCE = 2.d0 * GET_REAL_MISC (1)       ! twice twist          !15
                                                                 !16
!      STORE NUMBER OF POINTS FOR STRESS OR ERROR EST            !17
  CALL STORE_FLUX_POINT_COUNT ! Save LT_QP                       !18
                                                                 !19
!-->   NUMERICAL INTEGRATION LOOP                                !20
  DO IP = 1, LT_QP                                               !21
    H   = GET_H_AT_QP (IP)      ! INTERPOLATION FUNCTIONS        !22
    XYZ = MATMUL (H, COORD)     ! FIND POINT, ISOPARAMETRIC      !23
    DLH = GET_DLH_AT_QP (IP)    ! FIND LOCAL DERIVATIVES         !24
    AJ  = MATMUL (DLH, COORD) ! FIND JACOBIAN AT THE PT          !25
!     FORM INVERSE AND DETERMINATE OF JACOBIAN                   !26
    CALL INVERT_JACOBIAN (AJ, AJ_INV, DET, N_SPACE)              !27
    CONST  = DET * WT(IP)                                        !28
                                                                 !29
!     EVALUATE GLOBAL DERIVATIVES, B == DGH                      !30
    DGH = MATMUL (AJ_INV, DLH)                                   !31
    CALL APPLICATION_B_MATRIX (DGH, XYZ, B) ! for error est      !32
                                                                 !33
!     ELEMENT MATRICES: Stiffness, Source, Result integral       !34
    S = S + CONST * MATMUL ((MATMUL (TRANSPOSE (B), E)),B)       !35
    C = C + CONST * SOURCE * H    ! source                       !36
    H_INTG = H_INTG + H * CONST  ! for solution integral         !37
                                                                 !38
!-->  SAVE PT., E AND DERIVATIVES, FOR POST PROCESSING           !39
    CALL STORE_FLUX_POINT_DATA (XYZ, E, B)                       !40
  END DO                                                         !41
                                                                 !42
!-->     SAVE INTEGRAL OF INTERPOLATION FUNCTIONS                !43
  IF ( N_FILE1 > 0 )  WRITE (N_FILE1)  H_INTG ! post_el keyword  !44
!    ***  END ELEM_SQ_MATRIX PROBLEM DEPENDENT STATEMENTS ***    !45
```

Figure 11.90 *Element matrices for torsion*

```
! .......................................................        ! 1
!  ** POST_PROCESS_ELEM PROBLEM DEPENDENT STATEMENTS FOLLOW **   ! 2
!     Given: INTEGER,  INTENT(IN) :: N_FILE1, IE                 ! 3
!  REAL(DP), INTENT(IN) :: COORD (LT_N, N_SPACE), D (LT_FREE)    ! 4
! .......................................................        ! 5
! 2D TORSION: SHEAR STRESS & TORQUE, if keyword post_el is true  ! 6
!    (Stored as application source example number 205.)          ! 7
 LOGICAL, SAVE :: EACH = .false. ! list each or total ?          ! 8
                                                                 ! 9
   CALL LIST_ELEM_TORSION_STRESS  (IE)                           !10
   CALL LIST_ELEM_TORQUE_INTEGRAL (N_FILE1, IE, EACH)            !11
Contains ! the local methods cited above, in next two figures    !12
```

Figure 11.91 *Encapsulating the element stress and torque recovery* midpoints of the shortest sides of the rectangle. If we want to consider the torsion of a square bar then we could use the previous study of heat transfer of a square area with a constant internal source. As noted in Fig. 11.84 one can use a one-eighth symmetry model. From the analogy we expect the maximum shear stress to occur at point a.

Segerlind [22] presents a detailed solution of the torsion of a square bar shown in Fig. 11.84. He used two linear triangles combined with one bilinear square element. Here we will employ a slightly better mesh by employing four quadratic (six node) triangles over the one-eighth symmetry region. The MODEL data file is shown in Fig. 11.85. There the angle of twist per unit length is given in the last line because it is a global (miscellaneous) property that applies to all elements. The shear modulus, G, is given for each element to allow for applications involving bars of more than one material. The computed stress function amplitude is shown in Fig. 11.86 and the corresponding average error estimate is in Fig. 11.87. While the stress function may not be directly useful the shear stress vectors in the plane can be obtained from the physical derivatives and are shown at the quadrature points in Fig. 11.88, while their nodal average values are given in Fig. 11.89. Our analogy and handbook equations cite the mid-point a as the point of maximum shear stress. The handbook stress value is $\tau_{\max} = T/(1.664L^3)$ where T is the applied torque. The torque value is found by integrating the stress function in a post-processing step and this gives $T = 8(24.41\ Ncm) = 195.3\ Ncm$. Thus the estimated maximum shear stress is $938.9\ N/cm^2$ and the finite element prediction, at node 3, is $927.8\ N/cm^2$. The error of about one percent in the maximum stress for this crude mesh is quite reasonable (the three linear element model gave 11 percent error). In practice however, we generally know the applied torque, T, and not the twist. Thus we have to scale all these linear results to match the actual torque. For example, if the actual torque was $250\ Ncm$ we scale stress and twist angle by the ratio of T_{true}/T_{fea}, or $250./195.3 = 1.28$ to get the true maximum shear stress of $\tau_{\max} = 1,187.7\ N/cm^2$, and a true twist angle value of $0.0002234\ radians/cm$.

To be able to use any element in the interpolation library the solution has been implemented by numerical integration and the square matrix form is given in Fig. 11.90. To have the option to recover the physical gradients for stress calculation those data were saved to auxiliary storage in lines 18 and 40. Likewise, it is not unusual to need the integral of the solution so the MODEL system sets aside storage for the integral of the element interpolation functions, **H**, and calls it array H_INT. Line 44 of Fig. 11.90 allows for a user option to save those data for a later recovery of the torque, T. Keyword *post_el* in line 16 of Fig. 11.85 activated that storage as well as its recovery later on. The calculation of H_INT is quite cheap since **H** is already evaluated (at line 22) in the quadrature loop. In the next chapter we will see that most stress analysis problems are based on vector fields. Since data have been saved for post-processing we have supplied an INCLUDE file, my_post_el_inc, that is accessed when that keyword is present. Since two different recovery processes could be used we have included two special routines for this torsion problem. One to get only the shear stresses and one for the torque. They could be in a single code but it is best to use a modular programming style. The two methods are 'encapsulated' within the post-processing 'class' by placing them after the *CONTAINS* statement, as required by the language and noted in Fig. 11.91. The shear stress and resultant torque calculations are given in Figs. 11.92 and 93, respectively.

```
SUBROUTINE LIST_ELEM_TORSION_STRESS (IE)                                !16
! * * * * * * * * * * * * * * * * * * * * * * * * * * * * *             !17
!     LIST ELEMENT SHEAR STRESS AT QUADRATURE POINTS                    !18
! * * * * * * * * * * * * * * * * * * * * * * * * * * * * *             !19
Use System_Constants ! for DP, N_R_B, N_SPACE, E, XYZ                   !20
Use Elem_Type_Data   ! for LT_FREE, LT_N, D, DGH                        !21
 IMPLICIT NONE                                                          !22
 INTEGER,  INTENT(IN) :: IE                                             !23
!         Global Arrays                                                 !24
 REAL(DP) :: DGH (N_SPACE, LT_FREE), STRESS (N_R_B + 2), &              !25
             XYZ (N_SPACE), E (N_R_B, N_R_B)                            !26
                                                                        !27
 INTEGER,  SAVE :: TEST_E, TEST_P, J, N_IP ! for max value              !28
 REAL(DP), SAVE :: DERIV_MAX = -HUGE(1.d0) ! for max value              !29
                                                                        !30
!                 VARIABLES:                                            !31
! D       = NODAL PARAMETERS ASSOCIATED WITH AN ELEMENT                 !32
! E       = CONSTITUTIVE MATRIX                                         !33
! DGH     = GLOBAL DERIVATIVES INTERPOLATION FUNCTIONS                  !34
! IE      = CURRENT ELEMENT NUMBER                                      !35
! LT_N    = NUMBER OF NODES PER ELEMENT                                 !36
! LT_FREE = NUMBER OF DEGREES OF FREEDOM PER ELEMENT                    !37
! N_ELEMS = TOTAL NUMBER OF ELEMENTS                                    !38
! N_R_B   = NUMBER OF ROWS IN B AND E MATRICES                          !39
! N_SPACE = DIMENSION OF SPACE                                          !40
! STRESS  = STRESS OR GRADIENT VECTOR                                   !41
! XYZ     = SPACE COORDINATES AT A POINT                                !42
                                                                        !43
!-->     PRINT TITLES ON THE FIRST CALL                                 !44
  IF ( IE == 1 ) THEN ; WRITE (N_PRT, 5)                                !45
    5 FORMAT (/,'*** TORSIONAL SHEAR STRESSES ***',/,         &         !46
    ' ELEMENT, POINT,        X             Y ', /,            &         !47
    ' ELEMENT,          TAU_ZX        TAU_ZY        TAU', /)            !48
  END IF                                                                !49
                                                                        !50
  CALL READ_FLUX_POINT_COUNT (N_IP) ! NUMBER OF POINTS                  !51
                                                                        !52
  DO J = 1, N_IP                              ! QUADRATURE LOOP         !53
    CALL READ_FLUX_POINT_DATA (XYZ, E, B) ! RECOVER DATA                !54
                                                                        !55
!       CALCULATE SHEAR STRESSES,  STRESS = E*DGH*D                     !56
    STRESS (1:N_R_B) = MATMUL (DGH, D)                                  !57
    STRESS (N_R_B+1) = SQRT ( SUM (STRESS(1:N_R_B)**2) )                !58
                                                                        !59
!-->      PRINT COORDINATES AND GRADIENT AT THE POINT                   !60
    WRITE (N_PRT, 20) IE, J, XYZ                                        !61
    20 FORMAT ( I7, I6, 3(1PE13.5))                                     !62
    WRITE (N_PRT, 30) IE, STRESS(2), -STRESS(1), STRESS(3)              !63
    30 FORMAT ( I7, 6X, 4(1PE13.5) )                                    !64
    IF ( STRESS (N_R_B+1) > DERIV_MAX ) THEN                            !65
      DERIV_MAX = STRESS (N_R_B+1)     ! maximum value                  !66
      TEST_E    = IE ; TEST_P = J      ! maximum point                  !67
    END IF                                                              !68
  END DO ! integration                                                  !69
!-->   ARE CALCULATIONS COMPLETE FOR ALL ELEMENTS                       !70
  IF ( IE == N_ELEMS ) THEN ; WRITE (N_PRT,              &              !71
    "('LARGEST SHEAR STRESS = ', 1PE13.5)") DERIV_MAX                   !72
    WRITE (N_PRT, "('ELEM =', I6, ', POINT = ', I2)") &                 !73
  TEST_E, TEST_P ; END IF ! LAST ELEMENT                                !74
END SUBROUTINE LIST_ELEM_TORSION_STRESS                                 !75
```

Figure 11.92 *Element shear stress recovery option*

```
SUBROUTINE LIST_ELEM_TORQUE_INTEGRAL (N_FILE, IE, EACH)          ! 78
! * * * * * * * * * * * * * * * * * * * * * * * * * * *          ! 79
!  LIST INTEGRAL OF TORQUE FROM H INTEGRAL, ON N_FILE            ! 80
! * * * * * * * * * * * * * * * * * * * * * * * * * * *          ! 81
Use System_Constants  ! for DP                                   ! 82
Use Elem_Type_Data    ! for LT_FREE, LT_N, D, H_INTG             ! 83
 IMPLICIT NONE                                                   ! 84
 INTEGER,  INTENT(IN) :: N_FILE, IE  ! source of data            ! 85
 LOGICAL,  INTENT(IN) :: EACH        ! list each ?               ! 86
                                                                 ! 87
 REAL(DP), SAVE :: VALUE, TOTAL      ! integrals                 ! 88
 REAL(DP)       :: H_INTG (LT_FREE)                              ! 89
 INTEGER        :: EOF               ! End_Of_File               ! 90
                                                                 ! 91
!                 VARIABLES:                                     ! 92
! D       = NODAL PARAMETERS ASSOCIATED WITH ELEMENT             ! 93
! EACH    = TRUE IF ALL LISTED, ELSE JUST TOTAL                  ! 94
! H       = SOLUTION INTERPOLATION FUNCTIONS                     ! 95
! H_INTG  = INTEGRAL OF INTERPOLATION FUNCTIONS                  ! 96
! IE      = CURRENT ELEMENT NUMBER                               ! 97
! LT_FREE = NUMBER OF DEGREES OF FREEDOM PER ELEMENT             ! 98
! N_FILE  = UNIT FOR POST SOLUTION MATRICES STORAGE              ! 99
                                                                 !100
!-->     PRINT TITLES ON THE FIRST CALL AND INITIALIZE           !101
  IF ( IE == 1 ) THEN ; TOTAL = 0.d0                             !102
    IF ( EACH ) WRITE (N_PRT, 5) ; 5 FORMAT        &             !103
      (/,'** TORQUE INTEGRAL CONTRIBUTIONS **',/, &              !104
         'ELEMENT      TORQUE')                                  !105
  END IF                                                         !106
  IF ( IE <= N_ELEMS ) THEN ! ELEMENT RESULTS                    !107
    READ (N_FILE, IOSTAT = EOF) H_INTG   ! GET INTEGRAL          !108
    IF ( EOF /= 0 ) THEN ; PRINT *,                     &        !109
      'LIST_ELEM_TORQUE_INTEGRAL EOF AT ELEMENT ', IE            !110
      STOP 'ERROR, EOF IN LIST_ELEM_TORQUE_INTEGRAL'             !111
    END IF ! MISSING DATA                                        !112
                                                                 !113
!-->   CALCULATE ELEMENT CONTRIBUTION, VALUE = H_INTG*D          !114
    VALUE = DOT_PRODUCT (H_INTG, D) * 2.d0                       !115
    TOTAL = TOTAL + VALUE                                        !116
    IF ( EACH ) WRITE (N_PRT, '(I7, 1PE18.6)') IE, VALUE         !117
    IF ( IE == N_ELEMS ) WRITE (N_PRT, &                         !118
      "('TOTAL TORQUE INTEGRAL = ', 1PE16.6, /)")  TOTAL         !119
  END IF                                                         !120
END SUBROUTINE LIST_ELEM_TORQUE_INTEGRAL                         !121
```

Figure 11.93 *Element torque recovery option*

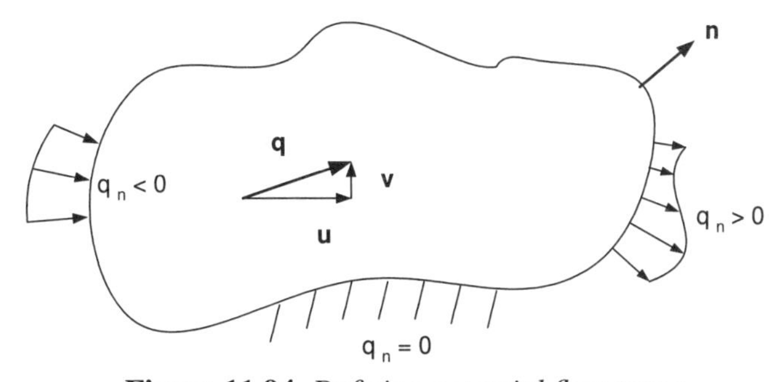

Figure 11.94 *Defining potential flow terms*

11.13 Introduction to linear flows

There are several classes of linear flow models that can be cast as a finite element model. Other flows are highly nonlinear and require much more advanced finite element methods and nonlinear equation solvers. Probably the most common example is the solution of the Navier-Stokes equations for fluid flow. There are several successful finite element formulations of computational fluid dynamics (CFD). Since they are nonlinear and require an iterative solution several implementations begin with a linear flow approximation that satisfies the continuity equation (mass conservation). Thus, there continues to be a need to have efficient linear flow models. Linear flow problems include a wide range of applications like potential flow, flow through porous media, lubrication flow, creeping viscous flows, all of which are elliptic in nature, and other classes like transonic potential flow that changes from elliptic to hyperbolic in nature as the Mach number increases. The subsonic flow of an ideal gas reduces to a nonlinear Poisson equation but it is often reduced to a linearized theory that gives a near singular Laplace equation usually known as the Prandl-Glauert equation. Here we will review some of the common examples of linear flows solved by finite element methods.

11.14 Potential flow

A common class of problem which can be formulated in terms of the Poisson equation is that of potential flow of ideal fluids. That is, we wish to model an invisid, irrotational, incompressible, steady state flow. Potential flow can be formulated in terms of the velocity potential, ϕ, or the stream function, ψ. The latter sometimes yields simpler boundary conditions, but ϕ will be utilized here since it can be easily extended to three dimensions while the stream function is quite difficult to generalize to three dimensions. For the velocity potential formulation the diffusion coefficients, K_x and K_y, become the fluid mass density, which is assumed to be constant. The source term, G, represents a source or sink term and is usually zero. In that common case the constant density term could be divided out so that the problem of potential flow is often presented as a solution of Laplace's equation rather than the Poisson equation. However, there are practical applications that merit retaining the Poisson form. The governing equation is

$$\rho\left(\frac{\partial^2\phi}{\partial x^2} + \frac{\partial^2\phi}{\partial y^2}\right) = Q, \tag{11.35}$$

where Q is a source or sink. The velocity potential, ϕ is usually of secondary interest and the analyst generally requires information on the velocity components. They are defined by the global derivatives of ϕ:

$$u \equiv \frac{\partial\phi}{\partial x}, \qquad v \equiv \frac{\partial\phi}{\partial y} \tag{11.36}$$

where u and v denote the x- and y-components of the velocity vector, $\mathbf{q}$, in the plane of analysis, as shown in Fig. 11.94. Thus, although the program will yield the nodal values of ϕ one must also calculate the above global derivatives. This can be done economically since these derivative quantities must be generated at each quadrature point during construction of the element square matrix, $\mathbf{S}$. Hence, one can simply store this derivative information, i.e., matrix **DGH**, and retrieve it later for calculating the global derivatives

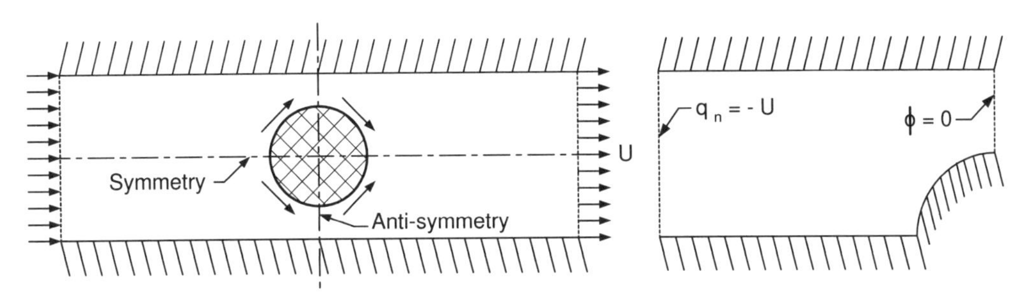

Figure 11.95 *Typical potential flow boundary condition considerations*

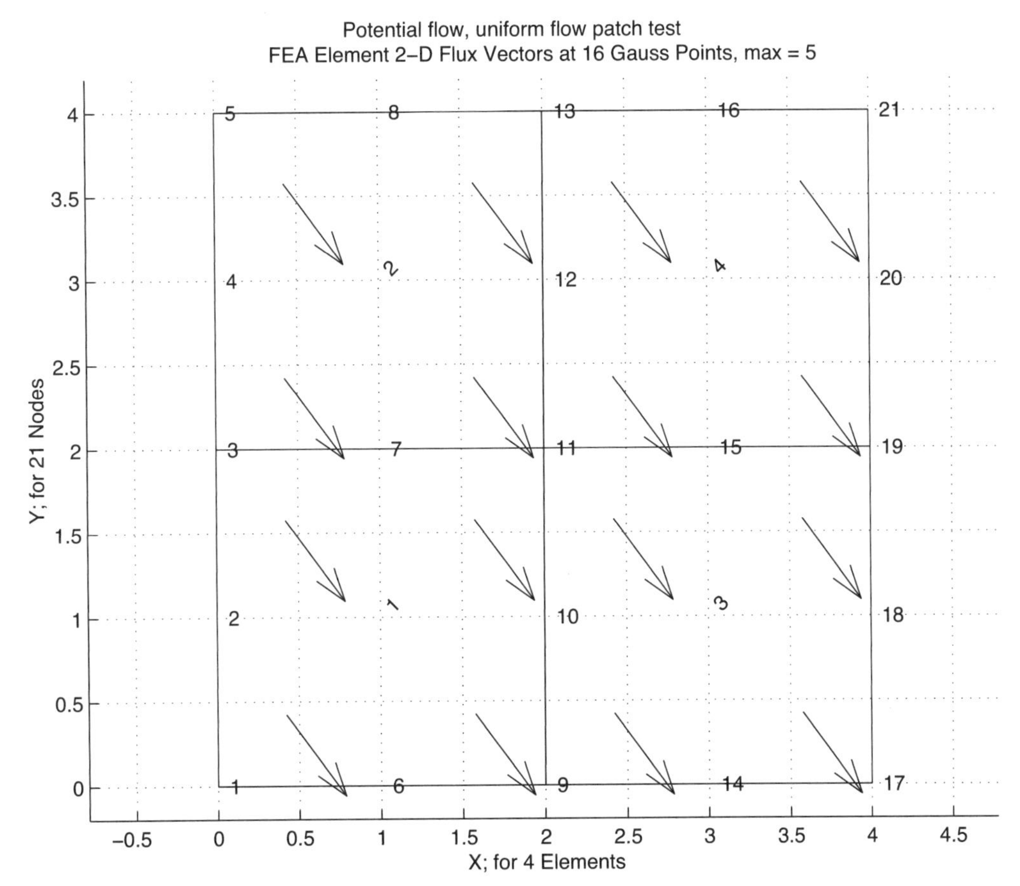

Figure 11.96 *A uniform potential flow patch test*

of ϕ. Another item of interest in Fig. 11.94 is the normal boundary flow into or out of the domain, $q_n = \partial\phi / \partial n$. The natural boundary condition is zero flux which represents an impervious wall. The use of these items to establish the boundary conditions in a simple flow domain is illustrated in Fig. 11.95. There on the vertical line of anti-symmetry we see that the vertical component of velocity, v, will be zero so $\partial \phi / \partial y = 0$ along the line of x constant, so we can set ϕ to an arbitrary constant.

11.14.1 Patch Test

We wish to test this Poisson equation solver by running a patch test. Originally based on engineering judgement, the *patch test* has been proven to be a mathematically valid convergence test [13, 14]. Consider a patch (or sub-assembly) of finite elements containing at least one internal node. An internal node is one completely surrounded by elements. Let the problem be formulated by an integral statement containing derivatives of order n. Assume an arbitrary function, $P(x)$, whose n^{th} order derivatives are constant. Use this function to prescribe the dependent variable on all external nodes of the patch (i.e., $\phi_e = P(x_e)$). Solve for the internal nodal values of the dependent variable, ϕ_I, and its nth order derivatives in each element. To be a convergent formulation:

1. The internal nodal values must agree with the assumed function evaluated at the internal points (i.e., $\phi_I = P(x_I)$?); and
2. The calculated n^{th} order derivatives *must* agree with the assumed constant values. In this application that means that all the velocity vectors are identical. Clearly that represents some inclined uniform flow state, as seen in Fig. 11.96.

It has been found that some non-conforming elements will yield convergent solutions for only one particular mesh pattern. The patch mesh should be completely arbitrary for a valid numerical test. The patch test is very important from the engineering point of view since it can be executed numerically. Thus, one obtains a numerical check of the entire program used to formulate the patch test.

The patch of four elements shown in Fig. 11.96 was utilized. It was assumed that the exact solution everywhere was

$$\phi(x, y) \equiv 1 + 3x - 4y \tag{11.37}$$

such that the derivatives $\phi_{,x} = 3$, $\phi_{,y} = -4$ are constant everywhere. All 16 points on the exterior boundary were assigned values by substituting their coordinates into the Eq. 11.37. That is, the boundary conditions that $\phi_1 \equiv \phi(x_1, y_1)$, etc., were applied on the exterior boundary. Then the problem was solved numerically to determine the value of ϕ at all of the interior points (7, 10, 11, 12, 15) and the values of its global derivatives at each integration point. The output results of this patch test are shown in Fig. 11.97. The output shows clearly that the global derivatives at all integration points have the assumed constant values. It is also easily verified that all the interior nodal values of ϕ are in exact agreement with the assumed form. Thus, the patch test is satisfied and the subroutines pass a necessary numerical test. It is also reassuring that the error estimator indicates that even this crude mesh does not need refinement.

```
TITLE: "Potential flow patch test, uniform flow"                    ! 1
                                                                    ! 2
***  REACTION RESULTANTS  ***                                       ! 3
PARAMETER,     SUM           POSITIVE     NEGATIVE                  ! 4
 DOF_1,      3.5527E-15   2.6000E+01  -2.6000E+01                   ! 5
                                                                    ! 6
***  OUTPUT OF RESULTS IN NODAL ORDER  ***                          ! 7
    NODE,  X-Coord,     Y-Coord,     DOF_1,                         ! 8
       1  0.0000E+00  0.0000E+00  1.0000E+00                        ! 9
    . . .                                                           !10
       7  1.0000E+00  2.0000E+00 -4.0000E+00                        !11
       8  1.0000E+00  4.0000E+00 -1.2000E+01                        !12
       9  2.0000E+00  0.0000E+00  7.0000E+00                        !13
      10  2.0000E+00  1.0000E+00  3.0000E+00                        !14
      11  2.0000E+00  2.0000E+00 -1.0000E+00                        !15
      12  2.0000E+00  3.0000E+00 -5.0000E+00                        !16
      13  2.0000E+00  4.0000E+00 -9.0000E+00                        !17
      14  3.0000E+00  0.0000E+00  1.0000E+01                        !18
      15  3.0000E+00  2.0000E+00  2.0000E+00                        !19
    . . .                                                           !20
      21  4.0000E+00  4.0000E+00 -3.0000E+00                        !21
                                                                    !22
*** FLUX COMPONENTS AT ELEMENT INTEGRATION POINTS ***               !23
ELEMENT, PT, X-Coord,    Y-Coord,     FLUX_1,     FLUX_2,           !24
      1  1  4.2265E-1  4.2265E-1  3.0000E+0 -4.0000E+0              !25
      1  2  1.5774E+0  4.2265E-1  3.0000E+0 -4.0000E+0              !26
      1  3  4.2265E-1  1.5774E+0  3.0000E+0 -4.0000E+0              !27
      1  4  1.5774E+0  1.5774E+0  3.0000E+0 -4.0000E+0              !28
    . . .                                                           !29
      4  1  2.4226E+0  2.4226E+0  3.0000E+0 -4.0000E+0              !30
      4  2  3.5774E+0  2.4226E+0  3.0000E+0 -4.0000E+0              !31
      4  3  2.4226E+0  3.5774E+0  3.0000E+0 -4.0000E+0              !32
      4  4  3.5774E+0  3.5774E+0  3.0000E+0 -4.0000E+0              !33
                                                                    !34
*** SUPER_CONVERGENT AVERAGED NODAL FLUXES ***                      !35
NODE,    X-Coord,     Y-Coord,     FLUX_1,      FLUX_2,             !36
     1  0.0000E+00  0.0000E+00  3.0000E+00 -4.0000E+00              !37
     7  1.0000E+00  2.0000E+00  3.0000E+00 -4.0000E+00              !38
    . . .                                                           !39
     8  1.0000E+00  4.0000E+00  3.0000E+00 -4.0000E+00              !40
     9  2.0000E+00  0.0000E+00  3.0000E+00 -4.0000E+00              !41
    10  2.0000E+00  1.0000E+00  3.0000E+00 -4.0000E+00              !42
    11  2.0000E+00  2.0000E+00  3.0000E+00 -4.0000E+00              !43
    12  2.0000E+00  3.0000E+00  3.0000E+00 -4.0000E+00              !44
    13  2.0000E+00  4.0000E+00  3.0000E+00 -4.0000E+00              !45
    14  3.0000E+00  0.0000E+00  3.0000E+00 -4.0000E+00              !46
    15  3.0000E+00  2.0000E+00  3.0000E+00 -4.0000E+00              !47
    . . .                                                           !48
    21  4.0000E+00  4.0000E+00  3.0000E+00 -4.0000E+00              !49
                                                                    !50
 ***  S_C_P  ENERGY  NORM  ERROR  ESTIMATE  DATA ***                !51
              ERROR IN         % ERROR IN       REFINEMENT          !52
 ELEMENT,     ENERGY_NORM,     ENERGY_NORM,     PARAMETER           !53
       1      4.5626E-15       2.2813E-14       4.5626E-14          !54
       2      7.2788E-15       3.6394E-14       7.2788E-14          !55
       3      6.5704E-15       3.2852E-14       6.5704E-14          !56
       4      8.2159E-15       4.1080E-14       8.2159E-14          !57
```

Figure 11.97 *Numerical results from standard patch test*

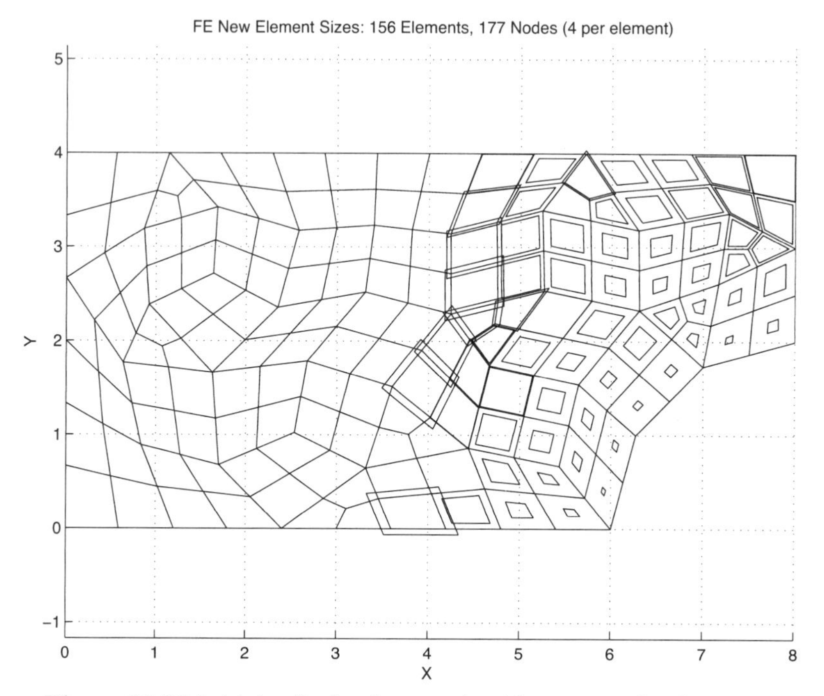

Figure 11.98 *Initial cylinder flow mesh with suggested refinements*

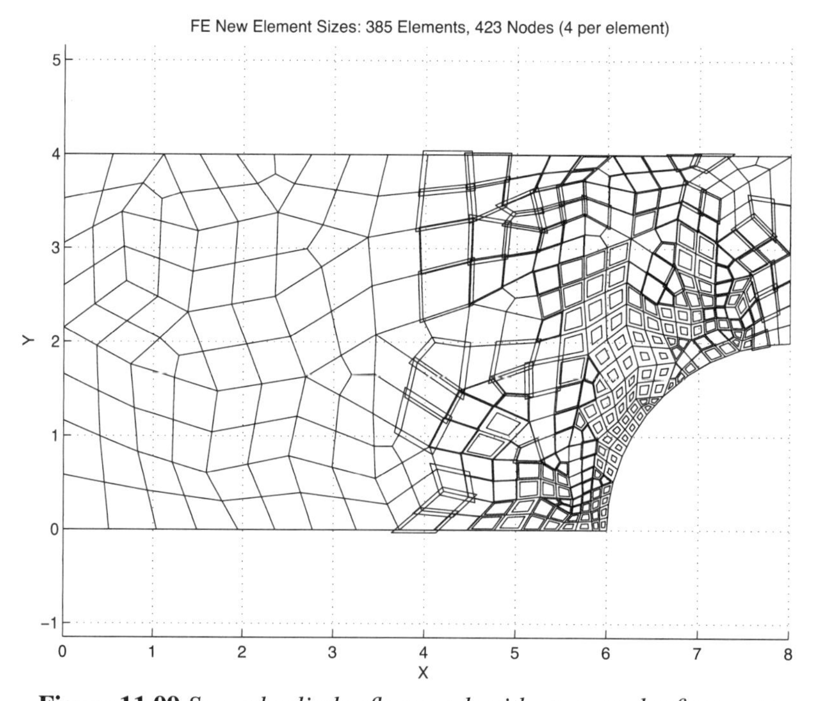

Figure 11.99 *Second cylinder flow mesh with suggested refinements*

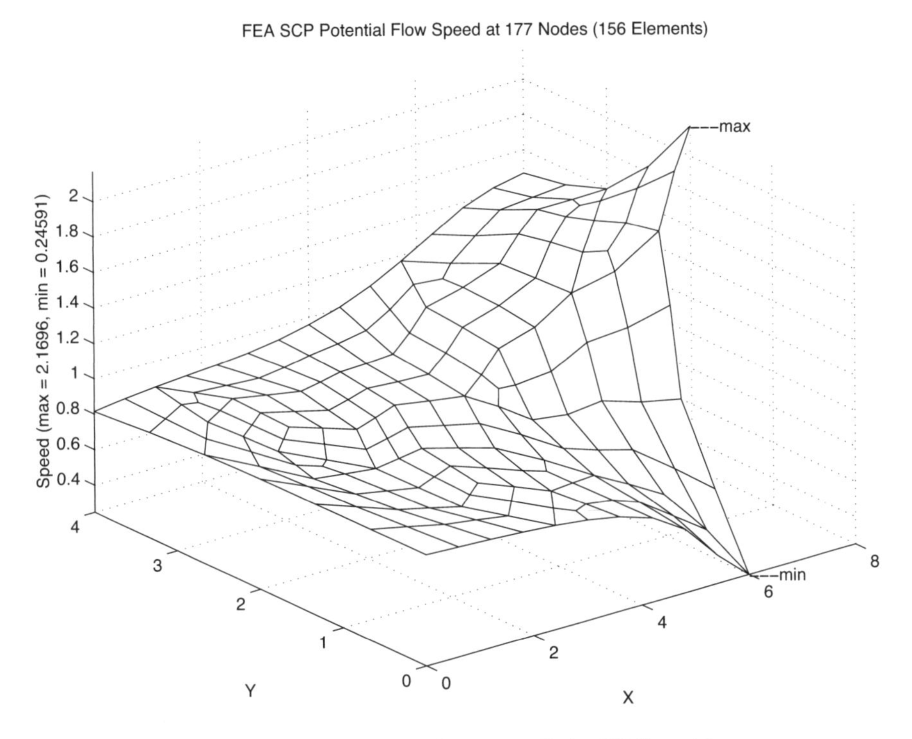

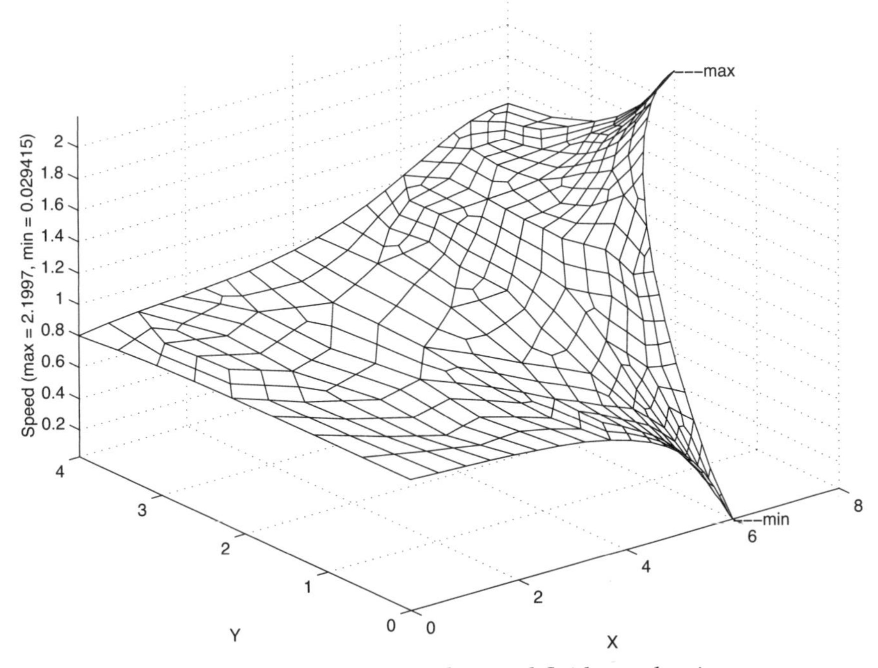

Figure 11.100 *First (top) and second fluid speed estimates*

11.14.2 Example Flow Around a Cylinder

Martin and Carey [17] were among the first to publish a numerical example of a finite element potential flow analysis. This same example is also discussed by others [6, 8]. The problem considers the flow around a cylinder in a finite rectangular channel with a uniform inlet flow. The geometry is shown in Fig. 11.95. By using centerline symmetry and midstream antisymmetry it is possible to employ only one fourth of the flow field. The stream function boundary conditions are discussed by Martin and Carey [17] and Chung [6]. For the velocity potential one has four sets of Neumann (boundary flux) conditions and one set of Dirichlet (nodal parameter) conditions. The first involve zero normal flow, $q_n = \phi_{,n} \equiv 0$, along the centerline ab and the solid surfaces bc and de, and a uniform unit inflow, $q_n \equiv -1$, along ad. At the mid-section, ce, antisymmetry requires that $v = 0$. Thus, $\phi_{,y} = 0$ so that $\phi = \phi(x)$, but in this special case x is constant along that line so we can set ϕ to any desired constant, say zero, along ce. A crude initial mesh of bilinear quadrilaterals was used to start the adaptation. That initial mesh and the suggested new element sizes are given in Fig. 11.98. The smaller element sizes were fed to a mesh generator to produce the first adaptive mesh. It is in Fig. 11.99 where its new suggested element sizes are also shown. Note that the first revised mesh also reduces the original geometric error by better matching the curved boundary of the cylinder. The computed fluid speed from both meshes is shown in Fig. 11.100. There we see the speed is reducing to zero at the stagnation point of the cylinder. The meshes are also refining mainly in that region because the velocity vectors are rapidly changing magnitudes and directions in that region. The minimum speed decreased from 0.246 to 0.029 with the first refinement while the maximum outflow speed increased from 2.17 to 2.20.

As before, the presence of a non-zero boundary flux, q_n, makes it necessary to evaluate the flux column matrix

$$\mathbf{C}^b = \int_{L^b} \mathbf{H}^{bT} q_n \, ds. \tag{11.38}$$

The variation of q_n along the boundary segment is assumed to be defined by the nodal (input) values and the segment interpolation equations, i.e.,

$$q_n(s) \equiv \mathbf{H}^b(s)\, \mathbf{q}_n^b.$$

Therefore, the segment column matrix (for a straight quadratic segment) becomes

$$\mathbf{C}^b = \int_{L^b} \mathbf{H}^{bT}\mathbf{H}^b \, ds \, \mathbf{q}_n^b = \frac{L^b}{30}\begin{bmatrix} 4 & 2 & -1 \\ 2 & 16 & 2 \\ -1 & 2 & 4 \end{bmatrix} \mathbf{q}_n^b. \tag{11.39}$$

The implementation of this segment calculation would be similar to that of Fig. 11.7 and the input segments would be similar to those in Fig. 11.23, but would involve three nodes for each element side subjected to the applied normal flux (fluid inflow).

11.15 Axisymmetric plasma equilibria *

Nuclear fusion is being developed as a future source of energy. The heart of the fusion reactors will be a device for confining the reacting plasma and heating it to thermonuclear temperatures. This confinement problem can be solved through the use of magnetic fields of the proper geometry which generate a so-called 'magnetic bottle'. The tokamak containment concept employs three magnetic field components to confine the plasma. An externally applied toroidal magnetic field, B_T, is obtained from coils through which the torus passes. A second field component is the polodial magnetic field, B_P, which is produced by a large current flowing in the plasma itself. This current is induced in the plasma by transformer action and assists in heating the plasma. Finally, a vertical (axial) field, B_V, is also applied. These typical fields are illustrated in Fig. 11.101. For many purposes a very good picture of the plasma behavior can be obtained by treating it as an ideal magnetohydrodynamic (MHD) media. The equations governing the steady state flow of an ideal MHD plasma are

$$grad\,\mathbf{B} = 0, \quad \nabla P = \mathbf{J} \times \mathbf{B}, \quad \text{curl}\,\mathbf{B} = \mu\mathbf{J}$$

where P is the pressure, **B** the magnetic flux density vector, **J** the current density vector, and μ a constant that depends on the system of units being employed. Consider an axisymmetric equilibria defined in cylindrical coordinates (r, z, θ) so that $\partial/\partial\theta = 0$. This implies the existence of a vector potential, **A**, such that curl **A** = **B**. Assuming that $\mathbf{A} = \mathbf{A}(r, z)$ and $A_\theta = \psi/r$, where ψ is a stream function, we obtain

$$B_r = -\psi_{,z}/r, \; B_z = \psi_{,r}/r, \; B_\theta = A_{r,z} - A_{z,r} = B_T$$

Therefore the governing equation simplifies to

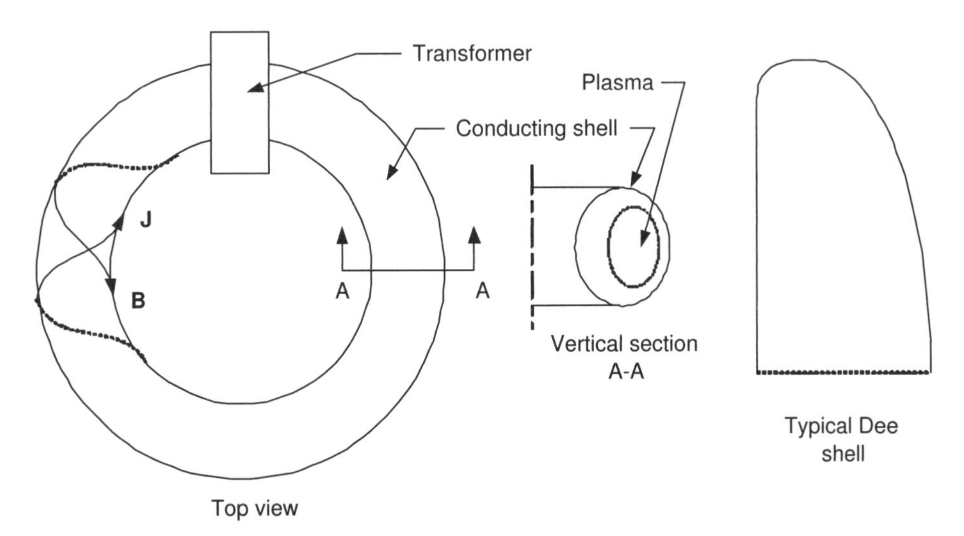

Figure 11.101 *Schematic of tokamak fields and currents*

$$\frac{\partial^2 \psi}{\partial r^2} - \frac{1}{r}\frac{\partial \psi}{\partial r} + \frac{\partial^2 \psi}{\partial z^2} = -\mu r^2 P' - X X' = r J_\theta, \tag{11.40}$$

where J_θ is the plasma current, P is the pressure, $X = r B_\theta$ and where P and X are functions of ψ alone. Both **J** and **B** are vectors that lie tangent to the surfaces of constant ψ. The above is the governing equation for the steady equilibrium flow of a plasma. For certain simple choices of P and B_θ, Eq. 11.40 will be linear but in general it is nonlinear. They are usually represented as a series in ψ as

$$P(\psi) = \alpha_0 + \alpha_1 \psi + \dots + \alpha_n \psi^n / n$$

$$X^2(\psi) = \beta_0 + \beta_1 \psi + \dots + \beta_n \psi^n / n$$

The essential boundary condition on the limiting surface, Γ_1, is

$$\psi = K + 1/2\, r^2 B_V \quad on\ \Gamma_1$$

where K is a constant and B_V is a superimposed direct current vertical (z) field. On planes of symmetry one also has vanishing normal gradients of ψ, i.e.,

$$\frac{\partial \psi}{\partial n} = 0\ on\ \Gamma_2.$$

The right-hand side of Eq. 11.40 can often be written as

$$r J_\theta = p\psi + q \tag{11.41}$$

where, for the above special cases, $p = p(r,\ z)$ and $q = q(r,\ z)$, but where in general q is a nonlinear function of ψ, i.e., $q = q(r,\ z, \psi)$. Equations 11.40 and 11.38 are those for which we wish to establish the finite element model.

A finite element formulation of this problem has been presented by Akin and Wooten [1]. They recast Eq. 11.40 in a self-adjoint form, applied the Galerkin criterion, and integrated by parts. This defines the governing variational statement

$$I = \iint_\Omega \left[\frac{1}{2} \{ (\psi_{,r})^2 + (\psi_{,z})^2 + p\psi^2 \} + q\psi \right] \frac{1}{r}\, dr\, dz \tag{11.42}$$

which, for the linear problem, yields Eq. 11.40 as the Euler equation when I is stationary, i.e., $\delta I = 0$. When $p = q = 0$, Eq. 11.42 also represents the case of axisymmetric inviscid fluid flow. Flow problems of this type were considered by Chung [6] using a similar procedure. For a typical element the element contributions for Eq. 11.41 are

$$\mathbf{S}^e = \int_{\Omega^e} [\mathbf{H}_{,r}^T \mathbf{H}_{,r} + \mathbf{H}_{,z}^T \mathbf{H}_{,z} + p(r,\ z)\mathbf{H}^T\mathbf{H}]\, \frac{1}{r}\, dr\, dz,$$

$$\mathbf{C}^e = \int_\Omega q(r,\ z)\mathbf{H}^T\, \frac{1}{r}\, dr\, dz. \tag{11.43}$$

These matrices are implemented in Fig. 11.102. Other applications of this model are given by Akin and Wooten [1]. The major advantage of the finite element formulation over other methods such as finite differences is that it allows the plasma physicist to study arbitrary geometries. Some feel that the fabrication of the toroidal field coils may require

```
!  ............................................................    ! 1
!     ***  ELEM_SQ_MATRIX PROBLEM DEPENDENT STATEMENTS FOLLOW ***   ! 2
!  ............................................................    ! 3
!  FOR AXISYMMETRIC MHD PLASMA EQUILIBRIUM,     P,Q  /=  0  I.E.   ! 4
!  L(U) = (U,R*1/R),R + U,ZZ*1/R - P*U/R - Q/R = 0                 ! 5
!  U = STREAM FUNCTION                                             ! 6
!         (Stored as source application example 217.)              ! 7
!  P = -C_1 * ALPHA_0 * R**2 - 0.5 * BETA_0, = 0 IF IDEAL PLASMA   ! 8
!  Q = -C_1 * ALPHA_1 * R**2 - 0.5 * BETA_1, = 0 IF IDEAL PLASMA   ! 9
                                                                   !10
! MISC PROPERTIES 1-5 ARE: C_1, ALPHA_0, ALPHA_1, BETA_0, BETA_1   !11
                                                                   !12
 REAL(DP), PARAMETER :: FOUR_PI = TWO_PI * 2.d0                    !13
 REAL(DP)            :: DET_WT, DET, P, Q, R                       !14
 REAL(DP), SAVE      :: ALPHA_0, ALPHA_1, BETA_0, BETA_1           !15
 INTEGER             :: IP, io_1                                   !16
                                                                   !17
!-->  DEFINE ELEMENT PROPERTIES (FIRST 2 TERMS IN POWERS OF U)      !18
  IF ( IE == 1 ) THEN ! first call                                 !19
    ALPHA_0 = GET_REAL_MISC (1)                                    !20
    ALPHA_1 = GET_REAL_MISC (2)                                    !21
    BETA_0  = GET_REAL_MISC (3)                                    !22
    BETA_1  = GET_REAL_MISC (4)                                    !23
  END IF                                                           !24
                                                                   !25
  CALL REAL_IDENTITY (N_R_B, E) !   DEFAULT TO IDENTITY MATRIX     !26
                                                                   !27
!      STORE NUMBER OF POINTS FOR FLUX CALCULATIONS                !28
  CALL STORE_FLUX_POINT_COUNT ! Save LT_QP                         !29
                                                                   !30
!-->   NUMERICAL INTEGRATION LOOP                                  !31
   DO IP = 1, LT_QP                                                !32
     H   = GET_H_AT_QP (IP)    ! EVALUATE INTERPOLATION FUNCTIONS  !33
     XYZ = MATMUL (H, COORD)   ! FIND GLOBAL COORD, (ISOPARAMETRIC) !34
     R   = XYZ (1)             ! CHANGE NOTATION                   !35
     DLH = GET_DLH_AT_QP (IP) ! FIND LOCAL DERIVATIVES             !36
     AJ  = MATMUL (DLH, COORD) ! FIND JACOBIAN AT THE POINT        !37
!        FORM INVERSE AND DETERMINATE OF JACOBIAN                  !38
     CALL INVERT_JACOBIAN (AJ, AJ_INV, DET, N_SPACE)               !39
     DET_WT  = DET * WT(IP) / R                                    !40
                                                                   !41
!        EVALUATE GLOBAL DERIVATIVES, DGH == B                     !42
     DGH = MATMUL (AJ_INV, DLH)                                    !43
     B   = COPY_DGH_INTO_B_MATRIX (DGH) ! B = DGH                  !44
                                                                   !45
!      EVALUATE CONTRIBUTIONS TO SQUARE AND COLUMN MATRICES        !46
     P = - FOUR_PI * R * R * ALPHA_1 - 0.5d0 * BETA_1              !47
     Q = - FOUR_PI * R * R * ALPHA_0 - 0.5d0 * BETA_0              !48
                                                                   !49
     S = S + ( MATMUL ((MATMUL (TRANSPOSE (B), E)), B)   &         !50
             + P * OUTER_PRODUCT (H, H) ) * DET_WT                 !51
     C = C - Q * H * DET_WT                                        !52
                                                                   !53
!-->  SAVE COORDS, E AND DERIVATIVE MATRIX, FOR POST PROCESSING     !54
     CALL STORE_FLUX_POINT_DATA (XYZ, E, B)                        !55
   END DO                                                          !56
                                                                   !57
!    ***  END ELEM_SQ_MATRIX PROBLEM DEPENDENT STATEMENTS ***      !58
```

Figure 11.102 *Plasma element matrices evaluations*

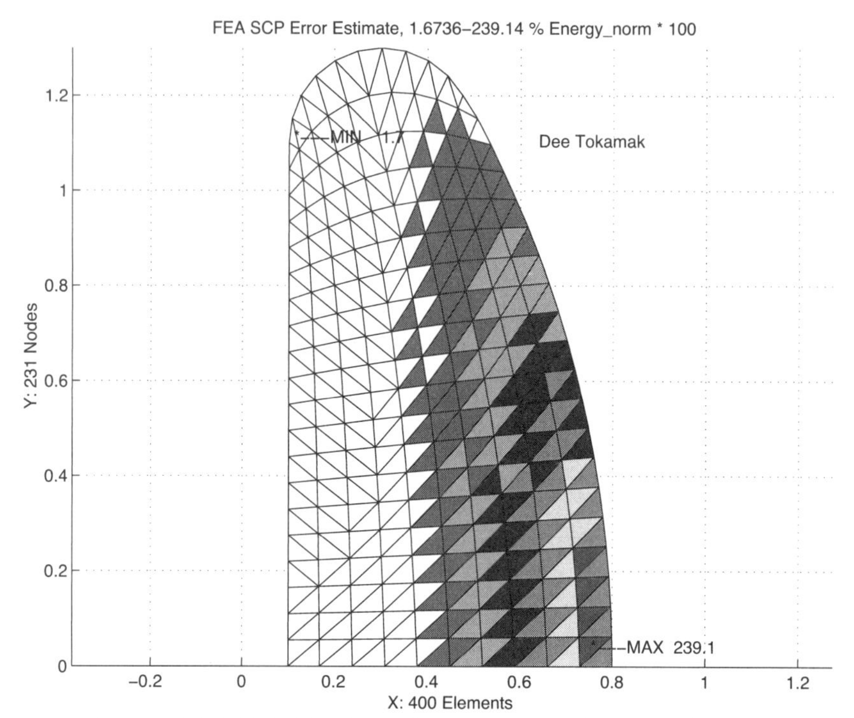

Figure 11.103 *Initial Dee plasma mesh and error estimates*

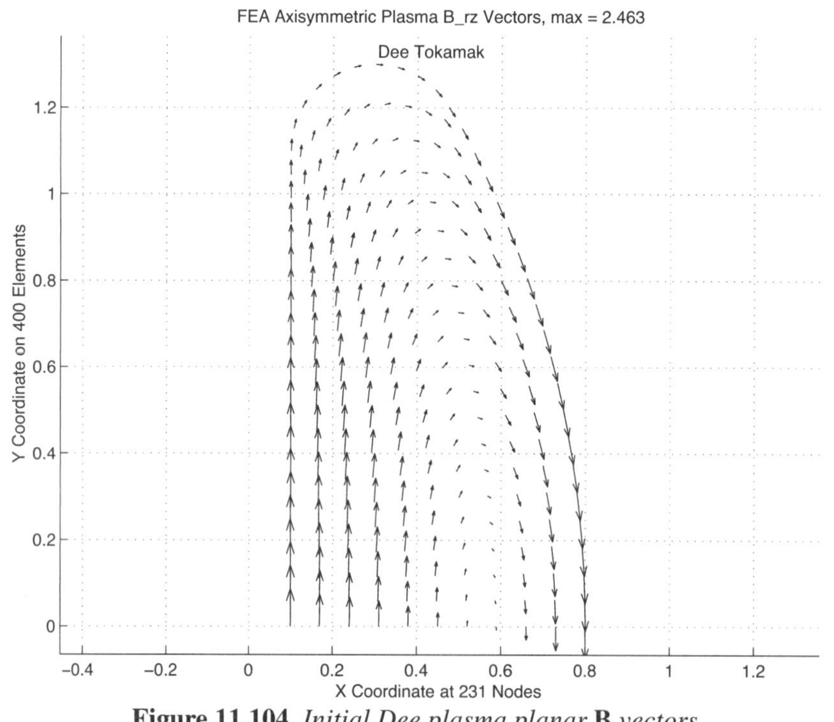

Figure 11.104 *Initial Dee plasma planar* **B** *vectors*

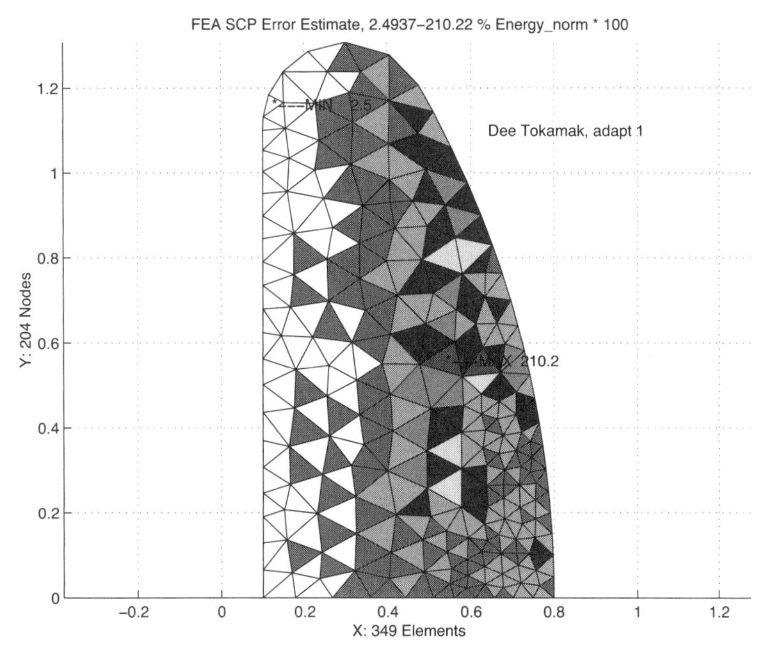

Figure 11.102 *Plasma error estimates in first adaptation (max 2.10 percent)*

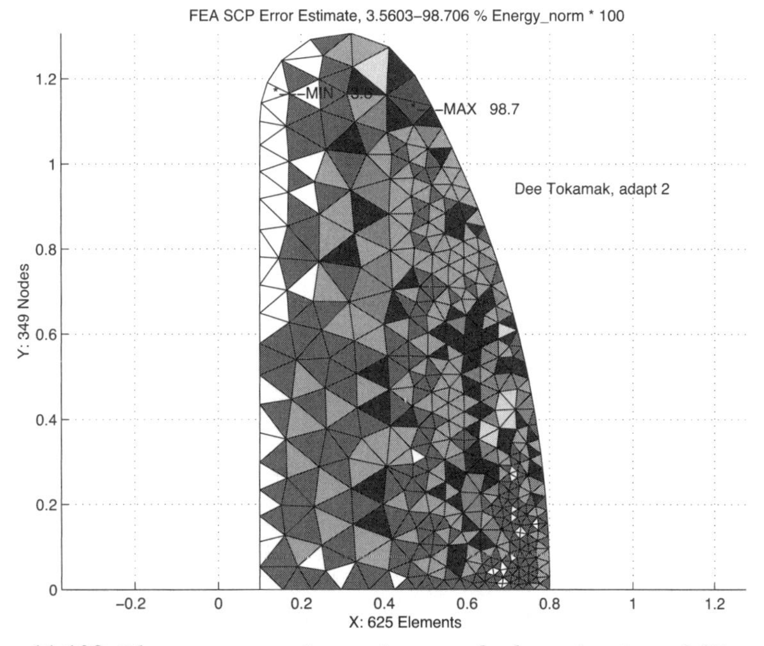

Figure 11.103 *Plasma error estimates in second adaptation (max 0.97 percent)*

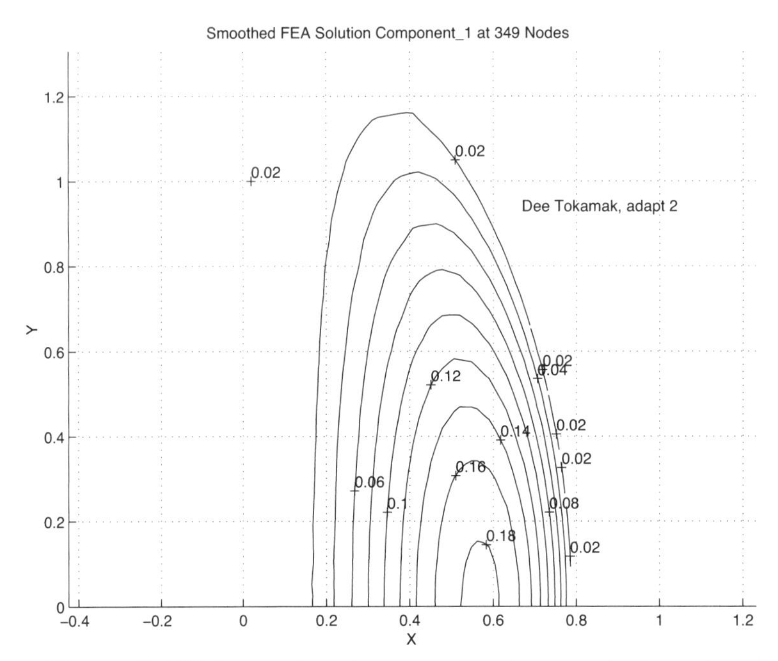

Figure 11.107 *Stream function, ψ, values in second adaptation*

the use of a circular plasma, while others recommend the use of dee-shaped plasmas. The current model has been applied to both of these geometries and the following figures illustrate typical results for a dee-shape torus cross-section where the linear triangle element was employed. Biquadratic or bicubic elements would be better for some formulations which require post-solution calculations using the first and second derivatives of ψ. Figure 11.103 shows the initial relatively uniform mesh and the element level error estimates in the energy norm. The corresponding plasma **B** vectors appear in Fig. 11.104. Note that the initial maximum error in the energy norm is about 2.4 percent so adaptive refinements were taken to reduce the maximum error level to less than 1 percent. The two adaptive meshes and error estimates are shown in Figs. 11.105 and 106. The stream function is of less interest but the final contours of ψ are in Fig 11.107. The initial and final surface plots of ψ are in Figs. 11.108 and 109. The above results have assumed no external vertical **B** field so ψ was assigned values of zero on Γ_1.

11.16 Slider bearing lubrication

Several references are available on the application of the method to lubrication problems. These include the early work of Reddi [21], a detailed analysis and computer program for the three node triangle by Allan [4], and a presentation of higher order elements by Wada and Hayashi [25]. The most extensive discussion is probably found in the text by Huebner [11]. These formulations are based on the Reynolds equation of lubrication. For simplicity a one-dimensional formulation will be presented here.

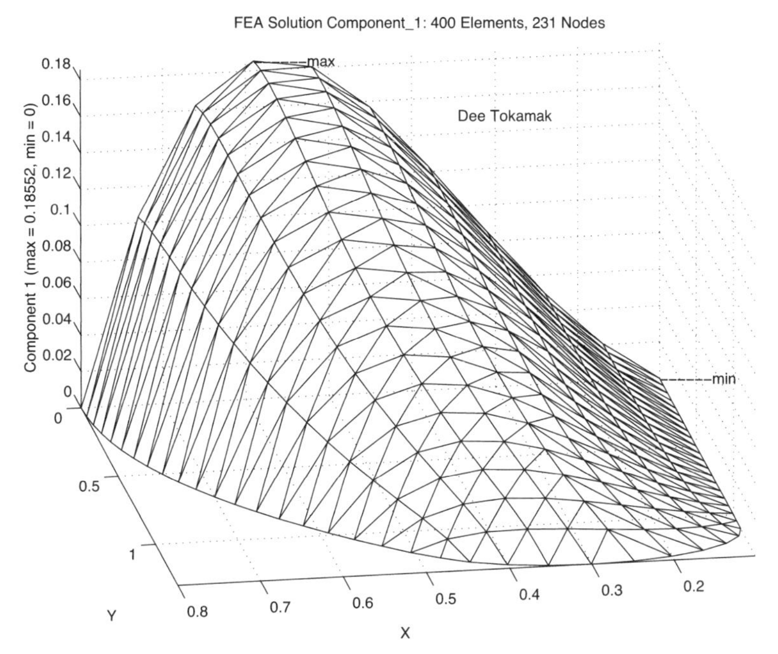

Figure 11.108 *Initial ψ surface for half symmetry model*

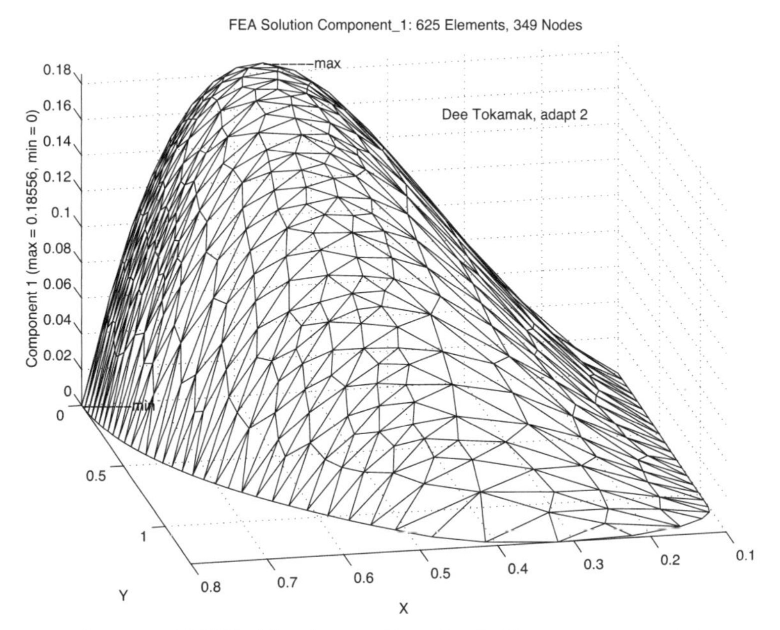

Figure 11.109 *Final ψ surface for half symmetry model*

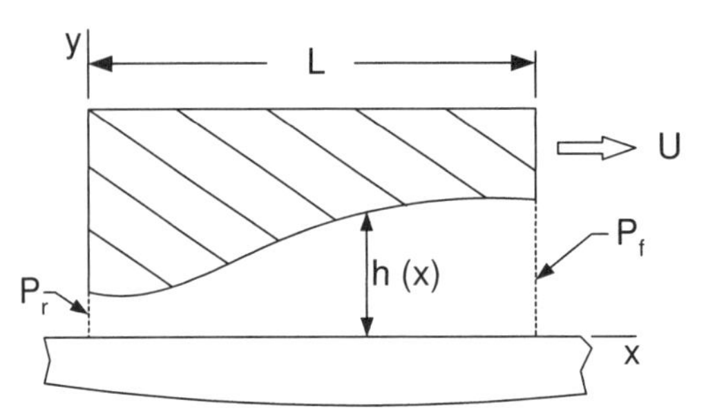

Figure 11.110 *Thin film slider bearing notation*

```
! ..............................................................   ! 1
!    ***  ELEM_SQ_MATRIX PROBLEM DEPENDENT STATEMENTS FOLLOW ***   ! 2
! ..............................................................   ! 3
!     APPLICATION: LINEAR SLIDER BEARING,    Example 102            ! 4
! N_SPACE = 1,  NOD_PER_EL = 2, N_G_DOF = 1                         ! 5
! N_EL_FRE = 2, MISC_FL = 2                                         ! 6
! FLT_MISC (1) = VISCOSITY, FLT_MISC (2) = VELOCITY                 ! 7
! EL_PROP (1) OR PRT_L_PT (K,1) = FILM THICKNESS                    ! 8
                                                                    ! 9
 INTEGER,   SAVE :: KALL = 1                                        !10
 REAL (DP), SAVE :: VIS, VEL, DL                                    !11
 REAL (DP)       :: THICK, CONST                                    !12
                                                                    !13
   IF ( KALL == 1 ) THEN ! GET GLOBAL REAL CONSTANTS                !14
     KALL = 0                                                       !15
     VIS = GET_REAL_MISC (1) ; VEL = GET_REAL_MISC (2)              !16
   END IF ! FIRST CALL                                              !17
                                                                    !18
!-->   DEFINE ELEMENT LENGTH AND ELEMENT THICKNESS                  !19
   DL    = COORD (2, 1) - COORD (1, 1)                              !20
   THICK = 0.d0                                                     !21
   IF ( EL_REAL  > 0 ) THICK = GET_REAL_LP (1)                      !22
                                                                    !23
!      CHECK FOR ALTERNATE AVERAGE NODE THICKNESS                   !24
   IF ( THICK == 0.d0 ) THEN  ! USE NODAL PROPERTY                  !25
     IF ( N_NP_FLO > 0 ) THEN ! DATA EXISTS                         !26
       THICK = 0.5d0 * (PRT_L_PT (1, 1) + PRT_L_PT (2, 1) )         !27
     ELSE                                                           !28
       STOP 'NO SLIDER BEARING THICKNESS DATA'                      !29
     END IF                                                         !30
   END IF ! NODAL THICKNESS DATA                                    !31
                                                                    !32
!-->  GENERATE ELEMENT SQUARE MATRIX & COLUMN MATRIX                !33
   CONST = THICK**3 / (6.0_DP * VIS * DL)                           !34
   S (1, 1) =  CONST    ; S (2, 2) =  CONST                         !35
   S (1, 2) = -CONST    ; S (2, 1) = -CONST                         !36
   C (1) =  VEL * THICK ; C (2) = -VEL * THICK                      !37
                                                                    !38
!-->   GENERATE DATA FOR LOAD CALCULATIONS AND STORE                !39
   H_INTG (1) = 0.5_DP * DL ; H_INTG (2) = 0.5_DP * DL              !40
   IF ( N_FILE1 > 0 ) WRITE (N_FILE1) H_INTG                        !41
                                                                    !42
!    ***  END ELEM_SQ_MATRIX PROBLEM DEPENDENT STATEMENTS ***       !43
```

Figure 11.111 *Slider bearing square and load matrices*

Consider the slider bearing shown in Fig. 11.110 which is assumed to extend to infinity out of the plane of the figure. It consists of a rigid bearing and a slider moving relative to the bearing with a velocity of U. The extremely thin gap between the bearing and the slider is filled with an incompressible lubricant having a viscosity of v. For the one-dimensional case the governing Reynolds equation reduces to

$$\frac{d}{dx}\left(\frac{h^3}{6v}\frac{dP}{dx}\right) = \frac{d}{dx}(Uh), \tag{11.44}$$

where $P(x)$ denotes the pressure and $h(x)$ denotes the thickness of the gap. The boundary conditions are that P must equal the known external pressures (usually zero) at the two ends of the bearing. It can be shown that the variational equivalent of the one-dimensional Reynolds equation requires the minimization of the functional

$$I = \int_0^L \left[\frac{h^3}{12v}\left(\frac{dP}{dx}\right)^2 + hU\left(\frac{dP}{dx}\right)\right] dx. \tag{11.45}$$

As a word of warning, it should be noted that, while the pressure P is continuous, the film thickness h is often discontinuous at one or more points on the bearing. Another related quantity of interest is the load capacity of the bearing. From statics one finds the resultant normal force per unit length in the z-direction, F_y, is

$$F_y = \int_0^L P\, dx. \tag{11.46}$$

This is a quantity which would be included in a typical set of post-solution calculations. Minimizing the above functional defines the element square and column matrices as

$$\mathbf{S}^e = \frac{1}{6v}\int_{x_i}^{x_j} h^3 \mathbf{H}^{eT}_{,x}\, \mathbf{H}^e_{,x}\, dx, \quad \mathbf{C}^e = -U\int_{x_i}^{x_j} h\, \mathbf{H}^{eT}_{,x}\, dx. \tag{11.47}$$

As a specific example of a finite element formulation, consider a linear element with two nodes ($n_n = 2$) and one pressure per node ($n_g = 1$). Thus, $P(x) = \mathbf{H}^e(x)\mathbf{P}^e$ where as before $\mathbf{P}^{eT} = [\, P_i \quad P_j]$ and the interpolation functions are $\mathbf{H}^e = [\,(x_j - x)\ (x - x_i)\,]/L$, where $L = (\, x_j - x_i\,)$ is the length of the element. For the element under consideration $\mathbf{H}^e$ is linear in x so that its first derivative will be constant. That is, $\mathbf{H}^e_{,x} = [\,-1 \quad 1\,]/L$ so that the element matrices simplify to

$$\mathbf{S}^e = \frac{1}{6vl^2}\begin{bmatrix} 1 & -1 \\ -1 & 1 \end{bmatrix}\int_{x_i}^{x_j} h^3\, dx, \quad \mathbf{C}^e = \frac{U}{l}\begin{Bmatrix} 1 \\ -1 \end{Bmatrix}\int_{x_i}^{x_j} h\, dx. \tag{11.48}$$

Thus, it is clear that the assumed thickness variation within the element has an important effect on the complexity of the element matrices. It should also be clear that the nodal points of the mesh must be located such that any discontinuity in h occurs at a node. The simplest assumption is that h is constant over the length of the element. In this case the latter two integrals reduce to $h^3 l$ and hl, respectively. One may wish to utilize this element to approximate a varying distribution of h by a series of constant steps. In this case, one could use an average thickness of $h = (\, h_i + h_j\,)/2$, where h_i and h_j denote the

```
!  .............................................................   ! 1
!   *** POST_PROCESS_ELEM PROBLEM DEPENDENT STATEMENTS FOLLOW ***  ! 2
!  .............................................................   ! 3
!       APPLICATION: LINEAR SLIDER BEARING, Example 102            ! 4
! N_SPACE = 1,   NOD_PER_EL = 2, N_G_DOF = 1                       ! 5
! N_EL_FRE = 2, MISC_FL = 2                                        ! 6
                                                                   ! 7
 INTEGER,  SAVE :: KALL = 1                                        ! 8
 REAL(DP), SAVE :: FORCE, TOTAL = 0.0_DP                           ! 9
                                                                   !10
   IF ( KALL == 1 ) THEN   !     PRINT TITLES ON THE FIRST CALL    !11
     KALL = 0 ; WRITE (6, 5)                                       !12
     5 FORMAT (/, '***   E L E M E N T   L O A D S   ***',/, &     !13
               'ELEMENT         LOAD              TOTAL')          !14
   END IF ! FIRST CALL                                             !15
                                                                   !16
!-->    CALCULATE LOADS ON THE ELEMENTS, F = H_INTG*D              !17
   READ (N_FILE1) H_INTG              ! RECOVER INTEGRAL OF H      !18
   FORCE = DOT_PRODUCT (H_INTG, D)    ! INTEGRAL OF PRESSURE       !19
   TOTAL = TOTAL + FORCE              ! SYSTEM UPDATE              !20
   WRITE (6, 10) IE, FORCE, TOTAL     ! LIST RESULTS               !21
   10 FORMAT (I5, 1PE16.5, 3X, 1PE16.5)                            !22
                                                                   !23
!   *** END POST_PROCESS_ELEM PROBLEM DEPENDENT STATEMENTS ***     !24
```

Figure 11.112 *Element bearing load calculations*

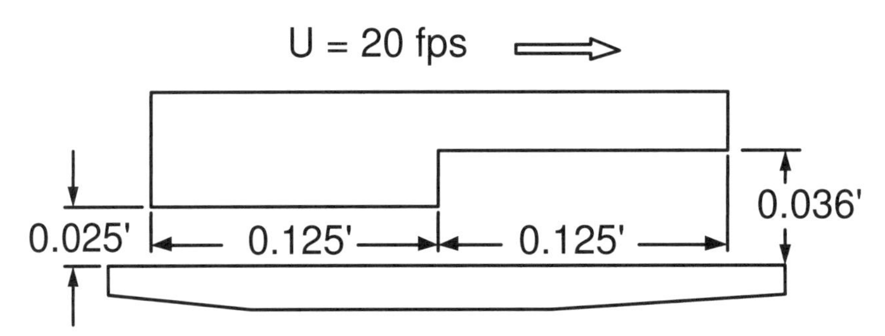

Figure 11.113 *Step bearing example geometry*

thickness at the nodal points of the element. Subroutine *ELEM_SQ_MATRIX* for this element is shown in Figs. 11.111.

Note that it allows for two methods of defining the film thickness, h, in each element. In the default option (keywords *el_real* 1, *pt_real* 0) the value of h is input as a floating point element property, i.e., $H = GET\_REAL\_LP(1)$. In the second option (keywords *el_real* 0, *pt_real* 1) the thickness is specific at each node as a floating point property. Note that for the latter option, *el_real* 0, causes $h = 0$. The program checks for this occurrence and then skips to the second definition of h.

The post-solution calculation function, *POST_PROCESS_ELEM*, is shown in Fig. 11.112. It evaluates the force, F_y^e, carried by each element. The load on a typical element is $F_y^e = \mathbf{Q}^e\mathbf{P}^e$ where $\mathbf{Q}^e = [l/2 \quad l/2]$. Subroutine *ELEM_SQ_MATRIX* also generates and stores $\mathbf{Q}^e$ for each element. Subroutine *POST_PROCESS_ELEM* carries

```
LINEAR SLIDER BEARING                                                    ! 1
                                                                         ! 2
NUMBER OF NODAL POINTS IN SYSTEM =..........           3                 ! 3
NUMBER OF ELEMENTS IN SYSTEM =..............           2                 ! 4
NUMBER OF NODES PER ELEMENT =...............           1                 ! 5
NUMBER OF PARAMETERS PER NODE =.............           2                 ! 6
DIMENSION OF SPACE =........................           1                 ! 7
NUMBER OF ITERATIONS TO BE RUN =............           1                 ! 8
NUMBER OF ROWS IN B MATRIX =................           1                 ! 9
ELEMENT SHAPE: LINE, TRI, QUAD, HEX, TET =...          1                 !10
NUMBER OF DIFFERENT ELEMENT TYPES =..........          1                 !11
STIFFNESS STORAGE MODE: SKY, BAND =..........          1                 !12
NUMBER OF REAL PROPERTIES PER ELEMENT =......          1                 !13
NUMBER OF REAL MISCELLANEOUS  PROPERTIES =...          2                 !14
OPTIONAL UNIT NUMBERS: N_FILE1 =   8                                     !15
*** NODAL POINT DATA ***                                                 !16
NODE, CONSTRAINT FLAG, 1 COORDINATES                                     !17
         1          1    0.0000                                          !18
         2          0    0.1250                                          !19
         3          1    0.2500                                          !20
*** ELEMENT CONNECTIVITY DATA ***                                        !21
ELEMENT NO., 2 NODAL INCIDENCES.                                         !22
    1    1    2                                                          !23
    2    2    3                                                          !24
***  NODAL PARAMETER CONSTRAINT LIST   ***                               !25
TYPE      EQUATIONS                                                      !26
   1          2                                                          !27
*** CONSTRAINT EQUATION DATA ***                                         !28
CONSTRAINT TYPE ONE                                                      !29
EQ. NO.   NODE1   PAR1          A1                                       !30
      1       1      1   0.00000E+00                                     !31
      2       3      1   0.00000E+00                                     !32
***  ELEMENT  PROPERTIES   ***                                           !33
ELEMENT  PROPERTY      VALUE                                             !34
    1          1    2.50000E-02                                          !35
    2          1    3.60000E-02                                          !36
***  MISCELLANEOUS SYSTEM PROPERTIES   ***                               !37
PROPERTY      VALUE                                                      !38
       1      2.00000E-03                                                !39
       2      2.00000E+01                                                !40
                                                                         !41
***  OUTPUT OF RESULTS  ***                                              !42
NODE, 1 COORDINATES, 1 PARAMETERS.                                       !43
    1  0.00000E+00  0.00000E+00                                          !44
    2  1.25000E-01  5.29857E+00                                          !45
    3  2.50000E-01  0.00000E+00                                          !46
                                                                         !47
***   E L E M E N T   L O A D S  ***                                     !48
ELEMENT          LOAD              TOTAL                                 !49
    1      3.31160E-01        3.31160E-01                                !50
    2      3.31160E-01        6.62321E-01                                !51
                                                                         !52
*** EXTREME VALUES OF THE NODAL PARAMETERS ***                           !53
PARAMETER    MAXIMUM, NODE       MINIMUM, NODE                           !54
     1   5.2986E+00,   2    0.0000E+00,    3                             !55
```

Figure 11.114 *Slider bearing results*

```
! .....................................................           ! 1
!    ***  ELEM_SQ_MATRIX PROBLEM DEPENDENT STATEMENTS FOLLOW ***    ! 2
!         NOW USING MASS MATRIX EL_M (LT_FREE, LT_FREE) ALSO        ! 3
! .....................................................           ! 4
!    TRANSIENT ANISOTROPIC POISSON EQUATION IN 1-, 2-, 3-D or       ! 5
!       AXISYMMETRIC VIA NUMERICALLY INTEGRATED ELEMENTS            ! 6
!  (K_ij * U,i),j + Q - Rho * U,t = 0;  1 <= (i,j) <= N_SPACE       ! 7
REAL(DP)         :: CONST, DET                       ! integration  ! 8
REAL(DP), SAVE :: SOURCE=0.d0, RHO=1.d0, THICK=1.d0 ! data          ! 9
INTEGER          :: IP                               ! counter      !10
!                   (Stored as source example 214.)                 !11
! 1-D properties, K_xx, Q, Thickness, Rho                           !12
! 2-D properties, K_xx, K_yy, K_xy, Q, Thickness, Rho               !13
! 3-D properties, K_xx, K_yy, K_zz, K_xy, K_xz, K_yz, Q, Rho        !14
 CALL POISSON_ANISOTROPIC_E_MATRIX (E) ! for 1-, 2-, or 3-D         !15
                                                                    !16
 IF ( SCALAR_SOURCE /= 0.d0 ) SOURCE = SCALAR_SOURCE                !17
 IF ( AREA_THICK    /= 1.d0 ) THICK  = AREA_THICK       ! if 2-D    !18
 IF ( EL_REAL > 0 ) THEN      ! Get local element constant values   !19
   SELECT CASE (N_SPACE)      ! for source, thickness, or rho       !20
     CASE (1) ; IF ( EL_REAL > 1 ) SOURCE = GET_REAL_LP (2)         !21
                IF ( EL_REAL > 2 ) THICK  = GET_REAL_LP (3)         !22
                IF ( EL_REAL > 3 ) RHO    = GET_REAL_LP (4)         !23
     CASE (2) ; IF ( EL_REAL > 3 ) SOURCE = GET_REAL_LP (4)         !24
                IF ( EL_REAL > 4 ) THICK  = GET_REAL_LP (5)         !25
                IF ( EL_REAL > 5 ) RHO    = GET_REAL_LP (6)         !26
     CASE (3) ; IF ( EL_REAL > 6 ) SOURCE = GET_REAL_LP (7)         !27
                IF ( EL_REAL > 7 ) RHO    = GET_REAL_LP (8)         !28
   END SELECT ! for spatial dimension                               !29
 END IF ! element data provided                                     !30
                                                                    !31
 CALL STORE_FLUX_POINT_COUNT ! Save LT_QP, for post-process         !32
                                                                    !33
 DO IP = 1, LT_QP           ! NUMERICAL INTEGRATION LOOP            !34
   H   = GET_H_AT_QP (IP)     ! EVALUATE INTERPOLATION FUNCTIONS    !35
   XYZ = MATMUL (H, COORD)    ! FIND GLOBAL COORD, ISOPARAMETRIC    !36
   DLH = GET_DLH_AT_QP (IP)   ! FIND LOCAL DERIVATIVES              !37
   AJ  = MATMUL (DLH, COORD) ! FIND JACOBIAN AT THE PT              !38
   CALL INVERT_JACOBIAN (AJ, AJ_INV, DET, N_SPACE) ! inverse        !39
   IF ( AXISYMMETRIC ) THICK = TWO_PI * XYZ (1)     ! axisymmetric  !40
   CONST = DET * WT(IP) * THICK                                     !41
   DGH   = MATMUL (AJ_INV, DLH) ; B = DGH      ! Physical gradient  !42
                                                                    !43
!   VARIABLE VOLUMETRIC SOURCE, via keyword use_exact_source        !44
!   Defaults to file my_exact_source_inc if no exact_case key       !45
   IF ( USE_EXACT_SOURCE )  CALL         & ! analytic Q             !46
        SELECT_EXACT_SOURCE (XYZ, SOURCE) ! via exact_case key      !47
   C = C + CONST * SOURCE * H            ! source resultant         !48
                                                                    !49
!    CONDUCTION SQUARE MATRIX, MASS (CAPACITY) MATRIX               !50
   S    = S + CONST * MATMUL ((MATMUL (TRANSPOSE (B), E)), B)       !51
   EL_M = EL_M + OUTER_PRODUCT (H, H) * CONST * RHO                 !52
                                                                    !53
!-->  SAVE COORDS, E AND DERIVATIVE MATRIX, FOR POST PROCESSING     !54
   CALL STORE_FLUX_POINT_DATA (XYZ, (E * THICK), B)                 !55
 END DO                                                             !56
!    ***  END ELEM_SQ_MATRIX PROBLEM DEPENDENT STATEMENTS ***       !57
```

Figure 11.115 *Including the mass matrix for transient applications* out the multiplication once the nodal pressures, $\mathbf{P}^e$, are known. In addition, it sums the force on each element to obtain the total load capacity of the bearing. It prints the element number and its load and the total load on the bearing. With the addition of a few extra post-solution calculations, one could also output the location of the resultant bearing force. Both U and v are constant along the entire length of the bearing. They are simply defined as floating point miscellaneous system properties (keyword *reals* 2).

As a numerical example consider the step bearing shown in Fig. 11.113, which has two constants gaps of different thicknesses but equal lengths. Select a mesh with three nodes (nodes 3) and two elements (elems 2). Let $L_1 = L_2 = 0.125$ ft, $U = 20$ ft/s, $v = 0.002$ lb s/ft^2, $h_1 = 0.025$ ft, and $h_2 = 0.036$ ft. The two boundary conditions are $P_1 = P_3 = 0$ and we desire to calculate the pressure, P_2, at the step. The calculated pressure is $P_2 = 5.299$ psf, which is the exact value, and the total force on the bearing is $F_y = 0.66$ ppf. The accuracy is not surprising since the exact solution for this problem gives a linear pressure variation over each of the two segments of the bearing. The typical output data are shown in Fig. 11.114.

11.17 Transient scalar fields

Transient (first order time dependent) studies are quite common in engineering analysis. All of the scalar field problems considered in this chapter can be easily extended to transient behavior. Huang and Usmani [10] have addressed error estimation in transient problems by creating new meshes at most time steps and interpolating the previous results to the new mesh. Wiberg [26] briefly outlines a similar approach. One could also apply the SCP error analysis to the thermal modes (eigenvectors) in expectation of improving the transient analysis. We will not go into transient adaptivity here. As seen in Eq. 2.53, to treat transient problems we just have to allow for a term involving the first partial derivative with respect to time. That also usually involves the input of one more material property and the definition of the symmetric consistent element mass matrix. The matrix form of the assembled transient problem is

$$\mathbf{M}\,\dot{\boldsymbol{\phi}}(t) + \mathbf{K}\,\boldsymbol{\phi}(t) = \mathbf{F}(t). \tag{11.49}$$

Of course, we also need to describe the initial condition of the primary unknown, and

```
title "Transient Square Conduction 1/8 Symmetry T3"                      ! 1
transient          ! Problem is first order in time                       ! 2
save_1248          ! Save after steps 1, 2, 4, 8, 16 ...                  ! 3
average_mass       ! Average consistent & diagonal mass matrices          ! 4
history_node   1   ! Node number for time-history graph output            ! 5
start_time    0.   ! A time history starting time                         ! 6
time_method    2   ! 1-Euler, 2-Crank-Nicolson, 3-Galerkin, etc           ! 7
time_step 4.0d-2   ! Time step size for time dependent solution           ! 8
time_steps    64   ! Number of time steps in each group                   ! 9
start_value   5.   ! Initial value of transient scalar everywhere         !10
```

Figure 11.116 *Modifying Fig. 11.10 for a transient study*

possibly allow for the Dirichlet and/or Neumann and/or Robin conditions to become functions of time. Often the initial conditions are a global constant, which can be input via a control keyword. Otherwise one has to provide source code (in include file *my_iter_start_inc*) to define its spatial distribution. A partial list of the more common keyword controls for semi-discrete transient integration methods was given in Table 2.1.

The classic finite element weighted residual formulation yields a full, symmetric consistent element mass matrix. This results in the assembled system mass matrix having the same sparsity as the system conduction square matrix. There are computational advantages in having a diagonal element, and system, mass matrix. They are often used to avoid physically impossible answers in the early stage of thermal shock applications. Good modeling practices, like having small element lengths in the direction normal to a thermal shock boundary, also avoid impossible answers but simply using diagonal mass matrices may be quicker. Hughes [12] compares several problems solved with both consistent and diagonal mass matrices. He also illustrates that the average of the consistent and diagonal mass matrices often gives a higher order of accuracy, especially for linear elements.

The storage for the element mass matrix (*EL_M*) is automatically dynamically allocated and its actual definition (in *ELEM_SQ_MATRIX*) is usually less than ten new lines of code. Mainly one needs to allow for a small group of new control keywords in the data stream to activate and control the additional matrix manipulations, and additional output for time history related plots. The additional minor programming steps, at the element level, are shown in Fig. 11.115 for the general anisotropic Poisson equation. Basically one only need recover the extra material time coefficient (mass density times specific heat), simply called ρ in that listing, (as done in line 23 or 26 or 28 or defaulted in line 9) and compute the integral of the element mass matrix by numerical integration (in line 52).

As a typical transient example let us return to the mesh shown in Fig. 11.9 where we considered the steady state temperature of a square having a constant internal heat source. The only input changes (if the density and thickness are defaulted) are based on the transient controls of Table 2.1 and are shown in Fig. 11.116. If we assume that the body was initially at a constant temperature of 5 (that matches the previous essential boundary condition) and pick ρc_p and a step size Δt so it reaches steady state in about 70 steps we compute the time history of the domain. If we display the temperature surface, over the one-eighth symmetry model, at time steps 1 (bottom), 4, 16, and 64 (top) we get a good feel for how the temperatures change with time. These are shown for selected times superimposed in Fig. 11.117. The center point (node 1) time-history is shown in Fig. 11.118.

A similar problem is the study of the cooling of a cylinder formed by revolving a 1:2 rectangle about the z-axis, Assume that it is everywhere at a constant uniform temperature of 100 when its outer cylindrical surface and two ends are suddenly changed to zero temperature. As you would expect, the center point will remain at its initial temperature for a while, but will eventually cool to zero. Such a time history for the center node is shown in Fig. 11.119 as obtained from a uniform mesh of 8 by 8 axisymmetric T3 elements.

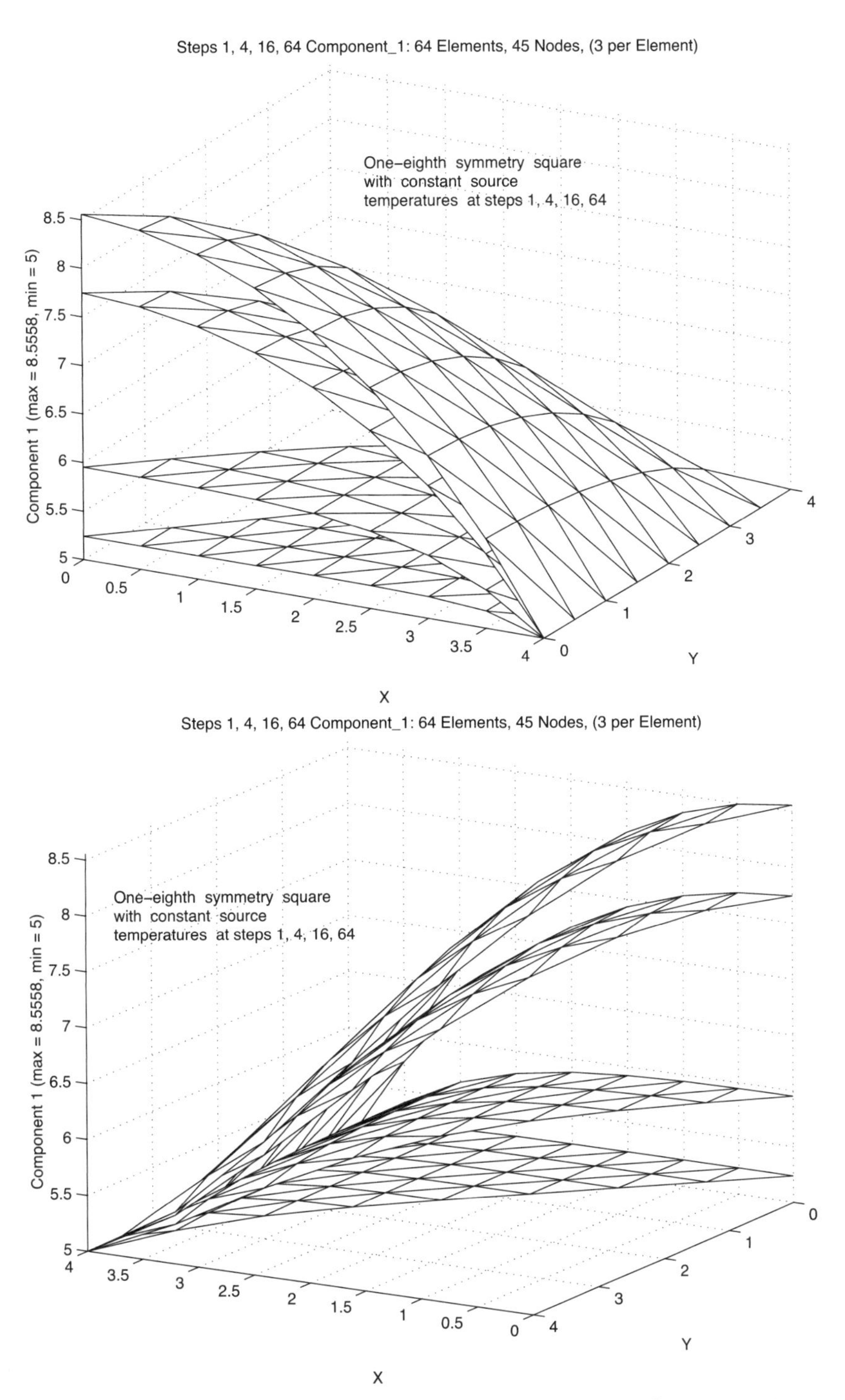

Figure 11.117 *Constant source square, temperature surfaces over time*

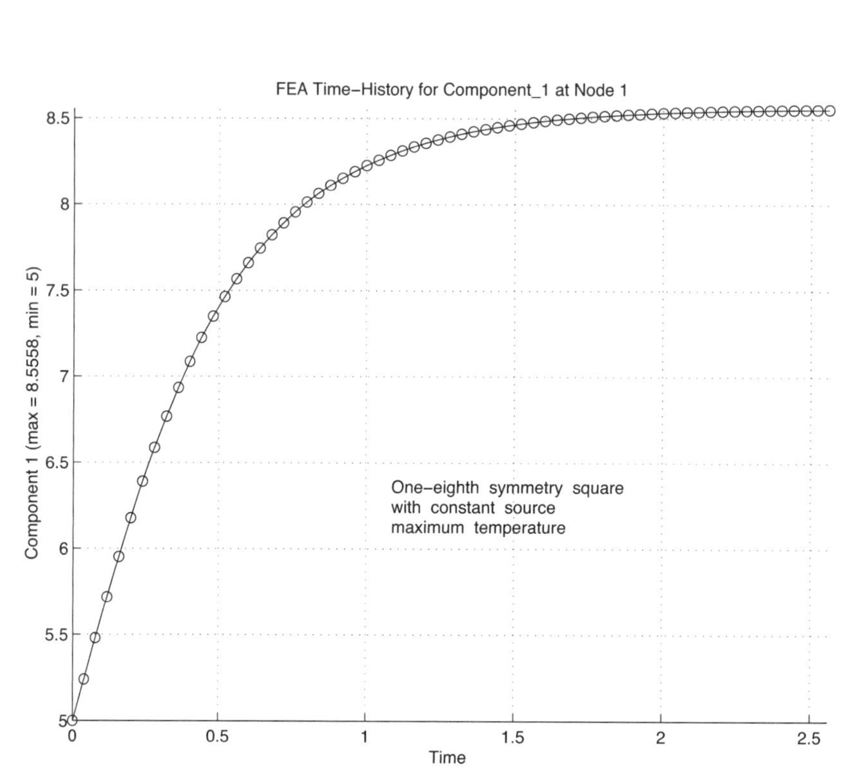

Figure 11.118 *Center point transient, constant source square*

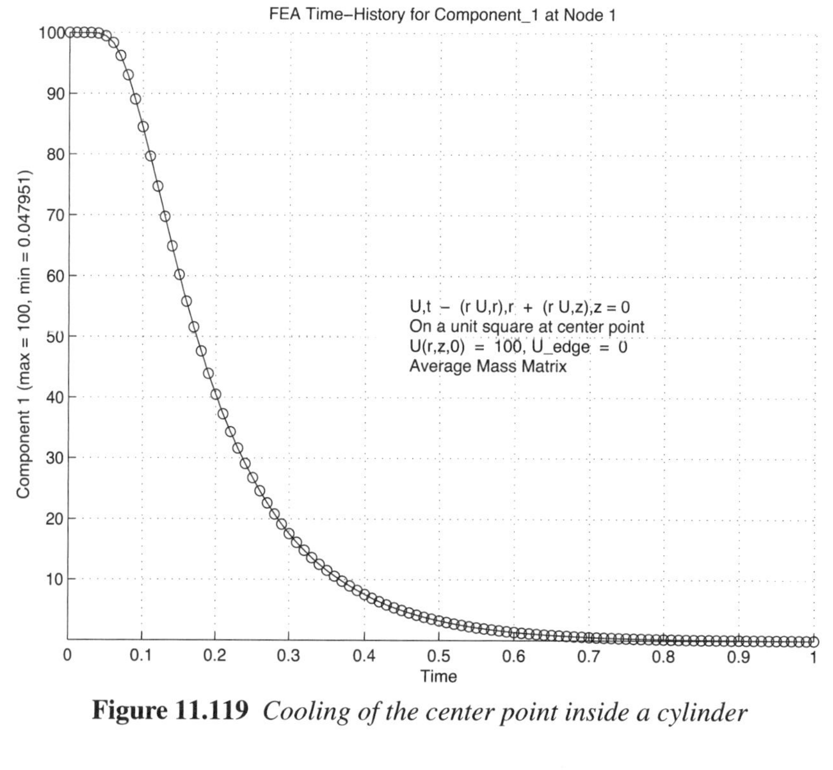

Figure 11.119 *Cooling of the center point inside a cylinder*

11.18 Exercises

1. The two-dimensional Laplace equation $\partial^2 u / \partial x^2 + \partial^2 u / \partial y^2 = 0$ is satisfied by the cubic $u(x, y) = -x^3 - y^3 + 3xy^2 + 3x^2 y$. It can be used to define the exact essential (Dirichlet) boundary conditions on the edges of any two-dimensional shape. For a unit square with a corner at the origin the boundary conditions are

$$u(0, y) = -y^3, \quad u(1, y) = -1 - y^3 + 3y^2 + 3y$$
$$u(x, 0) = -x^3, \quad u(x, 1) = -1 - x^3 + 3x^2 + 3x.$$

Obtain a finite element solution and sketch it along with the exact values along lines of constant x or y, or compare to *exact_case* 20 in the MODEL code. Solve using a domain of: a) the above unit square, or b) a rectangle over $0 \le x \le 3$ and $0 \le y \le 2$, or c) a quarter circle of radius 2 centered at the origin.

2. Modify the above problem to have normal flux (Neumann) boundary conditions on the two edges where y is constant (corresponding to the same exact solution):

$$u(0, y) = -y^3, \quad u(1, y) = -1 - y^3 + 3y^2 + 3y,$$

$$\frac{\partial u}{\partial y}(x, 0) = 3x^2, \quad \frac{\partial u}{\partial y}(x, 1) = -3 + 6x + 3x^2.$$

3. Consider a two-dimensional Poisson equation on a unit square $0 \le x \le 1, 0 \le y \le 1$ that contains a local high gradient peak on its interior centered at $x = \beta,\ y = \beta$. The amplitude of the peak is set by a parameter α. Let $A = (x - \beta)/\alpha$, $B = (y - \beta)/\alpha$, $C = (1 - \beta)/\alpha$, and $D = \beta/\alpha$. Then for the source, Q, defined by

$$\frac{\partial^2 u}{\partial x^2} + \frac{\partial^2 u}{\partial y^2} = Q = -6x - 6y - \frac{4}{\alpha^2}(1 + A^2 + B^2) \exp[-A^2 - B^2]$$

and edge Dirichlet and Neumann boundary conditions

$$u(0, y) = -y^3 + \exp[-D^2 - B^2], \quad u_{,y}(x, 0) = \frac{2\beta}{\alpha^2} \exp[-D^2 - A^2]$$

$$u(1, y) = -1 - y^3 + \exp[-C^2 - B^2], \quad u_{,y}(x, 1) = -3 - 2\frac{C}{\alpha} \exp[-A^2 - C^2]$$

the exact solution is $u = -x^3 - y^3 + \exp[-A^2 - B^2]$. Obtain a finite element solution and compare it to the exact values along the center lines $x = 1/2$, and $y = 1/2$. Assume $\beta = 0.5$ and $\alpha = 0.05$.

4. Consider the partial differential equation $\nabla \cdot (\mathbf{E} \nabla \phi) + \mathbf{v} \cdot \nabla \phi + F\phi + Q = 0$ *in* Ω. The second term is new. In 2-D it is $\mathbf{v} \cdot \nabla \phi = \begin{Bmatrix} v_x \\ v_y \end{Bmatrix} \begin{bmatrix} \frac{\partial \phi}{\partial x} & \frac{\partial \phi}{\partial y} \end{bmatrix}$. Apply the Galerkin method to get the additional matrix, say $\mathbf{U}_v^e$, that appears because of this term. State the result in matrix integral form. Is the result symmetric? The data $\mathbf{v}$ are often given at the nodes and we employ standard interpolation for those data:

$$\underset{1\,x\,n_s}{\mathbf{v}} = \underset{1\,x\,n_s}{[v_x \quad v_y]} = \underset{1\,x\,n_n}{\mathbf{H}^e} \underset{n_n\,x\,n_s}{[V_x^e \quad V_y^e]}.$$

Outline how you would numerically integrate $\mathbf{U}_v^e$ in that case. Give the linear line element matrix for a 1-D problem with v_x given constant data.

5. Sketch the boundary conditions for the ideal fluid flow around a cylinder given in Fig. 11.95 for an approach based on the use of a) the velocity potential, b) the stream function.

6. Solve the fin example of Fig. 11.26 by hand a) with insulated edges, b) with edge convection having the same convection coefficient.

7. For the solid conducting bar of Fig. 11.2 assume a length $L = 8$ cm, a width $2W = 4$ cm, and a thickness of $2H = 1$ cm. The material has a thermal conductivity of $k = 3\,W/cm\,C$, is surrounded by a fluid at a temperature of $\Theta_\infty = 20\,C$, and has a surface convection coefficient of $h = 0.1\,W/cm^2\,C$. Employ two equal length conduction elements and associated face, line, or point convection elements to obtain a solution for the temperature distribution and the convection heat loss. Use conduction elements that are: a) linear line elements, b) bilinear rectangular elements, c) trilinear brick elements.

8. How would you generalize the post-processing related to Eq. 11.46 to also locate the x position of the resultant force?

9. For the conduction mesh in Figs.11.9 and 11 prepare a combined surface plot of the element and SCP flux for: a) x, b) y, c) RMS values.

11.19 Bibliography

[1] Akin, J.E. and Wooten, J.W., "Tokamak Plasma Equilibria by Finite Elements," in *Finite Elements in Fluids III,* Chapter 21, ed. R.H. Gallagher, New York: John Wiley (1978).

[2] Akin, J.E., *Object-Oriented Programming Via Fortran 90/95*, Cambridge: Cambridge University Press (2003).

[3] Allaire, P.E., *Basics of the Finite Element Method*, Dubuque: Wm. C. Brown Pub. (1985).

[4] Allan, T., "The Application of Finite Element Analysis to Hydrodynamic and Externally Pressurized Pocket Bearings," *Wear*, **19**, pp. 169–206 (1972).

[5] Carslaw, H.S. and Jaeger, J.C., *Conduction of Heat in Solids*, Oxford: Oxford Press (1959).

[6] Chung, T.J., *Finite Element Analysis in Fluid Dynamics*, New York: McGraw-Hill (1978).

[7] Cook, R.D., Malkus, D.S., Plesha, N.E., and Witt, R.J., *Concepts and Applications of Finite Element Analysis*, New York: John Wiley (2002).

[8] Desai, C.S. and Kundu, T., *Introduction to the Finite Element Method*, Boca Raton: CRC Press (2001).

[9] Hinton, E. and Owen, D.R.J., *Finite Element Programming*, London: Academic Press (1977).

[10] Huang, H.C. and Usmani, A.S., in *Finite Element Analysis for Heat Transfer*, London: Springer-Verlag (1994).
[11] Huebner, K.H., Thornton, E.A., and Byrom, T.G., *Finite Element Method for Engineers*, New York: John Wiley (1994).
[12] Hughes, T.J.R., *The Finite Element Method*, Mineola: Dover Publications (2003).
[13] Irons, B.M. and Razzaque, A., "Experience with the Patch Test for Convergence of the Finite Element Method," pp. 557–587 in *Mathematical Foundation of the Finite Element Method*, ed. A.R. Aziz, New York: Academic Press (1972).
[14] Irons, B.M. and Ahmad, S., *Techniques of Finite Elements*, New York: John Wiley (1980).
[15] Kreyszig, E., *Advanced Engineering Mathematics*, New York: Wiley (1962).
[16] Kwon, Y.W. and Bang, H., *The Finite Element Method using Matlab*, Boca Raton: CRC Press (1997).
[17] Martin, H.C. and Carey, G.F., *Introduction to Finite Element Analysis*, New York: McGraw-Hill (1974).
[18] Meek, J.L., "Field Problems Solutions by Finite Element Methods," *Civil Eng. Trans., Inst. Eng. Aust.*, , pp. 173–180 (Oct. 1968).
[19] Myers, G.E., *Analytical Methods in Conduction Heat Transfer*, New York: McGraw-Hill (1971).
[20] Oden, J.T., Demkowicz, L., Strouboulis, T., and Devloo, P., "Adaptive Methods for Problems in Solid and Fluid Mechanics," pp. 249–280 in *Accuracy Estimates and Adaptive Refinements in Finite Element Computations*, ed. I. Babuska, O.C. Zienkiewicz, J. Gago, and E.R. de A. Oliveira, Chichester: John Wiley (1986).
[21] Reddi, M.M., "Finite Element Solution of the Incompressible Lubrication Problem," *J. Lubrication Technology*, **53**(3), pp. 524–532 (July 1969).
[22] Segerlind, L.J., *Applied Finite Element Analysis*, New York: John Wiley (1984).
[23] Shephard, M.S., "Approaches to the Automatic Generation and Control of Finite Element Meshes," *Appl. Mech. Rev.*, **41**(4), pp. 169–190 (Apr. 1988).
[24] Silvester, P.P. and Ferrari, R.L., *Finite Elements for Electrical Engineers*, Cambridge: Cambridge University Press (1983).
[25] Wada, S. and Hayashi, H., "Application of Finite Element Method to Hydrodynamic Lubrication Problems," *Bulletin of Japanese Soc. Mech. Eng.*, **14**(77), pp. 1222–1244 (1971).
[26] Wiberg, N.-E., "Superconvergent Patch Recovery – A Key to Quality Assessed FE Solutions," *Adv. Eng. Software*, **28**, pp. 85–95 (1997).
[27] Zienkiewicz, O.C. and Zhu, J.Z., "Superconvergent Patch Recovery Techniques and Adaptive Finite Element Refinement," *Comp. Meth. Appl. Mech. Eng.*, **101**, pp. 207–224 (1992).
[28] Zienkiewicz, O.C. and Taylor, R.L., *The Finite Element Method,* 5th Edition, London: Butterworth-Heinemann (2000).

Chapter 12

Vector fields

12.1 Introduction

Here we will extend the concepts from the previous chapter on scalar fields to those that involve vector unknowns. The most common (and historically the first) application is displacement based stress analysis. Other electrical field applications will only be noted through recent publications on those techniques which often require different 'edge based' interpolation methods not covered here. The main new considerations here are to select ordering for the vector unknowns and related vector or tensor quantities so that they can be cast in an expanded matrix notation.

12.2 Displacement based stress analysis summary

Let $\boldsymbol{u}$ denote a displacement vector at a point $\boldsymbol{x}$ in a solid. The finite element displacement formulation for stress analysis is based on the energy concept of finding $\boldsymbol{u}(\boldsymbol{x})$ that both satisfies the essential boundary conditions on $\boldsymbol{u}$ and minimizes the Total Potential Energy:

$$\Pi(\boldsymbol{u}) = U - W$$

where W is the external mechanical work due to supplied loading data with body forces per unit volume, $\boldsymbol{X}$, surface tractions per unit area, $\boldsymbol{T}$, and point loads, $\boldsymbol{P}_j$, at point $\boldsymbol{x}_j$ so

$$W(u) = \int_{\Omega} \boldsymbol{u}^T \boldsymbol{X} \, d\Omega + \int_{\Gamma} \boldsymbol{u}^T \boldsymbol{T} \, d\Gamma + \sum_j \boldsymbol{u}_j^T \boldsymbol{P}_j$$

and $U(\boldsymbol{u})$ consists of the strain energy due to the deformation, $\boldsymbol{u}$, of the material. It is defined as

$$U = \frac{1}{2} \int_{\Omega} \boldsymbol{\varepsilon}^T \boldsymbol{\sigma} \, d\Omega$$

where $\boldsymbol{\varepsilon}$ are the strain components resulting from the displacements, $\boldsymbol{u}$, and $\boldsymbol{\sigma}$ are the stress components that correspond to the strain components.

The strains, $\boldsymbol{\varepsilon}$, are defined by a 'Strain-Displacement Relation' for each class of stress analysis. This relation can be represented as a differential operator matrix, say $\boldsymbol{L}$, acting on the displacements $\boldsymbol{\varepsilon} = \boldsymbol{L}\boldsymbol{u}$. There are four commonly used classes that we will consider here. They are

1. Axial stress: $\boldsymbol{u} = u$, $\boldsymbol{L} = \partial/\partial x$, $\boldsymbol{\varepsilon} = \varepsilon_x = \partial u/\partial x$, $B_j = \partial H_j/\partial x$
2. Plane Stress and Plane Strain: $\boldsymbol{u}^T = \lfloor u \quad v \rfloor$

$$\boldsymbol{L} = \begin{bmatrix} \frac{\partial}{\partial x} & 0 \\ 0 & \frac{\partial}{\partial y} \\ \frac{\partial}{\partial y} & \frac{\partial}{\partial x} \end{bmatrix}, \quad \mathbf{B}_j = \begin{bmatrix} \frac{\partial H_j}{\partial x} & 0 \\ 0 & \frac{\partial H_j}{\partial y} \\ \frac{\partial H_j}{\partial y} & \frac{\partial H_j}{\partial x} \end{bmatrix}$$

$$\boldsymbol{\varepsilon}^T = [\varepsilon_x \quad \varepsilon_y \quad \gamma] = \begin{bmatrix} \frac{\partial u}{\partial x} & \frac{\partial v}{\partial y} & \left(\frac{\partial u}{\partial y} + \frac{\partial v}{\partial x}\right) \end{bmatrix}$$

3. Axisymmetric solid with radius r and axial position z : $\boldsymbol{u}^T = \lfloor u \quad v \rfloor$

$$\boldsymbol{L} = \begin{bmatrix} \frac{\partial}{\partial r} & 0 \\ 0 & \frac{\partial}{\partial z} \\ \frac{\partial}{\partial z} & \frac{\partial}{\partial r} \\ 1/r & 0 \end{bmatrix}, \quad \mathbf{B}_j = \begin{bmatrix} \frac{\partial H_j}{\partial r} & 0 \\ 0 & \frac{\partial H_j}{\partial z} \\ \frac{\partial H_j}{\partial z} & \frac{\partial H_j}{\partial r} \\ H_j/r & 0 \end{bmatrix},$$

$$\boldsymbol{\varepsilon}^T = \lfloor \varepsilon_r \quad \varepsilon_z \quad \gamma \quad \varepsilon_\theta \rfloor = \begin{vmatrix} \frac{\partial u}{\partial r} & \frac{\partial u}{\partial z} & (\frac{\partial u}{\partial z} + \frac{\partial v}{\partial r}) & u/r \end{vmatrix}$$

4. The full three-dimensional solid: $\boldsymbol{u}^T = [u \quad v \quad w]$

$$\boldsymbol{L} = \begin{bmatrix} \frac{\partial}{\partial x} & 0 & 0 \\ 0 & \frac{\partial}{\partial y} & 0 \\ 0 & 0 & \frac{\partial}{\partial z} \\ \frac{\partial}{\partial y} & \frac{\partial}{\partial x} & 0 \\ \frac{\partial}{\partial z} & 0 & \frac{\partial}{\partial x} \\ 0 & \frac{\partial}{\partial z} & \frac{\partial}{\partial y} \end{bmatrix}, \quad \mathbf{B}_j = \begin{bmatrix} \frac{\partial H_j}{\partial x} & 0 & 0 \\ 0 & \frac{\partial H_j}{\partial y} & 0 \\ 0 & 0 & \frac{\partial H_j}{\partial z} \\ \frac{\partial H_j}{\partial y} & \frac{\partial H_j}{\partial x} & 0 \\ \frac{\partial H_j}{\partial z} & 0 & \frac{\partial H_j}{\partial x} \\ 0 & \frac{\partial H_j}{\partial z} & \frac{\partial H_j}{\partial y} \end{bmatrix}$$

$$\boldsymbol{\varepsilon}^T = \left\lfloor \varepsilon_x \ \varepsilon_y \ \varepsilon_z \ \gamma_{xy} \ \gamma_{xz} \ \gamma_{yz} \right\rfloor = \left\lfloor \frac{\partial u}{\partial x} \ \frac{\partial v}{\partial y} \ \frac{\partial w}{\partial z} \ \left(\frac{\partial u}{\partial y} + \frac{\partial v}{\partial x}\right) \ \left(\frac{\partial u}{\partial z} + \frac{\partial w}{\partial y}\right) \ \left(\frac{\partial v}{\partial z} + \frac{\partial w}{\partial y}\right) \right\rfloor.$$

The above notation for strain is called the engineering definition. There is a more general definition, that we will not use, called the strain tensor which has the form $\varepsilon_{jk} = (u_{j,k} + u_{k,j})/2$. The classic elasticity definitions are expressed in a matrix form and do not yet include any finite element assumptions. Within any element we interpolate the displacement vector in terms of nodal displacement vector components, $\boldsymbol{\delta}^e$:

$$\mathbf{u}(\mathbf{x}) \equiv \mathbf{N}(x)\boldsymbol{\delta}^e \quad \mathbf{x} \in \Omega^e$$

where $\mathbf{N}$ denotes that spatial interpolation for vector components. Then, in all cases, we can define a common notation for the element strains in terms of the element displacement:

$$\boldsymbol{\varepsilon} = \mathbf{L}(\mathbf{x})\mathbf{N}^e(x)\boldsymbol{\delta}^e = \mathbf{B}^e(x)\boldsymbol{\delta}^e$$

where the differential operator action on the displacement vector interpolations defines the 'B-matrix':

$$\mathbf{B}^e(\mathbf{x}) = \mathbf{L}(\mathbf{x})\mathbf{N}^e(\mathbf{x})$$

Usually we use the same interpolation for each of the scalar components of the displacement vector, $\mathbf{u}$. For example

$$u(\mathbf{x}) = \mathbf{H}^e(\mathbf{x})\mathbf{u}^e, \; v(\mathbf{x}) = \mathbf{H}^e(\mathbf{x})v^e, \; w(\mathbf{x}) = \mathbf{H}^e(\mathbf{x})\boldsymbol{w}^e$$

and we order the unknown element displacements as $\boldsymbol{\delta}^{e^T} = [u_1 \; v_1 \; w_1 \; u_2 \; v_2 \; . \; . \; . \; v_m \; w_m]$ for an element with m nodes. Thus the vector interpolation matrix, $\mathbf{N}$, simply contains the $\mathbf{H}$ scalar interpolation functions and usually some zeros;

$$\mathbf{N}(\mathbf{x}) = \mathbf{N}(\mathbf{H}(\mathbf{x})) = [\mathbf{N}_1 \; \mathbf{N}_2 \; . \; . \; . \; \mathbf{N}_m],$$

where $\mathbf{N}_j$ is the matrix partition associated with the j-th node. It follows that the $\mathbf{B}^e$ matrix can also be partitioned into a set of nodal matrices

$$\mathbf{B}(\mathbf{x}) = \mathbf{B}(\mathbf{H}(\mathbf{x})) = [\mathbf{B}_1 \; \mathbf{B}_2 \; . \; . \; . \; \mathbf{B}_m],$$

where $\mathbf{B}_j = \mathbf{L}(\mathbf{x})\,\mathbf{I}\,H_j(\mathbf{x}) = [\mathbf{L}(\mathbf{x})\,H_j(\mathbf{x})]$. With the above choices for the order of the unknowns in $\boldsymbol{\delta}^e$ we note that for our four analysis classes the vector interpolation sub-matrices at node j are:

1. Axial stress: $\mathbf{N}_j(\mathbf{x}) = H_j(x)$
2. & 3. Two-dimensional and axisymmetric cases

$$\mathbf{N}_j(\mathbf{x}) = \begin{bmatrix} H_j(x) & 0 \\ 0 & H_j(x) \end{bmatrix}$$

4. Three-Dimensional Case

$$\mathbf{N}_j(\mathbf{x}) = \begin{bmatrix} H_j(x) & 0 & 0 \\ 0 & H_j(x) & 0 \\ 0 & 0 & H_j(x) \end{bmatrix}$$

In general we can use an identity matrix, $\boldsymbol{I}$, the size of the displacement vector, to define $\mathbf{N}_j(\mathbf{x}) = \boldsymbol{I}H_j(\mathbf{x})$. Returning to the elasticity notations, the corresponding stress components for our four cases are

1. Axial Stress: $\boldsymbol{\sigma} = \sigma_x$

2. Plane Stress ($\sigma_z = 0$), Plane Strain ($\varepsilon_z = 0$, plus a post-process for σ_z): $\boldsymbol{\sigma}^T = \lfloor \sigma_x \;\; \sigma_y \;\; \tau \rfloor$

3. Axisymmetric Solid: $\boldsymbol{\sigma}^T = \lfloor \sigma_r \; \sigma_z \; \tau \; \sigma_\theta \rfloor$

4. Three-dimensional Solid: $\boldsymbol{\sigma}^T = \lfloor \sigma_x \;\; \sigma_y \;\; \sigma_x \;\; \tau_{xy} \;\; \tau_{xz} \;\; \tau_{yz} \rfloor$

where the τ terms denote shear stresses and the other terms are normal stresses. For the state of plane strain σ_z is not zero, but is recovered in a post-processing operation. The mechanical stresses are related to the mechanical strains (ignoring initial effects) by the minimal form of a material constitutive law usually known as the basic Hooke's Law: $\boldsymbol{\sigma} = \mathbf{E}\,\boldsymbol{\varepsilon}$, where $\mathbf{E} = \mathbf{E}^T$ is a symmetric material properties matrix that relates the stress and strain components for each case. Actually, it is defined for the three-dimensional case and the others are special forms of it. For an isotropic material they are:

1. Axial stress: $\mathbf{E} = \mathrm{E}$, *Young's Modulus*
2. Plane Stress ($\sigma_z \equiv 0$)

$$\mathbf{E} = \frac{\mathrm{E}}{1 - \nu^2}\begin{bmatrix} 1 & \nu & 0 \\ \nu & 1 & 0 \\ 0 & 0 & (1-\nu)/2 \end{bmatrix}$$

$$\nu,\ \textit{Poisson's ratio},\ 0 \le v < 0.5$$

3. Plane Strain ($\varepsilon_z \equiv 0$)

$$\mathbf{E} = \frac{\mathrm{E}}{(1+\nu)(1-2\nu)}\begin{bmatrix} (1-\nu) & \nu & 0 \\ \nu & (1-\nu) & 0 \\ 0 & 0 & (1-2\nu)/2 \end{bmatrix}$$

4. Axisymmetric Solid

$$\mathbf{E} = \frac{\mathrm{E}}{(1+\nu)(1-2\nu)}\begin{bmatrix} (1-\nu) & \nu & 0 & \nu \\ \nu & (1-\nu) & 0 & \nu \\ 0 & 0 & (1-2\nu)/2 & 0 \\ \nu & \nu & 0 & (1-\nu) \end{bmatrix}$$

5. General Solid

$$E_{11} = E_{22} = E_{33} = (1-v)c$$

$$E_{12} = E_{13} = E_{23} = vc = E_{21} = E_{31} = E_{32}$$

$$E_{44} = E_{55} = E_{66} = G$$

where $\quad c = \dfrac{E}{(1+v)\,(1-2v)} \quad$ and $\quad G = \dfrac{E}{2(1+v)}$

Note that these data are independent of the element type and so their implementation for isotropic materials just depends on the input order of the material properties. Typical implementations for the plane stress, and axisymmetric assumptions are shown in Figs. 12.1 and 2, respectively. Note that if one has an *incompressible material*, such as rubber where $\nu = 1/2$, then division by zero would cause difficulties for the plane strain and general solid problems, as currently formulated.

```
SUBROUTINE E_PLANE_STRESS (E)                                    ! 1
! * * * * * * * * * * * * * * * * * * * * * * * * * * * * * *    ! 2
!      PLANE_STRESS CONSTITUTIVE MATRIX DEFINITION               ! 3
! STRESS & STRAIN COMPONENT ORDER: XX, YY, XY, SO N_R_B = 3      ! 4
! PROPERTY ORDER: 1-YOUNG'S MODULUS, 2-POISSON'S RATIO           ! 5
! * * * * * * * * * * * * * * * * * * * * * * * * * * * * * *    ! 6
Use System_Constants    ! for DP, N_R_B                          ! 7
Use Sys_Properties_Data ! for function GET_REAL_LP               ! 8
 IMPLICIT NONE                                                   ! 9
 REAL(DP), INTENT(OUT) :: E (N_R_B, N_R_B)                       !10
 REAL(DP) :: C_1, P_R, S_M, Y_M                                  !11
                                                                 !12
! E     = CONSTITUTIVE MATRIX                                    !13
! N_R_B = NUMBER OF ROWS IN B AND E MATRICES                     !14
! P_R   = POISSON'S RATIO                                        !15
! S_M   = SHEAR MODULUS                                          !16
! Y_M   = YOUNG'S MODULUS OF ELASTICITY                          !17
                                                                 !18
!      RECOVER ELEMENT PROPERTIES                                !19
  IF ( EL_REAL < 2 ) STOP 'el_real < 2 IN E_PLANE_STRESS'        !20
  Y_M = GET_REAL_LP (1)          ; P_R = GET_REAL_LP (2)         !21
  S_M   = 0.5d0 * Y_M / (1 + P_R) ; C_1 = Y_M / (1 - P_R * P_R)  !22
                                                                 !23
  E (1, 1) = C_1       ; E (2, 1) = C_1 * P_R ; E (3, 1) = 0.d0  !24
  E (1, 2) = C_1 * P_R ; E (2, 2) = C_1       ; E (3, 2) = 0.d0  !25
  E (1, 3) = 0.d0      ; E (2, 3) = 0.d0      ; E (3, 3) = S_M   !26
END SUBROUTINE E_PLANE_STRESS                                    !27
```

Figure 12.1 *Plane-stress isotropic constitutive law*

```
SUBROUTINE E_AXISYMMETRIC_STRESS (E)                             ! 1
! * * * * * * * * * * * * * * * * * * * * * * * * * * * * * *    ! 2
! AXISYMMETRIC CONSTITUTIVE MATRIX DEFINITION, N_R_B = 4         ! 3
! STRESS & STRAIN COMPONENT ORDER: RR, ZZ, RZ, AND TT            ! 4
! PROPERTY ORDER: 1-YOUNG'S MODULUS, 2-POISSON'S RATIO           ! 5
! * * * * * * * * * * * * * * * * * * * * * * * * * * * * * *    ! 6
Use System_Constants    ! for DP, N_R_B                          ! 7
Use Sys_Properties_Data ! for function GET_REAL_LP               ! 8
 IMPLICIT NONE                                                   ! 9
 REAL(DP), INTENT(OUT) :: E (N_R_B, N_R_B)  ! CONSTITUTIVE       !10
 REAL(DP) :: C_1, C_2, C_3, P_R, S_M, Y_M                        !11
                                                                 !12
! N_R_B = NUMBER OF ROWS IN B AND E MATRICES                     !13
! P_R   = POISSON'S RATIO                                        !14
! S_M   = SHEAR MODULUS                                          !15
! Y_M   = YOUNG'S MODULUS OF ELASTICITY                          !16
                                                                 !17
  IF (EL_REAL < 2) STOP 'el_real < 2 IN E_AXISYMMETRIC_STRESS'   !18
  Y_M = GET_REAL_LP (1) ; P_R = GET_REAL_LP (2)                  !19
  S_M = 0.5d0 * Y_M / ( 1 + P_R)                                 !20
  C_1 = Y_M /(1 + P_R)/(1 - P_R - P_R)                           !21
  C_2 = C_1 * P_R  ; C_3 = C_1 * (1 - P_R)                       !22
                                                                 !23
  E (:, 3) = 0.d0  ; E (3, :) = 0.d0                             !24
  E (1, 1) = C_3   ; E (2, 1) = C_2   ; E (4, 1) = C_2           !25
  E (1, 2) = C_2   ; E (2, 2) = C_3   ; E (4, 2) = C_2           !26
  E (1, 4) = C_2   ; E (2, 4) = C_2   ; E (4, 4) = C_3           !27
  E (3, 3) = S_M                                                 !28
END SUBROUTINE E_AXISYMMETRIC_STRESS                             !29
```

Figure 12.2 *Axisymmetric stress isotropic constitutive law*

The strain energy matrix definition becomes

$$U(\boldsymbol{u}) = \frac{1}{2}\int_{\Omega} \boldsymbol{\varepsilon}^T \mathbf{E} \boldsymbol{\varepsilon} d\Omega$$

which when evaluated in an element domain becomes

$$U^e = \frac{1}{2}\int_{\Omega^e} (\mathbf{B}^e \boldsymbol{\delta}^e)^T \mathbf{E}^e (\mathbf{B}^e \boldsymbol{\delta}^e) d\Omega = \frac{1}{2} \boldsymbol{\delta}^{e^T} \int_{\Omega^e} \mathbf{B}^{e^T}(\boldsymbol{x}) \mathbf{E}^e \mathbf{B}^e(\boldsymbol{x}) d\Omega \boldsymbol{\delta}^e = \frac{1}{2} \boldsymbol{\delta}^{e^T} \boldsymbol{S}^e \boldsymbol{\delta}^e$$

where the element stiffness matrix $\boldsymbol{S}^e$ has the same matrix form as seen in heat transfer:

$$\boldsymbol{S}^e = \int_{\Omega^e} \mathbf{B}^{e^T}(x) \mathbf{E}^e \mathbf{B}^e(x) d\Omega.$$

If additional data on 'initial strains', $\boldsymbol{\varepsilon}_0$, and/or 'initial stresses', $\boldsymbol{\sigma}_0$, are given then the stress-strain law must be generalized to:

$$\boldsymbol{\sigma} = \mathbf{E}(\boldsymbol{\varepsilon} - \boldsymbol{\varepsilon}_0) + \boldsymbol{\sigma}_0$$

where $\boldsymbol{\varepsilon}_0$ and $\boldsymbol{\sigma}_0$ are initial strains and stresses, respectively. If not zero they cause additional loading terms (source vectors) and require additional post-processing. The additional work terms due to the initial strains and stresses are

$$W_{\varepsilon_0} = \int_{\Omega} \boldsymbol{\sigma}^T \boldsymbol{\varepsilon}_0 d\Omega = \int_{\Omega} \boldsymbol{\varepsilon}^T \mathbf{E}\, \boldsymbol{\varepsilon}_0 d\Omega, \quad W_{\sigma_0} = -\int_{\Omega} \boldsymbol{\varepsilon}^T \boldsymbol{\sigma}_0 d\Omega.$$

There are several common causes of initial strains, such as thermal and swelling effects. It is unusual to know an initial stress state and it is usually assumed to be zero. $\Pi(\boldsymbol{\Delta}) = \boldsymbol{\Delta}^T \boldsymbol{S}\boldsymbol{\Delta}/2 - \boldsymbol{\Delta}^T \boldsymbol{C}$ so minimizing with respect to all the displacements, $\boldsymbol{\Delta}$, gives $\boldsymbol{S\Delta} - \boldsymbol{C} = \boldsymbol{O}$ which are known as the algebraic Equations of Equilibrium.

12.3 Planar models

The states of *plane stress* and *plane strain* are interesting and useful examples of stress analysis of a two-dimensional elastic solid (in the x-y plane). The assumption of plane stress implies that the component of all stresses normal to the plane are zero ($\sigma_z = \tau_{zx} = \tau_{zy} = 0$) whereas the plane strain assumption implies that the normal components of the strains are zero ($\varepsilon_z = \gamma_{zx} = \gamma_{zy} = 0$). The state of plane stress is commonly introduced in the first course of mechanics of materials. It was also the subject of some of the earliest finite element studies.

The assumption of plane stress means that the solid is very thin and that it is loaded only in the direction of its plane. At the other extreme, in the state of plane strain the dimension of the solid is very large in the z-direction. It is loaded perpendicular to the longitudinal (z) axis, and the loads do not vary in that direction. All cross-sections are assumed to be identical so any arbitrary x-y section can be used in the analysis. These two states are illustrated in Fig. 12.3. There are three common approaches to the variational formulation of the plane stress (or plane strain) problem: 1) Displacement formulation, 2) Stress formulation, and 3) Mixed formulation. We will select the common displacement method and utilize the total potential energy of the system. This can be proved to be equal to assuming a Galerkin weighted residual approach. In any event, note that it will be necessary to define all unknown quantities in terms of the displacements of the solid. Specifically, it will be necessary to relate the strains and stresses to the displacements as was illustrated in 1-D in Sec. 7.3.

The finite element form is based on the use of strain energy density, as discussed in Chapter 7. Since it is half the product of the stress and strain tensor components, we do not need, at this point, to consider either stresses or strains that are zero. This means that only three of six products will be used for a planar formulation.

Our notation will follow that commonly used in mechanics of materials. The displacements components parallel to the x- and y-axes will be denoted by $u(x, y)$ and $v(x, y)$, respectively. The normal stress acting parallel to the x- and y-axes are σ_x and σ_y, respectively. The shear stress acting parallel to the y-axis on a plane normal to the x-axis is τ_{xy}, or simply τ. The corresponding components of strain are ε_x, ε_y, and γ_{xy}, or simply γ. Figure 12.4 summarizes the engineering (versus tensor) strain notations that we will employ in our matrix definitions.

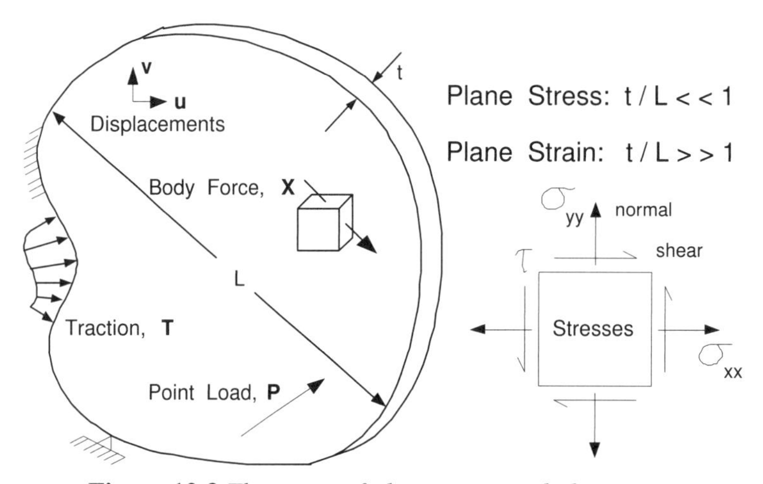

Figure 12.3 *The states of plane stress and plane strain*

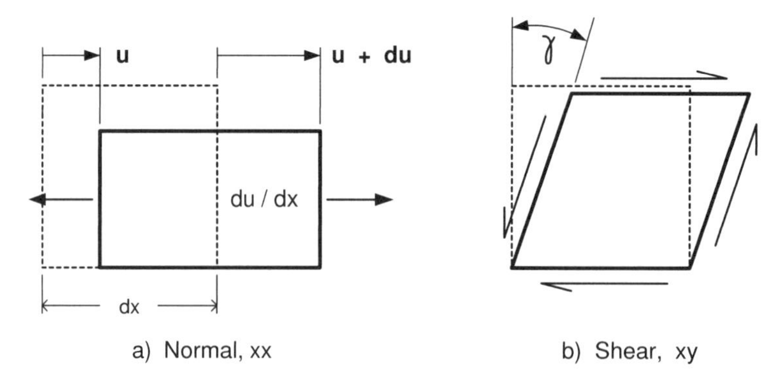

Figure 12.4 *Engineering notation for planar strains*

12.3.1 Minimum total potential energy

Plane stress analysis, like other elastic stress analysis problems, is governed by the principle of minimizing the total potential energy in the system. It is possible to write the generalized forms of the element matrices and boundary segment matrices defined in above. The symbolic forms are:

(1) Stiffness matrix

$$\mathbf{S}^e = \int_{V^e} \mathbf{B}^{e^T} \mathbf{E}^e \mathbf{B}^e(x, y)\, dV \tag{12.1}$$

$\mathbf{B}^e$ = element strain-displacement matrix, $\mathbf{E}^e$ = material constitutive matrix;

(2) Body Force Matrix

$$\mathbf{C}_x^e = \int_{V^e} \mathbf{N}^e(x, y)^T\, \mathbf{X}^e(x, y)\, dV \tag{12.2}$$

$\mathbf{N}^e$ = generalized interpolation matrix, $\mathbf{X}^e$ = body force vector per unit volume;

(3) Initial Strain Load Matrix

$$\mathbf{C}_o^e = \int_{V^e} \mathbf{B}^{e^T}(x, y)\, \mathbf{E}^e \boldsymbol{\varepsilon}_o^e(x, y)\, dV \tag{12.3}$$

$\boldsymbol{\varepsilon}_o^e$ = initial strain matrix;

(4) Surface Traction Load Matrix

$$\mathbf{C}_T^b = \int_{A^b} \mathbf{N}^b(x, y)^T\, \mathbf{T}^b(x, y)\, dA \tag{12.4}$$

$\mathbf{N}^b$ = boundary interpolation matrix, $\mathbf{T}^b$ = traction force vector per unit area;

and where V^e is the element volume, A^b is a boundary segment surface area, dV is a differential volume, and dA is a differential surface area. Now we will specialize these relations for plane stress (or strain). The two displacement components will be denoted by $\mathbf{u}^T = [\,u \quad v\,]$. At each node there are two displacement components to be determined ($n_g = 2$). The total list of element degrees of freedom is denoted by $\boldsymbol{\delta}^e$.

12.3.2 Displacement interpolations

As before, it is necessary to define the spatial approximation for the displacement field. Consider the x-displacement, u, at some point in an element. The simplest approximation of how it varies in space is to assume a complete linear polynomial. In two-dimensions a complete linear polynomial contains three constants. Thus, we select a triangular element with three nodes (see Figs. 3.2, 9.1, and 9.2) and assume u is to be computed at each node. Then

$$u(x, y) = \mathbf{H}^e\, \mathbf{u}^e = H_1^e\, u_1^e + H_2^e\, u_2^e + H_3^3\, u_3^e. \tag{12.5}$$

The interpolation can either be done in global (x, y) coordinates or in a local system. If global coordinates are utilized then, from Eq. 8.13, the form of the typical interpolation function is

$$H_i^e(x, y) = (\,a_i^e + b_i^e x + c_i^e\, y\,)/2A^e, \qquad 1 \le i \le 3 \tag{12.6}$$

where A^e is the area of the element and a_i^e, b_i^e, and c_i^e denote constants for node i that depend on the element geometry. Clearly, we could utilize the same interpolations for the y-displacement:

$$v(x, y) = \mathbf{H}^e\, \mathbf{v}^e. \tag{12.7}$$

To define the element dof vector $\boldsymbol{\delta}^e$ we chose to order these six constants such that

$$\boldsymbol{\delta}^{e^T} = \lfloor u_1 \quad v_1 \quad u_2 \quad v_2 \quad u_3 \quad v_3 \rfloor^e. \tag{12.8}$$

To refer to both displacement components at a point we employ a generalized rectangular element interpolation matrix, **N**, which is made up of the scalar interpolation functions for a node, H_j, and numerous zero elements. Then

$$\mathbf{u}(x, y) = \begin{Bmatrix} u(x, y) \\ v(x, y) \end{Bmatrix} = \begin{bmatrix} H_1 & 0 & H_2 & 0 & H_3 & 0 \\ 0 & H_1 & 0 & H_2 & 0 & H_3 \end{bmatrix}^e \boldsymbol{\delta}^e = \mathbf{N}^e \boldsymbol{\delta}^e. \tag{12.9}$$

More advanced polynomials could be selected to define the **H** or **N** matrices. This vector interpolation array, **N**, could be partitioned into typical contributions from each node. Since it is not efficient to multiply by a lot of zero elements you will sometimes prefer to simply use the scalar array interpolations, **H**, and the scatter the results into other arrays. (Logical masks are available in *f*95 and *Matlab* to assist with such efficiencies, but they are omitted here to avoid confusion.)

```
SUBROUTINE ELASTIC_B_PLANAR (DGH, B)                                    ! 1
! * * * * * * * * * * * * * * * * * * * * * * * * * * * * * *           ! 2
! 2-D ELASTICITY STRAIN-DISPLACEMENT RELATIONS  (B)                     ! 3
! STRESS & STRAIN COMPONENT ORDER: XX, YY, XY                           ! 4
! * * * * * * * * * * * * * * * * * * * * * * * * * * * * * *           ! 5
Use System_Constants ! for DP, N_R_B, N_G_DOF, N_SPACE                  ! 6
Use Elem_Type_Data   ! for LT_FREE, LT_N                                ! 7
 IMPLICIT NONE                                                          ! 8
 REAL(DP), INTENT(IN)  :: DGH (N_SPACE, LT_N)         ! Gradients ! 9
 REAL(DP), INTENT(OUT) :: B   (N_R_B, LT_N * N_G_DOF) ! Strains   !10
 INTEGER :: J, K, L                                   ! Loops     !11
                                                                        !12
! B       = STRAIN-DISPLACEMENT MATRIX (RETURNED)                       !13
! DGH     = GLOBAL DERIVATIVES OF ELEM INTERPOLATION FUNCTIONS          !14
! LT_N    = NUMBER OF NODES PER ELEMENT TYPE                            !15
! N_G_DOF = NUMBER OF PARAMETERS PER NODE = 2 (U & V)                   !16
! N_R_B   = NUMBER OF STRAINS (ROWS IN B) = 3: XX, YY, XY               !17
! N_SPACE = DIMENSION OF SPACE = 2 here                                 !18
                                                                        !19
  DO J = 1, LT_N           ! ROW NUMBER                                 !20
    K = N_G_DOF * (J - 1) + 1 ! FIRST COLUMN,  U                        !21
    L = K + 1                 ! SECOND COLUMN, V                        !22
    B (1, K) = DGH (1, J)     ! DU/DX  FOR XX NORMAL                    !23
    B (2, K) = 0.d0                                                     !24
    B (3, K) = DGH (2, J)     ! DU/DY  FOR XY SHEAR                     !25
    B (1, L) = 0.d0                                                     !26
    B (2, L) = DGH (2, J)     ! DV/DY  FOR YY NORMAL                    !27
    B (3, L) = DGH (1, J)     ! DV/DX  FOR XY SHEAR                     !28
  END DO                                                                !29
END SUBROUTINE ELASTIC_B_PLANAR                                         !30
```

Figure 12.5 *Strain-displacement matrix for planar elements*

12.3.3 Strain-displacement relations

From mechanics of materials we can define the strains in terms of the displacement. Order the three strain components so as to define $\boldsymbol{\varepsilon}^T = [\,\varepsilon_x \;\; \varepsilon_y \;\; \gamma\,]$. These terms are defined as:

$$\varepsilon_x = \frac{\partial u}{\partial x}, \quad \varepsilon_y = \frac{\partial v}{\partial y}, \quad \gamma = \left(\frac{\partial u}{\partial y} + \frac{\partial v}{\partial x} \right)$$

if the common engineering form is selected for the shear strain, γ. Two of these terms are illustrated in Fig. 12.3.2. From Eqs. 12.5 and 12.7 we note

$$\varepsilon_x = \frac{\partial \mathbf{H}^e}{\partial x}\mathbf{u}^e, \quad \varepsilon_y = \frac{\partial \mathbf{H}^e}{\partial y}\mathbf{v}^e, \quad \gamma = \frac{\partial \mathbf{H}^e}{\partial y}\mathbf{u}^e + \frac{\partial \mathbf{H}^e}{\partial x}\mathbf{v}^e. \tag{12.10}$$

These can be combined into a single matrix identity to define

$$\begin{Bmatrix} \varepsilon_x \\ \varepsilon_y \\ \gamma \end{Bmatrix}^e = \begin{bmatrix} H_{1,x} & 0 & H_{2,x} & 0 & \cdots & H_{n,x} & 0 \\ 0 & H_{1,y} & 0 & H_{2,y} & \cdots & 0 & H_{n,y} \\ H_{1,y} & H_{1,x} & H_{2,y} & H_{2,x} & \cdots & H_{n,y} & H_{n,x} \end{bmatrix}^e \boldsymbol{\delta}^e \tag{12.11}$$

or symbolically, $\boldsymbol{\varepsilon}^e = \mathbf{B}^e(x, y)\boldsymbol{\delta}^e$, where the shorthand notation $H_{j,x} = \partial H_j / \partial x$, etc. has been employed, and where we have assumed the element has n nodes. Thus, the $\mathbf{B}^e$ matrix size depends on the type of element being utilized. This defines the *element strain-displacement operator* $\mathbf{B}^e$ that would be used in Eqs. 12.1 and 12.3. Note that $\mathbf{B}$ could also be partitioned into 3×2 sub-partitions from each node on the element, as shown in the implementation in Fig. 12.5.

12.3.4 Stress-strain law

The stress-strain law (*constitutive relations*) between the strain components, $\boldsymbol{\varepsilon}$, and the corresponding stress components, $\boldsymbol{\sigma}^T = [\,\sigma_x \;\; \sigma_y \;\; \tau\,]$, is defined in mechanics of materials. For the case of an isotropic, and isothermal, material in plane stress these are listed in terms of the mechanical strains as

$$\sigma_x = \frac{E}{1-\nu^2}(\varepsilon_x + \nu\varepsilon_y), \quad \sigma_y = \frac{E}{1-\nu^2}(\varepsilon_y + \nu\varepsilon_x), \quad \tau = \frac{E}{2(1+\nu)}\gamma = G\gamma \tag{12.12}$$

where E is the elastic modulus, ν is Poisson's ratio, and G is the shear modulus. In theory, G is not an independent property. In practice it is sometimes treated as independent. Some references list the inverse relations since the strains are usually experimentally determined from the applied stresses. In the alternate inverse form the constitutive relations for the mechanical strains, $\boldsymbol{\varepsilon}$, are

$$\varepsilon_x = \frac{1}{E}(\sigma_x - \nu\sigma_y), \quad \varepsilon_y = \frac{1}{E}(\sigma_y - \nu\sigma_x), \quad \gamma = \frac{\tau}{G} = \tau\frac{2(1+\nu)}{E}. \tag{12.13}$$

We will write Eq. 12.13 in a more general matrix symbolic form

$$\boldsymbol{\sigma} = \mathbf{E}(\boldsymbol{\varepsilon} - \boldsymbol{\varepsilon}_o) \tag{12.14}$$

by allowing for the presence of initial strains, $\boldsymbol{\varepsilon}_o$, that are not usually included in mechanics of materials. Note that $\mathbf{E}$ (given above) is a symmetric matrix. This is almost always true. This observation shows that in general the element stiffness matrix,

Eq. 12.1, will also be symmetric.

The most common type of initial strain, ε_o, is that due to temperature changes. For an isotropic material these *thermal strains* are

$$\varepsilon^T{}_o = \alpha\,\Delta\theta \lfloor 1 \quad 1 \quad 0 \rfloor \tag{12.15}$$

where α is the coefficient of thermal expansion and $\Delta\theta = (\theta - \theta_o)$ is the temperature rise from a stress free temperature of θ_o. Usually the $\Delta\theta$ is supplied as piecewise constant element data, or θ_o is given as global data along with the nodal temperatures computed from the procedures in the previous chapter. At any point in the element the initial thermal strain is proportional to

$$\Delta\theta = \mathbf{H}^e(\mathbf{x})\boldsymbol{\Theta}^e - \theta_o \tag{12.16}$$

In other words the gathered answers from the thermal study, $\boldsymbol{\Theta}^e$, are loading data for a thermal stress problem. Notice that thermal strains in isotropic materials do not include thermal shear strains. If the above temperature changes were present then the additional loading effects could be included via Eq. 12.3. If the material is not isotropic the the initial thermal effects require local coordinate transformations and more input data (described in Sections 12.5 and 12.8, respectively). So the resultant nodal load due to thermal strains in the most general case becomes

$$\mathbf{C}^e_o = \int_{\Omega^e} \mathbf{B}^{e^T}(\mathbf{x})\,\mathbf{E}^e(\mathbf{x})\,\mathbf{t}^{-1}_{(ns)}(\mathbf{x})\,\varepsilon_{0\,(ns)}(\mathbf{x})\,d\Omega \tag{12.17}$$

where $\mathbf{t}_{(ns)}$ denotes a square strain transformation matrix from the local material principal coordinate direction, (ns), at local position $\mathbf{x}$ to the global coordinate axes.

At this point, we do not know the nodal displacements, $\boldsymbol{\delta}^e$, of the element. Once we do know them, we will wish to use the above arrays to get post-processing results for the stresses and, perhaps, for *failure criteria*. Therefore, for each element we usually store the arrays $\mathbf{B}^e$, $\mathbf{E}^e$, and ε^e_o so that we can execute the products in Eqs. 12.11 and 12.14 after the displacements are known.

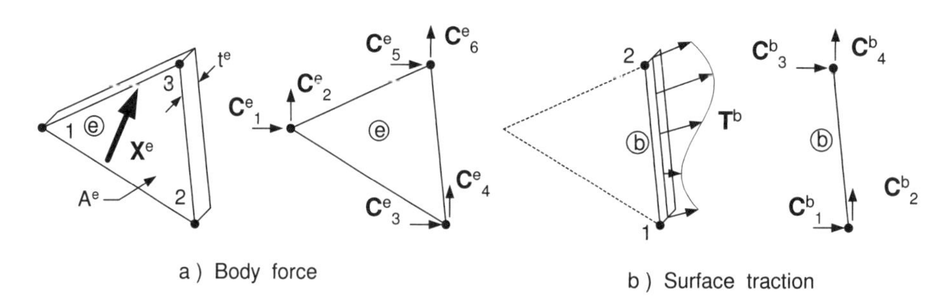

Figure 12.6 *Element loads and consistent resultants*

12.4 Matrices for the constant strain triangle (CST)

Beginning with the simple linear triangle displacement assumption we note that for a typical CST interpolation function $\partial H_i^e / \partial x = b_i^e / 2A^e$, and $\partial H_i^e / \partial y = c_i^e / 2A^e$. Therefore, from Eqs. 12.11 and 12.12, the strain components in the triangular element are constant. Specifically,

$$\mathbf{B}^e = \frac{1}{2A^e} \begin{bmatrix} b_1 & 0 & b_2 & 0 & b_3 & 0 \\ 0 & c_1 & 0 & c_2 & 0 & c_3 \\ c_1 & b_1 & c_2 & b_2 & c_3 & b_3 \end{bmatrix}^e . \tag{12.18}$$

For this reason this element is commonly known as the *constant strain triangle*, CST. Letting the material properties, E and ν, be constant in a typical element then the stiffness matrix in Eq. 12.1 simplifies to

$$\mathbf{S}^e = \mathbf{B}^{e^T} \mathbf{E}^e \mathbf{B}^e V^e \tag{12.19}$$

where the element volume is

$$V^e = \int_{V^e} dv = \int_{A^e} t^e(x, y)\, dx\, dy \tag{12.20}$$

where t^e is the element thickness. Usually the thickness of a typical element is constant so that $V^e = t^e A^e$. Of course, it would be possible to define the thickness at each node and to utilize the interpolation functions to approximate $t^e(x, y)$, and then the average thickness is $t^e = (t_1 + t_2 + t_3)/3$. Similarly if the temperature change in the element is also constant within the element then Eq. 12.3 defines the thermal load matrix

$$\mathbf{C}_o^e = \mathbf{B}^{e^T} \mathbf{E}^e \boldsymbol{\varepsilon}_o^e \mathbf{t}^e A^e . \tag{12.21}$$

It would be possible to be more detailed and input the temperature at each node and integrate its change over the element.

It is common for plane stress problems to include body force loads due to gravity, centrifugal acceleration, etc. For simplicity, assume that the body force vector $\mathbf{X}^e$, and the thickness, t^e, are constant. Then the body force vector in Eq. 12.2 simplifies to

$$\mathbf{C}_X^e = \mathbf{t}^e \int_{A^e} \mathbf{N}^{e^T}(x, y)\, dx\, dy\, \mathbf{X}^e . \tag{12.22}$$

From Eq. 12.9 it is noted that the non-zero terms in the integral typically involve scalar terms such as

$$I_i^e = \int_{A^e} H_i^e(x, y)\, da = \frac{1}{2A^e} \int_{A^e} (a_i^e + b_i^e x + c_i^e y)\, da . \tag{12.23}$$

These three terms can almost be integrated by inspection. The element geometric constants can be taken outside parts of the integrals. Then from the concepts of the first moment (centroid) of an area

$$a_i^e \int da = a_i^e A^e, \quad \int b_i^e x\, da = b_i^e \bar{x}^e A^e, \quad \int c_i^e y\, da = c_i^e \bar{y}^e A^e \tag{12.24}$$

where $\bar{x}$ and $\bar{y}$ denote the *centroid* coordinates of the triangle, $\bar{x}^e = (x_1 + x_2 + x_3)^e/3$, and $\bar{y}^e = (y_1 + y_2 + y_3)^e/3$. In view of Eq. 12.24, the integral in Eq. 12.23 becomes $I_i^e = A^e(a_i + b_i\bar{x} + c_i\bar{y})^e/2A^e$. Reducing the algebra to its simplest form, using Table 9.1, yields

$$I_i^e = A^e/3, \qquad 1 \le i \le 3.$$

Therefore, for the CST the expanded form of the body force resultant is

$$\mathbf{C}_X^e = \frac{t^e A^e}{3} \begin{bmatrix} 1 & 0 \\ 0 & 1 \\ 1 & 0 \\ 0 & 1 \\ 1 & 0 \\ 0 & 1 \end{bmatrix} \begin{Bmatrix} X_x \\ X_y \end{Bmatrix}^e = \frac{t^e A^e}{3} \begin{Bmatrix} X_x \\ X_y \\ X_x \\ X_y \\ X_x \\ X_y \end{Bmatrix}^e$$

where X_x and X_y denote the components of the body force vector. To assign a physical meaning to this result note that $t^e A^e X_x^e$ is the resultant force in the x-direction. Therefore, the above calculation has replaced the distributed load with a statically equivalent set of three nodal loads. Each of these loads is a third of the resultant load. These *consistent loads* are illustrated in Figs. 12.6.

A body force vector, $\mathbf{X}$, can arise from several important sources. An example is one due to acceleration (and gravity) loads. We have been treating only the case of equilibrium. When the acceleration is unknown, we have a dynamic system. Then, instead of using Newton's second law, $\sum \mathbf{F} = m\,\mathbf{a}$ where $\mathbf{a}(t)$ is the acceleration vector, we invoke the D'Alembert's principle and rewrite this as a pseudo-equilibrium problem $\sum \mathbf{F} - m\,\mathbf{a} = \mathbf{0}$, or $\sum \mathbf{F} + \mathbf{F}_I = \mathbf{0}$ where we have introduced an inertial body force vector due to the acceleration, that is, we use $\mathbf{X} = -\rho\,\mathbf{a}$ for the equilibrium integral form. Since the acceleration is the second time derivative of the displacement vector, we can write

$$\mathbf{a}(t) = \mathbf{N}^e(x,y)\,\ddot{\boldsymbol{\delta}}^e$$

in a typical element. The typical element inertial contribution is, therefore,

$$-\,\mathbf{m}^e\,\mathbf{a}^e = -\int_{\Omega^e} \mathbf{N}^{e^T} \varsigma\, \mathbf{N}^e\, d\,\Omega\, \ddot{\boldsymbol{\delta}}^e$$

where $\mathbf{m}^e$ is the element mass matrix. Since the acceleration vector is unknown, we move it to the LHS of the (undamped) system equations of motion:

$$\mathbf{M}\,\ddot{\boldsymbol{\delta}} + \mathbf{K}\,\boldsymbol{\delta} = \mathbf{F}. \tag{12.25}$$

Here, $\mathbf{M}$ is the assembled system mass matrix, and the above are the *structural dynamic equations.* This class of problem will be considered later. If we had free ($\mathbf{F} = 0$) *simple harmonic motion,* so that $\boldsymbol{\delta}(t) = \boldsymbol{\delta}_j \operatorname{Sin}(\omega_j t)$, then we get the alternate class known as the *eigen-problem,*

$$[\,\mathbf{K} - \omega_j^2\,\mathbf{M}\,]\,\boldsymbol{\delta}_j = \mathbf{0}, \tag{12.26}$$

where ω_j is the eigenvalue, or natural frequency, and $\boldsymbol{\delta}_j$ is the mode shape, or eigenvector. The computational approaches to eigen-problems are covered in detail in the texts by Bathe [4], and by others. As noted in Table 2.2, we employ the Jacobi iteration technique for relatively small eigen-problems. The use of SCP error estimators is discussed by Wiberg [25] for use in vibration problems.

Another type of body force that is usually difficult to visualize is that due to electromagnetic effects. In the past, they were usually small enough to be ignored. However, with the advances in superconducting materials, very high electrical current densities are possible, and they can lead to significant mechanical loads. Similar loads develop in medical scan devices, and in fusion energy reactors which are currently experimental. Recall from basic physics that the mechanical force, **F**, due to a current density vector, $\vec{\mathbf{j}}$, in a field with a magnetic flux density vector of $\vec{\mathbf{b}}$ is the vector cross product $\mathbf{F} = \vec{\mathbf{j}} \times \vec{\mathbf{b}}$. For a thin wire conductor, $\vec{\mathbf{j}}$ is easy to visualize since it is tangent to the conductor. However, the $\vec{\mathbf{b}}$ field, like the earth's gravity field, is difficult to visualize. This could lead to important forces on the system which might be overlooked.

The final load to be considered is that acting on a typical boundary segment. As indicated in Fig. 12.6, such a segment is one side of an element being loaded with a traction. In plane stress problems these pressures or distributed shears act on the edge of the solid. In other words, they are distributed over a length L^b that has a known thickness, t. Those two quantities define the surface area, A^b, on which the tractions in Eq. 12.4 are applied. Similarly, the differential surface area is $da = t\, dL$. We observe that such a segment would have two nodes. We can refer to them as local boundary nodes 1 and 2. Of course, they are a subset of the three element nodes and also a subset of the system nodes. Before Eq. 12.4 can be integrated to define the consistent loads on the two boundary nodes it is necessary to form the boundary interpolation, $\mathbf{N}^b$. That function defines the displacements, u and v, at all points on the boundary segment curve. By analogy with Eq. 12.8 we can denote the dof of the boundary segment as

$$\mathbf{a}^{b^T} = \lfloor u_1 \quad v_1 \quad u_2 \quad v_2 \rfloor^b .$$

Then the requirement that $\mathbf{u} = \mathbf{N}^b \boldsymbol{\delta}^b$, for all points on L^b, defines the required $\mathbf{N}^b$. There are actually two ways that its algebraic form can be derived:

1) Develop a consistent (linear) interpolation on the line between the nodal dof.
2) Degenerate the element function $\mathbf{N}^e$ in Eq. 12.9 by restricting the x and y coordinates to points on the boundary segment.

If the second option is selected then all the H_i^e vanish except for the two associated with the two boundary segment nodes. Those two H^e are simplified by the restriction and thus define the two H_i^b functions. While the result of this type of procedure may be obvious, the algebra is tedious in global coordinates. (For example, let $y^b = mx^b + n$ in Eq. 12.6.) It is much easier to get the desired results if local coordinates are used. (For example, set $s^b = 0$ in Eq. 9.6.) The net result is that one obtains a one-dimensional linear interpolation set for $\mathbf{H}^b$ that is analogous to Eq. 4.11. If we assume constant thickness, t^b, and constant tractions, $\mathbf{T}^b$, then Eq. 12.4 becomes

$$\mathbf{C}_T^b = t^b \int_{L^b} \mathbf{N}^{b^T} dL\, \mathbf{T}^b .$$

Repeating the procedure used for Eq. 5.5 a typical non-zero contribution is

$$I_i^b = \int_{L^b} H_i^b\, dL = L^b/2, \quad 1 \le i \le 2$$

and the final result for the four force components is

$$\mathbf{C}_T^{b^T} = \frac{t^b L^b}{2} \,[\, T_x \;\; T_y \;\; T_x \;\; T_y \,]^b \tag{12.27}$$

where T_x and T_y are the two components of T. Physically, this states that half of the

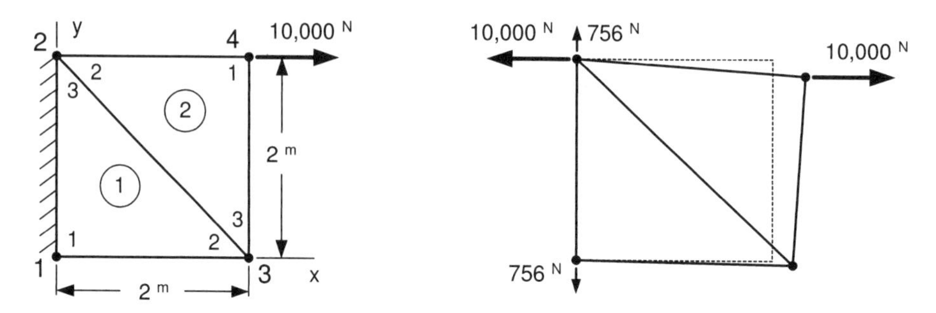

Figure 12.7 *An example plane stress structure*

```
title "Two element plane stress example"                 ! 1
area_thick 5e-3 ! Thickness of all planar elements       ! 2
nodes      4 ! Number of nodes in the mesh               ! 3
elems      2 ! Number of elements in the system          ! 4
dof        2 ! Number of unknowns per node               ! 5
el_nodes   3 ! Maximum number of nodes per element       ! 6
space      2 ! Solution space dimension                  ! 7
b_rows     3 ! Number of rows in the B matrix            ! 8
shape      2 ! Element shape, 1=line, 2=tri, 3=quad      ! 9
gauss      1 ! number of quadrature points               !10
el_real    3 ! Number of real properties per element     !11
el_homo      ! Element properties are homogeneous        !12
pt_list      ! List the answers at each node point       !13
loads        ! An initial source vector is input         !14
post_1       ! Require post-processing                   !15
example 201 ! Source library example number              !16
remarks    5 ! Number of user remarks                    !17
quit ! keyword input, remarks follow                     !18
        2------4 ---> P = 1e4 N                          !19
 Fixed  : \ (2):  width = height = 2 m,                  !20
 edge   :  \   :  E = 15e9 N/m^2, nu = 0.25,             !21
        : (1)\ :  thickness = 5e-3 m                     !22
        1------3                                         !23
1    11   0.0   0.0 ! node, U-V BC flag, x, y            !24
2    11   0.0   2.0 ! node, U-V BC flag, x, y            !25
3    00   2.0   0.0 ! node, U-V BC flag, x, y            !26
4    00   2.0   2.0 ! node, U-V BC flag, x, y            !27
1  1 3 2 ! element, 3 nodes                              !28
2  4 2 3 ! element, 3 nodes                              !29
 1  1    0.0 ! node, direction, BC value (U)             !30
 1  2    0.0 ! node, direction, BC value (V)             !31
 2  1    0.0 ! node, direction, BC value (U)             !32
 2  2    0.0 ! node, direction, BC value (V)             !33
1 15e9 0.25 10.5e6 ! elem, E, nu, yield stress           !34
4 1 10000.   ! node, direction, force                    !35
4 2 0.       ! terminate with last force in the system   !36
```

Figure 12.8 *Plane stress data file*

resultant x-force, $t^b\, L^b\, T_x^b$ is lumped at each of the two nodes. The resultant y-force is lumped in the same way as illustrated in Fig. 12.6.

There are times when it is desirable to rearrange the constitutive matrix, **E**, into two parts. One part, $\mathbf{E}_n$, is due to normal strain effects, and the other, $\mathbf{E}_s$, is related to the shear strains. Therefore, in general it is possible to write Eq. 12.5 as

$$\mathbf{E} = \mathbf{E}_n + \mathbf{E}_s. \tag{12.28}$$

In this case such a procedure simply makes it easier to write the CST stiffness matrix in closed form. Noting that substituting Eq. 12.28 into Eq. 12.1 allows the stiffness to be separated into parts $\mathbf{S}^e = \mathbf{S}_n^e + \mathbf{S}_s^e$, where

$$\mathbf{S}_n^e = \frac{EV}{4A^2(1-\nu^2)} \begin{bmatrix} b_1^2 & & & & & \text{sym} \\ \nu b_1 c_1 & c_1^2 & & & & \\ b_1 b_2 & \nu c_1 b_2 & b_2^2 & & & \\ \nu b_1 c_2 & c_1 c_2 & \nu b_2 c_2 & c_2^2 & & \\ b_1 b_3 & \nu c_1 b_3 & b_2 b_3 & \nu c_2 b_3 & b_3^2 & \\ \nu b_1 c_3 & c_1 c_3 & \nu b_2 c_3 & c_2 c_3 & \nu b_3 c_3 & c_3^2 \end{bmatrix}$$

$$\mathbf{S}_s^e = \frac{EV}{8A^2(1+\nu)} \begin{bmatrix} c_1^2 & & & & & \text{sym} \\ c_1 b_1 & b_1^2 & & & & \\ c_1 c_2 & b_1 c_2 & c_2^2 & & & \\ c_1 b_2 & b_1 b_2 & c_2 b_2 & b_2^2 & & \\ c_1 c_3 & b_1 c_3 & c_2 c_3 & b_2 c_3 & c_3^2 & \\ c_1 b_3 & b_1 b_3 & c_2 b_3 & b_2 b_3 & c_3 b_3 & b_3^2 \end{bmatrix}$$

and where V is the volume of the element. For constant thickness $V = At$.

The strain-displacement matrix $\mathbf{B}^e$ can always be partioned into sub-matrices associated with each node. Thus, the square stiffness matrix **S** can also be partitioned into square sub-matrices, since it is the product of $\mathbf{B}^T\,\mathbf{E}\,\mathbf{B}$. For local nodes j and k, they interact to give a contribution defined by:

$$\mathbf{S}_{jk} = \int_\Omega \mathbf{B}_j^T\, \mathbf{E}\, \mathbf{B}_k\, d\Omega.$$

If we choose to split **E** into two distinct parts, say $\mathbf{E} = \mathbf{E}_n + \mathbf{E}_s$, then we likewise have two contributions to the partitions of **S**, namely

$$\mathbf{S}_{jk}^n = \int_\Omega \mathbf{B}_j^T\, \mathbf{E}_n\, \mathbf{B}_k\, d\Omega\ , \qquad \mathbf{S}_{jk}^s = \int_\Omega \mathbf{B}_j^T\, \mathbf{E}_s\, \mathbf{B}_k\, d\Omega\ .$$

Sometimes we may use different numerical integration rules on these two parts. For the constant strain triangle, CST, we have the nodal partitions of $\mathbf{B}^{e_j}$:

$$\mathbf{B}_j^e = \frac{1}{2\,A^e} \begin{bmatrix} b_j & 0 \\ 0 & c_j \\ c_j & b_j \end{bmatrix}^e = \begin{bmatrix} d_{xj} & 0 \\ 0 & d_{yj} \\ d_{yj} & d_{xj} \end{bmatrix}^e$$

which involve the geometric constants defined earlier. Once we know the gradients of the scalar interpolation functions, **H** we can compute $\mathbf{B}^e$ from Eq. 12.11 as shown earlier in Fig. 12.5, for any planar element in our element library. For a constant isotropic **E**, the integral gives the partitions

$$\mathbf{S}^s_{jk} = \frac{t\,E_{33}}{4A}\begin{bmatrix} c_j\,c_k & c_j\,b_k \\ b_j\,c_k & b_j\,b_k \end{bmatrix}^e, \qquad \mathbf{S}^n_{jk} = \frac{t}{4A}\begin{bmatrix} E_{11}\,b_j\,b_k & E_{12}\,b_j\,c_k \\ E_{12}\,c_j\,b_k & E_{22}\,c_j\,c_k \end{bmatrix}^e$$

where E_{33} brings in the shear effects, and E_{11}, E_{12} couple the normal stress effects. If we allow j and k to range over the values 1, 2, 3, we would get the full 6×6 stiffness

$$\mathbf{S} = \begin{bmatrix} \mathbf{S}_{11} & \mathbf{S}_{12} & \mathbf{S}_{13} \\ \mathbf{S}_{21} & \mathbf{S}_{22} & \mathbf{S}_{23} \\ \mathbf{S}_{31} & \mathbf{S}_{32} & \mathbf{S}_{33} \end{bmatrix}.$$

Since $\mathbf{E}$ is symmetric, it should be clear that $\mathbf{S}_{jk} = \mathbf{S}^T_{kj}$. A similar split can be made utilizing the constitutive law in terms of the Lamé constants,

$$\lambda = K - \frac{2G}{3} = \frac{E\,\nu}{(1+\nu)(1-2\nu)}, \qquad \mu = G = \frac{E}{2(1+\nu)},$$

where K and G are the bulk modulus and the shear modulus, respectively. Then the plane strain $\mathbf{E}$ matrix can be split as

$$\mathbf{E} = \lambda\begin{bmatrix} 1 & 1 & 0 \\ 1 & 1 & 0 \\ 0 & 0 & 0 \end{bmatrix} + \mu\begin{bmatrix} 2 & 0 & 0 \\ 0 & 2 & 0 \\ 0 & 0 & 1 \end{bmatrix}$$

$$\mathbf{E} = K\begin{bmatrix} 1 & 1 & 0 \\ 1 & 1 & 0 \\ 0 & 0 & 0 \end{bmatrix} + \frac{G}{3}\begin{bmatrix} 4 & -2 & 0 \\ -2 & 4 & 0 \\ 0 & 0 & 3 \end{bmatrix}.$$

Likewise, for the full three-dimensional case with six strains and six stresses, we have a similar form

$$\mathbf{E} = \lambda\begin{bmatrix} 1 & 1 & 1 & 0 & 0 & 0 \\ 1 & 1 & 1 & 0 & 0 & 0 \\ 1 & 1 & 1 & 0 & 0 & 0 \\ 0 & 0 & 0 & 0 & 0 & 0 \\ 0 & 0 & 0 & 0 & 0 & 0 \\ 0 & 0 & 0 & 0 & 0 & 0 \end{bmatrix} + \mu\begin{bmatrix} 2 & 0 & 0 & 0 & 0 & 0 \\ 0 & 2 & 0 & 0 & 0 & 0 \\ 0 & 0 & 2 & 0 & 0 & 0 \\ 0 & 0 & 0 & 1 & 0 & 0 \\ 0 & 0 & 0 & 0 & 1 & 0 \\ 0 & 0 & 0 & 0 & 0 & 1 \end{bmatrix}.$$

This means we have split the strain energy into two corresponding parts: the distortional strain energy and the volumetric strain energy. We define an *incompressible material* as one that has no change in volume as it is deformed. For such a material $\nu = \frac{1}{2}$. For a nearly incompressible material, we note that as $\nu \to \frac{1}{2}$, we see that $\lambda \to \infty$ and $K \to \infty$. Since many rubber materials are nearly incompressible, we can expect to encounter this difficulty in practical problems. Since incompressibility means no volume change, it also means there is no volumetric strain. For plane strain we have the *incompressibility constraint:* $\partial u/\partial x + \partial v/\partial y \equiv 0$. In such a case, we must either use an alternate variational form that involves the displacements and the mean stress (pressure), or we must undertake numerical corrections to prevent the solution from *locking.*

```
TITLE: "Two element plane stress example"                          ! 1
                                                                    ! 2
***  ELEMENT  PROPERTIES   ***                                      ! 3
ELEMENT,  3 PROPERTY & REAL_VALUE PAIRS                             ! 4
   1  1  1.50000E+10  2  2.50000E-01  3  1.05000E+07                ! 5
                                                                    ! 6
NOTE: 2-D DOMAIN THICKNESS SET TO 5.000000000000E-03                ! 7
                                                                    ! 8
*** INITIAL FORCING VECTOR DATA ***                                 ! 9
NODE   PARAMETER     VALUE     EQUATION                             !10
   4          1   1.00000E+04        7                              !11
   4          2   0.00000E+00        8                              !12
                                                                    !13
*** REACTION RECOVERY ***                                           !14
NODE, PARAMETER,      REACTION, EQUATION                            !15
   1,  DOF_1,       6.8212E-13     1                                !16
   1,  DOF_2,       7.5630E+02     2                                !17
   2,  DOF_1,      -1.0000E+04     3                                !18
   2,  DOF_2,      -7.5630E+02     4                                !19
                                                                    !20
***  OUTPUT OF RESULTS IN NODAL ORDER  ***                          !21
NODE,  X-Coord,     Y-Coord,     DOF_1,       DOF_2,                !22
   1  0.0000E+00  0.0000E+00  0.0000E+00  0.0000E+00                !23
   2  0.0000E+00  2.0000E+00  0.0000E+00  0.0000E+00                !24
   3  2.0000E+00  0.0000E+00  2.5210E-05 -6.7227E-05                !25
   4  2.0000E+00  2.0000E+00  2.4650E-04 -1.5406E-04                !26
                                                                    !27
*** STRESSES AT INTEGRATION POINTS ***                              !28
           COORDINATES                 STRESSES                     !29
POINT     X            Y          XX            YY                  !30
POINT                             XY            EFFECTIVE           !31
         ELEMENT NUMBER  1                                          !32
  1  6.667E-01  6.667E-01   2.01681E+05   5.04202E+04               !33
  1                        -2.01681E+05   3.93794E+05               !34
         ELEMENT NUMBER  2                                          !35
  1  1.333E+00  1.333E+00   1.79832E+06  -2.01681E+05               !36
  1                         2.01681E+05   1.93890E+06               !37
```

Figure 12.9 *Selected CST results for the two element model*

As an example of the use of the CST, consider the thin structure shown in Fig. 12.7. Its elastic modulus is $15 \times 10^9 \ N/m^2$, Poisson's ratio is 0.25, and the yield stress is $10.5 \times 10^6 \ N/m^2$. The uniform thickness of the material is $5 \times 10^{-3} \ m$. The corresponding input data file is given in Fig. 12.8. From Eq. 9.14 the element geometric constants are

$e = 1$			$e = 2$		
i	b_i	c_i	i	b_i	c_i
1	−2	−2	1	+2	+2
2	+2	0	2	−2	0
3	0	+2	3	0	−2

For the given data the constants multiplying the $\mathbf{S}_n$ and $\mathbf{S}_s$ matrices are 1×10^7 and $6 \times 10^7 \ N/m$, respectively. For the first element the two contributions to the element stiffness matrix are

$$\mathbf{S}_n^e = \frac{1 \times 10^7}{2} \begin{bmatrix} +8 & & & & & \text{Sym.} \\ +2 & +8 & & & & \\ -8 & -2 & +8 & & & \\ 0 & 0 & 0 & 0 & & \\ 0 & 0 & 0 & 0 & 0 & \\ -2 & -8 & +2 & 0 & 0 & +8 \end{bmatrix}$$

$$\mathbf{S}_s^e = \frac{3 \times 10^7}{2} \begin{bmatrix} +1 & & & & & \text{Sym.} \\ +1 & +1 & & & & \\ 0 & 0 & 0 & & & \\ -1 & -1 & 0 & +1 & & \\ -1 & -1 & 0 & +1 & +1 & \\ 0 & 0 & 0 & 0 & 0 & 0 \end{bmatrix}$$

Thus, the first element stiffness is:

$$S^e = 5 \times 10^6 \begin{bmatrix} +11 & & & & & \text{Sym.} \\ +5 & +11 & & & & \\ -8 & -2 & +8 & & & \\ -3 & -3 & 0 & +3 & & \\ -3 & -3 & 0 & +3 & +3 & \\ -2 & -8 & +2 & 0 & 0 & +8 \end{bmatrix}$$

and its global and local degree of freedom numbers are the same. The second stiffness matrix happens to be the same due to its 180° rotation in space. Of course, its global dof numbers are different. That list is: 7, 8, 3, 4, 5, and 6. Since there are no body forces or surface tractions these matrices can be assembled to relate the system stiffness to the applied point load, P, and the support reactions. Applying the direct assembly procedure gives

$$\begin{matrix} 1 & 2 & 3 & 4 & 5 & 6 & 7 & 8 & \text{global} \end{matrix}$$

$$5 \times 10^6 \begin{bmatrix} +11 & & & & & & & \text{Sym.} \\ +5 & +11 & & & & & & \\ -3 & -3 & +11 & & & & & \\ -2 & -8 & 0 & +11 & & & & \\ -8 & -2 & 0 & +5 & +11 & & & \\ -3 & -3 & +5 & 0 & 0 & +11 & & \\ 0 & 0 & -8 & -3 & -3 & -2 & +11 & \\ 0 & 0 & -2 & -3 & -3 & -8 & +5 & +11 \end{bmatrix} \mathbf{\Delta} = \begin{Bmatrix} R_1 \\ R_2 \\ R_3 \\ R_4 \\ 0 \\ 0 \\ 10^4 \\ 0 \end{Bmatrix}.$$

Applying the conditions of zero displacement at nodes 1 and 2 reduces this set to

$$5 \times 10^6 \begin{bmatrix} 11 & & & \text{Sym.} \\ 0 & 11 & & \\ -3 & -2 & 11 & \\ -3 & -8 & 5 & 11 \end{bmatrix} \begin{Bmatrix} u_3 \\ v_3 \\ u_4 \\ v_4 \end{Bmatrix} = \begin{Bmatrix} 0 \\ 0 \\ 10^4 \\ 0 \end{Bmatrix}.$$

Inverting the matrix and solving gives the required displacement vector, transposed: $10^5 \times u_r = [\,2.52 \quad -6.72 \quad 24.65 \quad -15.41\,]\ m$. Substituting to find the reactions yields $\mathbf{R}_g^T = [\,-0.002 \quad -756.3 \quad -10{,}000 \quad -756.3\,]\ N$. The deformed shape and resulting reactions are also shown in Fig. 12.7. One should always check the equilibrium of the

```
!     ***  ELEM_SQ_MATRIX PROBLEM DEPENDENT STATEMENTS ***            ! 1
! PLANE_STRESS ANALYSIS, NON-ISOPARAMETRIC.  Example source 201.      ! 2
! STRESS AND STRAIN COMPONENT ORDER: XX, YY, XY, SO N_R_B = 3         ! 3
                                                                      ! 4
  INTEGER  :: IP                        ! loops                       ! 5
  REAL(DP) :: DET, DET_WT, THICK        ! volume                      ! 6
! PROPERTIES: 1-YOUNG'S MODULUS, 2-POISSON'S RATIO, AND               ! 7
!             3-YIELD STRESS, IF PRESENT                              ! 8
                                                                      ! 9
  CALL STORE_FLUX_POINT_COUNT ! Save LT_QP                            !10
                                                                      !11
  THICK = 1                           ! DEFINE CONSTANT PROPERTIES    !12
  IF ( AREA_THICK /= 1.d0 ) THICK = AREA_THICK                        !13
                                                                      !14
!   FORM THE CONSTITUTIVE MATRIX (OR GET_APPLICATION_E_MATRIX )       !15
  CALL E_PLANE_STRESS (E)                                             !16
                                                                      !17
  DO IP = 1, LT_QP      !      NUMERICAL INTEGRATION LOOP             !18
    G        = GET_G_AT_QP   (IP)         ! GEOMETRY INTERPOLATIONS   !19
    GEOMETRY = COORD (1:LT_GEOM,:)        ! GEOMETRY NODES            !20
    XYZ      = MATMUL ( G, GEOMETRY )     ! COORDINATES OF POINT      !21
                                                                      !22
    DLG      = GET_DLG_AT_QP (IP)         ! GEOMETRIC DERIVATIVES     !23
    AJ       = MATMUL (DLG, GEOMETRY (:, 1:LT_PARM)) ! JACOBIAN       !24
    CALL INVERT_2BY2 (AJ, AJ_INV, DET)    ! INVERSE, DET              !25
    DET_WT   = DET * WT (IP) * THICK                                  !26
                                                                      !27
    H   = GET_H_AT_QP   (IP)       ! SCALAR INTERPOLATIONS            !28
    DLH = GET_DLH_AT_QP (IP)       ! SCALAR DERIVATIVES               !29
    DGH = MATMUL ( AJ_INV, DLH )   ! PHYSICAL DERIVATIVES             !30
                                                                      !31
!---> FORM STRAIN-DISPLACEMENT, B (OR GET_APPLICATION_B_MATRIX)       !32
    CALL ELASTIC_B_PLANAR (DGH, B)                                    !33
                                                                      !34
!     EVALUATE ELEMENT MATRICES                                       !35
    S = S + DET_WT * MATMUL (TRANSPOSE(B), MATMUL (E, B))             !36
                                                                      !37
!     SAVE PT, CONSTITUTIVE & STRAIN_DISP FOR POST_PROCESS & SCP      !38
    CALL STORE_FLUX_POINT_DATA (XYZ, E, B)                            !39
  END DO  ! Over quadrature points                                    !40
!     ***  END ELEM_SQ_MATRIX PROBLEM DEPENDENT STATEMENTS ***        !41
```

Figure 12.10 *A general plane stress isotropic stiffness matrix*

reactions and applied loads. Checking $\sum \mathbf{F}_x = 0$, $\sum \mathbf{M} = 0$ does show minor errors in about the sixth significant figure. Thus, the results are reasonable.

At this point we can recover the displacements for each element, and then compute the strains and stress. The element dof vectors (in meters) are, respectively

```
!  *** POST_PROCESS_ELEM PROBLEM DEPENDENT STATEMENTS FOLLOW ***    ! 1
!        PLANE_STRESS ANALYSIS, using STRESS (N_R_B + 2)            ! 2
! STRESS AND STRAIN COMPONENT ORDER: XX, YY, XY, SO N_R_B = 3       ! 3
!            (Stored as application source example 201.)            ! 4
! PROPERTIES: 1-YOUNG'S MODULUS, 2-POISSON'S RATIO, AND             ! 5
!             3-YIELD STRESS, IF PRESENT                            ! 6
  INTEGER  :: J, N_IP                  ! LOOPS                      ! 7
  REAL(DP), SAVE :: YIELD              ! FAILURE DATA               ! 8
                                                                    ! 9
  IF ( IE == 1 ) THEN ! PRINT TITLES & INITIALIZE                   !10
    STRAIN = 0.d0 ; STRAIN_0 = 0.d0  ! INITIALIZE ALL OF "STRAIN"   !11
    IF ( EL_REAL > 2 ) THEN          ! INITIALIZE YIELD STRESS      !12
      YIELD = GET_REAL_LP (3)                                       !13
    ELSE ; YIELD = HUGE (1.d0) ; END IF ! YIELD DATA                !14
                                                                    !15
    WRITE (6, 50) ; 50 FORMAT ( /,                          &       !16
    '*** STRESSES AT INTEGRATION POINTS ***',          /, &         !17
    '            COORDINATES             STRESSES',    /, &         !18
    'POINT     X           Y          XX         YY',  /, &         !19
    'POINT                            XY         EFFECTIVE')        !20
  END IF ! NEW HEADINGS                                             !21
                                                                    !22
  WRITE (6,  * ) '        ELEMENT NUMBER ', IE                      !23
  CALL READ_FLUX_POINT_COUNT (N_IP) ! NUMBER OF QUADRATURE POINTS   !24
  DO J = 1, N_IP                    ! AT QUADRATURE POINTS          !25
                                                                    !26
    CALL READ_FLUX_POINT_DATA (XYZ, E, B) ! PT, PROP, STRAIN_DISP   !27
                                                                    !28
!     MECHANICAL STRAINS & STRESSES                                 !29
    STRAIN (1:N_R_B) = MATMUL (B, D) ! STRAINS AT THE POINT         !30
    STRESS = MATMUL (E, STRAIN)      ! CALCULATE STRESSES           !31
                                                                    !32
!     VON_MISES FAILURE CRITERION (EFFECTIVE STRESS, ADD TO END)    !33
    STRESS (4) = SQRT ( (STRESS (1) - STRESS (2) ) **2        &     !34
               + (STRESS (2)) **2 + (STRESS (1)) **2          &     !35
               + 6.d0 * STRESS (3) **2 ) * 0.7071068d0              !36
    IF ( STRESS (4) >= YIELD ) PRINT *,                       &     !37
      'WARNING: FAILURE CRITERION EXCEEDED IN ELEMENT -', IE        !38
                                                                    !39
!        LIST STRESSES AND FAILURE CRITERION AT POINT               !40
    WRITE (6, 52) J, XYZ (1:2), STRESS (1:2)                        !41
    WRITE (6, 51) J,            STRESS (3:4)                        !42
    52 FORMAT ( I3, 2(1PE11.3), 5(1PE14.5) )                        !43
    51 FORMAT ( I3, 22X,        5(1PE14.5) )                        !44
                                                                    !45
  END DO ! AT QUADRATURE POINTS                                     !46
!  *** END POST_PROCESS_ELEM PROBLEM DEPENDENT STATEMENTS ***       !47
```

Figure 12.11 *Plane stress mechanical stress recovery*

$$\boldsymbol{\delta}^{e^T} = \lfloor 0 \quad 0 \quad 2.521 \quad -6.723 \quad 0 \quad 0 \rfloor \times 10^{-5}$$

$$\boldsymbol{\delta}^{e^T} = \lfloor 24.650 \quad -15.406 \quad 0 \quad 0 \quad 2.521 \quad -6.723 \rfloor \times 10^{-5}$$

and the strain-displacement matrices, from Eq. 12.18 are

$$\mathbf{B}^e = \frac{1}{4}\begin{bmatrix} -2 & 0 & 2 & 0 & 0 & 0 \\ 0 & -2 & 0 & 0 & 0 & 2 \\ -2 & -2 & 0 & 2 & 2 & 2 \end{bmatrix}$$

for $e = 1$, while for $e = 2$

$$\mathbf{B}^e = \frac{1}{4}\begin{bmatrix} 2 & 0 & -2 & 0 & 0 & 0 \\ 0 & 2 & 0 & 0 & 0 & -2 \\ 2 & 2 & 0 & -2 & -2 & 0 \end{bmatrix}.$$

Recovering the element strains, $\boldsymbol{\varepsilon}^e = \mathbf{B}^e \boldsymbol{\delta}^e$ in meters/meter gives

$$e = 1, \quad \boldsymbol{\varepsilon}^{e^T} = 10^{-5} \lfloor 1.261 \quad 0.000 \quad -3.361 \rfloor$$

$$e = 1, \quad \boldsymbol{\varepsilon}^{e^T} = 10^{-5} \lfloor 12.325 \quad -4.342 \quad 3.361 \rfloor.$$

Utilizing the constitute law, with no initial strains, $\boldsymbol{\varepsilon}_o = \mathbf{0}$, gives

$$\mathbf{E}^e = \frac{15 \times 10^9}{(15/16)}\begin{bmatrix} 1 & 1/4 & 0 \\ 1/4 & 1 & 0 \\ 0 & 0 & 3/8 \end{bmatrix} = 2 \times 10^9 \begin{bmatrix} 8 & 2 & 0 \\ 2 & 8 & 0 \\ 0 & 0 & 3 \end{bmatrix},$$

and the element stresses, in Newtons/meter^2, are

$$e = 1, \quad \sigma^{e^T} = 10^4 \lfloor 20.17 \quad 5.04 \quad -20.17 \rfloor$$

$$e = 2, \quad \sigma^{e^T} = 10^4 \lfloor 179.83 \quad -20.17 \quad 20.17 \rfloor.$$

A good engineer should have an estimate of the desired solution before approaching the computer. For example, if the load had been at the center of the edge, then

$$\sigma_x = P/A = 10^4/(2)\,(5 \times 10^{-3}) = 10^6\ \text{N/m}^2,$$

and $\sigma_y = 0 = \tau$. The values are significantly different from the computed values. A better estimate would consider both the axial and bending effects so $\sigma_x = P/A \pm Mc/I$. At the centroid of these two elements ($y = 0.667$ and $y = 1.333$) the revised stress estimates are $\sigma_x = 0$ and $\sigma_x = 2 \times 10^6$ N/m^2, respectively. The revised difference between the maximum centroidial stress and our estimate is only 10 percent. Of course, with the insight gained from the mechanics of materials our mesh was not a good selection. We know that while an axial stress would be constant across the depth of the member, the bending effects would vary linearly with y. Thus, it was poor judgement to select a single element through the thickness. These hand calculations are validated in the selected output file shown in Fig. 12.9.

To select a better mesh we should imagine how the stress would vary through the member. Then we would decide how many constant steps are required to get a good fit to the curve. Similarly, if we employed linear stress triangles (LST) we would estimate the required number of piece-wise linear segments needed to fit the curve. For example,

```
title "2D STRESS PATCH TEST, T6 ESSENTIAL BC"                       ! 1
nodes       9 ! Number of nodes in the mesh                         ! 2
elems       2 ! Number of elements in the system                    ! 3
dof         2 ! Number of unknowns per node                         ! 4
el_nodes    6 ! Maximum number of nodes per element                 ! 5
space       2 ! Solution space dimension                            ! 6
b_rows      3 ! Number of rows in the B (operator) matrix           ! 7
shape       2 ! Element shape, 1=line, 2=tri, 3=quad, 4=hex         ! 8
gauss       4 ! number of quadrature points                         ! 9
el_real     2 ! Number of real properties per element               !10
el_homo       ! Element properties are homogeneous                  !11
post_el       ! Require element post-processing                     !12
example     201 ! Source library example number                     !13
data_set     01 ! Data set for example (this file)                  !14
exact_case   12 ! Exact analytic solution                           !15
list_exact      ! List given exact answers at nodes, etc            !16
list_exact_flux ! List given exact fluxes at nodes, etc             !17
remarks       8 ! Number of user remarks                            !18
quit ! keyword input                                                !19
Note: Patch test yields constant gradient and strains               !20
 3--6---9 Mesh to left. Exact solution u = 1 + 3x - 4y              !21
 :(2) / :                        du/dx = 3,     du/dy = -4          !22
 2  5   8             Exact solution v = 1 + 3x - 4y                !23
 : / (1):              dv/dx = 3,     dv/dy = -4                    !24
 1/--4--7  Thus answer at node 5 is u = v = -1                      !25
Strains: 3, -4, -1. Stresses: 3, -4, -0.5, for E=1, nu=0            !26
Stresses: 2.13333, -3.46667, -0.4, for E=1, nu=0.25                 !27
     1      11   0.0    0.0 ! begin: node, bc_flags, x, y           !28
     2      11   0.0    2.0                                         !29
     3      11   0.0    4.0                                         !30
     4      11   2.0    0.0                                         !31
     5      00   2.0    2.0  ! only unknown                         !32
     6      11   2.0    4.0                                         !33
     7      11   4.0    0.0                                         !34
     8      11   4.0    2.0                                         !35
     9      11   4.0    4.0                                         !36
  1  1  7  9  4  8  5 ! begin elements                              !37
  2  1  9  3  5  6  2                                               !38
     1      1   1.0 ! essential bc                                  !39
     2      1  -7.0                                                 !40
     3      1 -15.0                                                 !41
     4      1   7.0                                                 !42
     6      1  -9.0                                                 !43
     7      1  13.0                                                 !44
     8      1   5.0                                                 !45
     9      1  -3.0                                                 !46
     1      2   1.0 ! essential bc                                  !47
     2      2  -7.0                                                 !48
     3      2 -15.0                                                 !49
     4      2   7.0                                                 !50
     6      2  -9.0                                                 !51
     7      2  13.0                                                 !52
     8      2   5.0                                                 !53
     9      2  -3.0                                                 !54
1  1.0    0.25   0.0 ! el, E, Nu, yield                             !55
```

Figure 12.12 *Plane stress patch test data*

consider a cantilever beam subjected to a bending load at its end. We know the exact normal stress is linear through the thickness and the shear stress varies quadratically through the thickness. Thus, through the depth we would need several linear (CST), or a few quadratic (LST), or a single cubic triangle (QST).

Converting to such higher order elements is relatively simple if we employ numerical intergration. Extending the numerically integrated scalar element square matrix of Fig. 11.41 a plane stress formulation, independent of element type, is obtained as shown in Fig. 12.10. The corresponding element stress recovery at each quadrature point is given in Fig. 12.11. The stiffness matrix given in Fig. 12.10 is actually basically the form that would be needed for any 1-, 2-, 3-dimensional, or axisymmetric solid. For example, one would mainly need to change two calls and make them more general. Line 16 recovers the constitutive matrix. It could be replaced with a call, say to routine *E_ISOTROPIC_STRESS* (E), that had the necessary logic to treat the 5 cases cited above. The choices for $\mathbf{E}^e$ mainly depend on the dimension of the space (N_SPACE set by keyword) and whether the keyword *axisymmetric* is present.

Usually the state of plane stress is taken as the default in a 2-dimensional model (if not axisymmetric) so one would have to provide another logical control variable (say keyword *plane_strain*) to allow for activating that condition. Likewise, the strain-displacement call for the $\mathbf{B}^e$ matrix (at line 33) could be changed to a general form say, *ELASTIC_B_MATRIX* (DGH, H, XYZ, B), that works for any of the above 4 cases by allowing for an axisymmetric model to use the current point, XYZ, and the interpolation, H, needed to obtain the hoop strain. The choices for $\mathbf{B}^e$ depend only on the dimension of the space and whether the keyword *axisymmetric* is present.

It is always wise to test such implementations by means of a numerical patch test. Such a test is given in Fig. 12.12 where both the u and v displacements are prescribed, by the same equation, at all the boundary nodes as essential conditions. Then the one interior node displacement is computed. All displacements are then employed to obtain the generalized flux components (here mechanical stresses) and compared to the corresponding constant values assumed in the analytic expression picked to define the patch test. Here we see that the results for displacements and stresses, in Fig. 12.13, are everywhere exact and we pass the patch test and thus assume the programming is reasonably correct. In the latter figure the one interior node displacement vector actually computed is node 5 (line 13) while the other 8 node displacement vectors were prescribed as essential boundary conditions. The element post-processing results are given in lines 43-55. The flux (stress) components at the integration points (lines 19-29), and their smoothed values at the nodes (lines 31-41) are output when numerical integration is used (keyword *gauss* > 0) unless turned off by keywords *no_scp_ave*, or *no_error_est*.

12.5 Stress and strain transformations *

Having computed the global stress components at a point in an element, we may wish to find the stresses in another direction, that is, with respect to a different coordinate system. This can be done by employing the transformations associated with *Mohr's circle*. Mohr's circles of stress and strain are usually used to produce graphical solutions. However, here we wish to rely on automated numerical solutions. Thus, we will review the *stress transformation* laws. Refer to Fig. 12.14 where the quantities used in Mohr's transformation are defined. The alternate coordinate set (n, s) is used to describe the surfaces on which the normal stresses, σ_n and σ_s, and the shear stress, τ_{ns}, act. The

```
TITLE: "2D STRESS PATCH TEST, T6 ESSENTIAL BC"                          ! 1
                                                                        ! 2
 *** SYSTEM GEOMETRIC PROPERTIES ***                                    ! 3
VOLUME   =  1.60000E+01                                                 ! 4
CENTROID =  2.00000E+00  2.00000E+00                                    ! 5
                                                                        ! 6
***  OUTPUT OF RESULTS AND EXACT VALUES IN NODAL ORDER  ***             ! 7
NODE  X-Coord  Y-Coord    DOF_1       DOF_2       EXACT1      EXACT2    ! 8
   1  0.00E+0  0.00E+0  1.0000E+0  1.0000E+0  1.0000E+0  1.0000E+0 ! 9
   2  0.00E+0  2.00E+0 -7.0000E+0 -7.0000E+0 -7.0000E+0 -7.0000E+0 !10
   3  0.00E+0  4.00E+0 -1.5000E+1 -1.5000E+1 -1.5000E+1 -1.5000E+1 !11
   4  2.00E+0  0.00E+0  7.0000E+0  7.0000E+0  7.0000E+0  7.0000E+0 !12
   5  2.00E+0  2.00E+0 -1.0000E+0 -1.0000E+0 -1.0000E+0 -1.0000E+0 !13
   6  2.00E+0  4.00E+0 -9.0000E+0 -9.0000E+0 -9.0000E+0 -9.0000E+0 !14
   7  4.00E+0  0.00E+0  1.3000E+1  1.3000E+1  1.3000E+1  1.3000E+1 !15
   8  4.00E+0  2.00E+0  5.0000E+0  5.0000E+0  5.0000E+0  5.0000E+0 !16
   9  4.00E+0  4.00E+0 -3.0000E+0 -3.0000E+0 -3.0000E+0 -3.0000E+0 !17
                                                                   !18
*** FE AND EXACT FLUX COMPONENTS AT INTEGRATION POINTS ***         !19
EL X-Coord Y-Coord FLUX_1  FLUX_2  FLUX_3  EXACT1  EXACT2  EXACT3  !20
 1 2.67E+0 1.33E+0 2.13E0 -3.47E0 -4.00E-1 2.13E0 -3.47E0 -4.00E-1 !21
 1 3.20E+0 8.00E-1 2.13E0 -3.47E0 -4.00E-1 2.13E0 -3.47E0 -4.00E-1 !22
 1 3.20E+0 2.40E+0 2.13E0 -3.47E0 -4.00E-1 2.13E0 -3.47E0 -4.00E-1 !23
 1 1.60E+0 8.00E-1 2.13E0 -3.47E0 -4.00E-1 2.13E0 -3.47E0 -4.00E-1 !24
EL X-Coord Y-Coord FLUX_1  FLUX_2  FLUX_3  EXACT1  EXACT2  EXACT3  !25
 2 1.33E+0 2.67E+0 2.13E0 -3.47E0 -4.00E-1 2.13E0 -3.47E0 -4.00E-1 !26
 2 2.40E+0 3.20E+0 2.13E0 -3.47E0 -4.00E-1 2.13E0 -3.47E0 -4.00E-1 !27
 2 8.00E-1 3.20E+0 2.13E0 -3.47E0 -4.00E-1 2.13E0 -3.47E0 -4.00E-1 !28
 2 8.00E-1 1.60E+0 2.13E0 -3.47E0 -4.00E-1 2.13E0 -3.47E0 -4.00E-1 !29
                                                                   !30
** SUPER_CONVERGENT AVERAGED NODAL FLUXES & EXACT FLUXES **        !31
PT X-Coord Y-Coord FLUX_1  FLUX_2  FLUX_3  EXACT1  EXACT2  EXACT3  !32
 1 0.00E+0 0.00E+0 2.13E0 -3.47E0 -4.00E-1 2.13E0 -3.47E0 -4.00E-1 !33
 2 0.00E+0 2.00E+0 2.13E0 -3.47E0 -4.00E-1 2.13E0 -3.47E0 -4.00E-1 !34
 3 0.00E+0 4.00E+0 2.13E0 -3.47E0 -4.00E-1 2.13E0 -3.47E0 -4.00E-1 !35
 4 2.00E+0 0.00E+0 2.13E0 -3.47E0 -4.00E-1 2.13E0 -3.47E0 -4.00E-1 !36
 5 2.00E+0 2.00E+0 2.13E0 -3.47E0 -4.00E-1 2.13E0 -3.47E0 -4.00E-1 !37
 6 2.00E+0 4.00E+0 2.13E0 -3.47E0 -4.00E-1 2.13E0 -3.47E0 -4.00E-1 !38
 7 4.00E+0 0.00E+0 2.13E0 -3.47E0 -4.00E-1 2.13E0 -3.47E0 -4.00E-1 !39
 8 4.00E+0 2.00E+0 2.13E0 -3.47E0 -4.00E-1 2.13E0 -3.47E0 -4.00E-1 !40
 9 4.00E+0 4.00E+0 2.13E0 -3.47E0 -4.00E-1 2.13E0 -3.47E0 -4.00E-1 !41
                                                                   !42
*** STRESSES AT INTEGRATION POINTS ***                             !43
    COORDINATES                         STRESSES                   !44
PT   X          Y            XX          YY          XY        EFFECTIVE !45
                         ELEMENT NUMBER  1                         !46
 1 2.667E+0  1.333E+0   2.1333E+0 -3.4667E+0 -4.0000E-1 4.9441E+0  !47
 2 3.200E+0  8.000E-1   2.1333E+0 -3.4667E+0 -4.0000E-1 4.9441E+0  !48
 3 3.200E+0  2.400E+0   2.1333E+0 -3.4667E+0 -4.0000E-1 4.9441E+0  !49
 4 1.600E+0  8.000E-1   2.1333E+0 -3.4667E+0 -4.0000E-1 4.9441E+0  !50
                         ELEMENT NUMBER  2                         !51
 1 1.333E+0  2.667E+0   2.1333E+0 -3.4667E+0 -4.0000E-1 4.9441E+0  !52
 2 2.400E+0  3.200E+0   2.1333E+0 -3.4667E+0 -4.0000E-1 4.9441E+0  !53
 3 8.000E-1  3.200E+0   2.1333E+0 -3.4667E+0 -4.0000E-1 4.9441E+0  !54
 4 8.000E-1  1.600E+0   2.1333E+0 -3.4667E+0 -4.0000E-1 4.9441E+0  !55
```

Figure 12.13 *Correct plane stress patch test results*

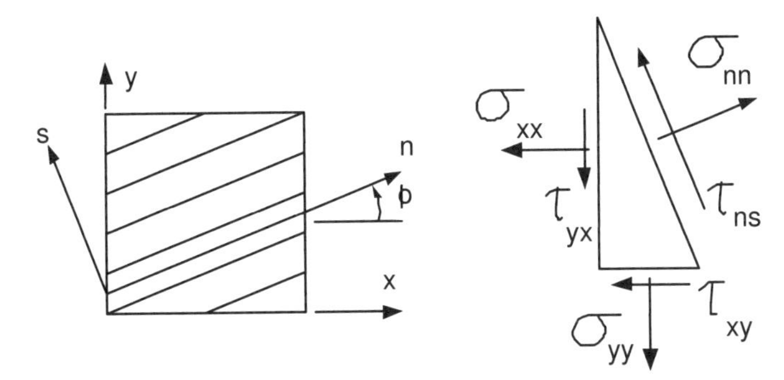

Figure 12.14 *Local material or stress axes*

n-axis is rotated from the x-axis by a positive (counter-clockwise) angle of β. By considering the equilibrium of the differential element, it is shown in mechanics of materials that

$$\sigma_n = \sigma_x \text{Cos}^2\beta + \sigma_y \text{Sin}^2\beta + 2\tau_{xy} \text{Sin}\,\beta \,\text{Cos}\,\beta. \tag{12.29}$$

Likewise the shear stress component is found to be

$$\tau_{ns} = -\sigma_x \text{Sin}\,\beta\,\text{Cos}\,\beta + \sigma_y \text{Sin}\,\beta\,\text{Cos}\,\beta + \tau_{xy}(\text{Cos}^2\beta - \text{Sin}^2\beta). \tag{12.30}$$

For Mohr's circle only these two stresses are usually plotted in the $\sigma_n - \tau_{ns}$ space. However, for a useful analytical statement we also need to define σ_s. Again from equilibrium considerations it is easy to show that

$$\sigma_s = \sigma_x \text{Sin}^2\beta + \sigma_y \text{Cos}^2\beta - 2\tau_{xy} \text{Sin}\,\beta\,\text{Cos}\,\beta. \tag{12.31}$$

Prior to this point we have employed matrix notation to represent the stress components. Then we were considering only the global coordinates. But now when we refer to the stress components it will be necessary to indicate which coordinate system is being utilized. We will employ the subscripts *xy* and *ns* to distinguish between the two systems. Thus, our previous stress component array will be denoted by

$$\boldsymbol{\sigma}^T = \boldsymbol{\sigma}^T_{(xy)} = [\sigma_x \quad \sigma_y \quad \tau_{xy}]$$

where we have introduced the notation that subscripts enclosed in parentheses denote the coordinate system employed. The stress components in the second coordinate system will be ordered in a similar manner and denoted by $\boldsymbol{\sigma}^T_{(ns)} = [\sigma_n \quad \sigma_s \quad \tau_{ns}]$. In this notation the stress transformation laws can be written as

$$\begin{Bmatrix} \sigma_n \\ \sigma_s \\ \tau_{ns} \end{Bmatrix} = \begin{bmatrix} +C^2 & +S^2 & +2SC \\ +S^2 & +C^2 & -2SC \\ -SC & +SC & (C^2 - S^2) \end{bmatrix} \begin{Bmatrix} \sigma_x \\ \sigma_y \\ \tau_{xy} \end{Bmatrix} \tag{12.32}$$

where $C \equiv \text{Cos}\,\beta$ and $S \equiv \text{Sin}\,\beta$ for simplicity. In symbolic matrix form this is

$$\boldsymbol{\sigma}_{(ns)} = \mathbf{T}(\beta)\,\boldsymbol{\sigma}_{(xy)} \tag{12.33}$$

where **T** will be defined as the stress transformation matrix. Clearly, if one wants to know the stresses on a given plane one specifies the angle β, forms **T**, and computes the results from Eq. 12.32. A similar procedure can be employed to express Mohr's circle of strain as a strain matrix transformation law. Denoting the new strains as $\boldsymbol{\varepsilon}^T_{(ns)} = [\varepsilon_n \quad \varepsilon_s \quad \gamma_{ns}]$ then the strain transformation law is

$$\begin{Bmatrix} \varepsilon_n \\ \varepsilon_s \\ \gamma_{ns} \end{Bmatrix} = \begin{bmatrix} +C^2 & +S^2 & +SC \\ +S^2 & +C^2 & -SC \\ -2SC & +2SC & (C^2 - S^2) \end{bmatrix} \begin{Bmatrix} \varepsilon_x \\ \varepsilon_y \\ \gamma_{xy} \end{Bmatrix} \tag{12.34}$$

or simply

$$\boldsymbol{\varepsilon}_{(ns)} = \mathbf{t}(\beta)\, \boldsymbol{\varepsilon}_{(xy)}. \tag{12.35}$$

Note that the two transformation matrices, **T** and **t**, are not identical. This is true because we have selected the engineering definition of the shear strain (instead of using the tensor definition). Also note that both of the transformation matrices are square. Therefore, the reverse relations can be found by inverting the transformations, that is,

$$\boldsymbol{\sigma}_{(xy)} = \mathbf{T}(\beta)^{-1} \boldsymbol{\sigma}_{(ns)}, \qquad \boldsymbol{\varepsilon}_{(xy)} = \mathbf{t}(\beta)^{-1} \boldsymbol{\varepsilon}_{(ns)}. \tag{12.36}$$

These two transformation matrices have the special property that the inverse of one is the transpose of the other, that is, it can be shown that

$$\mathbf{T}^{-1} = \mathbf{t}^T, \quad \mathbf{t}^{-1} = \mathbf{T}^T. \tag{12.37}$$

This property is also true when generalized to three-dimensional properties. Another generalization is to note that if we partition the matrices into normal and shear components, then

$$\mathbf{T} = \begin{bmatrix} \mathbf{T}_{11} & \mathbf{T}_{12} \\ \mathbf{T}_{21} & \mathbf{T}_{22} \end{bmatrix}, \quad \mathbf{t} = \begin{bmatrix} \mathbf{T}_{11} & \mathbf{T}_{12}/2 \\ 2\mathbf{T}_{21} & \mathbf{T}_{22} \end{bmatrix}.$$

In mechanics of materials it is shown that the principal normal stresses occur when the angle is given by $\mathrm{Tan}(2\beta_p) = 2\tau_{xy}/(\sigma_x - \sigma_y)$. Thus, if β_p were substituted into Eq. 12.31 one would compute the two principal normal stresses. In this case it may be easier to use the classical form that

$$\sigma_p = \frac{\sigma_x + \sigma_y}{2} \pm \left[\left(\frac{\sigma_x - \sigma_y}{2}\right)^2 + \tau_{xy}^2 \right]^{\frac{1}{2}}.$$

However, to illustrate the use of Eq. 12.31 we will use the results of the previous two element plane stress examples to find the maximum normal stress at the second element centroid. Then $\mathrm{Tan}(2\beta_p) = 2(20.17)/(179.83 - 20.17) = 0.2017$ so $\beta_p = 5.70°$, $\mathrm{Cos}\,\beta_p = 0.995$, $\mathrm{Sin}\,\beta_p = 0.099$, and the transformation is

$$\begin{Bmatrix} \sigma_n \\ \sigma_s \\ \tau_{ns} \end{Bmatrix} = \begin{bmatrix} 0.9901 & 0.0099 & 0.1977 \\ 0.0099 & 0.9901 & -0.1977 \\ -0.0989 & 0.0989 & 0.9803 \end{bmatrix} \begin{Bmatrix} 179.83 \\ -20.17 \\ 20.17 \end{Bmatrix}$$

or

$$\boldsymbol{\sigma}^T_{(ns)} = [\ 181.84 \quad -22.18 \quad -0.00\ ]\ \mathrm{N/m^2}.$$

The maximum shear stress is $\tau_{\max} = (\sigma_n - \sigma_s)^2/2 = 102.01\ \mathrm{N/m^2}$. These shear stresses occur on planes located at $(\beta_p \pm 45°)$. The classical form for $\tau_{\max}$ for two-dimensional problems is $\tau_{\max}^2 = [(\sigma_x - \sigma_y)/2]^2 + \tau_{xy}^2$.

The above example is not finished at this point. In practice, we probably would want to check the failure criterion for this material, and obtain an error estimate to begin an adaptive solution. There are many failure criteria. The three most common ones are the Maximum Principal Stress, the Maximum Shear Stress, and the Von Mises Strain Energy of Distortion. The latter is most common for ductile materials. It can be expressed in terms of a scalar measure known as the *Effective Stress*, σ_E:

$$\sigma_E = \frac{1}{\sqrt{2}} \sqrt{(\sigma_x - \sigma_y)^2 + (\sigma_x - \sigma_z)^2 + (\sigma_y - \sigma_z)^2 + 6(\tau_{xy}^2 + \tau_{xz}^2 + \tau_{yz}^2)}$$

$$\sigma_E = \frac{1}{\sqrt{2}} \sqrt{(\sigma_1 - \sigma_2)^2 + (\sigma_1 - \sigma_3)^2 + (\sigma_2 - \sigma_3)^2}$$

```
SUBROUTINE ELASTIC_B_AXISYMMETRIC (DGH, R, B)                          ! 1
! * * * * * * * * * * * * * * * * * * * * * * * * * * * * * * *       ! 2
! AXISYMMETRIC ELASTICITY STRAIN-DISPLACEMENT RELATIONS   (B)          ! 3
! STRESS & STRAIN COMPONENT ORDER: RR, ZZ, RZ, AND TT                  ! 4
! * * * * * * * * * * * * * * * * * * * * * * * * * * * * * * *       ! 5
Use System_Constants ! for DP, N_R_B, N_G_DOF, N_SPACE                 ! 6
Use Elem_Type_Data   ! for LT_FREE, LT_N, H (LT_N)                     ! 7
 IMPLICIT NONE                                                         ! 8
 REAL(DP), INTENT(IN)  :: DGH (N_SPACE, LT_N)          ! Gradients     ! 9
 REAL(DP), INTENT(IN)  :: R                            ! Radius        !10
 REAL(DP), INTENT(OUT) :: B   (N_R_B, LT_N * N_G_DOF) ! Strains        !11
 INTEGER :: J, K, L                                                    !12
                                                                       !13
! B       = STRAIN-DISPLACEMENT MATRIX (RETURNED)                      !14
! DGH     = GLOBAL DERIVATIVES OF H                                    !15
! H       = ELEMENT INTERPOLATION FUNCTIONS                            !16
! LT_N    = NUMBER OF NODES PER ELEMENT TYPE                           !17
! N_G_DOF = NUMBER OF PARAMETERS PER NODE = 2 here (U & V)             !18
! N_R_B   = NUMBER OF STRAINS (ROWS IN B) = 4: XX, YY, XY, HOOP        !19
! N_SPACE = DIMENSION OF SPACE = 2 here                                !20
! R,Z,T    DENOTE RADIAL, AXIAL, CIRCUMFERENCE                         !21
                                                                       !22
  B = 0.d0                                                             !23
  DO J = 1, LT_N                     ! ROW NUMBER                      !24
    K = N_G_DOF * (J - 1) + 1              ! FIRST COLUMN,  U          !25
    L = K + 1                              ! SECOND COLUMN, V          !26
                                                                       !27
    B (1, K) = DGH (1, J)                  ! DU/DX  FOR XX NORMAL      !28
    B (3, K) = DGH (2, J)                  ! DU/DY  FOR XY SHEAR       !29
    IF ( R <= 0.d0 ) STOP 'R=0, IN ELASTIC_B_AXISYMMETRIC'             !30
    B (4, K) = H (J) / R                   ! U/R HOOP, ZZ NORMAL       !31
    B (2, L) = DGH (2, J)                  ! DV/DY  FOR YY NORMAL      !32
    B (3, L) = DGH (1, J)                  ! DV/DX  FOR XY SHEAR       !33
  END DO                                                               !34
END SUBROUTINE ELASTIC_B_AXISYMMETRIC                                  !35
```

Figure 12.15 *Axisymmetric strain-displacement matrix*

in terms of the stress tensor components and principal stresses, respectively. For yielding in a simple tension test, $\sigma_x = \sigma_{yield}$, and all the other stresses are zero. Then, the effective stress becomes $\sigma_E = \sigma_{yield}$ which implies failure. This is the general test for ductile materials. For brittle materials, one may use the maximum stress criteria where failure occurs at $\sigma_1 = \sigma_{yield}$. The Tresca maximum shear stress criteria is also commonly used. With it failure occurs at $\tau_{max} = \sigma_{yield}/2$. For the plane stress state, all the z-components of the stress tensor are zero. However, in the state of plane strain, σ_z is not zero and must be recovered using the Poisson ratio effect. For an isotropic material (without an initial thermal strain) the result is $\sigma_z = \nu(\sigma_x + \sigma_y)$.

12.6 Axisymmetric solid stress *

There is another common elasticity problem class that can also be formulated as a two-dimensional problem involving two unknown displacement components. It is an axisymmetric solid subject to axisymmetric loads and axisymmetric supports. That is, the geometry, properties, loads, and supports do not have any variation around the circumference of the solid. The problem is usually discussed in terms of axial and radial position, and axial and radial displacements. The solid is defined by the shape in the radial-axial plane as it is completely revolved about the axis. Let (r, z) denote the coordinates in the plane of revolution, and u, v denote the corresponding radial and axial displacements at any point. This is an extension of the methods in Chapter 8 in that we now allow changes in the axial, z, direction. The axisymmetric solid has four stress and strain components, three of them the same as those in the state of plane stress. We simply replace the x, y subscripts with r, z.

The fourth strain is the so-called *hoop strain.* It arises because the material around the circumference changes length as it moves radially. The circumferential strain at a radial position, r, is defined in Fig. 8.10 as

$$\varepsilon_\Theta = \frac{\Delta L}{L} = \frac{2\pi(r+u) - 2\pi r}{2\pi r} = \frac{u}{r}.$$

This is a normal strain, and it is usually placed after the other two normal strains. Note that on the axis of revolution, $r = 0$. It can be shown that both $u = 0$ and $\varepsilon_\Theta = 0$ on the axis of revolution. However, one can encounter numerical problems if numerical integration is employed with a rule that has quadrature points on the edge of an element.

We typically order the strains as $\boldsymbol{\varepsilon}^T = [\,\varepsilon_r \;\; \varepsilon_z \;\; \varepsilon_\Theta \;\; \gamma_{rz}\,]$ and the corresponding stresses as $\boldsymbol{\sigma} = [\,\sigma_r \;\; \sigma_z \;\; \sigma_\Theta \;\; \tau_{rz}\,]$ where σ_Θ is the corresponding *hoop stress.* From the above definition we now see that the contribution of a typical node j to the strain-displacement relation for the matrix partition $\mathbf{B}_j^e = \mathbf{L}\,\mathbf{N}^e$ is given correctly near the beginning of Section 12.2. Therefore, we now see that in addition to the physical derivatives of the interpolation functions, we now must also include the actual interpolation functions (for the u contribution) as well as the radial coordinate ($r = \mathbf{G}^e\,\mathbf{r}^e$). Of course, the matrix is usually evaluated at a quadrature point, as in line 33 of Fig. 12.10. The implementation of the axisymmetric version is shown in Fig. 12.15, where it is assumed that the radial position is already known.

With the above changes and the observation that $d\,\Omega = 2\pi\, r\, da$, we note that the analytic integrals involve terms with $1/r$. These introduce logarithmic terms where we used to have only polynomial terms from exact analytic integration. Some of these become indeterminate at $r = 0$. For this and other practical considerations, one almost always employs numerical integration to form the element matrices. Clearly, one must interpolate from the given data to find the radial coordinate, r_q, at a quadrature point. That requires that we need to select quadrature rules that do not give points on the element boundary which does occur for some triangular rules and the Lobatto rules.

12.7 General solid stress *

For the completely general three-dimensional solid, there are three displacement components, $\mathbf{u}^T = \lfloor u \quad v \quad w \rfloor$, and the corresponding load vectors, at each node, have three components. There are six stresses, $\boldsymbol{\sigma}^T = [\, \sigma_x \;\; \sigma_y \;\; \sigma_z \;\; \tau_{xy} \;\; \tau_{xz} \;\; \tau_{yz} \,]$, and six corresponding strain components, $\boldsymbol{\varepsilon}^T = [\, \varepsilon_x \;\; \varepsilon_y \;\; \varepsilon_z \;\; \gamma_{xy} \;\; \gamma_{xz} \;\; \gamma_{yz} \,]$, and therefore the constitutive array $\mathbf{E}$ is 6 by 6 in size. The enginering strain-displacement relations are defined by a partition at node j as shown earlier in Section 12.2.

12.8 Anisotropic materials *

A material is defined to be *isotropic* if its material properties do not depend on direction. Otherwise it is called *anisotropic.* Most engineering materials are considered to be isotropic. However, there are many materials that are anisotropic. Examples of anisotropic materials include plywood, and filament wound fiber-glass. Probably the most common case is that of an *orthotropic material.* An orthotropic material has structural (or thermal) properties that can be defined in terms of two principal material axis directions. Let (n, s) be the principal material axis directions. For anisotropic materials it is usually easier to define the generalized compliance law in the form:

$$\boldsymbol{\varepsilon}_{(ns)} = \mathbf{E}^{-1}_{(ns)} \boldsymbol{\sigma}_{(ns)} + \boldsymbol{\varepsilon}_{0\,(ns)}, \tag{12.38}$$

where the inverse of the constitutive matrix, $\mathbf{E}^{-1}_{(ns)}$, is usually called the compliance matrix. Often the values in the compliance matrix can be determined from material experiments and that matrix is numerically inverted to form $\mathbf{E}_{(ns)}$. Note by way of comparison that Eq. 12.38 is written relative to the global coordinate axes. In Eq. 12.14 the square matrix contains the mechanical properties as experimentally measured relative to the principal material directions. For a two-dimensional orthotropic material the constitutive law is

$$\begin{Bmatrix} \varepsilon_n \\ \varepsilon_s \\ \gamma_{ns} \end{Bmatrix} = \begin{bmatrix} 1/E_n & -\nu_{sn}/E_s & 0 \\ -\nu_{ns}/E_n & 1/E_s & 0 \\ 0 & 0 & 1/G_{ns} \end{bmatrix} \begin{Bmatrix} \sigma_n \\ \sigma_s \\ \tau_{ns} \end{Bmatrix} + \boldsymbol{\varepsilon}_{0\,ns}. \tag{12.39}$$

Here the moduli of elasticity in the two principal directions are denoted by E_n and E_s. The shear modulus, G_{ns}, is independent of the elastic moduli. The two Poisson's ratios are defined by the notation:

$$\nu_{ij} = -\,\varepsilon_j \,/\, \varepsilon_i \tag{12.40}$$

where i denotes the direction of the load, ε_i is the normal strain in the load directions, and ε_j is the normal strain in the transverse (orthogonal) direction. Symmetry considerations require that

$$E_n \nu_{sn} = E_s \nu_{ns}. \tag{12.41}$$

Thus, four independent constants must be measured to define the orthotropic material mechanical properties. If the material is isotropic then $\nu = \nu_{ns} = \nu_{sn}$, $E = E_s = E_n$, and $G = G_{ns} = E/[2(1+\nu)]$. In that case only two constants (E and ν) are required and they can be measured in any direction. When the material is isotropic then the inverse of Eq. 12.39 reduces the plane stress model given at the beginning of Sec. 12.2.

The axisymmetric stress-strain law for an isotropic material was also given earlier. For an orthotropic axisymmetric material, we utilize the material properties in the principal material axis (n, s, Θ) direction. In that case, the compliance matrix, $\mathbf{E}_{ns}^{-1}$, and initial strain matrix, $\boldsymbol{\varepsilon}_{0(ns)}$ are

$$\begin{Bmatrix} \varepsilon_n \\ \varepsilon_s \\ \gamma_{ns} \\ \varepsilon_\Theta \end{Bmatrix} = \begin{bmatrix} \dfrac{1}{E_n} & \dfrac{-\nu_{sn}}{E_s} & \dfrac{-\nu_{\Theta n}}{E_\Theta} & 0 \\ \dfrac{-\nu_{ns}}{E_n} & \dfrac{1}{E_s} & \dfrac{-\nu_{\Theta s}}{E_\Theta} & 0 \\ \dfrac{-\nu_{n\Theta}}{E_n} & \dfrac{-\nu_{s\Theta}}{E_s} & \dfrac{1}{E_\Theta} & 0 \\ 0 & 0 & 0 & \dfrac{1}{G_{ns}} \end{bmatrix} \begin{Bmatrix} \sigma_n \\ \sigma_s \\ \tau_{ns} \\ \sigma_\Theta \end{Bmatrix} + \Delta\theta \begin{Bmatrix} \alpha_n \\ \alpha_s \\ 0 \\ \alpha_\Theta \end{Bmatrix}.$$

For the general anisotropic three-dimensional solid, there are nine independent material constants. However, due to the axisymmetry, there are only seven independent constants. When the material is also *transversely isotropic,* with the n - t plane being the plane of isotropic properties, then there are only five material constants and they are related by $E_n = E_\Theta$, $\nu_{n\Theta} = \nu_{\Theta n}$, and $E_\Theta \nu_{s\Theta} = E_s \nu_{ns}$. In practical design problems with anisotropic materials, it is difficult to get accurate material constant measurements. To be a physically possible material, both $\mathbf{E}$ and $\mathbf{E}^{-1}$ should have a positive determinant. When that is not the case, the program should issue a warning and terminate the analysis. The possible values of an orthotropic Poisson's ratio can be bounded by

$$|\nu_{sn}| < \sqrt{E_s/E_n}, \qquad -1 < \nu_{n\Theta} < 1 - 2E_n \nu_{sn}^2/E_s.$$

Orthotropic materials also have thermal properties that vary with direction. If $\Delta\theta$ denotes the temperature change from the stress free state then the local initial thermal strain is $\boldsymbol{\varepsilon}_{0(ns)}^T = \Delta\theta\,[\alpha_n \quad \alpha_s \quad 0]$ where α_n and α_s are the principal coefficients of thermal expansion. If one is given the orthotropic properties it is common to numerically invert $\mathbf{E}_{(ns)}^{-1}$ to give the form

$$\boldsymbol{\sigma}_{(ns)} = \mathbf{E}_{(ns)}(\boldsymbol{\varepsilon}_{(ns)} - \boldsymbol{\varepsilon}_{0(ns)}). \tag{12.42}$$

This is done since the form of $\mathbf{E}_{(ns)}$ is algebraically much more complicated than its inverse. Due to experimental error in measuring the anisotropic constants there is a potential difficulty with this concept. For a physically possible material it can be shown that both $\mathbf{E}$ and $\mathbf{E}^{-1}$ must be positive definite. This means that the determinant must be greater than zero. Due to experimental error it is not unusual for this condition to be violated. When this occurs the program should be designed to stop and require acceptable data. Then the user must adjust the experimental data to satisfy the condition that $(E_n - E_s \nu_{ns}^2) > 0$.

From the previous section on stress and strain transformations we know how to obtain $\boldsymbol{\varepsilon}_{(xy)}$ and $\boldsymbol{\sigma}_{(xy)}$ from given $\boldsymbol{\varepsilon}_{(ns)}$ and $\boldsymbol{\sigma}_{(ns)}$. But how do we obtain $\mathbf{E}_{(xy)}$ from $\mathbf{E}_{(ns)}$? We must have $\mathbf{E}_{(xy)}$ to form the stiffness matrix since it must be integrated relative to the x-y axes, that is,

$$\mathbf{K}_{(xy)} = \int \mathbf{B}^T_{(xy)}\, \mathbf{E}_{(xy)}\, \mathbf{B}_{(xy)}\, d\Omega.$$

Thus, the use of the $n - s$ axes, in Fig. 12.6, to define (input) the material properties requires that we define one more transformation law. It is the transformation from $\mathbf{E}_{(ns)}$ to $\mathbf{E}_{(xy)}$. There are various ways to derive the required transformation. One simple procedure is to note that the strain energy density is a scalar. Therefore, it must be the same in all coordinate systems. The strain energy density at a point is $dU = \boldsymbol{\sigma}^T \boldsymbol{\varepsilon}/2 = \boldsymbol{\varepsilon}^T \boldsymbol{\sigma}/2$. In the global axes it is

$$dU = \tfrac{1}{2}\, \boldsymbol{\sigma}^T_{(xy)}\, \boldsymbol{\varepsilon}_{(xy)} = \tfrac{1}{2}\, (\mathbf{E}_{(xy)}\, \boldsymbol{\varepsilon}_{(xy)})^T\, \boldsymbol{\varepsilon}^T_{(xy)} = \tfrac{1}{2}\, \boldsymbol{\varepsilon}^T_{(xy)}\, \mathbf{E}_{(xy)}\, \boldsymbol{\varepsilon}_{(xy)}.$$

In the principal material directions it is

$$dU = \tfrac{1}{2}\, \boldsymbol{\sigma}^T_{(ns)}\, \boldsymbol{\varepsilon}_{(ns)} = \tfrac{1}{2}\, (\mathbf{E}_{(ns)}\, \boldsymbol{\varepsilon}_{(ns)})^T\, \boldsymbol{\varepsilon}_{(ns)} = \tfrac{1}{2}\, \boldsymbol{\varepsilon}^T_{(ns)}\, \mathbf{E}_{(ns)}\, \boldsymbol{\varepsilon}_{(ns)}.$$

But from our Mohr's circle transformation for strain $\boldsymbol{\varepsilon}_{(ns)} = \mathbf{t}_{(ns)}\, \boldsymbol{\varepsilon}_{(xy)}$ so in the $n - s$ axes

$$dU = \tfrac{1}{2}\, (\mathbf{t}_{(ns)}\, \boldsymbol{\varepsilon}_{(xy)})^T\, \mathbf{E}_{(ns)} (\mathbf{t}_{(ns)}\, \boldsymbol{\varepsilon}_{(xy)}) = \tfrac{1}{2}\, \boldsymbol{\varepsilon}^T_{(xy)}\, (\mathbf{t}^T_{(ns)}\, \mathbf{E}_{(ns)}\, \mathbf{t}_{(ns)})\, \boldsymbol{\varepsilon}_{(xy)}. \tag{12.43}$$

Comparing the two forms of dU gives the *constitutive transformation* law that

$$\mathbf{E}_{(xy)} = \mathbf{t}^T_{(ns)}\, \mathbf{E}_{(ns)}\, \mathbf{t}_{(ns)}. \tag{12.44}$$

The same concept holds for general three-dimensional problems.

Before leaving the concept of anisotropic materials we should review the initial thermal strains. Recall that for an isotropic material or for an anisotropic material in principal axes a change in temperature does not induce an initial shear strain. However, an anisotropic material does have initial thermal shear strain in other coordinate directions. From Eqs. 12.36 and 12.37 we have

$$\boldsymbol{\varepsilon}_{0\,(xy)} = \mathbf{t}^{-1}_{(ns)}\, \boldsymbol{\varepsilon}_{0\,(ns)}, \qquad \begin{Bmatrix} \varepsilon^0_x \\ \varepsilon^0_y \\ \gamma^0_{xy} \end{Bmatrix} = \begin{bmatrix} +C^2 & +S^2 & -SC \\ +S^2 & +C^2 & +SC \\ +2SC & -2SC & (C^2 - S^2) \end{bmatrix} \begin{Bmatrix} \varepsilon^0_n \\ \varepsilon^0_s \\ 0 \end{Bmatrix}.$$

Thus, the thermal shear strain is $\gamma^0_{xy} = 2\,\text{Sin}\,\beta\,\text{Cos}\,\beta\,(\varepsilon^0_n - \varepsilon^0_s)$. This is not zero unless the two axes systems are the same ($\beta = 0$ or $\beta = \pi/2$). Therefore, one must replace the previous null terms in Eq. 12.17.

12.9 Circular hole in an infinite plate

A classical problem in elasticity is that of a two-dimensional solid having a traction free hole of radius a at its center and symmetrically loaded by a uniform stress of $\sigma_x = \sigma_\infty$ along the lines of $x = \pm\infty$. The analytic solution for the stresses and the displacements are known. Let θ be the angle from the x-axis and $r \geq a$ the radius from the center point. Then the displacement components are

$$u_x = \frac{\sigma_\infty a}{8G}\left[\frac{r}{a}(\kappa + 1)\,\text{Cos}\,\theta + 2\frac{a}{r}((1+\kappa)\,\text{Cos}\,\theta + \text{Cos}\,3\theta) - 2\frac{a^3}{r^3}\,\text{Cos}\,3\theta\right]$$

$$u_y = \frac{\sigma_\infty a}{8G}\left[\frac{r}{a}(\kappa - 3)\,\text{Sin}\,\theta + 2\frac{a}{r}((1-\kappa)\,\text{Sin}\,\theta + \text{Sin}\,3\theta) - 2\frac{a^3}{r^3}\,\text{Sin}\,3\theta\right]$$

where $G = E/(1 - 2\nu)$ is the shear modulus, ν is Poisson's ratio, E is the elastic modulus and $\kappa = (3 - 4\nu)$ for plain strain or $\kappa = (3 - \nu)/(1 + \nu)$ for plane stress. Note that the displacements depend on the material properties (E, ν), but the stresses given below do not depend on the material but depend on the geometry only. That is a common situation in solid mechanics and this serves as a reminder that one must always validate both the displacements and stresses before accepting the results of a stress analysis problem. We can use the above displacement components as essential boundary conditions on the boundary of a finite domain to compare a finite element solution to the exact results as one way to validate the program for stress analysis or error estimation.

As an aside remark note that the *MODEL* program allows the keyword *use_exact_bc* to employ a supplied analytic solution to override any user supplied essential boundary conditions. First one must identify the *exact_case* number assigned to the corresponding part of the exact source code library. Here for the circular hole in an infinite plate the value is 23. In this specific example the exact boundary conditions depend on material properties and the geometry. Therefore one must also supply data to the analytic solution by using the keyword *exact_reals* 3 and then supply those data (E, ν, a) immediately after the user remarks and before the usual finite element nodal data. (Most examples in this book that compare to analytic solutions do not have that additional data requirement since they usually depend only on location.)

In other words, if so desired one can generate a mesh that will use an analytic solution but it is only necessary to give data on where the essential boundary conditions occur and supply dummy null values which will be overwritten by the selected analytic solution, if the control keyword *use_exact_bc* is present in the data. If you want to use an analytic solution not built-in to the existing library (*exact_case* 0) then you must supply your own exact source code via the include file *my_exact_inc* (and re-compile).

The corresponding stress components in the infinite plate are

$$\sigma_x = \sigma_\infty\left[1 - \frac{a^2}{r^2}\left(\frac{3}{2}\,\text{Cos}\,2\theta + \text{Cos}\,4\theta\right) + \frac{3}{2}\frac{a^4}{r^4}\,\text{Cos}\,4\theta\right]$$

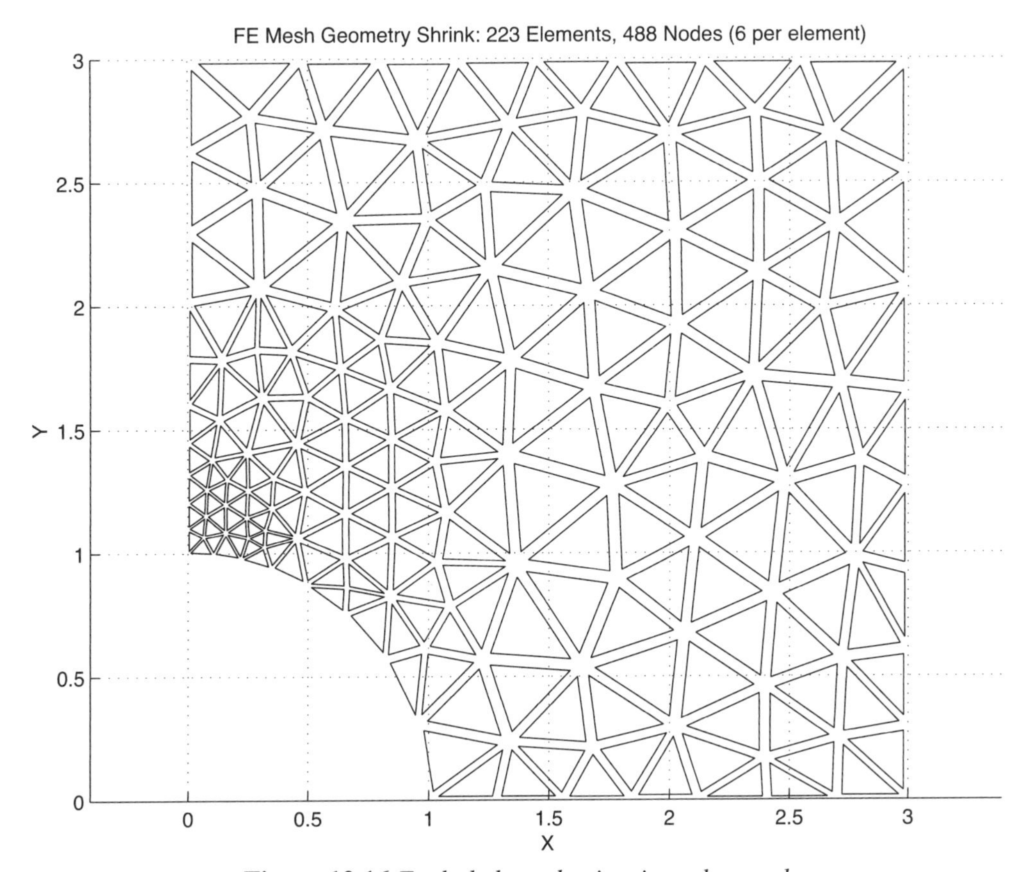

Figure 12.16 *Exploded quadratic triangular mesh*

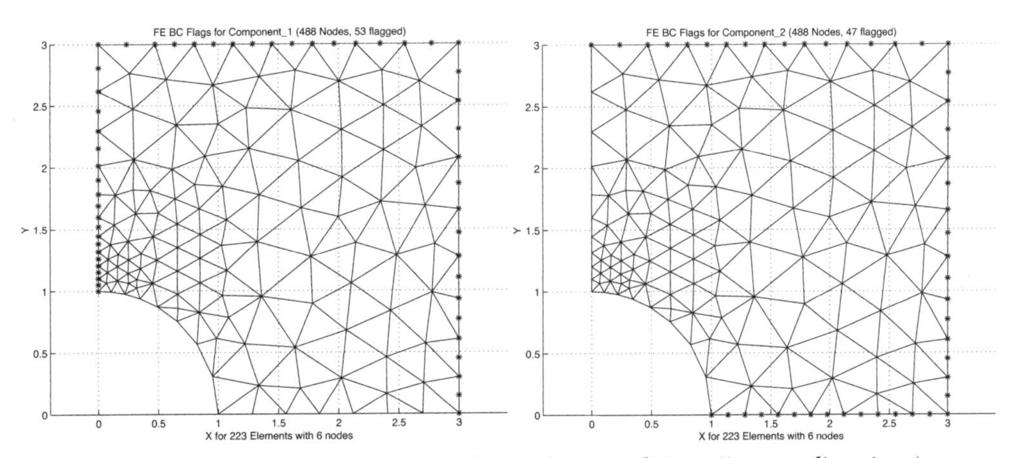

Figure 12.17 *Nodes with essential boundary conditions (in x-, y-directions)*

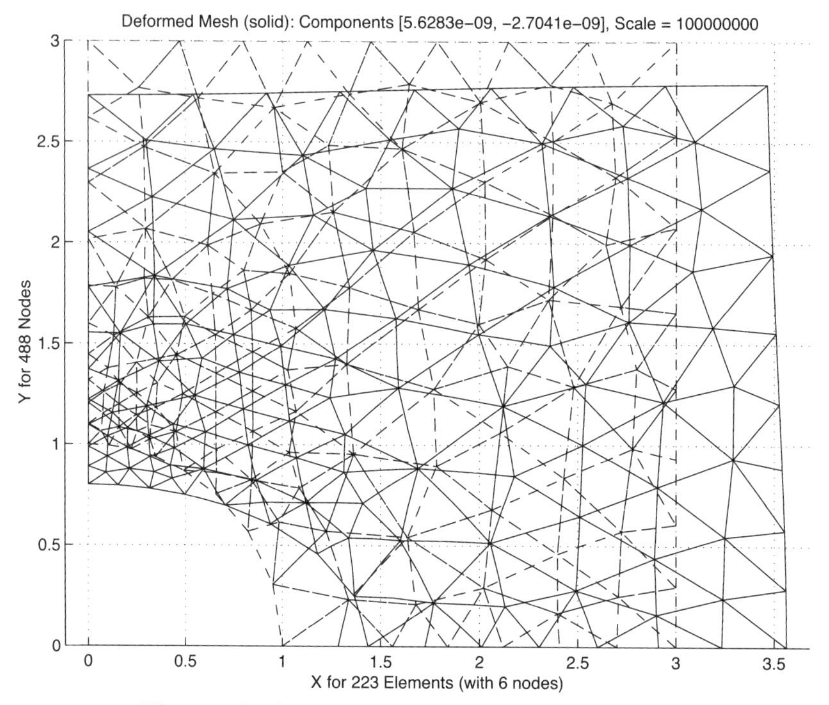

Figure 12.18 *Original and scaled deformed mesh*

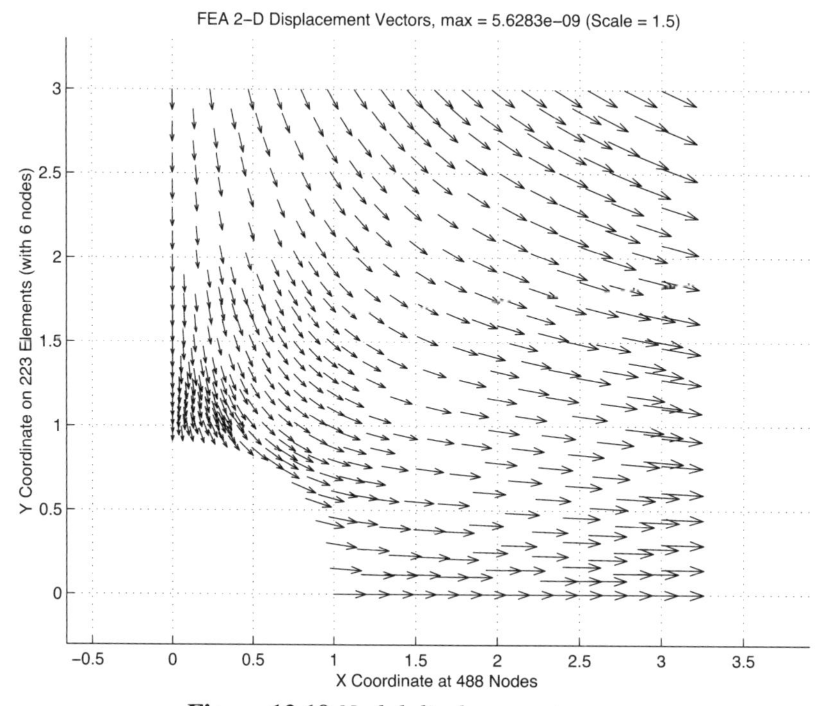

Figure 12.19 *Nodal displacement vectors*

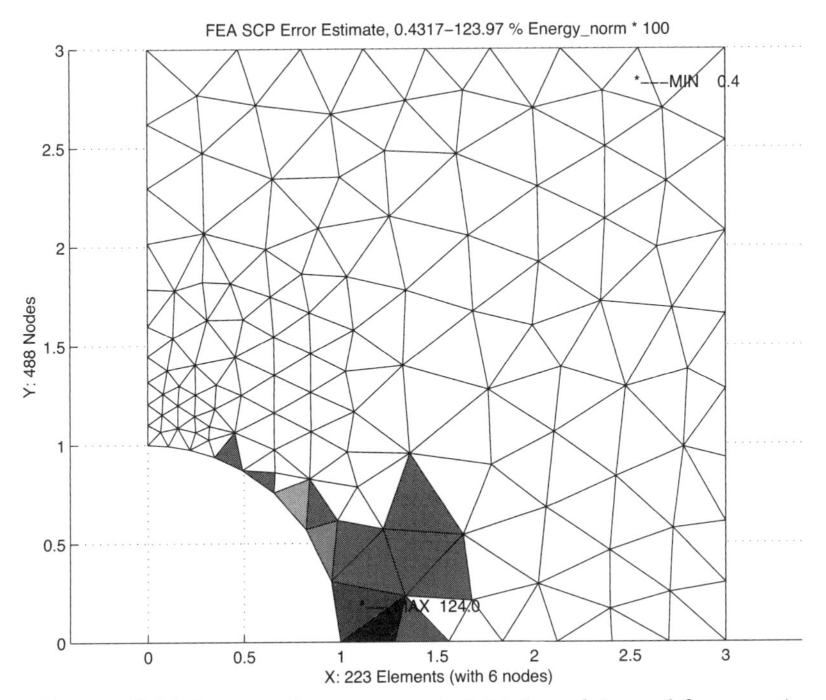

Figure 12.20 *Largest element error in initial mesh (max 1.2 percent)*

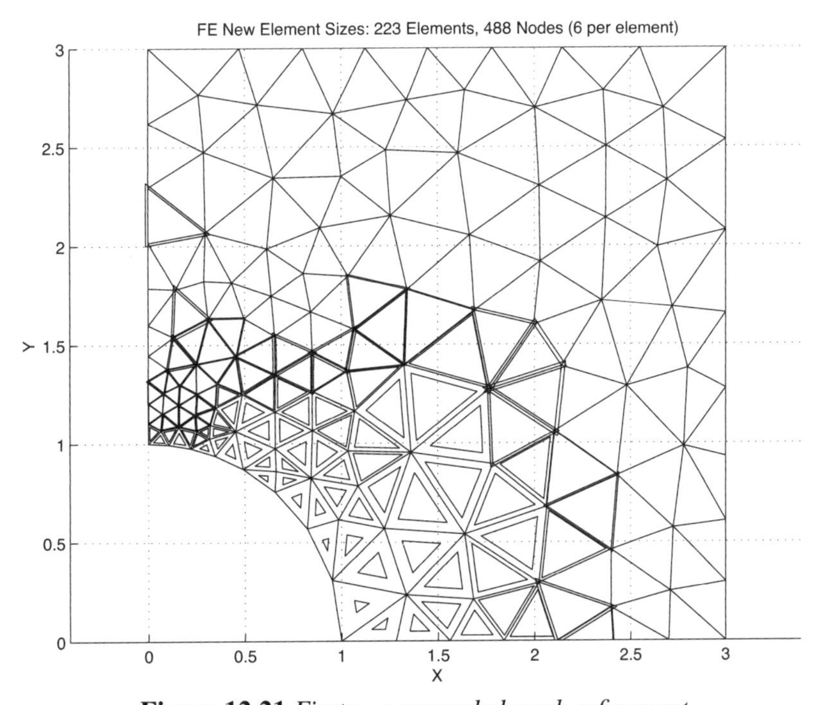

Figure 12.21 *First recommended mesh refinement*

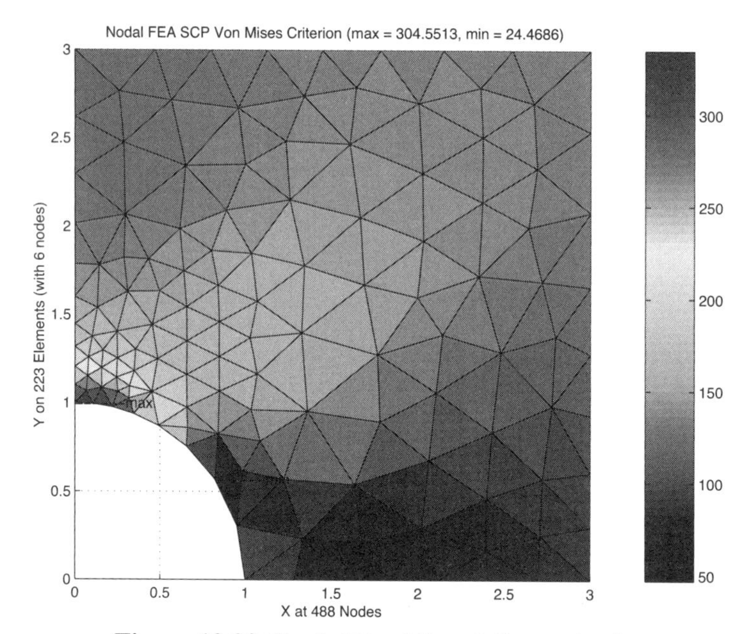

Figure 12.22 *Shaded Von Mises failure criterion*

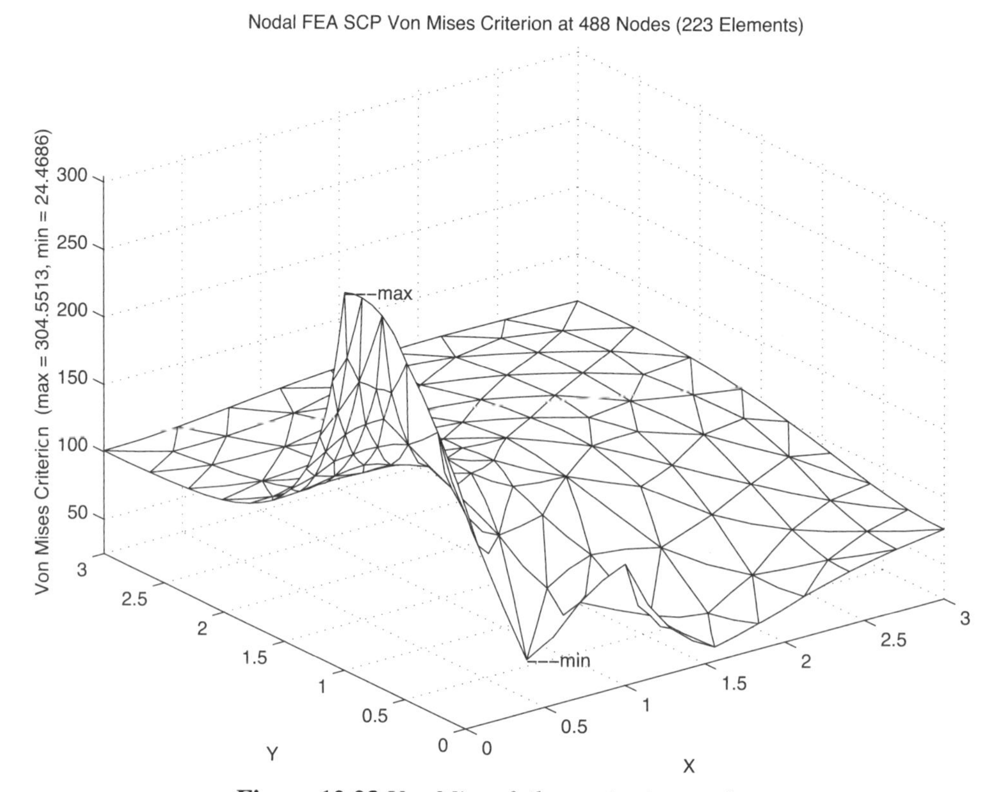

Figure 12.23 *Von Mises failure criterion surface*

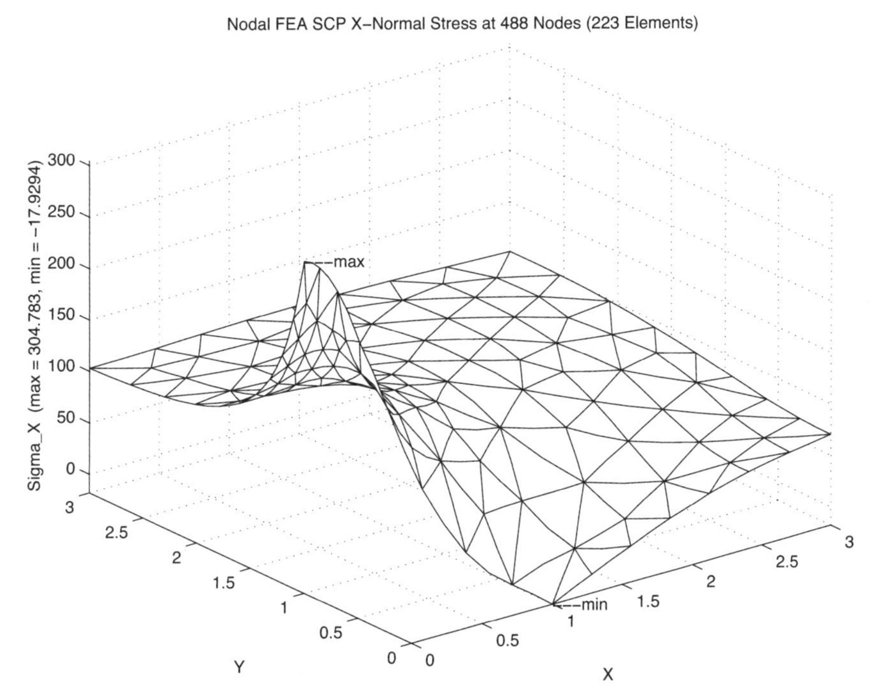

Figure 12.24 *Initial estimate of x-normal stress*

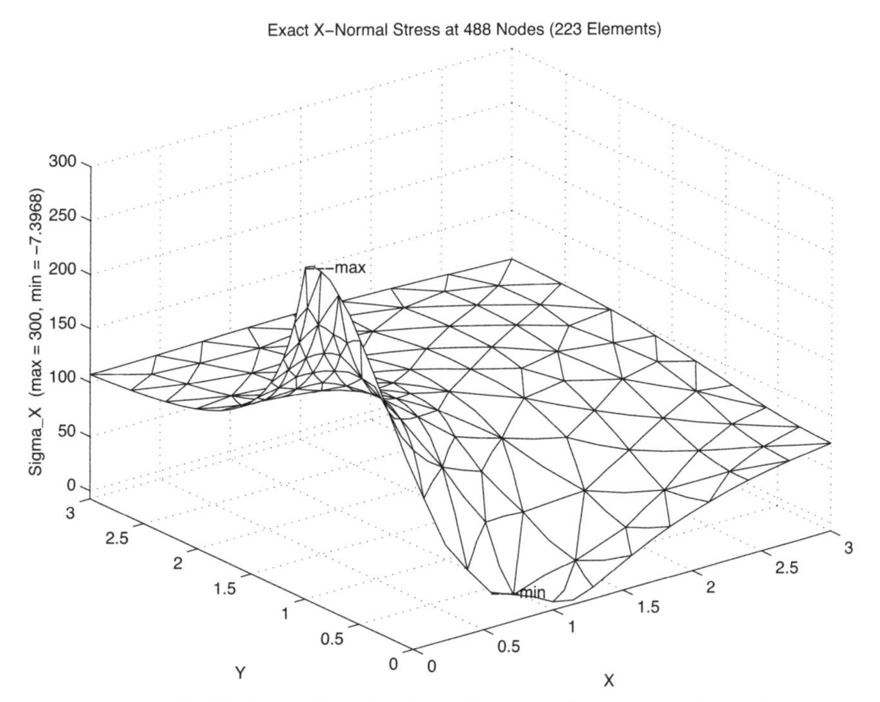

Figure 12.25 *Exact distribution of x-normal stress (at the nodes)*

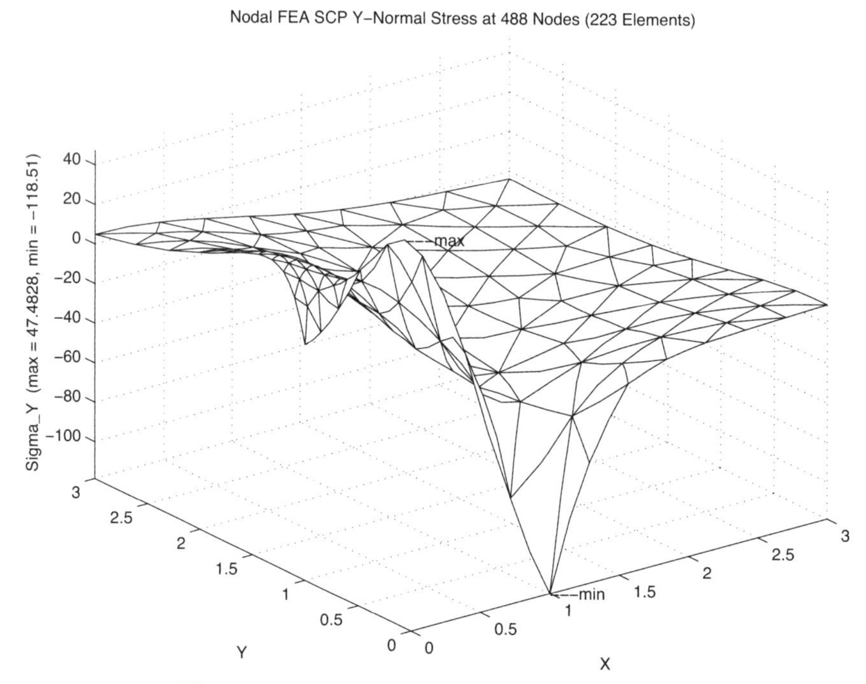

Figure 12.26 *Initial estimate of y-normal stress*

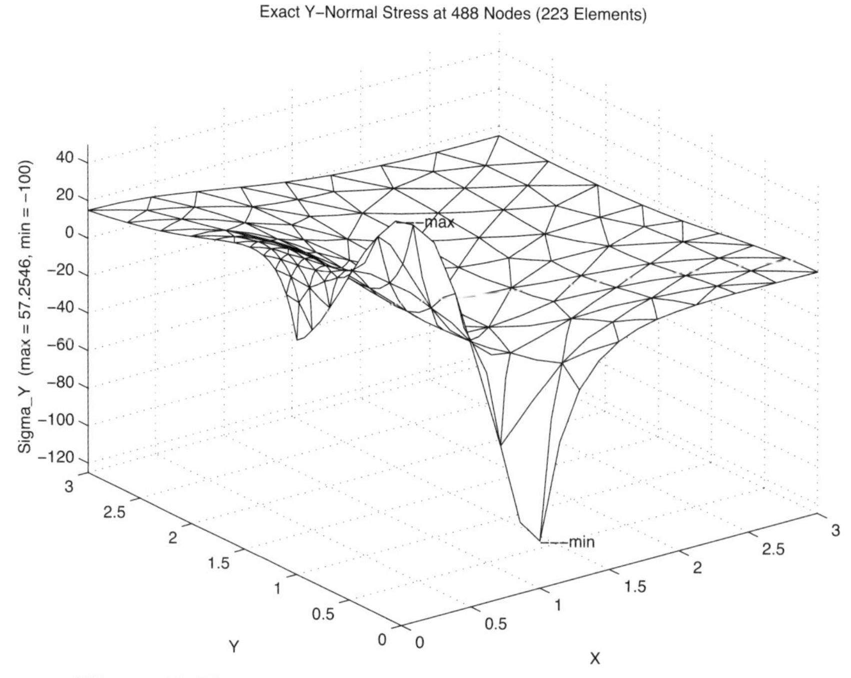

Figure 12.27 *Exact distribution of y-normal stress (at the nodes)*

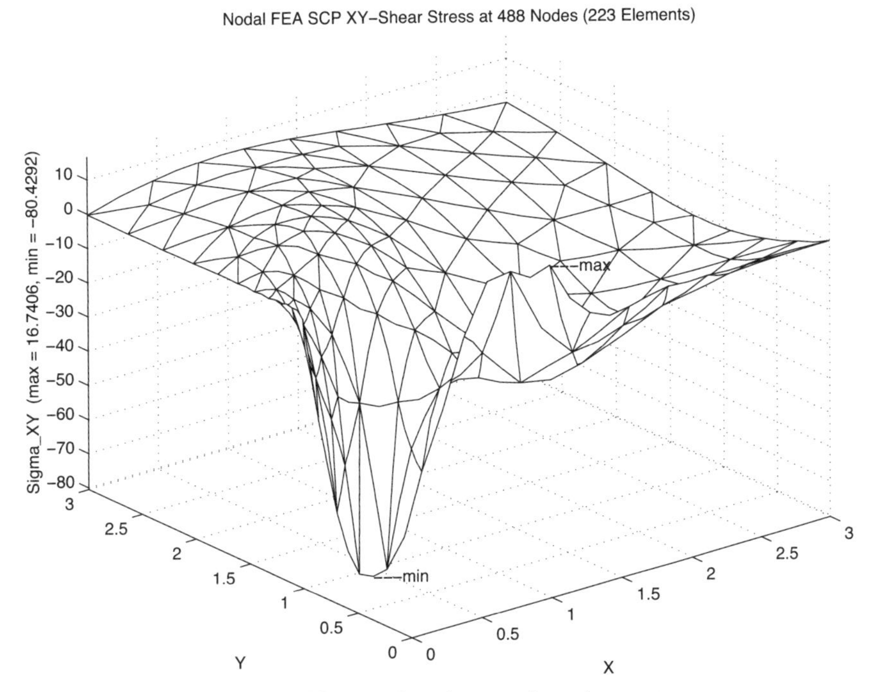

Figure 12.28 *Initial estimate of xy-shear stress*

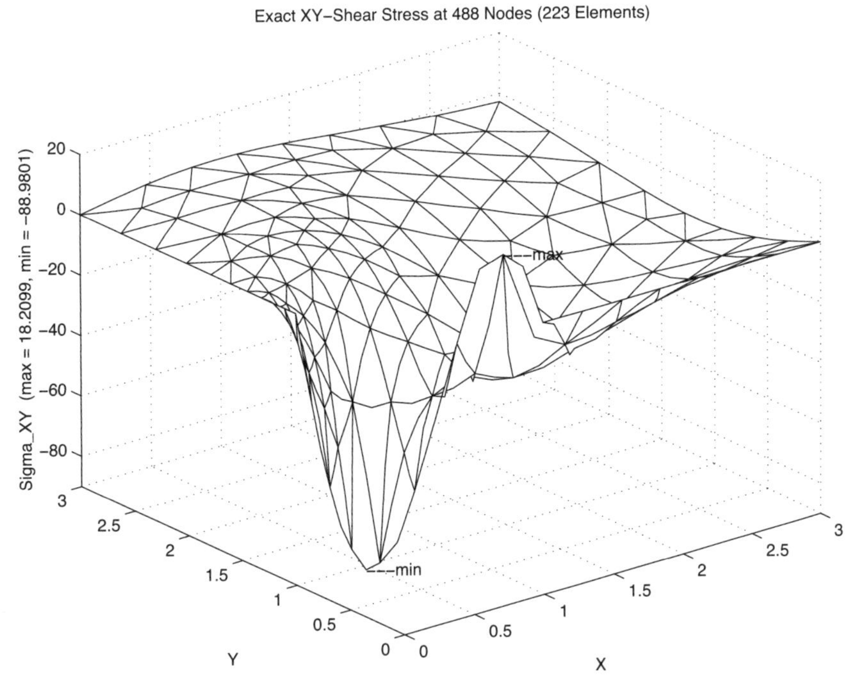

Figure 12.29 *Exact distribution of xy-shear stress (at the nodes)*

$$\sigma_y = \sigma_\infty \left[-\frac{a^2}{r_2}\left(\frac{1}{2}\text{Cos } 2\theta - \text{Cos } 4\theta\right) - \frac{3}{2}\frac{a^4}{r^4}\text{Cos } 4\theta \right]$$

$$\tau_{xy} = \sigma_\infty \left[-\frac{a^2}{r^2}\left(\frac{1}{2}\text{Sin } 2\theta + \text{Sin } 4\theta\right) + \frac{3}{2}\frac{a^4}{r^4}\text{Sin } 4\theta \right].$$

In polar components the radial and shear stresses on the hole surface $r = a$ are both zero, $\sigma_r = 0 = \tau_{r\theta}$ while the circumferential stress varies as $\sigma_\theta = \sigma_\infty(1 - 2\,\text{Cos } 2\theta)$. Thus it (and σ_x) has a maximum value of $3\sigma_\infty$ at $\theta = 90$ degrees. Likewise, it (and σ_y) has a minimum value of $-\sigma_\infty$ at $\theta = 0$.

If we solve a finite rectangular plate with $\sigma_x = \sigma_\infty$ along $x_{\max} >> a$ and $\sigma_y = 0$ along $y_{\max} > a$ then the peak stress values should be similar to those in the above special case. Of course, the maximum stresses will be somewhat higher since the hole takes up a bigger percentage of the center ($x = 0$) section.

Here we will present the common example of the hole in an infinite plate to illustrate finite element stress analysis and the usefulness of an error estimator. We will zoom-in on the region around the hole and use a radius of $a = 1\,m$ and bound the upper right quarter of the domain with a width and height of $h = 3\,m$ each. The material is taken to be aluminum ($E = 70\,GPa, \nu = 0.33$), and we assume a traction of $T_x = 100\,N/m$ at $x = \infty$. The analytic solution will specify both $x-$ and $y-$ displacements on the right and top edges of the mesh. The edges along the axes will invoke symmetry conditions ($v = 0$, and $u = 0$, respectively) which are consistent with the analytic solution. The edge of the circular arc is traction free and has no prescribed displacements. We expect the maximum stress concentration to occur at the top point of the arc so the initial mesh is slightly refined in that region. Here we employ the 6 noded quadratic triangular (T6) elements. The initial exploded view of the mesh is given in Fig. 12.16, while Fig. 12.17 shows an asterisk at nodes where one or more displacements are prescribed.

The scaled deformed geometry and displacements are shown in Figs. 12.18 and 19, respectively. The first error estimate determines that the largest error occurs in the elements shown in Fig. 12.20. They are the relatively large elements around the hole and in the region near the x-axis where we will see secondary stress components are varying very rapidly. The error range is cited in the top caption of that figure. As the mesh is refined we expect that the error will become more uniformly distributed over the mesh. The total system error was 0.118 percent in the enery norm.

The computed mesh refinement is shown in Fig.s 12.20 & 21 and will be passed to the mesh generation code for creating the next analysis. In that figure the estimated new local element size is plotted relative to the centroid of the current element that was used to compute the new size. In most cases one will note a refinement giving a smaller suggested element size. In a few cases a de-refinement gives a larger new element. In most elements the projected change in size is not noticeable.

Before going on to the next mesh stage we can compare the finite element stress estimates to the exact ones. Since the plate is made of a ductile material (aluminum) we begin with the Von Mises failure criterion which gives a positive value that is compared to the yield stress of the material (about 140 MPa). The shaded planar view and carpet

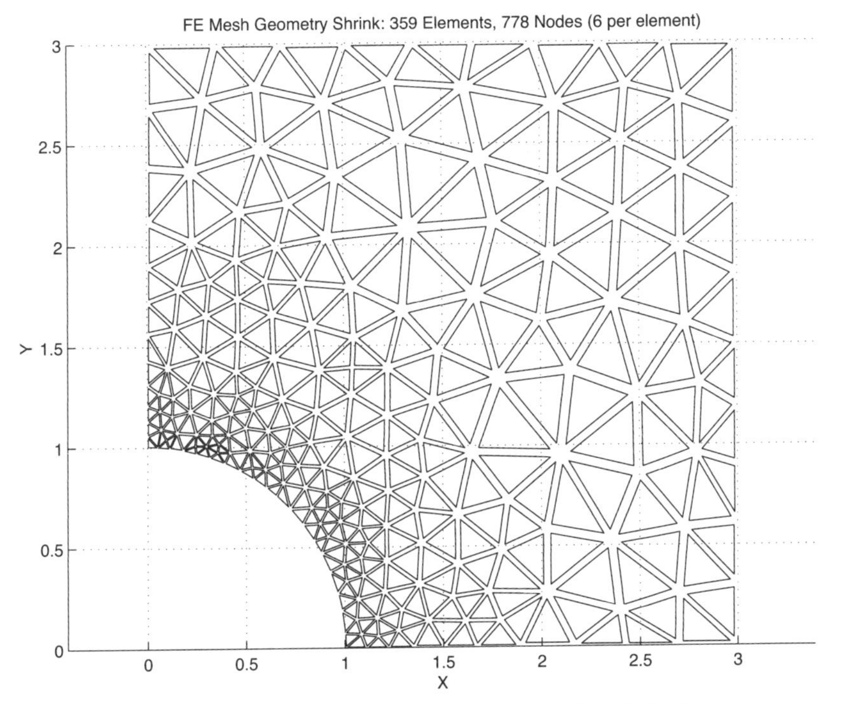

Figure 12.30 *Second mesh generation (exploded)*

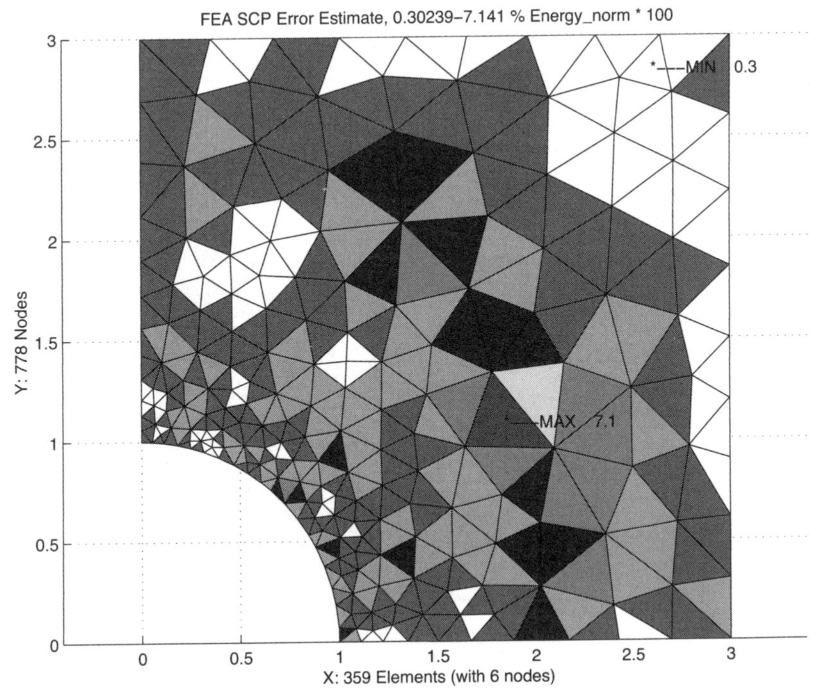

Figure 12.31 *Second error estimate distribution (0.07 percent max)*

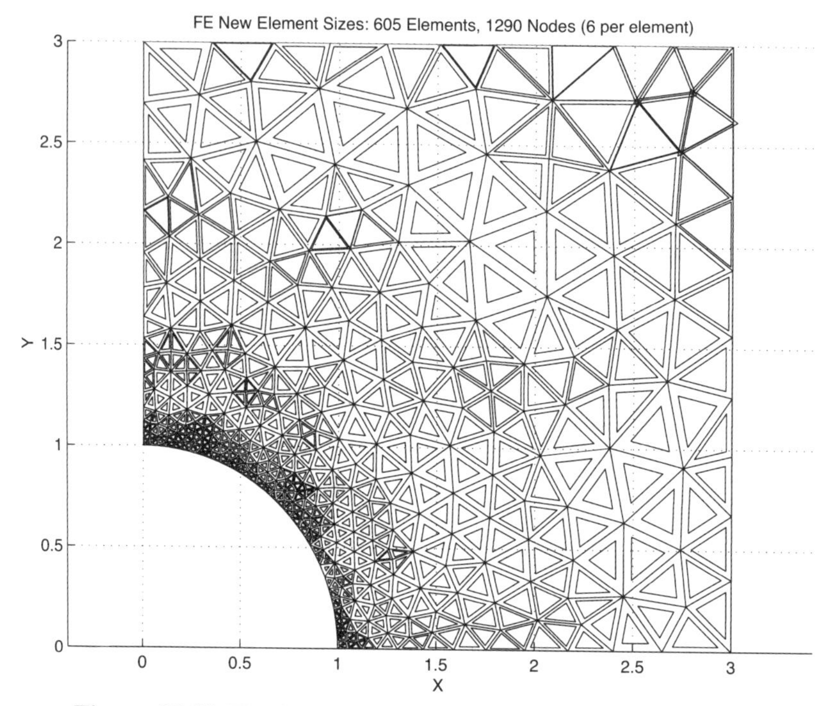

Figure 12.32 *Third mesh generation and its suggested revision*

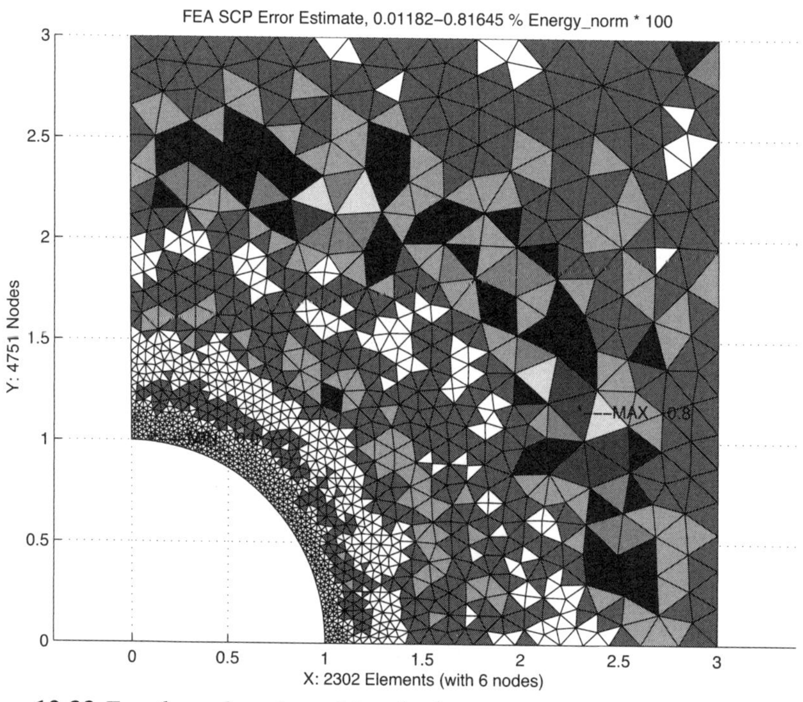

Figure 12.33 *Fourth mesh and resulting final error estimates (0.0001 percent max)*

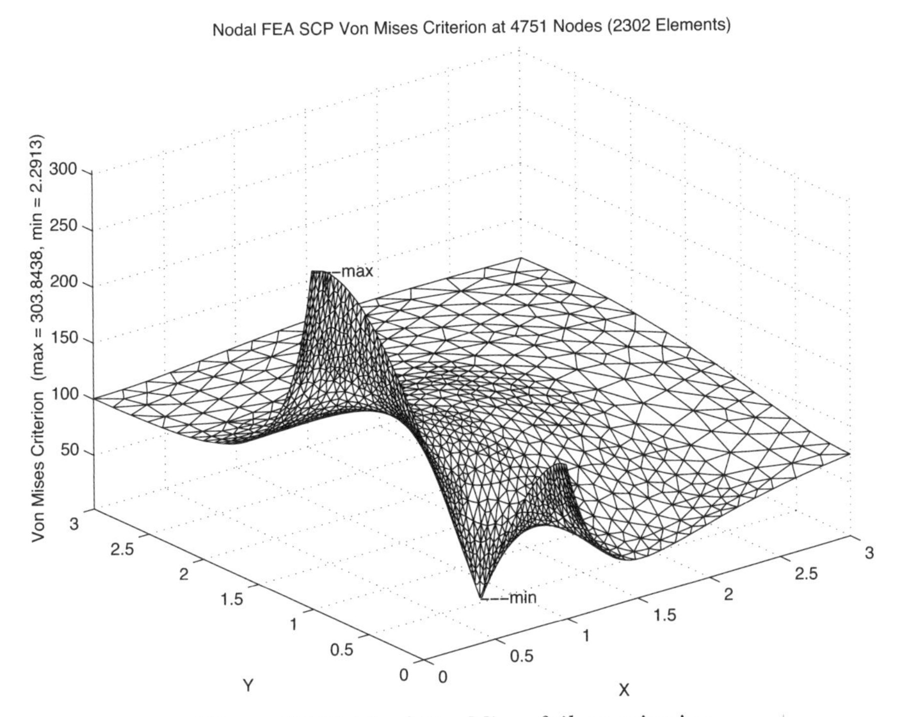

Figure 12.34 *Final Von Mises failure criterion*

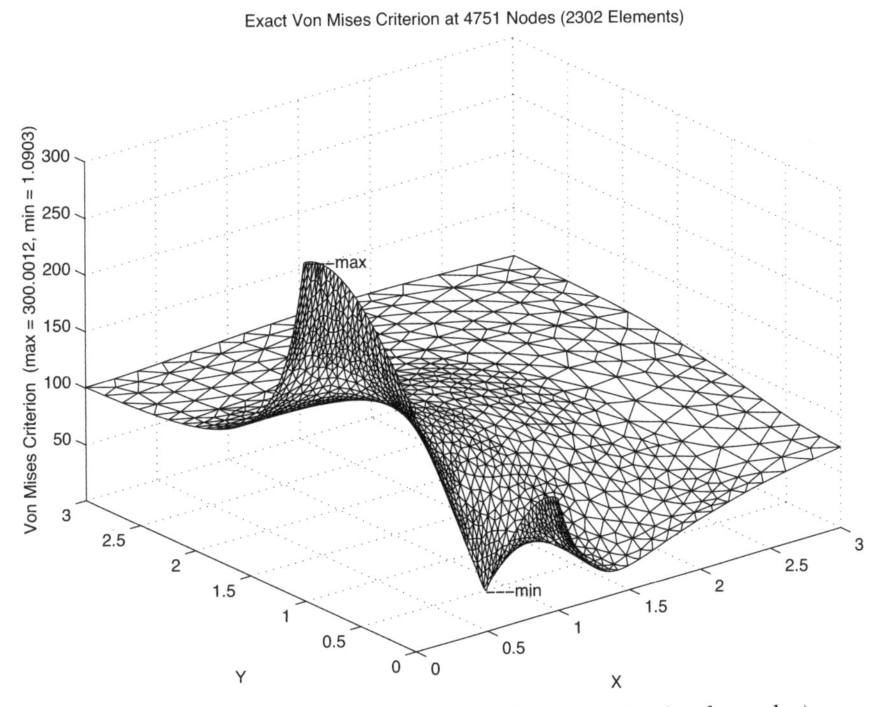

Figure 12.35 *Exact final Von Mises failure criterion (at the nodes)*

plot view of that criterion are shown in Figs. 12.22 and 23, respectively. The maximum value should be about 300.0 MPa, and it is.

The error estimates outlined in Fig. 12.18 are based on the nodal stress averages from the super-convergent patch processing (defined in Chapters 12 and 13). The finite element nodal averages for the horizontal normal stress σ_x are shown in Fig. 12.24 as a carpet plot and should be compared to the exact nodal distribution in Fig. 12.25. The vertical normal stresses, σ_y, finite element and exact values are shown in Figs. 12.26 and 27, respectively. Likewise, the comparisons for the shear stress component, τ_{xy}, are in Figs. 12.28 and 29. In general we see that the spatial distributions are similar but the amplitudes of the finite element stresses are a little high. Hopefully the mesh refinements directed by the error estimates will lead to an efficient and accurate solution for all the stresses.

The first mesh revision shown in Fig. 12.21, leads to the second mesh, in Fig. 12.30, and that in turn leads to a new distribution of error estimates as seen in Fig. 12.31. In the latter figure we see that the error is becoming more dispersed over the mesh. The system energy norm error has been cut almost in half to 0.0669 percent. Likewise the maximum error in a single element has been cut by about 40 percent and its location moved from near the hole edge to the largest interior element. Repeating this process gives the third mesh and its recommended mesh refinement given in Fig. 12.32. The refinement in element sizes is based on the error distribution in Fig. 12.31. There the lowest errors (in the energy norm) occur near the edge of the hole, where we expect the maximum stresses and failure criterion. In the second mesh the maximum single element error has been further reduced and is localized near the plate center where the stresses are low and slowly changing. The system wide error has been reduced to 0.051 percent. The fourth, and final, mesh is shown in Fig. 12.33 along with the final error estimates. The corresponding finite element and exact Von Mises values are given in Figs. 12.34 and 35. Clearly, one could continue this process but the resulting figures are too dense to show here. The third set of maximum single element error estimates is about a factor of 150 times smaller than the maximum element error in the first mesh.

12.10 Dynamics of solids

As noted in Eq. 12.25 we often encounter problems of the vibration of solids or their response to time varying loads. The assembled matrix equations thus involve the second time derivative of the displacements. If we include damping then the general form of the matrix equation of motion is

$$\mathbf{M}\,\ddot{\boldsymbol{\delta}} + \mathbf{D}\,\dot{\boldsymbol{\delta}} + \mathbf{K}\,\boldsymbol{\delta}(t) \;=\; \mathbf{F}(t)\,. \tag{12.45}$$

where the damping matrix, $\mathbf{D}$, is often taken to be of the form $\mathbf{D} = \alpha\mathbf{M} + \beta\mathbf{K}$. Given the initial conditions on $\dot{\boldsymbol{\delta}}(0)$ and $\boldsymbol{\delta}(0)$ one can determine the initial acceleration $\ddot{\boldsymbol{\delta}}(0)$ and then proceed to obtain the time-history of $\boldsymbol{\delta}(t)$ by direct integration in time. There are many algorithms for doing that. They differ in terms of stability, accuracy, storage requirements, etc. Several authors, such as Akin [1], Bathe [4], Hughes [14], Smith and Griffiths [20], Newmark [16], Petyt [18], and Zienkiewicz and Taylor [26] have given example algorithms for structural dynamics. Often the methods are close to each other and can be stated in terms of a small number of scalar coefficients that are multiplied times the various square matrices and vectors in Eq. 12.45 (see for example coefficients $a1$ through $a6$ in Fig. 12.36). Here we will illustrate one of the most popular algorithms called the Newmark Beta method. It is assumed here that the matrices **M**, **D**, and **K** are symmetric and have been assembled at the system level into a skyline sparse storage mode. It is also assumed that the resultant force vector **F** has been assembled and that it is multiplied by a know function of time. That scaling of the forces is done by subroutine *FORCE_AT_TIME* which is application dependent. Of course, the initial conditions on $\boldsymbol{\delta}$ and $\dot{\boldsymbol{\delta}}$ must be given. Figure 12.36 gives the implementation details of a Newmark integrator. For simplicity it assumes that any essential boundary conditions on the components of $\boldsymbol{\delta}$ are zero. In other words, it only allows for fixed supports in the form shown there.

```
SUBROUTINE NEWMARK_METHOD_SYM (I_DIAG, A_SKY, B_SKY, C_SKY, P_IN, & ! 1
                               R, DR, DT, N_STEPS, I_PRINT, N_BC, & ! 2
                               NULL_BC)                             ! 3
! * * * * * * * * * * * * * * * * * * * * * * * * * * * * * * *     ! 4
!    NEWMARK (BETA, GAMMA) CONDITIONALLY STABLE METHOD OF            ! 5
!     STEP BY STEP TIME INTEGRATION OF MATRIX EQUATIONS:             ! 6
!        A*D2R(T)/DT2 + B*DR(T)/DT + C*R(T) = P(T)                   ! 7
!      ALSO INCLUDES HUGHES AND TRAPEZOIDAL METHODS                  ! 8
!    INITIAL VALUES OF R AND DR ARE PASSED THRU ARGUMENTS            ! 9
! * * * * * * * * * * * * * * * * * * * * * * * * * * * * * * *     !10
Use System_Constants ! N_COEFF, N_D_FRE, NEWMARK_METHOD, FROM_REST, !11
                     ! HUGHES_METHOD, MASS_DAMPING, STIF_DAMPING    !12
 IMPLICIT NONE                                                      !13
 INTEGER,   INTENT (IN)    :: N_STEPS, I_PRINT, N_BC                !14
 INTEGER,   INTENT (IN)    :: I_DIAG (N_D_FRE)                      !15
 INTEGER,   INTENT (IN)    :: NULL_BC (N_BC)                        !16
 REAL(DP),  INTENT (IN)    :: DT                                    !17
 REAL(DP),  INTENT (INOUT) :: A_SKY (N_COEFF), C_SKY (N_COEFF)      !18
 REAL(DP),  INTENT (INOUT) :: B_SKY (N_COEFF)                       !19
 REAL(DP),  INTENT (IN)    :: P_IN  (N_D_FRE)                       !20
 REAL(DP),  INTENT (INOUT) :: R     (N_D_FRE), DR (N_D_FRE)         !21
 REAL(DP),  PARAMETER      :: ZERO = 0.d0                           !22
                                                                    !23
 INTEGER  :: I, K, I_STEP, I_COUNT                                  !24
 REAL(DP) :: a1, a2, a3, a4, a5, a6, BETA, GAMMA, T                 !25
                                                                    !26
!                 Automatic Work Arrays                             !27
 REAL(DP) :: R_PLUS (N_D_FRE), D2R    (N_D_FRE), P (N_D_FRE)        !28
 REAL(DP) :: WORK_1 (N_D_FRE), WORK_2 (N_D_FRE)                     !29
                                                                    !30
! DT, T     = TIME STEP SIZE, CURRENT TIME                          !31
! N_COEFF   = TOTAL NUMBER OF TERMS IN SKYLINE                      !32
! N_BC      = NUMBER OF  D.O.F. WITH SPECIFIED VALUES OF ZERO       !33
! NULL_BC   = ARRAY CONTAINING THE N_BC DOF NUMBERS WITH ZERO BC    !34
! R,DR,D2R  = 0,1,2 ORDER DERIVATIVE OF R W.R.T. T AT TIME=T        !35
! R_PLUS    = VALUE OF R AT TIME = T + DELT                         !36
! N_STEPS   = NO. OF INTEGRATION STEPS                              !37
! I_PRINT   = NO. OF INTEGRATION STEPS BETWEEN PRINTING             !38
                                                                    !39
```

Figure 12.36a *Interface for the Newmark method*

```
!                ** INITIAL  CALCULATIONS **                             !40
 IF ( NEWMARK_METHOD ) THEN                                              !41
    WRITE (6, '("NEWMARK STEP BY STEP INTEGRATION",/)')                  !42
    BETA = NEWMARK_BETA          ; GAMMA = NEWMARK_GAMMA                 !43
 ELSEIF ( HUGHES_METHOD ) THEN                                           !44
    WRITE (6, '( "HUGHES STEP BY STEP INTEGRATION",/)')                  !45
    BETA  = 0.25d0 * (1.d0 - HUGHES_ALPHA) **2                           !46
    GAMMA = 0.5d0 * (1.d0 - 2 * HUGHES_ALPHA)                            !47
 ELSE ! default to average acceleration method                           !48
    WRITE (6, '( "TRAPEZOIDAL STEP BY STEP INTEGRATION",/)')             !49
    BETA  = 0.25d0 ; GAMMA = 0.5d0                                       !50
 END IF                                                                  !51
                                                                         !52
 IF ( N_BC < 1 ) PRINT *,'NOTE: NO CONSTRAINTS IN NEWMARK_METHOD'        !53
 P = 0.d0 ; WORK_1 = 0.d0 ; WORK_2 = 0.d0 ; T = 0.d0                     !54
                                                                         !55
 IF ( .NOT. ELEMENT_DAMPING ) THEN                                       !56
   B_SKY = 0.d0                                                          !57
   IF ( MASS_DAMPING > 0.d0 ) B_SKY = MASS_DAMPING * A_SKY                !58
   IF ( STIF_DAMPING > 0.d0 ) B_SKY = B_SKY + STIF_DAMPING * C_SKY       !59
 END IF ! damping defined at element level                               !60
                                                                         !61
 IF ( FROM_REST ) THEN ! vel = acc = 0                                   !62
   DR = 0.d0 ; D2R = 0.d0                                                !63
 ELSE ! approx initial acc from diagonal scaled mass                     !64
   WORK_1 = A_SKY (I_DIAG) ! extract diagonal of mass matrix             !65
   D2R    = (SUM (A_SKY) / SUM (WORK_1)) * WORK_1 ! scaled M             !66
   CALL FORCE_AT_TIME (T, P_IN, P)               ! initial force         !67
   CALL VECTOR_MULT_SKY_SYM (B_SKY, DR, I_DIAG, WORK_1)                  !68
   CALL VECTOR_MULT_SKY_SYM (C_SKY,  R, I_DIAG, WORK_2)                  !69
   D2R = (P - WORK_1 - WORK_2) / D2R  ! approx initial acc               !70
 END IF ! need initial acc                                               !71
                                                                         !72
!       Print and/or save initial conditions                             !73
 I_STEP = 0 ; WRITE (6, 5020) I_STEP, ZERO                               !74
 5020 FORMAT ( /,' STEP NUMBER = ',I5,5X,'TIME = ',E14.8,/, &            !75
      & '     I          R(I)              DR/DT           ', &          !76
      & 'D2R/DT2')                                                       !77
 WRITE (6, 5030) (K, R (K), DR (K), D2R (K), K = 1, N_D_FRE)             !78
 5030 FORMAT ( I10, 2X, E14.8, 2X, E14.8, 2X, E14.8 )                    !79
                                                                         !80
! Form time integration constants                                        !81
 a1 = 1.d0 / (BETA * DT **2) ; a2 = 1.d0 / (BETA * DT)                   !82
 a3 = 0.5d0/BETA - 1.d0      ; a4 = GAMMA * a2                           !83
 a5 = GAMMA / BETA - 1.d0    ; a6 = 0.5d0 * GAMMA / BETA - 1.d0          !84
                                                                         !85
! Form system work array to factor for each time step group              !86
 C_SKY = C_SKY + a1 * A_SKY + a4 * B_SKY                                 !87
                                                                         !88
! ** APPLY BOUNDARY CONDITIONS (Zero C_SKY rows, cols) **                !89
 DO I = 1, N_BC                                                          !90
   CALL SKY_TYPE_1_SYM (NULL_BC (I), ZERO, C_SKY, P, I_DIAG)             !91
 END DO ! initial test of bc                                             !92
                                                                         !93
```

Figure 12.36b *Initialize, combine square matrices and apply BC*

```
!       *** TRIANGULARIZE C_SKY, in C_SKY * R = P ***                    ! 94
 CALL SKY_SOLVE_SYM (C_SKY, P, R, I_DIAG, .TRUE., .FALSE.)               ! 95
!       *** END OF INITIAL CALCULATIONS ***                               ! 96
                                                                          ! 97
!       *** CALCULATE SOLUTION AT TIME T ***                              ! 98
 I_COUNT = I_PRINT - 1                                                    ! 99
 DO I_STEP = 1, N_STEPS                                                   !100
   I_COUNT = I_COUNT + 1                                                  !101
   T       = DT * I_STEP                                                  !102
   IF ( I_COUNT == I_PRINT ) WRITE (6, 5020) I_STEP, T                    !103
                                                                          !104
!    FORM MODIFIED FORCING FUNCTION AT T + DELT                           !105
   CALL FORCE_AT_TIME (T, P_IN, P)                                        !106
   CALL VECTOR_MULT_SKY_SYM (A_SKY, (a1*R + a2*DR + a3*D2R), &            !107
                             I_DIAG, WORK_1)                              !108
   CALL VECTOR_MULT_SKY_SYM (B_SKY, (a4*R + a5*DR + a6*D2R), &            !109
                             I_DIAG, WORK_2)                              !110
   P = P + WORK_1 + WORK_2                                                !111
                                                                          !112
!       ** APPLY NULL BOUNDARY CONDITIONS ( TO P ) **                     !113
   P ( NULL_BC ) = 0.d0 ! vector subscripts                               !114
                                                                          !115
!           SOLVE FOR DR_PLUS AT TIME T+DELT                              !116
   CALL SKY_SOLVE_SYM (C_SKY, P, R_PLUS, I_DIAG, .FALSE., .TRUE.)         !117
                                                                          !118
!      UPDATE KINEMATICS OF D2R(+), DR(+), R(+)                           !119
   WORK_1 = a1*(R_PLUS - R - dt*DR) - a3*D2R      ! D2R(+)                !120
   WORK_2 = a4*(R_PLUS - R) - a5*DR - dt*a6*D2R ! DR(+)                   !121
   R      = R_PLUS ! R(+)                                                 !122
   DR     = WORK_2 ! DR(+)                                                !123
   D2R    = WORK_1 ! D2R(+)                                               !124
                                                                          !125
!             OUTPUT RESULTS FOR TIME T                                   !126
   IF ( I_COUNT /= I_PRINT ) CYCLE ! to next time step                    !127
     WRITE (6, 5030) (K, R (K), DR (K), D2R (K), K = 1, N_D_FRE)          !128
     I_COUNT = 0                                                          !129
 END DO ! over I_STEP for time history                                    !130
END SUBROUTINE NEWMARK_METHOD_SYM                                         !131
```

Figure 12.36c *Factor once and march through time*

The details of implementing such time integrators are usually fairly similar. Figure 12.36a shows a typical interface segment to a main program which has already dynamically allocated the necessary storage space for the initial system arrays (*A_SKY*, *B_SKY*, *C_SKY*, *P_IN* for **M**, **D**, **K**, **F**, respectively) and the initial conditions (*R*, *DR* for $\boldsymbol{\delta}$, $\dot{\boldsymbol{\delta}}$, respectively). The damping matrix may have been filled at the element level, or it may be filled here (lines 56-61) as a proportional damping matrix. The time integration requires storage of additional working arrays (lines 27-29) that are dynamically allocated for the scope of this routine only then released. One of these (*D2R*, for $\ddot{\boldsymbol{\delta}}$) holds the current value of the acceleration vector. Often input data keywords allow the user to select the specific algorithm to be employed as noted in Fig. 12.36b (lines 41-51). If the system does not start from rest then the initial acceleration must be computed once (lines 64-71). The scalar time integration constants are also determined once from the control data (lines 81-84) and then the effective square matrix overwrites the damping array to save storage.

The effective square matrix must be modified to include the zero value essential boundary condition at the supports (lines 89-92). After that it can be factored once (Fig. 12.36c, line 95) for future back-substitutions at each time step. It should be noted here that operations that are cited as done only once must actually be done again if the time step (*DT*) is changed. To avoid those relatively expensive changes one tries to keep the time step constant, or stop and do a restart from the last valid answer. The calculations for each time step (lines 105-130) basically determine the effective forcing function at the current time (lines 106-114) perform a forward and backward substitution to get the solution at the next time step (*R_PLUS* for $\delta(t + \Delta t)$) and use it along with the previous time step answers to update the velocity and acceleration vectors (lines 119-124). At that point the results are output to a sequential file for later post-processing and/or printed.

At any time step of interest we can recover the system displacement vector. That of course allows us to recover the strain and stresses as we did above for the static examples. Thus we can also carry out an error estimation. If that error estimation requires a new mesh to be created then one must use the displacements and velocities from the last acceptable time solution to interpolate the initial conditions on the new mesh that are required to restart the full time history process, including assembly. One should also consider the kinetic energy in such a process. Wiberg and Li [24] discuss the additional contributions to the error and use the above Newmark integration approach as a specific example.

To test such a time integrator it is best to begin with a simple system with a known analytic solution. Biggs [7] presented the analytic time history for a simple structure consisting of the linear springs and three lumped masses. One end is fixed while the other three joints are subjected to given forces having triangular time pulses that drop to zero. The Newmark and analytic (dashed) solutions for the displacement, and velocity at node 3 (of 4) are given in Fig. 12.37. They basically fall on top of each other for the chosen time step size. The popular Wilson Theta method gives similar, but less accurate, results as shown in Fig. 12.38.

Smith and Griffiths [20] also present a few examples more closely related to the present chapter in that they solve plane strain examples. They also present the so called 'element by element' approach, originally developed by Hughes [14], that saves storage by avoiding the assembly of the system square matrices. Instead, a product approximation of the square matrix assemblies is made so that the inversion of the effective matrix is replaced by sequential inversions of smaller element level square matrices. Both Bathe and Hughes give comparisons of the most popular algorithms in terms of their stability, accuracy, overshoot, and damping. They also give the time histories of a few test cases. One should review such comparisons before selecting an algorithm for a dynamic time-history solution.

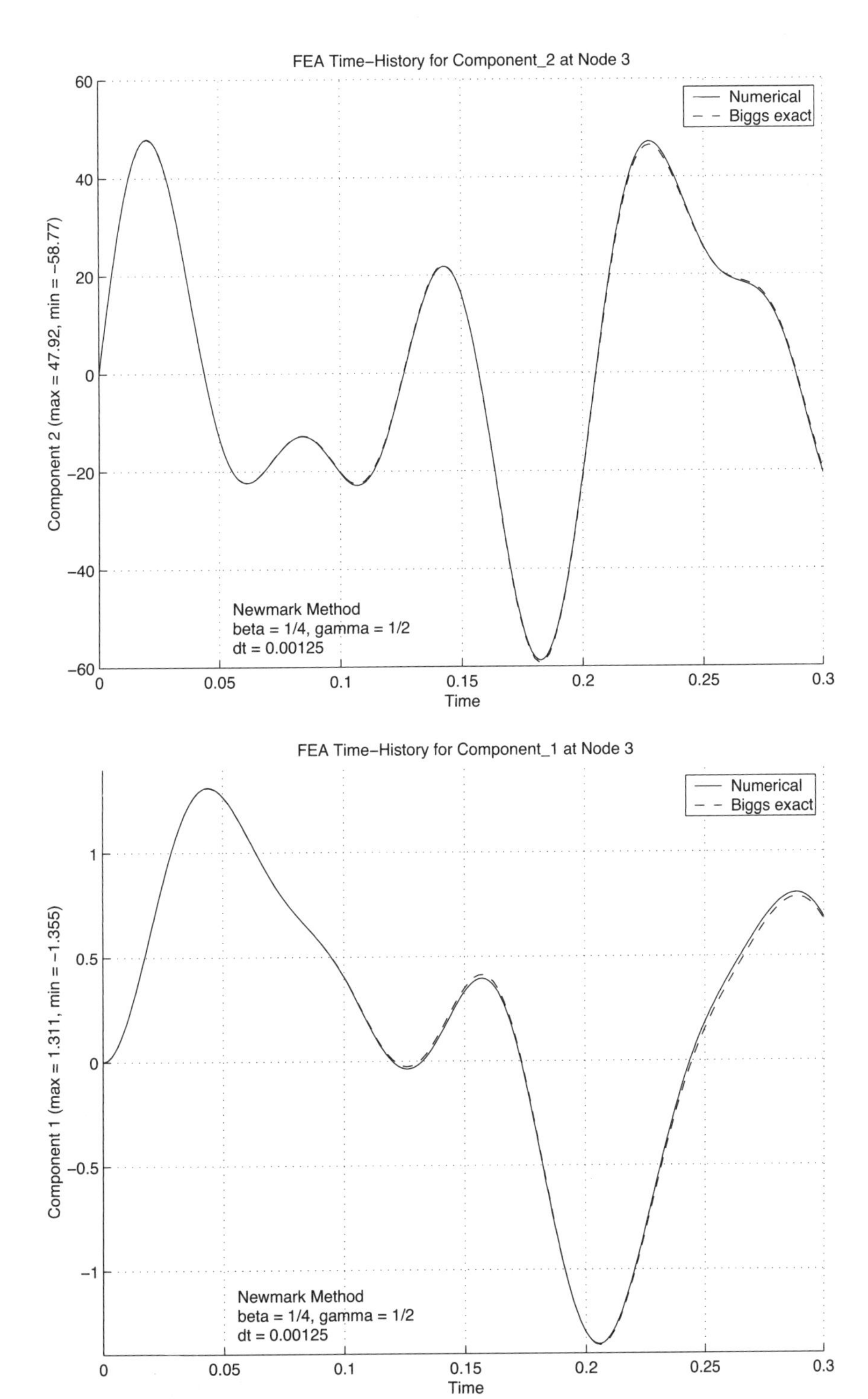

Figure 12.37 *Newmark method velocity (top), and displacement time history*

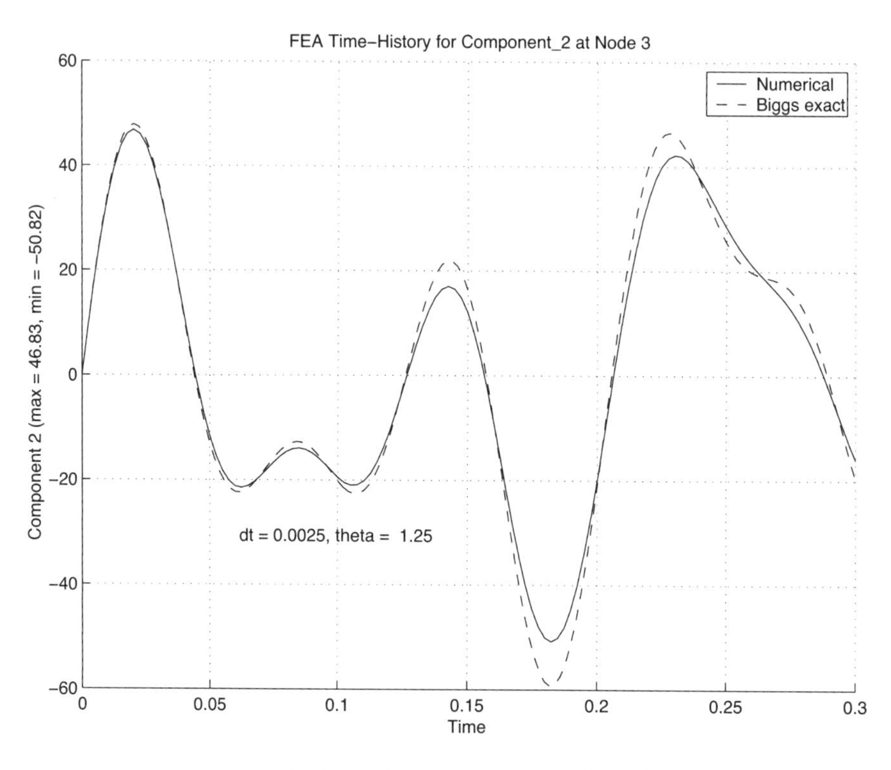

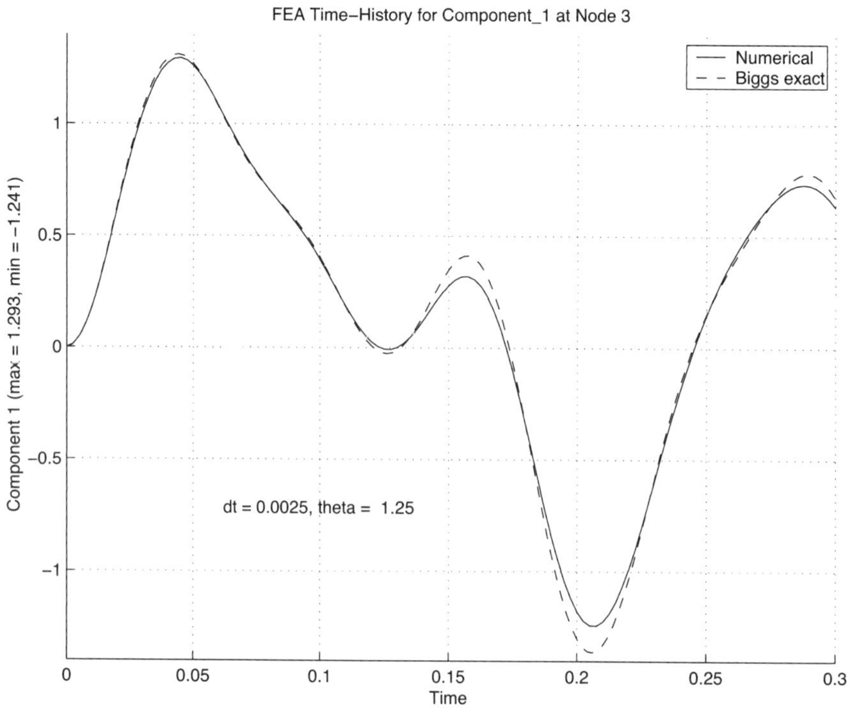

Figure 12.38 *Wilson method velocity (top), and displacement time history*

12.11 Exercises

1. Implement the isotropic $\mathbf{E}^e$ matrix for the plane strain assumption. Assume the stress components are in the *xx*, *yy*, *xy* order.
2. Implement the isotropic $\mathbf{E}^e$ matrix for the general solid. Assume the stress components are in the *xx*, *yy*, *xy*, *zz*, *xz*, *yz* order.
3. Give an implementation of the $\mathbf{B}^e$ matrix for a general solid. Assume the strain components are in the *xx*, *yy*, *xy*, *zz*, *xz*, *yz* order.
4. Review the scalar implementation for the square matrix for the T3 element, given in Fig. 11.6, and extend it to the case of a two-dimensional vector displacement model (that will double the matrix size).
5. Implement a general evaluation of the elasticity $\mathbf{E}^e$ matrix that will handle all the 1-, 2-, and 3-dimensional cases.
6. Implement a general evaluation of the elasticity $\mathbf{B}^e$ matrix that will handle all the 1-, 2-, and 3-dimensional cases.
7. Modify the plane stress implementation shown in Figs. 12.10 and 11 to solve a plane strain problem. Remember that an additional normal stress must be considered in the post-processing stage. Validate the results with a patch test.
8. Use the definition of the kinetic energy of a differential mass, *dm*, at a point to develop the element kinetic energy and thereby the consistent mass matrix for an element.
9. For the three node plane stress triangle use the previous scalar integrals of Eq. 11.11 and Fig. 11.8 to write, by inspection, the 6×6 element mass matrix. How much of the mass is associated with the x- and y-directions?

12.12 Bibliography

[1] Akin, J.E., *Finite Elements for Analysis and Design*, London: Academic Press (1994).

[2] Akin, J.E., *Object-Oriented Programming Via Fortran 90/95*, Cambridge: Cambridge University Press (2003).

[3] Babuska, I. and Strouboulis, T., *The Finite Element Method and its Reliability*, Oxford: Oxford University Press (2001).

[4] Bathe, K.J., *Finite Element Procedures*, Englewood Cliffs: Prentice Hall (1996).

[5] Becker, E.B., Carey, G.F., and Oden, J.T., *Finite Elements – An Introduction*, Englewood Cliffs: Prentice Hall (1981).

[6] Belytschko, T., Liu, W.K., and Moran, B., *Nonlinear Finite Elements for Continua and Structures*, New York: John Wiley (2000).

[7] Biggs, J.M., *Introduction to Structural Dynamics*, New York: McGraw-Hill (1964).

[8] Bonet, J. and Wood, R.D., *Nonlinear Continuum Mechanics for Finite Element Analysis*, Cambridge: Cambridge University Press (1997).

[9] Cook, R.D., Malkus, D.S., Plesha, N.E., and Witt, R.J., *Concepts and Applications of Finite Element Analysis*, New York: John Wiley (2002).

[10] Crisfield, M.A., *Non-linear Finite Element Analysis of Solids and Structures, Vol. 1*, Chichester: John Wiley (1997).

[11] Desai, C.S. and Abel, J.F., *Introduction to the Finite Element Method*, New York: Van Nostrand-Reinhold (1972).

[12] Gupta, K.K. and Meek, J.L., *Finite Element Multidisciplinary Analysis*, Reston: AIAA (2000).

[13] Hinton, E. and Owen, D.R.J., *Finite Element Programming*, London: Academic Press (1977).

[14] Hughes, T.J.R., *The Finite Element Method*, Mineola: Dover Publications (2003).

[15] Meek, J.L., *Computer Methods in Structural Analysis*, London: E & F.N. Spon (1991).

[16] Newmark, N.M., "A Method of Computation for Structural Dynamics," *ASCE J. Eng. Mech. Div.*, **85**(EM3), pp. 67–94 (July 1959).

[17] Oden, J.T., *Finite Elements of Nonlinear Continua*, New York: McGraw-Hill (1972).

[18] Petyt, M., *Introduction to Finite Element Vibration Analysis*, Cambridge: Cambridge University Press (1998).

[19] Segerlind, L.J., *Applied Finite Element Analysis*, New York: John Wiley (1987).

[20] Smith, I.M. and Griffiths, D.V., *Programming the Finite Element Method,* 3rd Edition, Chichester: John Wiley (1998).

[21] Stein, E., *Error Controlled Adaptive Finite Elements in Solid Mechanics*, Chichester: John Wiley (2003).

[22] Szabo, B. and Babuska, I., *Finite Element Analysis*, New York: John Wiley (1991).

[23] Weaver, W.F., Jr. and Johnston, P.R., *Finite Elements for Structural Analysis*, Englewood Cliffs: Prentice Hall (1984).

[24] Wiberg, N.-E. and Li, X.D., "A Postprocessed Error Estimate and an Adaptive Procedure for the Semidiscrete Finite Element Method in Dynamic Analysis," *Int. J. Num. Meth. Eng*, **37**, pp. 3585–3603 (1994).

[25] Wiberg, N.-E., "Superconvergent Patch Recovery – A Key to Quality Assessed FE Solutions," *Adv. Eng. Software*, **28**, pp. 85–95 (1997).

[26] Zienkiewicz, O.C. and Taylor, R.L., *The Finite Element Method,* 5th Edition, London: Butterworth-Heinemann (2000).

Index